Springer Complexity

Springer Complexity is an interdisciplinary program publishing the best research and academic-level teaching on both fundamental and applied aspects of complex systems – cutting across all traditional disciplines of the natural and life sciences, engineering, economics, medicine, neuroscience, social and computer science.

Complex Systems are systems that comprise many interacting parts with the ability to generate a new quality of macroscopic collective behavior the manifestations of which are the spontaneous formation of distinctive temporal, spatial or functional structures. Models of such systems can be successfully mapped onto quite diverse "real-life" situations like the climate, the coherent emission of light from lasers, chemical reaction-diffusion systems, biological cellular networks, the dynamics of stock markets and of the internet, earthquake statistics and prediction, freeway traffic, the human brain, or the formation of opinions in social systems, to name just some of the popular applications.

Although their scope and methodologies overlap somewhat, one can distinguish the following main concepts and tools: self-organization, nonlinear dynamics, synergetics, turbulence, dynamical systems, catastrophes, instabilities, stochastic processes, chaos, graphs and networks, cellular automata, adaptive systems, genetic algorithms and computational intelligence.

The two major book publication platforms of the Springer Complexity program are the monograph series "Understanding Complex Systems" focusing on the various applications of complexity, and the "Springer Series in Synergetics", which is devoted to the quantitative theoretical and methodological foundations. In addition to the books in these two core series, the program also incorporates individual titles ranging from textbooks to major reference works.

Understanding Complex Systems

Founding Editor: J.A. Scott Kelso

Future scientific and technological developments in many fields will necessarily depend upon coming to grips with complex systems. Such systems are complex in both their composition – typically many different kinds of components interacting simultaneously and nonlinearly with each other and their environments on multiple levels – and in the rich diversity of behavior of which they are capable.

The Springer Series in Understanding Complex Systems series (UCS) promotes new strategies and paradigms for understanding and realizing applications of complex systems research in a wide variety of fields and endeavors. UCS is explicitly transdisciplinary. It has three main goals: First, to elaborate the concepts, methods and tools of complex systems at all levels of description and in all scientific fields, especially newly emerging areas within the life, social, behavioral, economic, neuro- and cognitive sciences (and derivatives thereof); second, to encourage novel applications of these ideas in various fields of engineering and computation such as robotics, nano-technology and informatics; third, to provide a single forum within which commonalities and differences in the workings of complex systems may be discerned, hence leading to deeper insight and understanding.

UCS will publish monographs, lecture notes and selected edited contributions aimed at communicating new findings to a large multidisciplinary audience.

Alfons G. Hoekstra · Jiří Kroc · Peter M.A. Sloot
Editors

Simulating Complex Systems by Cellular Automata

Springer

Editors

Dr. Alfons G. Hoekstra
University of Amsterdam
Computational Science
Faculty of Science
Science Park 107
1098 XG Amsterdam
Netherlands
a.g.hoekstra@uva.nl

Dr. Jiří Kroc
Havlickova 482
CZ 33203 Stahlavy
Czech Republic
kroc@c-mail.cz

Prof. Dr. Peter M.A. Sloot
Universiteit Amsterdam
Computational Science
Faculty of Science
Science Park 107
1098 XG Amsterdam
Netherlands
p.m.a.sloot@uva.nl

ISSN 1860-0832 e-ISSN 1860-0840
ISBN 978-3-642-26367-5 ISBN 978-3-642-12203-3(eBook)
DOI 10.1007/978-3-642-12203-3
Springer Heidelberg Dordrecht London New York

Cover design: Integra Software Services Pvt. Ltd., Pondicherry

Printed on acid-free paper

Springer is part of Springer Science+Business Media (www.springer.com)

Simulation based understanding of complex systems with cellular automata.

Foreword

What are Cellular Automata good for?
What you always wanted to ask about them
but where afraid of the answer.

Cellular automata (CA) are a paradigm of *fine-grained*, *uniform*, *parallel* computation. This mode of computing is supposedly one that is most naturally and efficiently supported by physics, since physics itself is at bottom a uniform parallel computing medium (thence the appeal of "cellular automata machines" and all that). Obviously, then, if you have a complex system (such as an urban traffic network or a simultaneous system of chemical reactions) and manage to represent it as a cellular automaton, then you can "run" it on CA hardware – which you'll be using effectively as a "numerical integrator," see what happens, and hopefully develop a better understanding of your systems structure and function. Just like with experimental mathematics, there is nothing wrong with this prescription – as long as it is taken only in *trace amounts*. But if you think that, just because you've managed to dress your system in the garb of a CA, then running it and watching its evolution will make you *understand* it – well, you are in for a big disappointment, and so will your readers.

Even though their distinguishing characteristic – that they lend themselves to efficient and *exact* numerical integration – plays a major role in our story (unlike partial differential equations, CA are not chronically beset by issues of approximation, convergence, and stability – and one does not have to be an expert in calculus and functional analysis to use one), nonetheless CA are primarily a *conceptual* tool. In this sense, they play a role, for understanding and taming a complex system, similar to that played by *mathematical physics* in making us understand physics itself.

As 50 years of development consistently testify, the best contribution that a CA approach can give to the understanding of a complex system is at the stage of *developing the model* – not running an industrial version of it. *A successful CA model is the seed of its own demise!* The reason is simple. A CA model is a plausible *microscopic* dynamics (no mysteries to it, no infinitesimals, no indeterminism) that yields in a well-understood way the emergence of a desired *mesoscopic* behavior. In principle, an astronomically large implementation of that model will also give us the full-fledged *macroscopics*. Suppose we have a CA that "predicts" the formation of water droplets or ice needles by means of a simple combinatorics of tokens. If we had the resources, by scaling the model billions of billions of times we could model fog, clouds, etc., all the way up to global weather. But once we have derived, from the CA, the bulk properties of a water droplet, we can feed these numerical parameters to a higher-level model (a finite-element model or a differential equation), in which all that counts is just droplets per cubic meter, temperature, and so forth, and *much more practically and efficiently* model a whole cloud. At a higher aggregation level, we can model regional – and ultimatelly global – weather. That is, once we have learned from a CA how a water droplet emerges and behaves, there is no point in re-running the entire process from scratch for every single droplet of a rainstorm – even though in principle we might! A better allocation of our computational budget is (a) to use CA to learn whether, under what conditions, and at what level there emerge recognizable *mesoscopic laws* out of simple microscopic mechanisms; and then (b) use these mesoscopic laws as a basis for higher-level analytical models ("equations") or numerical models ("simulations").

The present collection gives several examples of this *maieutic* role of CA.

In addition to that, recent developments in the theory of CA and their "lattice gas" variants suggest that these structures may play an even more blatantly conceptual modeling role, analogous to that of analytical mechanics. That is, after the arithmetization of physics (Galileo), the mechanization of physics (Newton, Faraday, Maxwell), and the geometrization of physics (Poincaré, Einstein), we may envisage the *informatization* of physics not only at the statistical level (Boltzmann, Gibbs) but also at the fundamental level.

According to *Hamiltons* "least action" *principle*, among all conceivable trajectories of a physical systems the effective ones are those whose *action* is stationary with respect to infinitesimal variations of the trajectory. According to *Noethers theorem*, in the dynamical systems typically dealt with by physics "to every continuous one-parameter group of symmetries there correspond a conserved quantity." (E.g., the existence of a conserved a quantity called *energy* comes from the fact that a systems laws do not change with *time*.) However, in spite of their sweeping generality, these principles are predicated only for physical systems that are continous, differentiable, invertible, and symplectic. Does that mean that nothing at all of these principles is left if one turns to systems that, like CA, share many properties with physics but are *discrete*?

To salvage some aspects of those principles for CA, one has to transliterate concepts, as far as possible, from the continouous to the discrete; to know how to do that, one first has to ask "What is the essential, most likely combinatorial, and inescapably tautological nature of those principles?", "What kind of *accounting* is it that they are doing?", or, "What is it that they are really trying to tell us?" (think of Boltzmann and his intuition that entropy has to do with *number* of states). More importantly for the advancement of science, we can reverse the roles of means and goal in the above endeavor. That is, let us work at fitting (as much as posible of) those principles into a CA context, hoping that that will reveal to all of us (CA aficionados and physicists alike) what those somewhat mystical and teleological principles "really mean." By using a CA as a *discrete* "kitchen experiment" of analytic mechanics, we bring "magical" aspects of theoretical physics down to earth. After the fair's magic show, the man in the street – any one of us, really – will then go home and tell his children (and, what's more important, to himself), "Oh, there was nothing to it! Here is how it must work – let me show you. . . ."

In conclusion, the discipline of modeling with cellular automata is an excellent way to identify, without being distracted by irrelevant technicalities, those element of a system that are *truly essential* – namely, those that are both necessary and sufficient to yield a *certain kind* of behavior. "Keep things simple!", or, in Donald Knuth's words, "Premature optimization is the root of all evil."

Boston University, Boston, MA, USA Tommaso Toffoli (tt@bu.edu)

Preface

Deeply rooted in fundamental research in Mathematics and Computer Science, Cellular Automata are recognized as an intuitive modeling paradigm for Complex Systems. Very basic Cellular Automata, with extremely simple micro dynamics such as the Game of Life, show an almost endless display of complex emergent behavior. By modeling natural or man-made Complex Systems with Cellular Automata we usually dissect the system to it's most fundamental and minimal properties and interactions, such that the simulated dynamics *mimics* the emergent behavior of the real Complex System, leading to a true *understanding* of the fundamental properties of the system under study.

For instance, Cellular Automata models of vehicle traffic are a beautiful example. A few simple rules relating to acceleration, deceleration, and maximum speed of vehicles in a one-dimensional Cellular Automata are sufficient to display all different types of motions that cars on a freeway can have (free flow, stop-and-go, jamming), as well as showing backward traveling density waves in stop-and-go traffic and reproducing the fundamental diagram of car throughput on a freeway as a function of car density.

Vice-versa, Cellular Automata can also be designed to *produce* a desired emergent behavior, using theoretical methodologies or using e.g. evolutionary techniques to find Cellular Automata rules that produce specified characteristics.

Cellular Automata can also actually *reproduce* the dynamics of Complex Systems *qualitatively*. For instance, Lattice Gas Cellular Automata are a class of Cellular Automata that reproduce many of the intricate dynamics in fluids. Likewise, other fundamental physical systems, such as Reaction–Diffusion or Advection–Diffusion can be qualitatively modeled. These Cellular Automata models can actually be used to *predict* the behavior of Complex Systems under many different circumstances. Nowadays there are many applications of Cellular Automata models in Computational Physics or – Chemistry, but also in for instance Systems Biology (e.g. models for diffusion limited gene regulatory networks).

Over the last decade or so, there has been a tremendous progress in studying Complex Systems with Cellular Automata. They are not only being used within their originating disciplines (say Physics, Computer Science, Mathematics), but are also applied in quite different disciplines such as epidemiology, immunology, sociology, and finance. Cellular Automata are quite successful in for instance modeling immune response after HIV infection, both the short term effects, as well as the long term effect that finally lead to the development of AIDS. Cellular Automata are also used to study the dynamics of crowds, for instance in situations where a crowd must escape from a confined space through a small door.

In this context of fast and impressive progress in the field the idea to compose this book emerged. Moreover, another experience convinced us that we should embark on this project that in the end resulted in this book. When teaching Complex Systems Simulations to Master students in the Amsterdam Master program on Computational Science, we always experience the great appeal that Cellular Automata have on the students, and we are always impressed by the deep understanding that our students – but we as well – obtain of a large range of complex systems they try to model and understand using Cellular Automata. These students come from many disciplines, as broad as the application areas of Cellular Automata mentioned earlier.

For us it became evident that an edited book focusing on all aspects of modeling Complex Systems with Cellular Automata was needed, as a welcome overview of the field for its practitioners, as well as a good starting point for detailed study on the graduate and post-graduate level. While Jiří Kroc was a visiting scientist in Amsterdam, in the period September 2007 to September 2008, the idea materialized and the "book project" went into high gear.

The book contains three parts, two major parts on theory and applications, and a smaller part on software. The theory part contains fundamental chapters on how to design and/or apply Cellular Automata for many different areas. This should give the readers a representative overview and strong background on many aspects related to modeling with Cellular Automata. In the applications part a number of representative examples of really using Cellular Automata in a large range of disciplines is presented. By providing a large set of examples, this part should give readers a good idea of the real strength of this kind of modeling and challenge them to apply Cellular Automata in their own field of study. Finally, we included a smaller section on software, to highlight the important work that has been done to create high quality problem solving environments that allow to quickly and relatively easily implement a Cellular Automata model and run simulations, both on the desktop and if needed, on High Performance Computing infrastructures.

We are very proud and happy that many prominent scientists agreed to join this project and prepared a chapter for this book. We are also very pleased to see it materialize in a way as we originally envisioned. We hope that this book will be a source of inspiration to the readers. We certainly challenge students on the graduate and post-graduate level to study this book in detail, learn from it, grasp the fundamental ideas behind modeling Complex Systems with Cellular Automata, and apply it to solve their own problems. For scientists working in many different fields we believe that this book will provide a representative state-of-the-art overview of this field. It not only shows what we *can* do, it also shows current gaps in our knowledge, open issues and suggestions for further study.

We wish all readers a fruitful time reading this book, and wish they experience the same excitement as we did – and still do – when using Cellular Automata for modeling complex systems.

Amsterdam, The Netherlands
May 2010

Alfons G. Hoekstra
Jiří Kroc
Peter M.A. Sloot

Acknowledgements

We wish to acknowledge all authors of this book for their continuing efforts writing high quality chapters and bearing with us during this project. We wish to acknowledge the professional support by Gabriele Hakuba and Chris Caron at Springer. We acknowledge financial support by the European Commission, through the DynaNets project (www.dynanets.org, EU-FP7-IST-FET Contract 233847) and through the COAST project (www.complex-automata.org, EU-FP6-IST-FET Contract 033664).

Contents

Contributors

Andrew Adamatzky University of the West of England, Bristol, UK, andrew.adamatzky@uwe.ac.uk

Olga Bandman Supercomputer Software Department, ICM&MG, Siberian Branch Russian Academy of Sciences, Novosibirsk, 630090, Russia, bandman@ssd.sscc.ru

Alfonso Caiazzo INRIA Rocquencourt – BP 105, F-78153 Le Chesnay Cedex, France, alfonso.caiazzo@inria.fr

Bastien Chopard Department of Computer Science, University of Geneva, 7 route de Drize, 1227 Carouge, Switzerland, bastien.chopard@unige.ch

Debashish Chowdhury Department of Physics, Indian Institute of Technology, Kanpur 208016, India, debch@iitk.ac.in

Andreas Deutsch Center for Information Services and High Performance Computing, Technische Universität Dresden, Nöthnitzerstr. 46, 01069 Dresden, Germany, andreas.deutsch@tu-dresden.de

Adam Dunn Centre for Health Informatics, University of New South Wales UNSW, Sydney NSW 2052, Australia; Alcoa Research Centre for Stronger Communities, Curtin University of Technology, PO Box U1985, Perth WA 6845, Australia, a.dunn@curtin.edu.au; a.dunn@unsw.edu.au

Jean-Luc Falcone Department of Computer Science, University of Geneva, 7 route de Drize, 1227 Carouge, Switzerland, jean-luc.falcone@unige.ch

Martin Grube Institute of Plant Sciences, Karl-Franzens-Universität Graz, Graz, martin.grube@uni-graz.at

Zafer Gürdal Faculty of Aerospace Engineering, Delft University of Technology, Delft, The Netherlands, z.gurdal@tudelft.nl

Haralambos Hatzikirou Center for Information Services and High Performance Computing, Technische Universität Dresden, Nöthnitzerstr. 46, 01069 Dresden, Germany, haralambos.hatzikirou@tu-dresden.de

Dirk Helbing ETH Zurich, CLU E1, Clausiusstr. 50, 8092 Zurich, Switzerland, dhelbing@ethz.ch

Alfons G. Hoekstra Computational Science, Faculty of Science, University of Amsterdam, Science Park 107, 1098 XG, Amsterdam, The Netherlands, a.g.hoekstra@uva.nl

Paulien Hogeweg Theoretical Biology and Bioinformatics Group, Utrecht University, Padualaan 8, 3584 CH, Utrecht, The Netherlands, p.hogeweg@bio.uu.nl

Jiří Kroc Havlíčkova 482, 332 03 Štáhlavy, The Czech Republic, jiri.kroc@gmail.com

Halim Kusumaatmaja Max Planck Institute of Colloids and Interfaces, Science Park Golm, 14424 Potsdam, Germany; The Rudolf Peierls Centre for Theoretical Physics, Oxford University, 1 Keble Road, Oxford OX1 3NP, UK, kusumaatmaja@gmail.com

Eric Lorenz Computational Science Group, Faculty of Science, University of Amsterdam, Science Park 107, 1098 XG, Amsterdam, The Netherlands, e.lorenz@uva.nl

Lev Naumov Computational Science Group, Faculty of Science, University of Amsterdam, Science Park 107, 1098 XG, Amsterdam, The Netherlands, levnaumov@gmail.com

Katsuhiro Nishinari Research Center for Advanced Science and Technology, The University of Tokyo, Komaba 4-6-1, Meguro-ku, Tokyo, 153-8904, Japan, tknishi@mail.ecc.u-tokyo.ac.jp

Zhijian Pan IBM Pervasive Computing Lab, 1997 Annapolis Exchange Pkwy, Annapolis, MD 21401, USA, edzpan@yahoo.com

James A. Reggia Computer Science Department, University of Maryland, A. V. Williams Building, College Park, MD 20742, USA, reggia@cs.umd.edu

Andreas Schadschneider Institut für Theoretische Physik, Universität zu Köln, 50937 Köln, Germany, as@thp.uni-koeln.de

Peter M.A. Sloot Computational Science, Faculty of Science, University of Amsterdam, Science Park 107, 1098 XG, Amsterdam, The Netherlands, p.m.a.sloot@uva.nl

Domenico Talia DEIS, University of Calabria, Rende, Italy, talia@deis.unical.it

Tommaso Toffoli ECE Department, Boston University, 8 Saint Mary's St, Boston, MA, USA, tt@bu.edu

Marco Tomassini Information Systems Department, University of Lausanne, Lausanne, Switzerland, marco.tomassini@unil.ch

Hiroshi Umeo University of Osaka Electro-Communication, Neyagawa-shi, Hatsu-cho, 18-8, Osaka, 572-8530, Japan, umeo@cyt.osakac.ac.jp

Julia M. Yeomans The Rudolf Peierls Centre for Theoretical Physics, Oxford University, 1 Keble Road, Oxford OX1 3NP, UK, j.yeomans1@physics.ox.ac.uk

Wenjian Yu ETH Zurich, CLU C4, Clausiusstr. 50, 8092 Zurich, Switzerland, yuwen@ethz.ch

Ramzi Zakhama Faculty of Aerospace Engineering, Delft University of Technology, Delft, The Netherlands, r.zakhama@tudelft.nl

Chapter 1
Introduction to Modeling of Complex Systems Using Cellular Automata

Alfons G. Hoekstra, Jiří Kroc, and Peter M.A. Sloot

"The real purpose of scientific method is to make sure Nature hasn't misled you into thinking you know something you don't actually know."

Robert M. Pirsig
Zen and the Art of Motorcycle Maintenance, 1974.

Since the sixteenth century there have been two main paradigms in the methodology of doing science. The first one is referred to as "the experimental" paradigm. During an experiment we observe, measure, and quantify natural phenomena in order to solve a specific problem, answer a question, or to decide whether a hypothesis is true or false. The second paradigm is known as "the theoretical" paradigm. A theory is generally understood as a fundamental, for instance logical and/or mathematical explanation of an observed natural phenomenon. Theory can be supported or falsified through experimentation.

The roots of the experimental and theoretical paradigms occurred much earlier and were already used by Pythagoras, Euclid, Archimedes and others (e.g. during ancient times in Greece, China, Egypt and other cultures thousands years BC). Since that time, the systematic use of those two paradigms has enabled us to understand and quantify some bits and pieces of Nature.

Since the Second World War, a third scientific paradigm appeared on the scene. This one is usually referred to as "the computational" paradigm, in which we study Nature through computer simulations. The first theoretical results dealing with this paradigm can be attributed to Alan Turing [24] in the 1930s. Computation is neither theory nor experiment. Computations can be implemented by many means: mechanically, electro-mechanically (for example the Bombe machine – used to decipher the German Enigma coding-machine), using electrical circuits (the first two programmable digital electronic computers Colossus and ENIAC), electronically (nowadays built in silico computers), chemically, biochemically, using DNA, quantum mechanically and in many other ways not mentioned here.

A.G. Hoekstra (✉)
Computational Science, Faculty of Science, University of Amsterdam,
Science Park 107, 1098 XG, Amsterdam, The Netherlands
e-mail: a.g.hoekstra@uva.nl

A.G. Hoekstra et al. (eds.), *Simulating Complex Systems by Cellular Automata*,
Understanding Complex Systems, DOI 10.1007/978-3-642-12203-3_1,

1.1 The Computational Paradigm

The computational paradigm uses computation to describe systems and natural phenomena employing a computer. The computational paradigm plays a fundamental role in situations where analytical descriptions of the observed phenomena are not tractable and/or out of reach of direct experimentation. Outputs from computations are often validated against experimental data and against simplified analytical models. The power of the computational paradigm has been demonstrated in physics, mathematics, chemistry, engineering and its importance is continuously increasing in such fields such as biology, medicine, sociology and psychology.

The invention of the computer enabled the solution of analytical models in terms of their numerical implementations, where by numerical we usually mean discretization of space and time and keeping continuous variables. In this way, for example, Partial Differential Equations (PDEs) were solved by the Finite Element Method (FEM). It brought a big breakthrough in scientific and engineering solutions of various problems. Finally, scientists realized that sometimes it is not necessary, or even possible, to design an analytical model describing a given natural phenomenon. In such cases, the observed phenomena can often be directly implemented using a discrete model.

1.2 Modeling

An abstract model is a construct incorporating a number of variables, processes and relationships among them. A model in this sense usually means that details about the original, modeled phenomenon are excluded from the model itself. An abstract model employing mathematical methods and tools to describe a natural phenomenon or system is called a mathematical model. Mathematical models are used within physics, engineering, biology (natural sciences), sociology, psychology, political science, economics (social sciences). Mathematical models employ many different types of mathematical disciplines and methods as statistics, differential equations, integro-difference equations, dynamical systems, game theory, particle systems, systems of difference equations, cellular automata, and many other not mentioned here.

A computer model (often called a computer simulation or computational model) is an abstract model implemented into a computer program. Computer models of natural systems (as a part of mathematical modeling) are widely used in physics, biology, social sciences, economics and engineering.

Mathematical and computational models can be grouped according to the use of continuous or discrete values in *space*, *state variables* and *time* into several computationally different classes (for details see Table 1.1).

Toffoli and others [22, 25] pointed out the following sequence of approximations typically used during modeling of natural phenomena where differential equations are used. Initially, there is a naturally observed phenomenon. This phenomenon is

Table 1.1 Analytical and numerical methods used to model natural phenomena where C stands for continuous and D for discrete *state*, *space*, or *time* [3, 7, 21]

Type of model	State	Space	Time
Partial differential equations (PDEs)	C	C	C
Integro-difference equations	C	C	D
Coupled ordinary differential equations (ODEs)	C	D	C
Interacting particle systems	D	D	C
Coupled map lattices (CMLs) and systems of difference equations lattice Boltzmann equations (LBEs),	C	D	D
Cellular automata (CAs) and lattice gas automata (LGAs)	D	D	D

observed and studied through experimentation. Then a mathematical model is build. In physics, it often means that suitable (set of) differential equations is used (e.g. partial differential equations). This represents the first-level approximation. In the second-level of approximation, the mathematical model is discretized into a numerical one. Typically, such numerical schemes end up with a numerical error.

The third-level approximation is associated with the representation of real numbers on computers. In general computers approximate most real numbers by using nearby rational numbers. In some cases irrational numbers can be treated exactly by computers (e.g. sqrt(2) function). Computers use either floating point or fixed-point number representations, resulting in rounding errors.

Contrary to the three previously listed approximations, we identify only one approximation in computational modeling with cellular automata [22, 25]. The natural phenomenon is directly implemented in the form of a cellular automaton. This is a potential advantage compared to the previous approaches. Due to the presence of only one approximation, the expressivity of cellular automata is in many cases higher as compared to other techniques.

1.3 Complex Systems

In complex systems we observe group or macroscopic behavior emerging from individual actions and interactions. One of the first biological observations of complex systems, if not the first one, is linked to ant-colony behavior [9], see Fig. 1.1. Each ant is simply following its internally encoded reactions to external stimuli. It is a well known fact that ant species may have a set of 20 up to 40 reactions on any given external stimulus. There is no ant "leader" which tells other ants what they should do nor a hierarchy of such leaders. Despite the lack of controlling entities, an ant-colony builds its ant-hill, feeds it, protects it, attacks other colonies, follows a foraging strategy, etc. This emergent behavior observed in ant-colonies is prototypical for many other complex systems.

We now know that a carefully selected set of reactions on external stimuli of large number of identical copies of several generic, mutually interacting entities creates a complex system often displaying self-organization and/or emergent behavior [1].

(a) (b)

Fig. 1.1 One of the biggest challenges of the current level of scientific understanding of Nature is to find out principles behind self-organizing and emergent systems. An excellent example is the ant-colony. One ant (**a**) (species *Formica*) in isolation is a "simple" system, 1000s ants working together are capable to build complex structures like an ant-colony without any central control (**b**). The solution of the backward problem is "relatively" easy (i.e. if we know ant-colony behavior then we could find out local rules through the observation of ants). The forward problem is in general untractable by all currently known techniques. The question is which set of local rules will lead to the desired global response

Unfortunately, we do not know a generic procedure or algorithm to design such systems. It is a tremendous task to find a set of local interactions which produce a desired global response. The opposite direction, which might be called a backward (or deconstructive) one, is relatively easy to perform.

Another striking biological example of self-organization is the existence of "V-formations" spontaneously occurring within flocks of flying birds (e.g. geese) [15]. The model describing this phenomenon takes into account line of sight and aerodynamic advantages of flocking birds due to an upwash behind their wing tips creating extra lift. Each bird controls its actual position according to its nearest neighbors with respect to aerodynamics and vision. The final "V-like formation" a flock spontaneously occurs due to this self-organization, regardless of the initial positions of birds.

These two examples lead us to the concept of complex systems. Complex systems were independently and often simultaneously (re)discovered in many scientific fields [1, 4, 6, 17]. This is an indirect indication of their universality. A typical complex systems consists of a vast number of identical copies of several generic and mutually interacting processes. Despite the lack of global control complex systems often express (macroscopic) self-organization and emergence, which are typically driven by dissipation of energy and/or information.

In general, self-organization often results from a competition between counteracting processes. A good example is provided by self-organized criticality observed in Earthquakes where the system is constantly fed by elastic energy which is released abruptly in the form of avalanches of various size and intensity. Self organised criticality has been observed in many natural and man-made systems [19]. Self-organization generates very robust solutions, which are intrinsically resistant to any kind of external and/or internal noise and perturbations. Self-organization is often accompanied by emergence and vice versa.

Emergence is defined as the occurrence of a new quality operating at a higher level than where the system units are operating. Going back to the example of ant-colonies, an ant hill is an emergent property arising from the local interactions of ants. A whole hierarchy of emergents can exist, like e.g. in the human body where we see a cascade from DNA, amino acids, polypeptides, proteins, cell structures, cells, tissues, organs, bodies, societies, ecosystems, etc.

Complex system behaviour is observed within Earthquakes, volcano activities, stock market behavior, social networks, traffic flow, etc. Complex systems like that express self-organized criticallity. There is a group of complex systems that show percolation behavior of some property such as, e.g., conductivity of a mixture of powders of an isolator and a metal, forest fires, opinion development within societies, and so on. It is worth stressing that no generally accepted definitions exist for complex systems, self-organization, and emergent behavior. There is a variety of techniques to model complex systems. In the following parts of the introduction as well as in the whole book, the attention is focussed to one of those computational techniques called “cellular automata”.

1.4 Cellular Automata

The notion of *cellular automata* has a long, living history going back to Stanislav Ulam and John von Neumann, see for instance [21]. Von Neumann’s cellular automaton (in plural cellular automata) [16] represents a direct predecessor of cellular automata (CAs). Shortly, von Neumann introduced cellular automata as a computational medium for machine self-replication motivated by a simple question: “What kind of logical organization is sufficient for an automaton to be able to reproduce itself?”. In another words “Can we reproduce computationally and in silico what living cells do?” Unluckily, he left his work unfinished. In principle, self-replication is possible to solve but, so far, the problem is far from being finished. Self-replication still attracts the attention of many researchers. There are two main streams in the design of self-replicating structures. The first one uses hand made local rules and topologies whereas the second one employs evolutionary algorithms to explore self-replicating structures and rules. You will find examples of both approaches in this book.

After the very promising start of cellular automata many theoretical and practical applications and models of natural phenomena were successfully described and solved by computational models using classical cellular automata. Some of them are presented in this book. There are still a vast number of prospective applications of cellular automata in almost all scientific fields. As you will see in this book, cellular automata themselves are undergoing a development as a computational method as well.

1.5 Classical Cellular Automata

The oldest versions of cellular automata have a simple and straightforward implementation. It is based on the use of a regular array of cells (often implemented as a matrix in computer simulations), their local variables, and a transition function working over a neighborhood. Those components of classical cellular automaton employing a von Neumann neighborhood are depicted in Fig. 1.2.

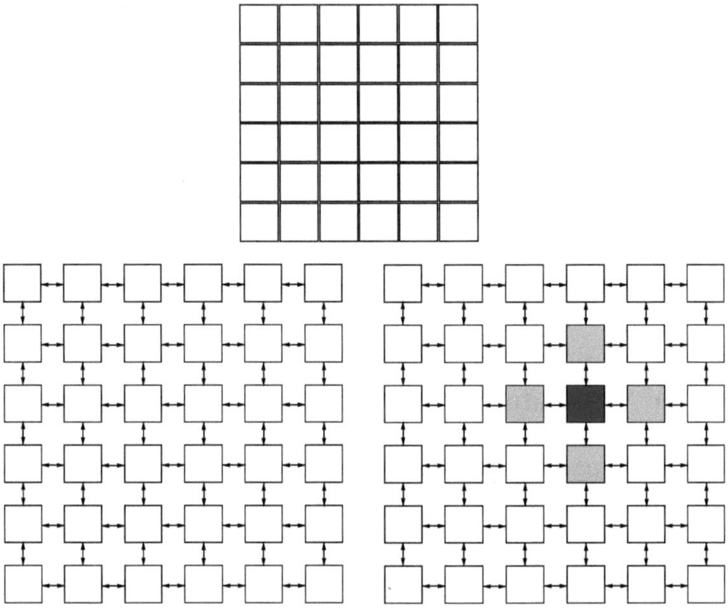

Fig. 1.2 A two-dimensional lattice defining cellular automaton having a size of 6×6 cells is shown (*top*). The spatially expanded lattice with explicitly displayed links (*lines with arrows*) between neighboring cells (*bottom-left*). Such topology, where only the nearest neighbors are linked, defines a von Neumann neighborhood with a radius $r = 1$ (*bottom-right*), an updated cell (*dark gray*) and neighbors (*light gray*). In general, different neighborhoods are defined by different sets of links

Let us define cellular automata as follows:

- *A discrete cellular state space* $\mathcal{L}$ is a set of locally interconnected finite state automata (FSAs) that is typically created by a regular d-dimensional lattice of FSAs. For example, in two-dimensions, a cellular automaton consists of a lattice of $\mathcal{L} = m \times n$ squares where each square, in the literature called a cell, is represented by one FSA.
- *A local value space* $\sum$ defines all possible states for each FSA. The state σ of each FSA can be in one of the finite number of states $\sigma \in \sum \equiv \{0, 1, 2, 3, 4, \ldots, k-1, k\}$. The composition of all possible states of all cells create a state space of the whole cellular automaton.

- A *neighborhood* $\mathcal{N}$ is created by a set of N topologically neighboring cells, which are influencing a change of the state of each updated cell in the next simulation step. Typically, homogeneous neighborhoods (composed of nearest neighbors) are used, see Fig. 1.3 for examples of von Neumann and Moore neighborhoods.
- *Boundary conditions* can be periodic, fixed or reflecting among others. The most important are periodic boundary conditions, which are used to simulate the infinite lattice using a finite one. Periodic boundary conditions are implemented as a torus in two dimensions.
- A *transition rule* ϕ: $\overbrace{\sum \times \sum \times \cdots \times \sum}^{\mathrm{N}} \rightarrow \sum$ describing the change of each updated cell from its current state value to a new one, operating over the neighborhood (having size N) – is defined.

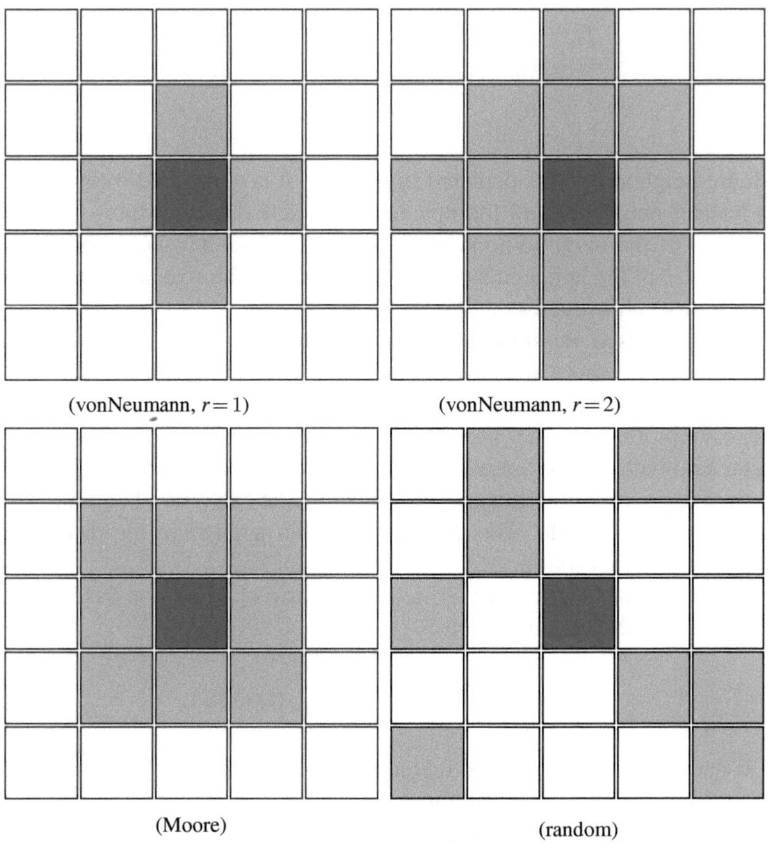

Fig. 1.3 Various types of neighborhoods are shown: von Neumann (radius 1 and 2), Moore, and a random one. Variables of cells within a given neighborhood (*light gray* plus *dark gray*) are used to evaluate new values of the updated cell (*dark gray*) using transition function

- All cells change their state synchronously at an externally provided clock step. This is usually called one iteration step.

A definition of a neighborhood is often provided in the following way. In the case of a von Neumann neighborhood having radius equal to one

$$\begin{aligned}\mathcal{N}_{\mathrm{N}}^{1}(i,j) &= \{\sigma_{k,l} \mid |i-k| + |j-l| \le 1\} \\ &= \{\sigma_{i,j}, \sigma_{i-1,j}, \sigma_{i,j-1}, \sigma_{i+1,j}, \sigma_{i,j+1}\}\,. \end{aligned} \tag{1.1}$$

This von Neumann neighborhood (with radius equal to one) is depicted in Fig. 1.3. Only the nearest neighbors of the updated cell are involved in this neighborhood (i.e. four plus one in total). It is worth to mention out that the updated cell itself (i.e. the one with indexes (i, j)) is involved in the neighborhood as well. The other type of neighborhood is Moore neighborhood having a radius r equal to one

$$\begin{aligned}\mathcal{N}_{\mathrm{M}}^{1}(i,j) &= \{\sigma_{k,l} \mid |i-k| \le 1, |j-l| \le 1\} \\ &= \{\sigma_{i,j}, \sigma_{i-1,j}, \sigma_{i-1,j-1}, \sigma_{i,j-1}, \sigma_{i+1,j-1}, \\ &\quad\ \sigma_{i+1,j}, \sigma_{i+1,j+1}, \sigma_{i,j+1}, \sigma_{i-1,j+1}\}\,. \end{aligned} \tag{1.2}$$

This Moore neighborhood is depicted in Fig. 1.3. It is composed from the first and second nearest neighbors and the updated cell itself (i.e. eight plus one in total). All members of the neighborhood are numbered anti-clockwise in both cases. In general, the size of the lattice must be much larger than the size of the neighborhood otherwise every cell becomes dependent on other cells.

A commonly used definition of cellular automata states that they represent dynamical systems where space, variables, and time are discrete. Such definition enables to employ techniques developed in statistical physics that are used to predict average, asymptotic and other properties of simulated natural phenomena. Results of cellular automata models are often compared to results of analytical models using a mean field approximation. It is a known fact that the behavior of cellular automata is in general unpredictable. This property is often applied in the design cellular automata used for encryption.

Any transition rule $\phi(t)$ can be written in the form of a function $\phi(t)$ using states $\sigma(t)$ of all cells in the neighborhood

$$\sigma_{i,j}(t+1) = \phi(\sigma_{k,l}(t) \mid \sigma_{k,l}(t) \in \mathcal{N}), \tag{1.3}$$

where $\mathcal{N}$ defines the neighborhood including the cell itself, see Eqs. (1.1) and (1.2), and Fig. 1.3 for examples of neighborhoods. The transition rule ϕ can be any combination of arithmetical and logical operations, including functions. In real applications, the typical transition rule has either the form of a transition table (defining for which input values a certain output value is taken) or the form of a computer program.

In order to provide an illustrative example, the general form of a transition rule for a von Neumann neighborhood $\mathcal{N}_N^1$ is given by

$$\sigma_{i,j}(t+1) = \phi(\sigma_{i,j}(t), \sigma_{i-1,j}(t), \sigma_{i,j-1}(t), \sigma_{i+1,j}(t), \sigma_{i,j+1}(t)), \tag{1.4}$$

which contains all five neighbors including the updated cell itself.

The number of all possible local rules N_r depends on the number of states σ and the number of neighbors n for a given cellular automaton in the following way

$$N_r = \sigma^{\sigma^n}. \tag{1.5}$$

As shown in Table 1.2, the number of rules dramatically increases with the number of neighbors n and states σ.

In general, the most difficult task in the design of cellular automata is to find a transition rule that will describe the temporal evolution of a modeled system, i.e. which leads to desired global response of the system. As the number of possible rules dramatically increases with a number of states σ and the size of the neighborhood n, it is usually non-trivial to find correct transition rules describing the system being modeled.

In order to narrow down the huge space of all possible rules of a cellular automaton for a given number of states σ and the size of neighborhood n, attempts were made to classify rules according to their complexity. Wolfram [26] proposed four classes of cellular automata behavior: (1) almost all configurations relax to a fixed configuration, (2) almost all configurations relax to either a fixed point or a periodic cycle according to the initial configuration, (3) almost all configurations relax to chaotic behavior, (4) sometimes initial configurations produce complex structures that might be persisting or long-living.

Table 1.2 The total number of local rules for a cellular automaton having a number of states σ and number of neighbors n. It is evident that even for automata with a relatively small number of states and neighbors the number of all possible rules increases dramatically with the number of states and neighbors

Number of states σ	Number of neighbors n	σ^{σ^n}	Number of rules N_r
2	2	2^{2^2}	16
2	3	2^{2^3}	256
2	5	2^{2^5}	4 294 967 296
2	10	$2^{2^{10}}$	$1.797 \cdot 10^{308}$
5	2	5^{5^2}	$2.98 \cdot 10^{17}$
5	3	5^{5^3}	$2.35 \cdot 10^{87}$
5	5	5^{5^5}	$1.91 \cdot 10^{2184}$
10	2	10^{10^2}	10^{100}
10	3	10^{10^3}	10^{1000}
10	5	10^{10^5}	10^{100000}

Langton studied the average type of behavior with respect to a statistic provided by the parameter λ of each rule table [13]. In a binary cellular automata, the parameter λ is the fraction of "ones" within the output column of a transition table. Advantages and disadvantages of both classification methods are discussed in detail elsewhere [14]. There have been a wide number of attempts to classify cellular automata rules without substantial success. All known classification techniques are not reliable (they misclassify rules quite easily). A search for a reliable classification technique is still needed for the design of new cellular automata rules having desired properties. Currently, the lack of such classification techniques is typically overcome by use of, e.g., genetic programming, genetic algorithms or any other evolutionary algorithm.

1.6 The Game of Life

Before we proceed let us emphasize that not all complex systems can be described mathematically. One of the best known examples of such a complex system is the Game of Life [8]. Due to its simplicity, it became a prototypical example of a complex system formulated by a cellular automaton.

The Game of Life represents a simple, vital, widely studied example displaying an extremely complex behavior arising from an extraordinary simple local rule [8]. This example is, either directly or indirectly, motivating many applications within such diverse fields as game theory, sociology, ecology, fire spreading, market behavior, etc. The transition rule consists of the evaluation of four independent logical conditions for each updated cell separately:

1. a living cell having less than two living neighbors dies (loneliness);
2. a living cell having more than three living neighbors dies (overcrowding);
3. a living cell having two or three living neighbors stay living (ideal living conditions);
4. a dead cell having exactly three living neighbors becomes alive (offspring).

When we take these simple rules and apply them to a randomly chosen initial configuration of living and dead cells within a lattice of $N \times M$ cells, amazing patterns start to appear. We observe configurations made of living cells transversing through space and persisting over time. Such configurations are called "gliders", see Fig. 1.4 for a sequence of snapshots. We might also find another configuration generating those "gliders" (called "glider-guns") or configurations destroying them (called "glider-eaters"), or oscillating structures. A lot of effort has been spent on the discovery of various structures emerging from the Game of Life.

Figure 1.5 shows two types of structures, a "glider-gun" and "gliders" produced by this "glider-gun". The "glider-gun" located at the bottom-left is continuously creating persistent, moving, living structures called "gliders" also shown in Fig. 1.4.

It is possible to create a whole set of logical and arithmetic operations using NOT, AND, and OR gates implemented within the Game of Life employing such building

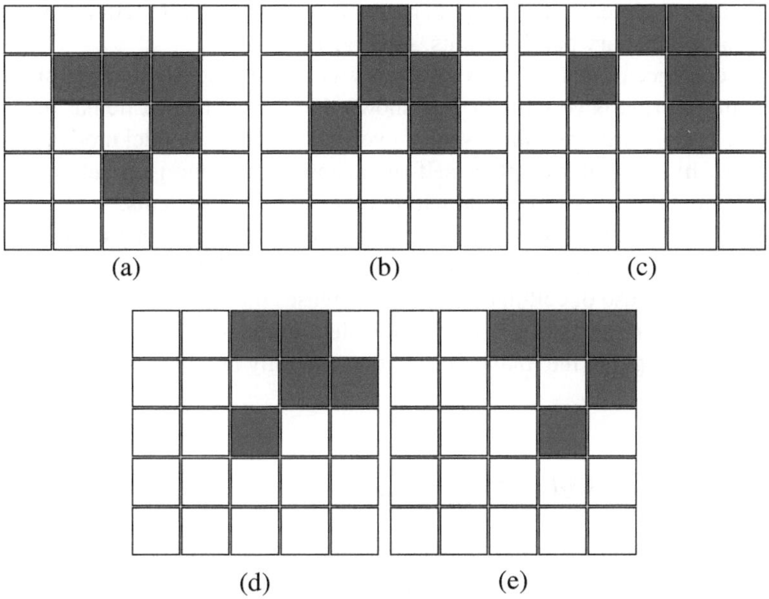

Fig. 1.4 A sequence of snapshots (**a**)–(**d**) depicting the propagation of initial configuration of one "glider" through a cellular space observed within the Game of Life. Coordinates of all cells are kept the same within all sub-figures. The "glider" moves by shifting itself one cell upwards and one cell to the right within four time steps and so on. There are four diagonally symmetrical directions of glider propagation

Fig. 1.5 One type of glider-gun (**bottom**) creating gliders (moving from the **left-bottom** corner to the **right-top** one) observed within the Game of Life. The whole logics and arithmetics can be build from them (such operations as NOT, AND, OR, etc.). Any recursive function can be build using such operations. Therefore the Game of Life is universal

units as gliders, glider-guns, and glider-eaters. Any recursive function can be build using those elementary building units, see Fig. 1.5.

The occurrence of real complex structures within a randomly initiated simulation of the Game of Life is very likely. It is shown that the Game of Life has the universality of a Turing machine. Surprisingly, even such a simple model motivated by the behavior of living matter displays self-organization and emergent behavior. These two fundamental processes operate within many complex systems.

There is no efficient way to "simulate" the game of life through a set of equations. In this case the efficiency of the cellular automata is much higher than any classical method, since the use of cellular automata decreases the level of dissipation of local information within the system. This is equivalent to the well-known Church-Turing hypothesis, stating that the final configuration can only be obtained through explicit simulation.

1.7 Advanced Cellular Automata

Classical cellular automata can be generalized in many ways. The best known and the most important generalizations, like the lattice Boltzmann method, networked automata, complex automata, asynchronous automata, quantum cellular automata, wetware, and real valued cellular automata are briefly introduced.

The lattice Boltzmann method was developed to simulate fluid flow as an alternative to the Navier–Stokes equations. It is based on the use of a discrete Boltzmann equation operating on a lattice of edges where virtual particles are moving. Particles have a distribution of velocities at each lattice point. They move along those edges and undergo collisions. Collisions are designed in a such way that time-averaged motion of particles is consistent with the Navier-Stokes equations. The lattice Boltzmann method is discussed in more detail in the chapter by Kusumaatmaja and Yeomans in this book. The lattice Boltzmann method is used to model a wide range of applications for single- and multiphase flows within complex geometries including porous media [5].

Networked automata appeared on the scene along with understanding that the lattice itself on which a particular cellular automaton operates can be understood as a network. The question is how a change of a regular lattice (i.e. a regular network) to more general types of networks influence the overall behavior of a specific cellular automaton under study. The answer is in many cases surprising. Such complicated networks as scale-free and small-world networks have in most cases a large, positive impact on the efficiency and robustness of the computation. What is even more important, such networks more precisely reflect natural phenomena which are described by cellular automata (e.g. social networks, gene regulatory networks, the Internet, etc.).

Complex automata (CxA) employ the idea that a multi-scale system can be decomposed into a number of single-scale automata, which are mutually interacting. The evolution of the system is driven by evolution at single-scale levels which are influenced by evolution at the other scale levels. There is a whole number of possible implementations of multi-scale models using cellular automata. Updating of the multi-scale model could be done in a variety of ways. This new type of cellular automata are introduced in this book by Hoekstra et al.

For a long time it was assumed that synchronous updating of the whole lattice of cells is fundamental for computations employing cellular automata. This assumption is proven to be too strict and can be weakened to *asynchronous automata*. Simply said, it is possible to develop asynchronous cellular automata rules, which perform the same tasks as synchronous cellular automata. The asynchronous updating mode of computation occurs in many real applications. It is known that an asynchronous CA for a particular task is more robust and error resistant than a synchronous equivalent.

Quantum cellular automata (QCA) proposed by Richard P. Feynman employ quantum mechanical phenomena in the design of computational models. They represents the quantum mechanical analogy of classical cellular automata. In many cases, they employ completely different computational principles.

Wetware represents one of the prospective media to implement cellular automata. This area of research remains mostly open for future discoveries. Employing chemical and/or biochemical reactions (going even up to the level of cells and tissues) may bring new computational tools (called wetware) potentially having a tremendous computational capacity (e.g. DNA computing). The term wetware stands for an abstraction used for hardware and software. Wetware is build on completely different computational principles compared to the ones currently used in silico circuits.

Some authors are using the notion of a cellular automaton even when the associated variables have real values. This is valuable in many models of naturally observed phenomena (e.g. recrystallization in solid state physics). Cellular automata using real values are often referred to as coupled map lattices (especially in physics).

1.8 Book Organization

The book is organized in three parts: (1) Theory of Cellular Automata; (2) Applications of Cellular Automata; and (3) Cellular Automata Software. The theory part contains fundamental chapters on different aspects of modeling complex systems with cellular automata. The applications part presents a number of representative examples of applications of Cellular Automata in a large range of scientific disciplines. The part on software highlights the important work that has been done to create high quality software environments that allow to quickly and relatively easily implement a Cellular Automata model and run simulations.

The first part on theory contains eight chapters, discussing fundamental issues related to complex systems modeling with Cellular Automata. An important class of complex systems are multi-level systems, where interactions between multiple levels to a large extend dictate their properties. In three chapters different aspects of this class of Cellular Automata Models are discussed. In Chap. 2 Hogeweg introduces multilevel Cellular Automata, where the states and rules of lower levels, that produce patterns on higher levels, can by adjusted by those very patterns in a feedback loop. Applications in Bioinformatic systems are reviewed. Hoekstra et al., in Chap. 3, propose Complex Automata as a formalism to model multi-scale complex systems. The idea is to decompose a multi-scale system into a set of single scale models, where each single scale model is represented by a CA. While Complex Automata assume a form of scale separation, Dunn in Chap. 4 introduces a strongly coupled hierarchical Cellular Automata to model multiscale, multiresolution systems, with special projection operators to formalize the coupling between the levels in the hierarchy. Applications in landscape ecology serve as an example.

Bandman, in Chap. 5, introduces a formalism for Cellular Automata as a model of spatially extended dynamical systems, focussing on composition techniques to develop and formally analyze advanced Cellular Automata models for such dynamical systems.

The great asset of Cellular Automata has always been to use them as minimal models capturing the main features of the dynamics of a system under study. In two chapters examples of such minimal Cellular Automata models are further explored. Chapter 6, by Umeo, discusses in detail 1-bit communication Cellular Automata, where inter-cell communication per step is restricted to one bit, applied to a number of applications. Adamatzky and Grube consider minimal Cellular Automata models for population dynamics, introducing only six very basic types of interaction between two species, and then exploring the behavior of their models as a function of these interactions.

Tomassini, in Chap. 8, introduces Cellular Evolutionary Algorithms, a class of probabilistic Cellular Automata that can be used to solve complex optimization problems, of which a number of examples are shown. Finally, in Chap. 9, Pan and Reggia continue the line of research that actually started of the field of Cellular Automata, that is, Cellular Automata for self-replicating structures. They give a historical overview, and discuss novel results using evolutionary algorithms to discover new self-replicating structures.

Part II contains 5 chapters showing impressive examples of applications of Cellular Automata to model complex systems in a large range of different scientific disciplines. Yu and Helbing discuss in Chap. 10 applications in Sociology, introducing migration games and studying in detail their behavior under a range of assumptions and parameters. On a quite different stance, Kusumaatmaja and Yeomans introduce in Chap. 11 the Lattice Boltzmann Method and its application to modeling complex fluids, specifically dynamics of drops and wetting dynamics.

Chowdhury, Nishinari, Schadschneider introduce in Chap. 12 single-and two-lane CA models for ant trails, and their validation against empirical data. Next, in Chap. 13, Hatzikirou and Deutsch discuss how the Lattice Gas Cellular Automaton can be applied as a model for interacting biological cells. They study pattern formation of growing cellular populations, and apply their models to simulate growth of solid tumors.

Gürdal and Zakhma show how Cellular Automata can be applied in an engineering context. In Chap. 14 they discuss in detail how Cellular Automata can be applied to topology design optimization and provide three challenging examples.

Finally Part III contains one chapter, by Talia and Naumov, on Cellular Automata software. They review a range of problem solving environments for Cellular Automata modeling and simulation, and discuss in some detail systems that facilitate parallel cellular programming.

1.9 Additional Resources

A wide number of high quality sources providing necessary starting information about cellular automata, their implementation, and their use in modeling of complex systems exist; among others there are [10, 23, 22, 25, 27, 18, 21, 12]. In the case you have no prior experience with cellular automata, it might be good to start with reading of the book of Mitchel Resnick [17] and use the StarLogo programable modeling environment [18], which is suitable for description of decentralized systems (e.g. ant colonies and market economies).

It is also recommended to study the book of Toffoli and Margolus [23] which presents a wide number of cellular automata models implemented in their specially dedicated computer with a computational environment called Cellular Automata Machine (CAM). Many more cellular automata software packages are available. We suggest to visit the web page of the IFIP Working Group 1.5 on Cellular Automata and Machines [11] and consult the review provided in Chap. 15 of this book.

A series of bi-annual cellular automata conferences [28, 20, 2] named ACRI presents every second year a wide number a valuable, up-to-date overview of cellular automata models. If you intend to use cellular automata to model a naturally observed phenomenon then this source – beside this book – might be a good starting point of your search for cellular automata models that are solving the same or similar problems.

References

1. P. Bak, *How Nature Works: The Science of Self-Organized Criticality*. (Springer, New York, NY, 1996)
2. S. Bandini, B. Chopard, M. Tomassini (eds.), Cellular Automata, 5th International Conference on Cellular Automata for Research and Industry, ACRI 2002, Geneva, Switzerland, October 9–11, 2002, Proceedings Lecture Notes in Computer Science, vol. 2493 (Springer, Heidelberg, 2002)
3. L. Berec, Techniques of spatially explicit individual-based models: Construction, simulation and mean-field analysis. Ecol. Model. **150**, 55–81 (2002)
4. N. Boccara, *Modeling Complex Systems* (Springer, Heidelberg, 2004)
5. B. Chopard, M. Droz, *Cellular Automata Modeling of Physical Systems*, (Cambridge University Press, Cambridge, 2005)
6. K. Christensen, N. Moloney, *Complexity and Criticality* (Imperial College Press, London, 2005)
7. A. Deutch, S. Dormann, *Cellular Automaton Modeling of Biological Pattern Formation* (Birkhauser, Basel, 2004)

8. M. Gardner, The fantastic combinations of John Conway's new solitaire game "life". Sci. Am. **223**, 120–123 (1970)
9. B. Hölldobler, E. Wilson, *Journey to the Ants: A Story of Scientific Exploration*, 3rd edn. (Harvard University Press, Cambridge, MA, 1995)
10. A. Ilachinski, *Cellular Automata: A Discrete Universe* (World Scientific Publishing Co. Pte. Ltd., London, 2001)
11. International Federation for Information Processing (IFIP), Working Group 1.5 on Cellular Automata and Machines, http://liinwww.ira.uka.de/ca/software/index.html: A list of software packages for Cellular Automata, http://liinwww.ira.uka.de/ca/software/index.html
12. J. Kroc, Special issue on modelling of complex systems by cellular automata 2007: Guest editors' introduction. Adv. Compl. Syst. **10**(1 supp), 1–3 (2007)
13. C. Langton, Computation at the edge of chaos: Phase transitions and emergent computation. Physica D **42**, 12–27 (1990)
14. M. Mitchell, *Nonstandard Computation*, chap. Computation in cellular automata: a selected review (VCH, Weinheim Verlagsgesellschaft, 1998) pp. 95–140
15. A. Nathan, V. Barbosa, V-like formations in flocks of artificial birds. ArXiv Computer Science e-prints (2006), http://arxiv.org/abs/cs/0611032
16. J. von Neumann, A. Burks, *Theory of Self-Reproducing Automata* (University of Illinois Press, Urbana, IL 1966)
17. M. Resnick, *Turtles, Termites, and Traffic Jams – Explorations in Massively Parallel Microworlds* (The MIT Press, Cambridge, MA, 1997)
18. M. Resnick, StarLogo – programmable environment for exploring decentralized systems flocks, traffic jams, termite and ant colonies. Tech. rep., MIT (2006), http://education.mit.edu/starlogo/
19. P.M.A. Sloot, B.J. Overeinder, A. Schoneveld, Self organized criticality in simulated correlated systems. Comput. Phys. Comm. **142**, 66–81 (2001)
20. P. Sloot, B. Chopard, A. Hoekstra (eds.), Cellular Automata, 6th International Conference on Cellular Automata for Research and Industry, ACRI 2004, Amsterdam, The Netherlands, October 25–28, 2004, Proceedings. Lecture Notes in Computer Science, vol. 3305 (Springer, Heidelberg, 2004)
21. P.M.A. Sloot, A.G. Hoekstra, Modeling dynamic systems with cellular automata, ed. by P.A. Fishwick *Handbook of Dynamic System Modelling* chapter 21 (Chapman and Hall London, 2007)
22. T. Toffoli, Cellular automata as an alternative to (rather than an approximation of) differential equations in modelling physics. Physica **D17**, 117–127 (1984)
23. T. Toffoli, N. Margolus, *Cellular Automata Machines: A New Environment for Modeling* (MIT Press, Cambridge, MA 1987)
24. A.M. Turing, *The Essential Turing: Seminal Writings in Computing, Logic, Philosophy, Artificial Intelligence, and Artificial Life plus The Secrets of Enigma* (Oxford University Press, New York, NY, 2004)
25. G. Vichniac, Simulating physics with cellular automata. Physica **D17**, 96–116 (1984)
26. S. Wolfram, Universality and complexity in cellular automata. Physica D 1–35 (1984)
27. S. Wolfram, *A New Kind of Science* (Wolfram Media Inc., Champaign, Il 2002)
28. S.E. Yacoubi, B. Chopard, S. Bandini, (eds.) Cellular Automata, 7th International Conference on Cellular Automata, for Research and Industry, ACRI 2006, Perpignan, France, September 20–23, 2006, Proceedings Lecture Notes in Computer Science, vol. 4173 (Springer, Heidelberg, 2006)

Part I
Theory of Cellular Automata

Chapter 2
Multilevel Cellular Automata as a Tool for Studying Bioinformatic Processes

Paulien Hogeweg

2.1 Introduction: *"one more soul"*

The signature feature of Cellular Automata is the realization that "simple rules can give rise to complex behavior". In particular how fixed "rock-bottom" simple rules can give rise to multiple levels of organization. Here we describe Multilevel Cellular Automata, in which the microscopic entities (states) and their transition rules themselves are adjusted by the mesoscale patterns that they themselves generate. Thus we study the feedback of higher levels of organization on the lower levels. Such an approach is preeminently important for studying bioinformatic systems. We will here focus on an evolutionary approach to formalize such Multilevel Cellular Automata, and review examples of studies that use them.

At the 2004 meeting on which the current book is based, Toffoli nicely discussed how science has annihilated one by one the "souls" around us, by bringing the phenomena they represent, which were deemed to lie outside the scientific realm, into the scientific discourse. Cellular Automata (CA) have played their part in this pursuit e.g. by von Neumann's existence proof on self-reproduction [19]. More generally CA have been the preeminent environment to demonstrate that simple local interactions can lead to complex "emergent" behavior at different scales of observation. Therewith it has not only eliminated "souls" but also many more mundane but too complex explanations of complex behavior. It may be argued, however, that CA have left one "soul" in tact: the rules themselves as an externally given rock-bottom. In this paper I will describe approaches to eliminate (or at least soften) this last "soul" in CA-like models. Moreover I will argue that for understanding biological complexity it is essential to allow not only for higher level emergent properties, but also to allow the lower level "entities" to become emergent properties of the system.

P. Hogeweg (✉)
Theoretical Biology & Bioinformatics Group, Utrecht University, Padualaan 8, 3584 CH, Utrecht, The Netherlands
e-mail: p.hogeweg@bio.uu.nl

A.G. Hoekstra et al. (eds.), *Simulating Complex Systems by Cellular Automata*, Understanding Complex Systems, DOI 10.1007/978-3-642-12203-3_2,

2.2 Modeling Bioinformatic Systems

A pre-eminent feature of biological systems is the intertwining of many levels of organization, spanning many orders of magnitude. For example changes of one nucleotide can change the shape of a leaf, which means a blow-up in scale of 10^{10}, as Enrico Coen recently[1] vividly reminded his audience by arguing that a human cannot have such a large effect, even by exploding a nuclear bomb. The percolation of effects through different scales is even more impressive when combined with the robustness the system exhibits. To understand the interactions between different scales, at whatever level, is a pre-eminent goal of bioinformatic research. As mentioned above, CA play an important role to understand how apparent complex phenomena at multiple scales can result form simple local interactions. The multiple scale nature of even the simplest CA has been beautifully demonstrated by Crutchfield and his collaborators ([9] and Crutchfield, this meeting) in his *tour the force* analysis of the elementary CA in terms of higher level "virtual particles" and their (reaction kinetic like) interactions, where the particles are the "defects" in the most frequent patterns formed by the CA. This illustrates the power of CA in "bottom up" studies (from lower level to higher level behavior) which make them an important tool in bioinformatic research. However for bioinformatic research we have to go beyond this because we can neither work only "bottom up" nor "top down" but have to work "middle out". This is because when we start at an intermediate scale the lower level entities cannot be treated simply as new rock-bottom micro entities for the following reasons.

- The micro entities we deal with are often not invariant, e.g. when we take cells as our micro level, their properties change through transcription regulation or mutation, and also directly through stress and strain caused by differential adhesion between cells.
- The micro entities, and the "rules" that govern their behavior, are not necessarily as simple as possible. As Zhou et al. [23] formulated it, living systems have a high "design degrees of freedom (DDOF)" and therefore may "choose" a solution that is quite different than the simple solution that many of us would stipulate. A great challenge is therefore to design modeling approaches that enable us to get insight not only in simple to complex mappings but also in complex to complex mappings.
- The "rules" that govern the behavior of the entities are neither "arbitrary" nor "some universal invariants" but are evolved. An evolutionary perspective is essential for understanding present day organisms. But also the ultimate aim of bioinformatic research is not only to understand the information transmission from genetic information to higher levels but also how the genetic information itself was generated over evolutionary time, and therewith "why it is as it is".

---

[1] BSDB autumn meeting 2007, Sheffield, UK.

Therefore we need a "middle-out" approach that allows for the feedback from mesoscale entities to the microscale entities which generate them so that both become emergent properties of the system. Here we take an evolutionary approach to the challenging problem of defining multilevel CA. Note however that several other multilevel CA have been formulated. For example in urban planning studies where different features of the landscape are modeled in different layers of the CA, and the rules include inter-level interactions and also may include rules for modifying the rules on the basis of the dynamics of other layers, e.g. [4]. Another interesting example is the so-called CPM formalism, used in e.g. foam physics and biological development. In CPM the mesoscale entities (cells) are priory defined and the feedback to the microscale is indirectly defined through an energy minimization process [8, 7, 17]; the interaction between micro and mesoscale entities generates very rich behavior and many scales (e.g [18]) and, it can be combined with an evolutionary approach (e.g. [14]).

The evolutionary approach I discuss here, allows the coupling the CA paradigm of "simple rules gives rise to complex behavior" to the famous observation of Dobzhansky "Nothing in biology makes sense except in the light of evolution"[6].

2.3 Multiscale Processes in Standard CA Models: Examples from Ecology

Ecology studies interactions between various species. Classical ecological models do this by defining interactions between populations, where all individuals of the population are assumed to be the same (or at least in an equilibrium distribution of variants) and not changing over time. Using CA as a paradigm, ecological modeling can instead take individuals as basic units. In such models the state of one automaton (cell[2]) in the CA represents the presence of an individual of a certain species at the corresponding location, whereas state 0 means that no individual is present (this encoding assumes that only one individual can be present at such a location). The (probabilistic) CA rules encode the interactions between individuals. Such interactions are thus assumed to be local as they necessarily are in real ecosystems, making CA an important tool in ecological research [11]. Modeling ecological interactions in this way will lead in many cases to a patterned distribution of the species over space. For example predator-prey or host-parasite interactions will for large parameter ranges lead to large scale wave-like patterns, whereas (interference) competition leads to static large scale patterns similar to those of voting rules. The formation of such mesoscale patterns (i.e. patterns at some scale in between the micro scale on which the rules of the CA operate and the CA as a whole) may influence the dynamics and persistence of the ecosystem as a whole as in als well as of each of the species. Within this framework we can study the fate of mutants, i.e.

---

[2] In biological context I will avoid to use the word cell for the automaton to prevent confusion with biological cells.

new individuals created by a mutation event (modeled as new states with slightly modified transition rules). by studying whether they can invade and/or persist in the ecosystem. In this way Boerlijst and Hogeweg [2, 3], showed clearly that mesoscale patterns drastically alter generally held expectations about invasions and persistence of mutants. They showed, for example, that a mutant with higher death rate, but equal in all other aspects to the resident population can invade and even replace the resident population. The explanation of this counter-intuitive phenomenon is that the competition shifts from the level of individuals to the level of mesoscale patterns. In the case mentioned these mesoscale patterns are spiral waves. The spiral waves with the species with the higher death rate rotate faster and therefore take over the domain of those with the species of the lower death rate, which rotate slower. All this fits in the classical CA formalism. It indeed exemplifies the "simple rules to complex phenomena" paradigm in a very nice way: not only do "emergent phenomena" arise, but they do, in fact, influence the dynamics of the system at both the microlevel and the level of the entire automaton, i.e. on macroscopic variables as which states (species) are present in equilibrium. These features of CA models have contributed to our understanding of ecological and invasion dynamics.

However, individuals within population are not genetically, or developmentally invariant, and the distribution of these variations is not invariant over time. Many strikingly fast changes in the properties have been reported, over time-frames shorter than equilibration of the ecological dynamics (see also [16]). In other words in order to understand ecological systems we should widen our scope to include evolutionary processes, i.e. we should shift from studying ecological dynamics *sensu strictu* to studying eco-evolutionary dynamics and eco-informatics [15]. This can be done by extending CAs to allow for expansion and/or change of the prior defined set of states and their transition rules, generating new "states" "on the fly", as described below.

2.4 Emergent Microscale Entities in Evolutionary CA Models: An Example

As an example of pattern generation (emergent properties) at multiple scales, including the scale of micro entities, I review here our work on "Evolution of Complexity in RNA-like replicators" [22]. The biological setting is pre-biotic evolution, and the question is how in early evolution the so-called information threshold could be circumvented. Here I emphasize the "middle-out" aspect of this work in the sense explained above.

A prerequisite for generating new microscale entities on the fly, is that their structure allows interaction to be computed.[3] Thus we have to switch from micro states

[3] One could argue, however, that strictly speaking our evolutionary CAs are normal CAs except with a huge number of states for each automaton where mesoscale pattern select a subset of those. Such an argument is however equivalent to saying that we do not need the concept to CA itself, because we can reformulate every (finite) CA as a single finite state machine with a huge number of states as CA are just a restricted subset of such huge finite state machines.

to micro entities, terms we used intermingled above. In the middle out approach these do not have to be very simple.

Here we use RNA sequences, i.e. strings of four bases, A,C,G,U, of length 50 as our microscale entities. Binding strength between the 4 bases are defined. From this follows a minimum energy secondary structure (folding). The generalized interaction rules are defined in terms of the secondary structure and the binding of the so-called 5′ and 3′ (open) ends (i.e. the left and right ends) of two of "molecules". A specific secondary structure (which can be realized by many, but a small subset of the sequences) defines a catalyst. Binding to the 5′ end a catalyst can lead to the replication: i.e. an automaton in state 0 can get the state of the complementary string of the replicating sequence. (so two rounds of "replication" are needed to produce the same sequence again). Binding to a non-catalyst has no effect (except implicitly by not being available for binding to a catalyst). Thus, a sequence (genotype), to structure (phenotype), to interaction mapping is defined (Fig. 2.1) In addition "mutations" can take place, i.e. a base of a sequence changes into another base (several mutations can occur at one replication step). Thus, new micro-entities are generated, and their interactions with other (present) micro-entities can be computed from the genotype-phenotype-interaction mapping. Finally any non-zero state has a certain probability to change into a zero state (i.e. decay of the molecules). For more details see [22]. Note that in this evolutionary model no external fitness criterion is defined: as in biological evolution only survival matters. The questions studied by this approach are therefore not how can we mimic some predefined (observed) behavior, but how/when does such a non-supervised evolutionary process give rise to complexity at various levels.

The upper part of Fig. 2.1 shows the feedback of the mesoscale to the microscale in such a system. Through mutations a single initial starting sequence generates a polymorphic so-called "quasi-species", i.e. a set of strings in its mutational neighborhood. Through the differential interaction between the different sequence/structures spatial patterns emerge, in this case mostly chaotic waves. Binding probabilities and chaotic waves together determine which sequences can "survive" (i.e. which sequences (states) occur). This process can lead to "speciation", the emergence of several lineages of replicators. The (long-term) occurrence of these lineages is not because of the mutational process (as is the case for quasi-species polymorphism) but because of their separate "functions" in the ecosystem.

These processes are shown in Fig. 2.2. The upper left panel shows the absence of any spatial structure in the initial, non-evolutionary CA: One particular replicator consisting of a catalytic sequence and its complement (colored the same) maintains

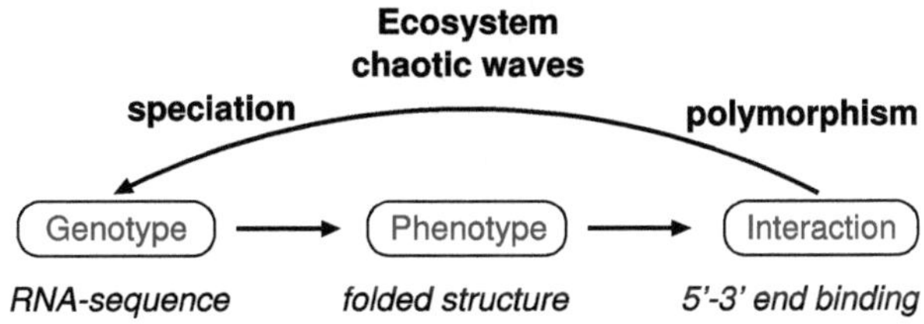

Fig. 2.1 Scheme of feedback of mesoscale patterns to microscale entities

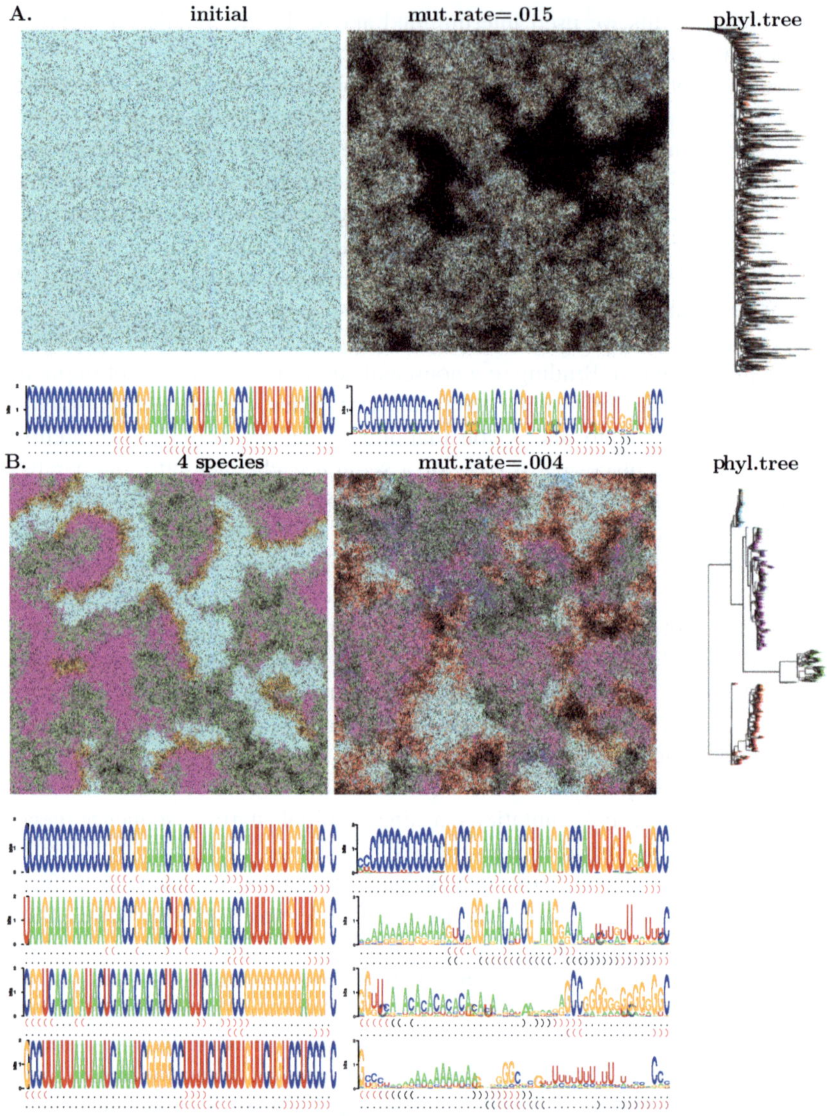

Fig. 2.2 Multilevel evolutionary CA: mutually dependent patterns and mesoscale and microscale. (**a**) Initial ecosystem (mutation rate 0), and eco-evolutionary system (mutation rate 0.015) with phylogenetic tree, showing one quasi-species. (**b**) Evolved ecosystem and eco-evolutionary system (mutation rate 0.004) with phylogenetic tree showing 4 quasi-species. Below the snapshots of the CA are the Sequence(logo): the larger the letter the more conserved that position is in the system. Below the sequence logo is the consensus secondary structure for each (quasi) species in *bracket* notation: corresponding *open* and *closing parentheses* indicate a 5′ to 3′ end base-pairing, *dots* are unbound bases. For a colored version of this figure see: http://www-binf.bio.uu.nl/ph/figacri.pdf

itself in the system, with empty places because of decay. At high mutation rates (upper right panel) a high degree of polymorphism emerges leading to spatial pattern formation, i.e. chaotic waves of non-zero and zero states. The polymorphism is "non-structured" as seen in the phylogenetic tree (clustering) on the right and consist of relatively close mutants of the initial sequence. In other words no speciation occurs. Nevertheless these close mutants may fold differently, and may bind stronger (with both complementary strands) to the catalyst, locally out-competing it. This process leads to the spatial pattern formation. When mutation rate is set to 0, the system as initially defined is recovered, with identical sequences and no spatial pattern formation. Survival of the system including mutations depends on the spatial pattern formation: when spatial pattern formation is prevented by mixing the entities(states) between every reproduction step, the system quickly dies out (all 0 state) because the catalysts goes extinct due to competition with mutants. This extinction happens in well mixed system for any mutation rate. We further only consider the non-mixed system.

In contrast to the results at high mutations rates, at somewhat lower mutation rates, speciation does occur in the full system. This can be clearly seen from the phylogenetic tree on the right: there are 4 quasi-species. Moreover, if mutation is stopped 4 mono-morphic species (i.e. 2×4 sequences because of the complementarity) survive indefinitely. There are 2 catalytic species (cyan, corresponding to the initial sequence, and magenta) and 2 non-catalytic species (parasites) (red and green). The binding probabilities are given in the Table 2.1

Table 2.1 Complex formation happens with a probability p as indicated in table when the two types of "molecules" are neighbors and interact. Complexes dissociate with a probability $(1 - p)$ Replication happens when a complex and an empty spot (state==0) interact; the complement of the string binding to the catalyst is produced and occupies the empty spot; the complex dissociates if replication happens. All non-zero states have a decay probability of 0.03, i.e. to become state 0

initial 1 species ecosystem
2 states and transition rules

	C-catalyst	
	catal.str.	comp.
C-cat	0.528	0.878

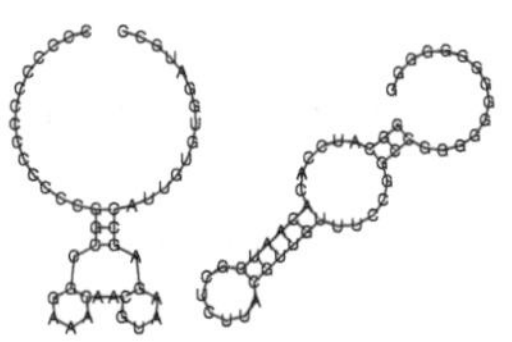

structure of catalyst and its compl. string

evolved 4 species ecosystem
8 states and transition rules

	C-catalyst CYAN		A-catalyst MAGENTA		G-parasite RED		U-parasite GREEN	
	catal.str.	comp.	catal.str.	comp.	logostr.	comp.	logostr.	comp.
C-cat	0.528	0.878	0.362	0.451	0.808	0.650	0.259	0.362
A-cat	0.393	0.049	0.503	0.765	0.139	0.478	0.632	0.551

The red parasite is strongly catalyzed by the cyan catalyst, the green one by the magenta one. Accordingly the emergent mesocale patterns in space show a succession of cyan, red, magenta and green waves, as red out-competes cyan, while the weaker catalyst (magenta), which catalyzes itself more than red succeeds it, but is itself out-competed by green, after which cyan invades again. In the full eco-evolutionary system the same succession of quasi-species can be seen. The different levels are tightly interwoven: the sequence, the folding structure and the spatial patterns can only be understood in terms of each other. Repeated runs converge to a similar organization: 2 catalysts and 2 parasites with similar relative binding strength, with the catalysts having many C or A's in the $5'$ end and the parasites having many G's or U's in a very long $3'$ end. The emerging folding structure of parasites and complementary strands can in hindsight be understood in terms of interaction avoidance.

In conclusion, in the evolutionary CA described above micro-scale and mesoscale patterns emerge in mutual dependence. The emerging interaction topology was very surprising: despite the fact that many interaction structures have been studied in ecology in general and prebiotic evolution in particular, no-one has ever proposed the one that emerged in the present study. All interaction topologies which were proposed proved to be vulnerable to high mutation rates and invasion of parasites, leading to extinction of the entire system by extinction of catalysts. The remarkable robustness of the systems depends on this topology, but not on the topology alone: also on the chosen coding structure of the sequences and folding structure which shape the mutational landscape, as well as the spatial patterns. Neither arbitrary nor "designed" microscale interaction are likely to exhibit the robustness of the system that emerged in our experiments. However rare the robustness may be in possibility space, it is repeatedly found in independent runs, i.e. they are both rare and likely to occur. It is very satisfying to find that the simple multiscale evolutionary CA approach allows us to investigate such rare but likely cases as indeed biological systems seem to belong to that category!

2.5 Evolutionary CA and Evolutionary Computation

In the field of evolutionary computation, CA have been combined with evolution in mainly three different ways, i.e.

(1) The evolution of a CA with specific properties, e.g. density classification [5] or diversity in replicator systems [12].
(2) The evolution of micro-scale entities that solve problems by co-evolution of solvers and problems in a CA setting (e.g the evolution of sorting algorithms [10] or function approximation [20, 1]).
(3) Both approaches can also be combined as we showed by evolving density classifiers in a co-evolutionary manner, improving performance as well as altering the type of solutions found: from mostly majority based solutions to the more intricate particle based solutions [21].

These approaches differ from the one described above in that there is an externally imposed fitness criterion. Thus what comes out has been much more predetermined. Nevertheless such systems, especially when the fitness criterion is very general (e.g. diversity) or high fitness can we attained in many different ways (given the basic building blocks), can share to a large extend the mutual multilevel properties described above. For example, although in the co-evolutionary CAs two different "species" are defined (host parasitoids) with predefined types of interactions, further speciation within the host and parasitoid populations occurs through indirect interactions and pattern formation. Indeed these mutual multilevel properties are essential for attaining the global performance fitness [1]. Recently we have shown that through a speciation process ecosystem based problem solving can be obtained. In this case the eco-evolutionary process automatically decomposes each of the problems in sub problems that are solved by different species (de Boer and Hogeweg submitted). I conjecture that designing such systems so as to maximize the degrees of freedom of the multilevel interactions may improve the performance.

2.6 Conclusion

We have defined evolutionary multilevel CA and we have shown how microscale and mesoscale entities emerge in consort. In doing so we deviate from the usual CA approach in which the microscale is the rock-bottom on which only higher level entities emerge. We have seen that through this approach we can zoom on "rare but likely" cases with, in some sense, superior properties (e.g. robustness). Moreover it allows us to go beyond the "simple rules give complex behavior" to begin studying in a meaningful and relatively simple way "how complex rules give rise to complex behavior" (and vice versa) (see also [13]) . Doing so is necessary for studying bioinformatic processes keeping in mind Einsteins famous dictum *"Everything should be made as simple as possible, but no simpler"*.

Acknowledgments I thank Nobuto Takeuchi for advancing the modeling methodology described here, for providing the figures and for critical discussions on the manuscript.

References

1. F. de Boer, P. Hogeweg, The role of speciation in spatial coevolutionary function approximation. In *Proceedings of the 2007 GECCO Conference Companion on Genetic and Evolutionary Computation*, http://doi.acm.org/10.1145/1274000.1274007 (2007) pp. 2437–2441
2. M.C. Boerlijst, P. Hogeweg, Selfstructuring and selection: Spiral waves as a substrate for prebiotic evolution, ed. by C.G. Langton, C. Taylor, J.D. Farmer, S. Rasmussen, *Artificial Life II* (Addison Wesley, 1991), pp. 255–276
3. M.C. Boerlijst, P. Hogeweg, Spiral wave structure in pre-biotic evolution: Hypercycles stable against parasites. Physica D, **48**(1), 17–28 (1991)
4. K.C. Clarke, S. Hoppen, L. Gaydos, A self-modifying cellular automaton model of historical urbanization in the San Francisco Bay area. *Environ. Plann. B* **24**, 247–262 (1997)

5. J.P. Crutchfield, M. Mitchell, The evolution of emergent computation. Proc. Natl. Acad. Sci. **92**(23), 10742–10746 (1995)
6. T. Dobzhansky, Nothing in biology makes sense except in the light of evolution. Am. Biol. Teach., **35**, 125–129 (1973)
7. J.A. Glazier, A. Balter, N.J. Poplawski, II. 1 Magnetization to morphogenesis: A brief history of the Glazier-Graner-Hogeweg Model. *Single-Cell-Based Models in Biology and Medicine* (Birkhauser Verlag, Basel/Switzerland, 2007), p. 79
8. J.A. Glazier, F. Graner, Simulation of the differential adhesion driven rearrangement of biological cells. Phy. Rev. E **47**(3), 2128–2154 (1993)
9. J.E. Hanson, J.P. Crutchfield, Computational mechanics of cellular automata: An example. Physica D, **103**(1–4), 169–189 (1997)
10. W.D. Hillis, Co-evolving parasites improve simulated evolution as an optimization procedure. Physica D **42**(1–3), 228–234 (1990)
11. P. Hogeweg, Cellular automata as a paradigm for ecological modeling. Appl. Math. Comput. **27**(1), 81–100, 1988
12. P. Hogeweg, Multilevel evolution: replicators and the evolution of diversity. Physica D **75**(1–3), 275–291 (1994)
13. P. Hogeweg, On searching generic properties of non generic phenomena: An approach to bioinformatic theory formation. In *Artificial Life VI: Proceedings of the Sixth International Conference on Artificial Life*, MIT Press, Cambridge, MA (1998), p. 286
14. P. Hogeweg, Evolving mechanisms of morphogenesis: On the interplay between differential adhesion and cell differentiation. *J. Theor. Biol.* **203**, 317–333 (2000)
15. P. Hogeweg, From population dynamics to ecoinformatics: Ecosystems as multilevel information processing systems. *Ecol. Inform.* **2**(2), 103–111 (2007)
16. J.D. van der Laan, P. Hogeweg, Predator-prey coevolution: Interactions across different timescales. In *Proceedings: Biological Sciences*, (1995) pp. 35–42
17. A.F.M. Marée, V.A. Grieneisen, P. Hogeweg, II. 2 The cellular potts model and biophysical properties of cells, tissues and morphogenesis. *Single-Cell-Based Models in Biology and Medicine*, (Birkhauser Verlag, Basel/Switzerland, 2007), p. 107
18. A.F.M. Marée, P. Hogeweg, How amoeboids self-organize into a fruiting body: Multicellular coordination in Dictyostelium discoideum. Proc. Natl. Acad. Sci. **98**(7), 3879 (2001)
19. J. von Neumann, A.W. Burks *Theory of Self-Reproducing Automata* (University of Illinois Press, Urbana, IL, 1966)
20. L. Pagie, P. Hogeweg, Evolutionary consequences of coevolving targets. Evol. Comput. **5**(4):401–418 (1997)
21. L. Pagie, P. Hogeweg, Information integration and red queen dynamics in coevolutionary optimization. In *Proceedings of the 2000 Congress on Evolutionary Computation, CEC*, IEEE Press, (2000), http://ieeexplore.ieee.org/xpl/tocresult.jsp? pp. 1260–1267
22. N. Takeuchi, P. Hogeweg, Evolution of complexity in RNA-like replicator systems. Biol. Direct **3**, 11 (2008)
23. T. Zhou, J.M. Carlson, J. Doyle, Mutation, specialization, and hypersensitivity in highly optimized tolerance. Proc. Natl. Acad. Sci. **99**(4), 2049 (2002)

Chapter 3
Complex Automata: Multi-scale Modeling with Coupled Cellular Automata

Alfons G. Hoekstra, Alfonso Caiazzo, Eric Lorenz, Jean-Luc Falcone, and Bastien Chopard

3.1 Multi-scale Modeling

3.1.1 Introduction

Cellular Automata (CA) are generally acknowledged to be a powerful way to describe and model natural phenomena [1–3]. There are even tempting claims that nature itself is one big (quantum) information processing system, e.g. [4], and that CA may actually be nature's way to do this processing [5–7]. We will not embark on this philosophical road, but ask ourselves a more mundane question. Can we use CA to model the inherently multi-scale processes in nature *and* use these models for efficient simulations on digital computers?

The ever increasing availability of experimental data on every scale, from "atom to material" or from "gene to health", in combination with the likewise ever increasing computational power [8, 9], facilitate the modeling and simulation of natural phenomena taking into account all the required spatial and temporal scales (see e.g. [10]). Multi-scale modeling and simulation, as a paradigm in Computational Science, is becoming more and more important, as witnessed by e.g. dedicated special issues [11] and thematic journals [12, 13].

Consider for example the field of physiology. The sequence from the genome, proteome, metabolome, physiome to health comprises multi-scale, multi-science systems [14, 15]. Studying biological sub-systems, their organization, and their mutual interactions, through an interplay between laboratory experiments and modeling and simulation, should lead to an understanding of biological function and to a prediction of the effects of perturbations (e.g. genetic mutations or presence of drugs) [16]. The concept "from genes to health" is the vision of the Physiome [17] and ViroLab [18] projects, where multi-scale modeling and simulation of aspects of human physiology is the ultimate goal. Modeling such systems is a challenging

A.G. Hoekstra (✉)
Computational Science, Faculty of Science, University of Amsterdam,
Science Park 107, 1098 XG, Amsterdam, The Netherlands
e-mail: a.g.hoekstra@uva.nl

A.G. Hoekstra et al. (eds.), *Simulating Complex Systems by Cellular Automata*,
Understanding Complex Systems, DOI 10.1007/978-3-642-12203-3_3,

problem but has the potential to improve our understanding of key interactions. The inherent complexity of biomedical systems is now beginning to be appreciated fully; they are multi-scale, multi-science systems, covering a range of phenomena from molecular and cellular biology, via physics and medicine, to engineering and crossing many orders of magnitude with regard to temporal and spatial scales [19].

Despite the widely acknowledged need for multi-scale modeling and simulation, there is a scarcity of underpinning literature on methodology and generic description of the process. There are many excellent papers that present multi-scale models, but few methodological papers on multi-scale modeling (such as [20, 21]) have appeared.

When using Cellular Automata to model a natural process, the lattice spacing and time step have a clear meaning in relation to the corresponding physical space and time of the process. We denote by $A(\Delta x, \Delta t, L, T)$ the spatio-temporal domain of a CA, whose spatial domain is made of cells of size Δx and it spans a region of size L, while the quantity Δt is the time step and T is the end of the simulated time interval. Therefore, processes with time scales between Δt and T can be represented and spatial scales ranging from Δx to L can be resolved. When executing such CA on a digital computer we note that the execution time T_{ex} scales as

$$T_{\text{ex}} \sim \frac{T}{\Delta t}\left(\frac{L}{\Delta x}\right)^{D}, \tag{3.1}$$

where D is the spatial dimension of the simulated domain. Trying to model a multi-scale system with a single CA would require to choose Δx and Δt in such a way that the smallest microscopic details and fastest dynamical response of the system are captured, yet the overall system size (L) and slowest dynamical time scale (T) need to be covered. For instance, in modeling human physiology the relevant range of spatial scales is from nanometer to meter (i.e. a factor 10^9) whereas temporal scale is from microseconds to human lifetime (i.e. a factor 10^{15}). These numbers, in combination with Eq. (3.1) immediately show that one will probably never be able to simulate multi-scale systems with a single CA spanning such a wide range of scales.

The literature on using Cellular Automata to model multi-scale phenomena is relatively small, maybe with the exception of using CA to model land usage and geographical systems (e.g. [22]). Furthermore, many papers exist that use CA in multi-scale modeling, but there CA is typically coupled to other types of models (e.g. [23]). The bulk of CA multi-scale attempts are grid refinement methods, also termed multi-blocks. The idea is to adapt the local grid size to the local process scale, i.e. using a fine grid in regions where small scale processes occur and a coarse grid where larger scales are sufficient. A common approach is to couple grids of different scales with an overlap region [24].

Other ways of coupling multi-scale CA come from two theoretical frameworks. The first one is based on higher-order CA [25]. In this framework, the CA rules are not only able to change the cell state, but also the rules themselves, the neighborhood and the topology. Moreover, these models are also able to take into account hierarchical CA where higher level cells are connected to one or more lower level cells. The second one results from the work of Israeli and Goldenfeld [26] who have shown that it is possible to coarse-grain 1D nearest-neighbor CA, by defining a macroscopic CA whose behavior is similar to a microscopic CA. That is an important result because the authors have achieved the coarse-graining of CA known to be irreducible.

We developed a multi-scale, multi-science framework, coined Complex Automata (CxA), for modeling and simulation of multi-scale complex systems [27–29]. The key idea is that a multi-scale system can be decomposed into N single-scale CA that mutually interact across the scales.[1] Decomposition is facilitated by building a Scale Separation Map (SSM) on which each single-scale system can be represented as an area according to its spatial and temporal scales. Processes having well-separated scales are easily identified as the components of the multi-scale model. We validate the CxA approach by building a large set of exemplary applications, and applying it to the challenging clinical problem of in-stent restenosis (ISR) [30]. The CxA approach was developed within the context of the COAST project [31].

In this chapter we will review the current state of development of Complex Automata and explore the possibilities that are offered by Cellular Automata (CA) for multi-scale Modeling and Simulation.

3.2 Complex Automata

3.2.1 A Definition

Formally, we shall define a CA as a tuple

$$\mathcal{C} = \{A(\Delta x, L, \Delta t, T), \mathbb{F}, \Phi, f_{\text{init}} \in \mathbb{F}, \mathbf{u}, \mathbf{O}\}. \quad (3.2)$$

A is the domain, made of spatial cells of size Δx and spanning a region of size L, while the quantity Δt is the time step and $T/\Delta t$ is the number of iterations during which the CA will be run. Therefore, processes with time scales between Δt and T can be represented and spatial scales ranging from Δx to L can be resolved. The state of the CA is described by an element of $\mathbb{F}$ (space of states) and it evolves according to the update rule $\Phi : \mathbb{F} \to \mathbb{F}$ (note that formally both $\mathbb{F}$ and Φ depend on the discretizations $(\Delta x, \Delta t)$). Additionally, we constrain the update rule to be in the form of *collision+propagation*, such that the operator Φ can be decomposed as a

$$\Phi = \mathbf{PCB}, \quad (3.3)$$

[1] Note that our approach is not limited to CA but also includes extensions such as lattice Boltzmann models and agent based models, because they can all be described by a generic update rule discussed in Sect. 3.2.3.

i.e. into a boundary condition, a propagation, and a collision operator, each depending, possibly, on the field $\mathbf{u}$ (see also Sect. 3.2.3 for more details). The terminology collision-propagation is borrowed from the lattice gas automata framework (see e.g. [1]). This is equivalent to the more classical Gather-Update CA paradigm, as was formally demonstrated recently [32]. The initial condition (f_{init}) is a particular element of the space of states. At the spatial boundaries of A, additional information is needed (boundary conditions).

In definition 3.2, we introduced additional elements. The field $\mathbf{u}$ collects the external information exchanged at each iteration between the CA and its environment. The functional $\mathbf{O} : \mathbb{F} \to \mathbb{R}^d$, the *observable*, specifies the quantity we are interested in.

A CxA can be viewed as a collection of interacting CA. Definition 3.2 suggests that a CxA can be represented as a graph $\mathcal{X} = (V, E)$ where V is the set of vertexes and E the set of edges with the following properties

- Each vertex is a CA $\mathcal{C}_i = \{A_i(\Delta x_i, L_i, \Delta t_i, T_i), \mathbb{F}_i, \Phi_i, f_{\text{init,i}} \in \mathbb{F}_i, \mathbf{u}_i, \mathbf{O}_i\}$
- each edge E_{ij} is a coupling procedure describing the interaction between $\mathcal{C}_i$ and $\mathcal{C}_j$. In practice, E_{ij} will define how and when information is exchanged between the two subsystems.

During the initialization phase, this problem-dependent graph is built according to the modeler's specifications.

3.2.2 The Scale Separation Map

A key idea behind CxA is that a multi-scale system can be decomposed into N single-scale Cellular Automata that mutually interact across the scales. The decomposition is achieved by building a Scale Separation Map (SSM) on which each system can be represented as an area according to its spatial and temporal scales. Processes having well separated scales are easily identified as the components of the multi-scale model.

Figure 3.1 shows a SSM, where the horizontal axis represents the temporal scales and the vertical axis the spatial scales. On the left a CA with spatio-temporal domain $A(\Delta x, \Delta t, L, T)$ is represented on the SSM. Assuming that the process to be simulated is really multi-scale in the sense that it contains relevant sub-processes on a wide range of scales, simulations based on the finest discretizations are not really feasible (recall Eq. (3.1)), the approach we propose in CxA modeling is to try to split the original CA into a number of single-scale CA and let these CA exchange information in such a way that the dynamical behavior of the multi-scale process is mimicked as accurately as possible. This is shown schematically in the right part in Fig. 3.1. The subsystem in the lower left part operates on small spatial scales, and short time scales, the one at the upper right part operates at large scales, and the other three at intermediate scales. This could e.g. be processes operating at the micro-, meso-, and macro scale.

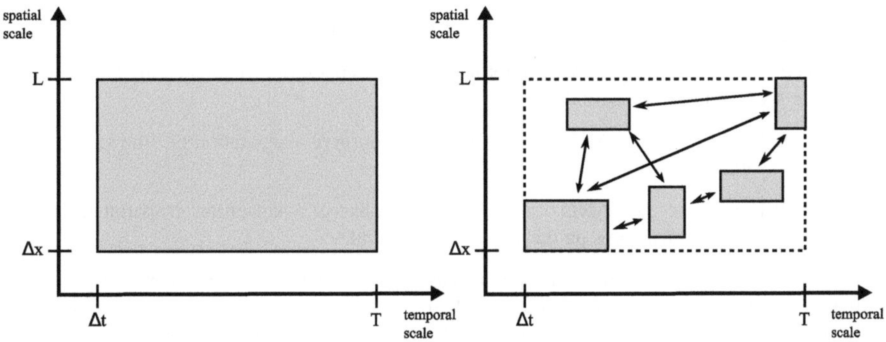

Fig. 3.1 The scale separation map with *left* a single CA and *right* a hypothetical CxA with 5 single scale CA modeling the same process

After identifying all subsystems and placing them on the scale map, coupling between subsystems is then represented by edges on the map. For instance, a process can be coupled with another through a lumped parameter or through detailed spatially and temporally resolved signals, in which case they would typically share a boundary and synchronously exchange information. The distance between subsystems on the map indicates which model embedding method to use to simulate the overall system. In the worst case, one is forced to use the smallest scales everywhere, probably resulting in intractable simulations. On the other hand, if the subsystems are well separated and the smallest scale subsystems are in quasi-equilibrium, then they can be solved separately, although infrequent (possibly event-driven) feedback between the subsystems will still be required.

Consider two processes A and B with their own specific spatial – and temporal scale, denoted by ξ_i and τ_i respectively ($i \in \{A, B\}$). Assume that A has the largest spatial scale. In case the spatial scales are the same, A has the largest temporal scale. In other words, $(\xi_B < \xi_A)$ OR $(\xi_B = \xi_A$ AND $\tau_B < \tau_A)$. We can now place A on the scale map and then investigate the different possibilities of placing B on the map relative to A. This will lead to a classification of types of multi-scale coupling, as in Fig. 3.2.

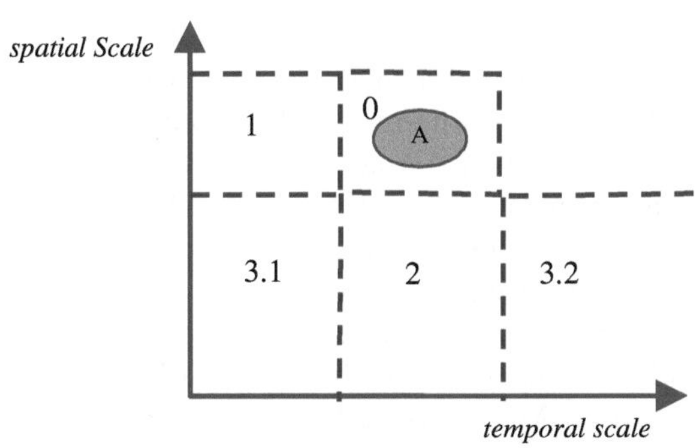

Fig. 3.2 Interaction regions on the scale map

Depending on where B is, we find the following regions:

Region 0: A and B overlap, so we do not have a scale separation, we are dealing here with a single-scale multi-science model.

Region 1: Here $\xi_B = \xi_A$ AND $\tau_B < \tau_A$, so we observe a separation of time scales at the same spatial scale.

Region 2: Here $\xi_B < \xi_A$ AND $\tau_B = \tau_A$, so we observe a separation in spatial scales, like coarse and fine structures on the same temporal scale.

Region 3: Separation in time – and spatial scales. Region 3.1 is the well-known micro $\Leftrightarrow$ macro coupling, so fast processes on a small spatial scale coupled to slow processes on a large spatial scale. This type of multi-scale model has received most attention in the literature. In region 3.2 we have the reversed situation, a slow process on small spatial scales coupled to a fast process on large spatial scales. We believe that this region is very relevant in for instance coupling of biological with physical processes, where the biological process is e.g. the slow response of cells to a faster physical process on a larger scale (e.g. blood flow in arteries).

Note that we do not have to consider other regions of the scale map, because then the role of A and B just reverses, and we fall back to one of the five cases identified above.

Next we address the question of the area that processes A and B occupy on the SSM. As discussed earlier, a 1D CA is characterized by a spatial discretization Δx and a system size L. We assume that Δx and L have been chosen such that the spatial scale of the process is well represented on this CA, so at least we will have $\Delta x < \xi < L$. We define $N^{(x)}$ as the number of CA cells that extend the full domain, i.e. $N^{(x)} = L/\Delta x$. Next assume that the discretization has been chosen such that the spatial scale is represented by $10^{\delta^{(x)}}$ cells (i.e. $\Delta x = \xi/10^{\delta^{(x)}}$) and the spatial extension of the CA is $10^{\eta^{(x)}}$ times the spatial scale, i.e. $L = \xi\, 10^{\eta^{(x)}}$, and therefore $N^{(x)} = 10^{\eta^{(x)}+\delta^{(x)}}$. Likewise for the temporal domain, i.e. a single scale CA has a time step Δt and the CA is simulated over a time span T, and we have $\Delta t < \tau < T$. The number of time steps $N^{(t)} = T/\Delta t$. The discretization has been chosen such that the temporal scale is represented by $10^{\delta^{(t)}}$ time steps (i.e. $\Delta t = \tau/10^{\delta^{(t)}}$) and that simulation time of the CA is $10^{\eta^{(t)}}$ times the temporal scale, i.e. $T = \tau\, 10^{\eta^{(t)}}$ and $N^{(t)} = 10^{\eta^{(t)}+\delta^{(t)}}$.

A process position on the scale map is now fully determined by the tuple $\{\xi, \delta^{(x)}, \eta^{(x)}; \tau, \delta^{(t)}, \eta^{(t)}\}$, and is drawn in Fig. 3.3, where the axes are now on a logarithmic scale. On such logarithmic SSM the process is rectangular with an area $(\delta^{(t)} + \eta^{(t)}) \times (\delta^{(x)} + \eta^{(x)})$ asymmetrically centered around the point $(\log(\tau), \log(\xi))$. In the special case that $\delta^{(x)} = \eta^{(x)} = \delta^{(t)} = \eta^{(t)} = 1$ (a reasonable first order assumption) we see that the process is symmetrically centered around $(\log(\tau), \log(\xi))$ and that the size of the box extends 2 decades in each dimension.

In Fig. 3.3 we show the extension of Fig. 3.2, where regions $1-3$ now have well defined positions and size. Depending on the location of process B, that is the point $(\log(\tau_B), \log(\xi_B))$ on the SSM, and with all information on the spatial and temporal extensions of processes A and B, we can unambiguously find in which region of the scale map they are located with respect to each other. The scale separation between

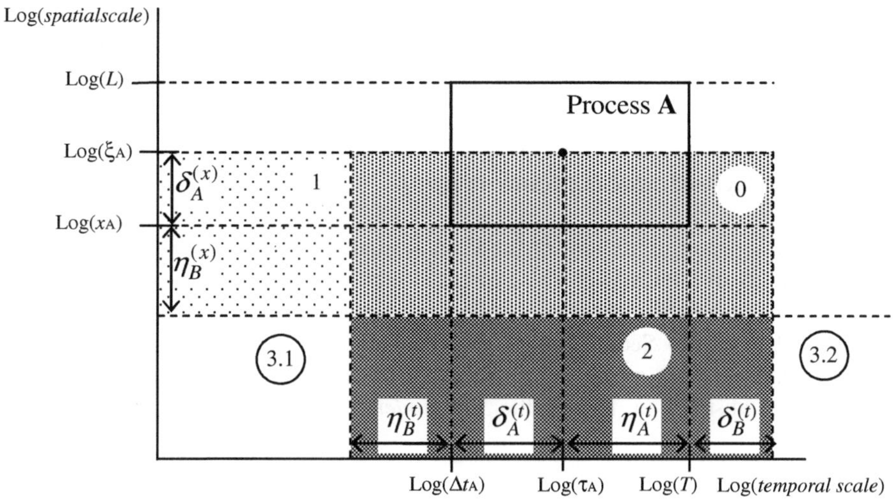

Fig. 3.3 Position of a process A with parameters $\{\xi, \delta^{(x)}, \eta^{(x)}; \tau, \delta^{(t)}, \eta^{(t)}\}$ and the interaction regions on the logarithmic scale map

two processes can now clearly be defined in terms of a distance on the SSM, and this can then become an important measure to determine errors that are induced by scale splitting procedures. This is further elaborated in Sect. 3.3.

Consider once more region 3, where there is a separation in time and length scales. In region 3.1 we find that $L_B < \Delta x_A$ and $T_B < \Delta t_A$. As said earlier, this is the classical micro $\Leftrightarrow$ macro coupling, and in our language this means the *full* spatio-temporal extend $T_B \times L_B$ of process B is smaller than one single spatio-temporal step $\Delta t_A \times \Delta x_A$ of process A. A number of modeling and simulation paradigms have been developed for this type of multi-scale systems (see e.g. [21]).

Region 3.2 also exhibits separation of time and length scales, but now the situation is quite different. We find that, just like in region 3.1, $L_B < \Delta x_A$. So, the spatial extend of process B is smaller than the grid spacing of process A. However, now we find that $T_A < \Delta t_B$. In other words, the full time scale of process A is smaller then the time step in process B. This will result in other modeling and simulation paradigms than in region 3.1. Typically, the coupling between A and B will involve time averages of the dynamics of the fast process A.

Let us now turn our attention to the regions where there is overlap on the temporal – or spatial scales, or both (regions 0, 1, and 2, in Fig. 3.3). In all these cases we can argue that we have partial or full overlap of the scales, giving rise to different types of (multi-scale) modeling and simulation. We say that the scales fully overlap if the point $(\log(\tau_B), \log(\xi_B))$ falls within (one of) the scales spanned by process A. On the other hand, there is partial overlap if $(\log(\tau_B), \log(\xi_B))$ falls *outside* (one of) the scales spanned by process A, but the rectangular area of process B still overlaps with (one of) the scales spanned by process A. The region of partial scale overlap can also be considered as a region of gradual scale separation, a boundary region between the scale separated regions 1, 2 and 3 and region 0. Simulations of this kind of multi-scale system would typically involve CxA's with local grid refinements, or multiple time stepping approaches, or a combination of both.

3.2.3 The Sub-Model Execution Loop

A second important ingredient of the CxA formalism is the observation that each CA (i.e. vertex of the CxA) can be expressed with a common instruction flow. This gives a way to identify generic coupling templates and achieve a precise execution model (see also Sect. 3.2.6). Using the specific collision+propagation form of the update rule, as introduced in Sect. 3.2.1, we represent the workflow with a pseudo-code abstraction, termed the *Sub-model Execution Loop* (SEL), as shown below.

```
D := D_init          /* initialization of the domain */
f := f_init          /* initialization of state variables */
t := 0               /* initialization of time */
    While Not EC
        t += Δt      /* increase time with one timestep t */
        D := U(D)    /* update the domain */
        f := B(f)    /* apply boundary conditions */
        f := C(f)    /* collision, update state of cells */
        f := P(f)    /* propagation, send information to neighbors */
        O_i(f)       /* compute observables from new state */
    End
O_f(f)               /* compute observables from final state */
```

Note that in the SEL, operators are written in bold and (state) variables as plain characters. The CA operates on a computing domain `D`, being the lattice of cells and the boundaries. Each cell in a CA has a set of state variables `f`. At the start of the SEL the domain and the state variables are initialized by the operators $\mathbf{D}_{\text{init}}$ and $\mathbf{f}_{\text{init}}$ respectively. The simulation time `t` is set to an initial value (`0` in this case). After initialization the CA enters into an iteration loop, whose termination is controlled by an end condition computed by **EC**. The end condition can simply be a fixed number of iterations, but could also be some convergence criterion depending upon the state variables. Within the main iteration loop, the time is first increased with a time step Δ`t`. Next the domain is updated by the operator **U**. If the domain is static, this operator is just the identity operator **I**. However, in many models the domain is dynamic. For instance, new cells can be created or existing cells removed (e.g. due to the movement of the boundary). In all these cases **U** will execute these domain updates. Next, the sequence **PCB**(`f`) is executed. First, the operator **B** applies the boundary conditions. This means that missing information is constructed that is needed for the actual state updates by **C** (see below) of the cells lying at the boundary of the domain D. For instance, if the state variables represent a concentration of some species, the boundary condition could specify a flux of those species into the domain, and from that missing information on the domain boundary cells is computed. Next the actual state change of all cells is computed by the Collision operator **C**. Finally, information is sent to neighboring cells or agents by the Propagation operator **P**. The CA is now updated for the current time step, and the simulation can proceed to the next iteration. However, before doing so an intermediate observation operator $\mathbf{O}_i$ computes observables from the state variables f. After termination of the main iteration loop a final observation is done of the state variables with the $\mathbf{O}_f$ operator.

3.2.4 CxA Multi-scale Coupling

Despite the growing literature there is not a well accepted generic methodology, nor a well-defined nomenclature of multi-scale modeling. A few authors have proposed different typologies of multi-scale models. Weinan E et al. [21] have proposed 4 types of multi-scale problems and 4 general strategies. Despite the many examples given by them the relevance of their classification is not always clear, because they single out, in all their examples, one specific item from their classification, and do not further discuss the relevance or completeness of the other classes. Another proposition for a multi-scale modeling methodology is that of Ingram. Working on chemical engineering simulations, Ingram et al. [20] have defined five types of macro-micro scale coupling. Ingram et al. present simulation examples for three types of coupling, showing that different strategies may be used to solve the same problem. The choice of coupling has an influence on both computational efficiency and accuracy. The fact that it is not always easy or possible to make the correspondence between the approaches by Ingram et al. and Weinan et al. indicates that the topic of multi-scale modeling lacks consensus. This lack of consensus on terminology and methodology can be attributed to the fact that actual coupling methodologies were mixed with classifications of the computational domain and/or with the type of scale separation (temporal, spatial, or both).

In the following discussion we try to clarify the situation, in the framework of the CxA formalism. However, we believe that this is also relevant to multi-scale modeling in general. Based on the discussion on the SSM in Sect. 3.2.2, we identified 5 different types of scale separation. We call them *Interaction Regions* on the SSM, and they are shown in Fig. 3.4. Another important parameter to distinguish multi-scale models is the Domain type. We distinguish between *single Domain* (sD) and *multi-Domain* (mD) types. In case of sD processes A and B can access the whole simulated domain and communication can occur everywhere, whereas in case of mD each process is restricted to a different physical region and communication can only occur across an interface or small overlap region.

For each combination of interaction region and domain type we can now try to identify a multi-scale coupling. We will base our approach on the SEL discussed in Sect. 3.2.3, and show which operators from the SEL are coupled to each other.

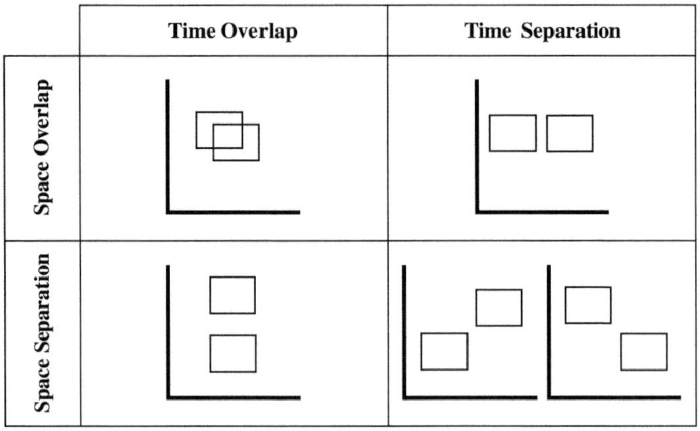

Fig. 3.4 Interaction regions on the SSM

We call this *Coupling Templates*. As an example consider Weinan E's Heterogeneous Multi-scale Method [21]. On close inspection we must conclude that this is a Coupling Template for single Domain processes in interaction region 3.1. In terms of the SEL of the macroscopic process A and the microscopic process B we find as Coupling Template $\mathbf{O}_f^B \rightarrow \mathbf{C}^A$; $\mathbf{O}_i^A \rightarrow \mathbf{f}_{\text{init}}^B$ (see also Fig. 3.8). At each time step of the macroscopic process B a microscopic process A is initialized using macroscopic information. The microscopic model then runs to completion and sends final information to the collision operator of the macroscopic process.

We are currently investigating many examples of multi-scale models, their mapping to a CxA model, and resulting coupling templates. A detailed discussion of the results will be reported later. As a summary, some of the examples are indicated in Fig. 3.5. At this stage we can extract two observations:

- In the case of time scale overlap, the coupling will occur inside the inner iteration loop. In contrast, in the case of time scale separation, coupling is realized outside the inner loop through the initialization operators and the final observation operator.
- Single-domain models are coupled through the collision operator. Multi-domain models are coupled through the domain update or the boundary operators.

Based on our current set of examples, we hypothesize that for each type of multiscale model, classified in terms of domain type and interaction region, only a very small set of coupling templates exists. If this is true, this would lead the way to a powerful CxA multi-scale modeling and simulation strategy, including a multi-scale modeling language, generic simulation software and a mathematical framework to analyze errors involved in CxA modeling. In what follows we will further elaborate on these ideas, sketching the contours of such a generic CxA based multi-scale modeling approach.

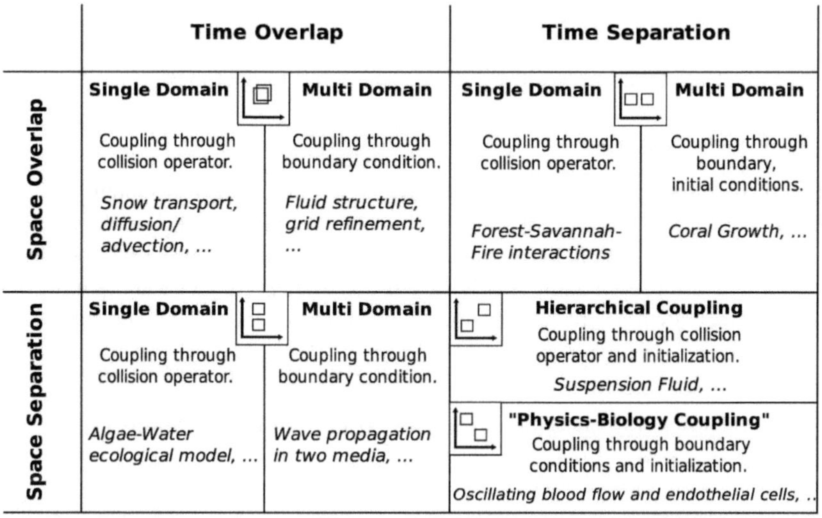

Fig. 3.5 Our classification of multiscale problems, for systems that can be reduced to two single-scale processes. This classification is based on the five interaction regions given by the SSM, and the domain type (sD or mD). For each class, the generic coupling template is indicated, in terms of the CxA operators. Examples of specific applications belonging to the given categories are indicated in *italic*

3.2.5 Multiscale Modeling Strategies

A key question when dealing with a multiscale system is how to decompose it in several coupled single-scale sub-processes. This decomposition is certainly not unique and a good knowledge of the system may be required. Once the sub-processes are chosen, this specifies the relation between the computational domains and the interaction regions on the SSM. Then, our classification scheme indicates the expected coupling templates.

We have observed several strategies that can be used to deal with systems having a broad range of scales and to reduce their area on the scale separation map. They are briefly discussed below.

3.2.5.1 Time Splitting

This approach is appropriate when two processes act at different time scales. Let us assume we have a sD problem described with a propagation operator P and a collision operator C that is the product of two operators

$$P_{\Delta t}C_{\Delta t} = P_{\Delta t}C^{(1)}_{\Delta t}C^{(2)}_{\Delta t}$$

where Δt specifies the finer scale of the process. Then, if $C^{(1)}_{\Delta t}$ acts at a longer time scale than $C^{(2)}_{\Delta t}$ we can approximate M iterations of the dynamics as

$$[P_{\Delta t} C_{\Delta t}]^M \approx P_{M\Delta t} C^{(1)}_{M\Delta t} [C^{(2)}_{\Delta t}]^M$$

We will illustrate this time-splitting strategy in detail in Sect. 3.3.

3.2.5.2 Coarse Graining

The goal of coarse graining is to express the dynamic of a given system at a larger temporal and/or spatial scale in some part of the computational domain where less accuracy is needed. After coarse graining we obtain a new process, specified by new collision and propagation operators and occupying a reduced area on the SSM. Within our formalism, a space-time coarse graining of a factor 2 can be expressed as

$$[P_{\Delta x} C_{\Delta x}]^n \approx \Gamma^{-1} [P_{2\Delta x} C_{2\Delta x}]^{n/2} \Gamma$$

where Γ is a projection operator, Δx the fine scale, and n is the number of iterations needed to simulate the problem.

3.2.5.3 Amplification

This strategy can be used to reduce the larger time scale of a process. For instance, we can consider a process acting with low intensity but for a long time, in a time periodic environment, such as a growth process in a pulsatile flow.

Within our formalism, let us consider two coupled (mD) processes which are iterated $n >> 1$ times

$$[P^{(1)} C^{(1)}]^n \qquad \text{and} \qquad [P^{(2)} C^{(2)}(k)]^n$$

where k expresses the intensity of the coupling of process 1 to process 2.

If the $C^{(1)}$ is periodic with period $m << n$, we can approximate the above evolution as

$$[P^{(1)} C^{(1)}]^m \qquad \text{and} \qquad [P^{(2)} C^{(2)}(k')]^m$$

with k' the new effective intensity of the coupling. For a linear coupling we would have $k' = (n/m)k$.

3.2.6 Execution Model

Coupling several sub-models, using coupling templates raises implementation issues. A typical situation is shown in Fig. 3.6 for the problem of coral growth. The growth of branching corals is modeled with the aim to understand the influence of abiotic factors (transport of nutrients by flow and diffusion) on the morphology. This is work performed under the supervision of Dr. Jaap Kaandorp, and for biological context and background we refer to his recent book [33] and to [34, 35]. In short, this model works as follows: the fluid flow is transporting nutrients that are needed by the coral to grow. There is a clear time scale separation that can be exploited. Fluid flow establishes at a few seconds whereas the coral grows at a much slower pace.

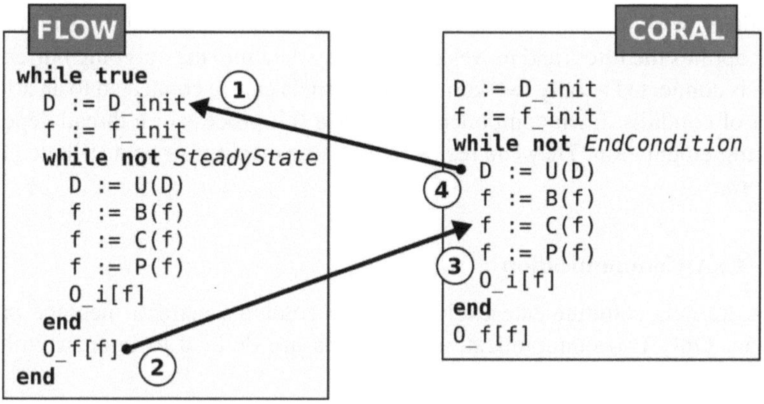

Fig. 3.6 Coupling template for the so-called coral growth model. *Numbers* corresponds to the communication operation described in Fig. 3.7

According to the coupling template shown in Fig. 3.6, the fluid solver is run until steady state and the resulting flow field is passed to the coral solver for calculating the growth rate. The new geometry of the coral is then used to build a new initial condition for the flow solver. The process stops when enough iterations of the coral solver have been performed.

Using this example we will explain the main concepts of our proposed execution model for CxA, which is compatible with the asynchronous channel actor-model framework [36]. A computer implementation for a CxA simulation environment, implementing this execution model, has been realized [37] and a public domain release is available.[2]

3.2.6.1 CxA Components

For the sake of the present discussion, CxA can be described as directed bipartite graphs whose edges represent a single direction communication channel and the vertexes are either *kernels* or *conduits*. The kernels are the main computational units of a CxA. Generally, kernels are the single-scale sub-model solvers as described above. However, when needed, they can also execute other tasks such as measurements or complex data mappings. The conduits are s̈mart c̈ommunication channels. Each conduit connects a pair of kernels together in an oriented fashion and, in principle, only one quantity is transported per conduit. These conduits are composed of three parts:

(1) an incoming buffer (the entrance)
(2) an outgoing buffer (the exit)
(3) (optional) one or several data *filters* between different scales (to perform interpolation, restriction, discretization, etc.)

[2] see http://www.complex-automata.org

Conduits work in a purely reactive way: when data is copied at the entrance, the conduit applies the filters and moves the resulting data into the outgoing buffer. Each conduit is connected to only two kernels, but kernels can be connected to an arbitrary number of conduits. Each component is either a full process or a thread depending on the implementation. They can reside in the same machine or be distributed across a network.

3.2.6.2 CxA Communication

In CxA, kernels communicate exclusively via conduits, using a message passing paradigm. Only two communication primitives are defined to interact with conduits:

1. `send(data)`: this primitive sends a data vector from a kernel to a conduit entrance. It is non-blocking, since it returns as soon as the data is sent to the conduit, whether or not the destination process has read the data. This corresponds to a push communication.
2. `receive()`: this primitive allows a kernel to receive data from a conduit exit. This primitive is blocking, it will return only when the desired data exist in the conduit. The receiving kernel will then simply wait until the data is available before resuming its computations. This corresponds to a pull communication.

Conduits entrances and exits are supposed to have large buffers, able to store several large data structures. These buffers act as FIFO ("first in, first out") where each entry is a reference to a date-structure. So, if the sending kernel is faster than the receiving one, several data vectors will be stored in the exit buffer, waiting for a receive() call from the destination kernel. The FIFO nature of the buffer ensures that the data are always read in the correct time order. The actual communication can be either a memory copy if the kernel and conduit reside in the same processor, or a network communication if both components reside on different machines. Note that the conduit could also be used to implement a mutex coordination primitive in case of shared memory execution.

Let us consider again the example of the coral growth. The coral SEL represented in Fig. 3.6, can be rewritten as follows, to include the two communication primitives explicitly:

```
While Not EC
    D := U(D)
    DomainConduit.send( D )
    f := B(f)
    velocityMap := VelocityConduit.receive()
    f := C(f,velocityMap)
    f := P(f)
End
```

3.2.6.3 CxA Initialization and Start

CxA initialization occurs in a semi-decentralized way. First, each conduit and kernel is spawned (possibly on several machines). Then a special process, termed *plumber*, is responsible for connecting each kernel with the entrances and exits of the relevant conduits. The plumber terminates as soon as this basic task is finished. The rest of the initialization process is then fully decentralized:

1. As soon as a kernel is fully connected with the required conduits, it starts its computations. If it is sending data to a yet unconnected kernel, the data will be kept in the conduit until the receiver is active and reading. On the other hand, if a conduit tries to receive data originating from an unconnected kernel, it will hang on until the sending kernel connects and transmits data.
2. For conduits the situation is even simpler. Since they are purely reactive components, nothing will happen in an unconnected conduit. Similarly, if only the conduit exit is connected, the conduit will do nothing. In contrast, if only the conduit entrance is connected, the conduit will simply process incoming data which will be accumulated in the exit buffer. Therefore, the conduit is always in a valid state (assuming it has enough internal memory).

3.2.6.4 CxA Synchronization

CxA graphs are usually cyclic. Even the basic examples with just two single-scale models (see Fig. 3.6) will display a communication cycle if both models can influence one another. Moreover CxA are multiscale systems and kernels can thus function at different time scales, maybe in an adaptive way. These properties make a central scheduler approach impractical. However, the fact that the receive primitive is blocking and the send is non-blocking, allows a data-driven synchronization to occur naturally. Indeed, kernels will just wait until information is available before continuing their computation. An example of such synchronization is shown in Fig. 3.7 for the coral model.

The main problem with this method are possible deadlock situations. However, such issues can be easily prevented with the CxA execution model. In the coral example, deadlock is avoided by having a model (the coral) which sends before receiving. This allows the flow model to continue its computations to produce the data that will unlock the coral, etc. In contrast, the situation presented in Fig. 3.8 will produce a deadlock because both models try to receive before sending anything. This problem is easily solved by moving the observation $\mathbf{O}_i$ at the beginning to the inner loop, or adding initial send instructions before entering the submodel execution loop.

Furthermore, the fact that communication is pairwise and that the conduits use buffers, makes race conditions impossible. Data are meant to be read by only one process, data sent in a conduit entrance will be processed only by that conduit and data moved to conduit exits will concern only a single kernel.

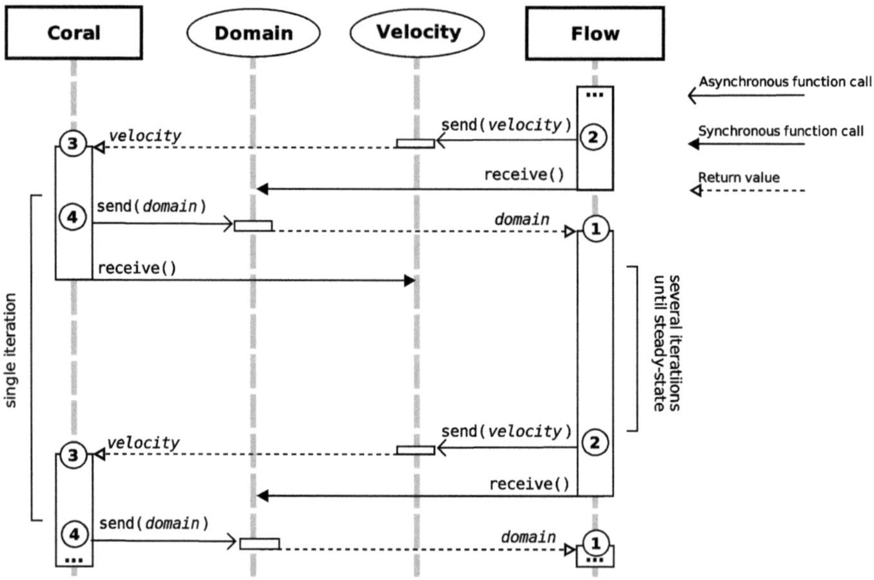

Fig. 3.7 UML sequence diagram of the CxA shown in Fig. 3.6. The *vertical lines* represent the "life-line" of the process: the kernels are represented by *rectangles* and the conduits by *ovals*. When a process is active, the *gray* life line is replaced by a *vertical white rectangle*. The *arrows* represent interaction. *Solid arrows* with *triangular heads* are blocking interactions and *solid arrows* with *thin heads* represent non-blocking interactions. The return values are indicated by *dashed arrows*. The *circled numbers* correspond to Fig. 3.6

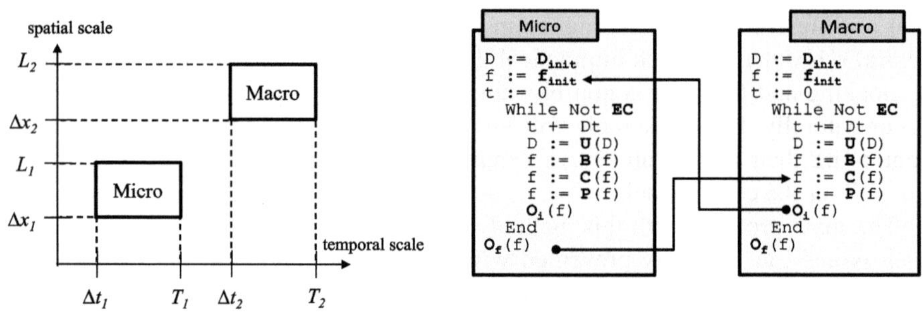

Fig. 3.8 Micro-macro coupling example. *Left*: SSM. *Right*: coupling template

3.2.6.5 CxA Termination

The termination of the whole CxA is also designed to be fully decentralized: when a kernel finishes its computations (because of e.g. a preset maximum time or a steady state condition), it first notifies all its conduits and then it terminates itself. Similarly, when a conduit receives termination notifications from all connected kernels, it can terminate itself. While the conduit termination rule is always safe (a conduit stops when no kernel is connected anymore), the kernel termination rule needs an extra mechanism. Otherwise, a problem occurs if a kernel is waiting for information from an already terminated kernel.

For instance, in the coral example (Fig. 3.6) the flow model will hang on for the domain update, even after the coral model termination. To solve this issue a stop signal is introduced which is able to release a kernel blocked in the receive primitive. This signal is propagated by a kernel through the existing conduits, using a third primitive: stop(): this primitive sends the stop signal through a conduit. The receive primitive is then modified slightly. It works exactly as seen above but can return either the expected data or the stop signal.

Therefore a kernel waiting for data can be released by a stop signal. Kernels are then responsible to send, process and propagate stop signals. Generally a kernel receiving a stop signal should:

1. Abort the submodel execution loop.
2. Send some final data, if required.
3. Propagate the stop signal to each connected conduit entrances.
4. Notify each connected entrance and exit.
5. Terminate itself gracefully.

With this termination scheme, all kernels which need data from the rest of the CxA will thus stop. The stop signal can originate from any kernel, and this approach also works if two (or more) kernels reach a stop condition at the same time.

As an illustration we can add a stop mechanism to the example of Fig. 3.6, as follows:

1. Coral submodel

```
While Not EC
   D := U(D)
   DomainConduit.send( D )
   f := B(f)
   velocityMap :=
      VelocityConduit.receive()
   f := C(f,velocityMap)
   f := P(f)
End
DomainConduit.stop()
myStop()
```

2. Flow submodel

```
While True
   domain :=
      DomainConduit.receive()
   If domain == STOP_SIGNAL
      myStop()
   D := domain
   f := f_init
   While Not Steady_State
      [SEL]
   End
End
```

where myStop() is a user-defined function which terminates the kernel. But, before, if needed, it: (i) saves results, (ii) propagates the stop signal, (iii) notifies the connected conduits.

3.2.6.6 Parallelization

With the execution model described above, our framework is compatible with a distributed or GRID computing approach, in which each submodel could run on a different core or, alternatively, as a different threads on the same core. The actual support for parallelization depends on the chosen implementation of our framework.

For instance, the MUSCLE library[3] offers an easy but manual parallelization. On the other hand the CxA-lite library[4] only allows a multithread execution in which all the submodels share the same memory space.

So far, we did not address the question of load balancing. This is clearly a separate issue and no tools have been yet developed to assist the user in distributing the computation in equal pieces over several processor.

3.2.7 Formalism

The concept of a CxA as a set of coupled CA's, where the coupling is expressed in terms of input–output relations between operators of the SEL of the coupled CA's is not just a concept that allows us to classify multi-scale models, as discussed in Sect. 3.2.4, or a powerful concept to built CxA simulation software, see Sect. 3.2.6, but it is also amenable to mathematical formalism and analysis. This section will introduce some of the formalism, which will be further used in one of the examples of Sect. 3.3.

Recalling (3.2), the state of a CA at a certain time t is described by a $f^t \in \mathbb{F}$, denoting the numerical solution at the time step t, which evolves according to

$$\begin{aligned} f^0 &= f_{\text{init}}[\mathbf{u}_0], \text{ initialcondition} \\ f^{t+\Delta t} &= \Phi[\mathbf{u}; f^t] \end{aligned} \tag{3.4}$$

where $\mathbf{u}_0$ is an external field connected to the initial condition. As previously discussed, we constrain the update rule Φ to the form

$$\Phi[\mathbf{u}; f] = (\mathbf{B}[\mathbf{u}_B] \circ \mathbf{P} \circ \mathbf{C}[\mathbf{u}_C])\,[f], \tag{3.5}$$

i.e. written as a composition of three operators: *collision* $\mathbf{C}[\mathbf{u}_C]$, depending on external parameters $\mathbf{u}_C$, *propagation* $\mathbf{P}$, depending on the topology of the domain, and *boundary condition* $\mathbf{B}[\mathbf{u}_B]$, depending on external parameters denoted by $\mathbf{u}_B$.

More precisely, the space of the states $\mathbb{F}$ and the update rule Φ depend in general on the discretization parameters Δx and Δt. For simplicity, in what follows, we let the definition of CA depend also on a (small) parameter h, related to spatial and temporal discretizations (for example $\Delta x_h = h$, $\Delta t_h = \alpha h$). Accordingly, considering the CA $\mathcal{C}_h$, the evolution space and the update rule can be denoted as: $\Phi_h : \mathbb{F}_h \to \mathbb{F}_h$. Shortly, we will call f_h the numerical outcome of the CA $\mathcal{C}_h$.

To begin with, as in the left diagram in Fig. 3.1, we consider a multi-scale system represented as a single $\mathcal{C}_h$ defined as in (3.2). Building a CxA, instead of describing the system with a single f_h, we lower the dimension of the problem and the computational complexity, introducing coarser temporal and/or spatial discretizations

$$H = (h_1, \ldots, h_M) \tag{3.6}$$

[3] http://developer.berlios.de/projects/muscle

[4] http://github.com/paradigmatic/CxALite/

and building a corresponding Complex Automaton

$$\mathrm{CxA}_H = (\mathcal{C}_{h_1}, \ldots, \mathcal{C}_{h_M}), \tag{3.7}$$

where each $\mathcal{C}_{h_m}$ is an object as in (3.2).

Formally, the definition of a CxA can be summarized in two steps. First, a projection of the space of states $\mathbb{F}$ on a product of spaces is considered

$$\Pi_{Hh} : \mathbb{F}_h \to \mathbb{F}_{h_1} \times \cdots \times \mathbb{F}_{h_M}, \tag{3.8}$$

each describing the evolution of a *single scale model* (on different discretizations). Second, a *rescaling* of the update rule is performed, according to the new discretizations h_i, on each space $\mathbb{F}_{h_i}$, for $i = 1, \ldots, M$, depending on the multiscale technique used. Due to the form of the execution model of each CA, the rescaling can be easily expressed in terms of operations on the operators **P**, **C**, **B**. Note that the spaces $\mathbb{F}_{h_i}$ are not necessarily disjoint, i.e. part of a single scale evolution space could be shared by several CA, in case of space overlap and single domain coupling.

Let us denote with f_{CxA} the numerical outcome of the complex automata simulation and with f_{h_m} (or f_m) the state variable of the single CAs. To be able to compare the results of the CxA versus the original multiscale algorithm, we associate an observable $\mathbf{O}_{\mathrm{CxA}}$ to the Complex Automata, which projects the result f_H on the space of $\mathbf{O}(f_h)$. A sketch of the relevant spaces and operators is drawn below.

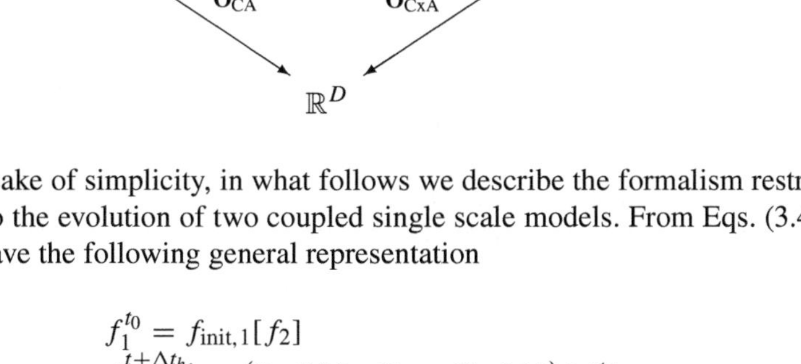

For the sake of simplicity, in what follows we describe the formalism restricting ourselves to the evolution of two coupled single scale models. From Eqs. (3.4) and (3.5), we have the following general representation

$$\begin{aligned} f_1^{t_0} &= f_{\mathrm{init},1}[f_2] \\ f_1^{t+\Delta t_{h_1}} &= \left(\mathbf{B}_{h_1}[f_2] \circ \mathbf{P}_{h_1} \circ \mathbf{C}_{h_1}[f_2]\right)[f_1^t], \\ f_2^{t_0} &= f_{\mathrm{init},2}[f_1] \\ f_2^{t+\Delta t_{h_2}} &= \left(\mathbf{B}_{h_2}[f_1] \circ \mathbf{P}_{h_2} \circ \mathbf{C}_{h_2}[f_1]\right)[f_2^t], \end{aligned} \tag{3.9}$$

where two CAs are fully coupled in all the components. In detail,

- $f_{\text{init},1}[f_2]$ denotes a coupling through initial conditions (i.e. the initial condition of 1 depends on the results of 2)
- $\mathbf{B}_i[f_j]$ expresses coupling through boundary conditions,
- $\mathbf{C}_i[f_j]$ expresses the coupling through collision operator.

In general, for different situations (multidomain/singledomain, time/space separation/overlap) we can restrict the set of possible couplings to a well-specified coupling template. Consider the example of a microscopic fast process coupled to a macroscopic slow process (micro-macro coupling), as introduced earlier in Sect. 3.2.4. The macroscopic process takes input from explicit simulations of microscopic processes at each time step and on each lattice site of the macroscopic process. The microscopic processes run to completion, assuming that they are much faster than the macroscopic process and therefore are in quasi-equilibrium on the macroscopic time scales (this approach is known in the literature as the Heterogeneous Multi-scale Method, see [21]). The macroscopic process could e.g. be a fluid flow with takes its viscosity from an underlying microscopic process (e.g. explicit suspension model).

In Fig. 3.8 we show for this example of micro-macro coupling the SSM (left) and the coupling template (right). The later is defined in [27] and shows how the operators as defined in (3.5) are coupled to each other. A close inspection of this coupling template shows indeed that, upon each iteration of the macroscopic process, the microscopic process executes a complete simulation, taking input from the macroscopic process. In turn, the output from the microscopic process is fed into the collision operator of the macroscopic process.

We can formulate the CxA dynamics as follows (based on Eq. (3.9))

$$\begin{aligned} f_1^{t_0} &= f_{\text{init},1}[f_2] \\ f_1^{t_1+\Delta t_1} &= (\mathbf{B}_1 \circ \mathbf{P}_1 \circ \mathbf{C}_1)\,[f_1^{t_1}], \\ \\ f_2^{t_0} &= f_{\text{init},2} \\ f_2^{t_2+\Delta t_2} &= (\mathbf{B}_2 \circ \mathbf{P}_2 \circ \mathbf{C}_2[f_1])\,[f_2^{t_2}], \end{aligned} \tag{3.10}$$

where 1 refers to the micro-scale and 2 to the macro-scale. The micro-scale model 1 is run until completion (i.e. until the final time T_1), then a single time step Δt_{h_2} is performed for the macro-scale model.

We can now compare an estimation of the execution time of the CxA model of Fig. 3.8 with that of using a single CA for the same system, as in the left part of Fig. 3.1. For the single CA the execution time would be $T_{CA} = k_{CA}\frac{T_2}{\Delta t_1}\left(\frac{L_2}{\Delta x_1}\right)^D$, which is (3.1) using the subscripts as introduced in Fig. 3.8. For the CxA, the execution time becomes

$$T_{\text{CxA}} = \frac{T_2}{\Delta t_2}\left(\frac{L_2}{\Delta x_2}\right)^D \left(k_2 + k_1\frac{T_1}{\Delta t_1}\left(\frac{L_1}{\Delta x_1}\right)^D\right). \tag{3.11}$$

where k_{CA}, k_1 and k_2 are the CPU times to update one spatial cell for one time step, respectively for the full scale CA, the micro and the macro submodels.

Next one can compute a speedup, comparing the single scale CA formulation and the CxA formulation as $S = T_{\text{CA}}/T_{\text{CxA}}$. After some algebra we find

$$S = \left(k_{CA}\frac{\Delta t_2}{\Delta t_1}\left(\frac{\Delta x_2}{\Delta x_1}\right)^D\right) \Big/ \left(k_2 + k_1\frac{T_1}{\Delta t_1}\left(\frac{L_1}{\Delta x_1}\right)^D\right). \tag{3.12}$$

Under the reasonable assumption that the execution time for a full micro scale simulation needs much more time than a single iteration of the macro scale model, i.e. when $k_1 \frac{T_1}{\Delta t_1}\left(\frac{L_1}{\Delta x_1}\right)^D >> k_2$, Eq. (3.12) reduces to $S = \frac{k_{CA}}{k_1}\frac{\Delta t_2}{T_1}\left(\frac{\Delta x_2}{L_1}\right)^D$. Note that $\frac{\Delta t_2}{T_1} > 1$ and $\frac{\Delta x_2}{L_1} > 1$, and can be interpreted as the distance on the SSM (Fig. 3.8). So, if the scale separation is large enough, the obtained speedups can be huge, principally rendering a CxA simulation feasible.

3.2.8 Scale-Splitting Error

The above arguments demonstrate the improvements in computational efficiency offered by the CxA formulation. On the other hand, replacing the original multi-scale model with many coupled single-scale algorithms, we face a partial loss of precision. A possible measure of this lowering in accuracy can be obtained considering the difference in the numerical results of the original $\mathcal{C}_h$ and the Complex Automaton CxA$_H$, which we call *scale-splitting error*.

This error is measured according to the observables, i.e. the quantity of interest, formally resulting from the observable operators:

$$E^{\mathcal{C}_h \to \text{CxA}} = \|\mathbf{O}_{\text{CA}}(f_h) - \mathbf{O}_{\text{CxA}}(f_H)\| \tag{3.13}$$

in an opportune norm. The scale-splitting error has a direct interpretation in terms of accuracy. In fact, calling $E^{\text{CxA,EX}}$ the absolute error of the CxA model with respect to an exact reference solution, and $E^{\mathcal{C}_h,\text{EX}}$ the error of the model itself, we have

$$\left\|E^{\text{CxA,EX}}\right\| \leq \left\|E^{\mathcal{C}_h,\text{EX}}\right\| + \left\|E^{\mathcal{C}_h \to \text{CxA}}\right\|. \tag{3.14}$$

If we heuristically assume that the original *fine-scale* algorithm has a high accuracy, the scale splitting error is a measure of the error of the CxA model.

In general, a detailed and rigorous investigation of the scale-splitting error requires a good base knowledge of the single scale CA and of the full multiscale algorithm. Case by case, error estimates can be derived using the properties of the algorithms, the operators involved in the update rule and in the coarse-graining procedure. An example of error investigation using the formalism for a simple CxA model can be found in Sect. 3.3.

3.3 Examples

3.3.1 Reaction Diffusion

Let us consider a reaction-diffusion process for a concentration field $\rho = \rho(t, x)$ described by the equation

$$\begin{aligned} &\partial_t \rho = d\partial_{xx}\rho + \kappa(\rho_\lambda - \rho),\ t \in (0, T_{end}],\ x \in (0, L] \\ &\rho(0, x) = \rho_0(x) \end{aligned} \tag{3.15}$$

with periodic boundary conditions in x, ρ_0 being the initial condition and $\rho_\lambda(x)$ a given function. To consider a multiscale model, we assume the reaction to be characterized by a typical time scale faster than the diffusion., i.e. $\|k\| \gg \|d\|$.

Numerically, problem (3.15) can be solved employing a lattice Boltzmann method (LBM) (see for example [1, 10, 38, 39] and the references therein, as well as the chapter by Kusumaatmaja and Yeomans of the present book), discretizing the space interval with a regular grid $\mathcal{G}_h = \{0, \ldots, N_x - 1\}$ of step size $\Delta x_h = h$ and associating each node $j \in \mathcal{G}_h$ with two variables, f_1 and f_{-1} representing the density of probabilities of populations traveling with discrete velocities $c_i \in \{-1, 1\}$. The collision+propagation update has the form

$$f_i^{t_n+\Delta t}(j + c_i) = f_i^{t_n}(j) + \frac{1}{\tau}\left(\frac{\hat{\rho}^{t_n}}{2} - f_i^{t_n}(j)\right) + \Delta t_h \frac{1}{2} R(\hat{\rho}^{t_n}(j)). \tag{3.16}$$

here $R(\hat{\rho}(j)) = \kappa(\rho_\lambda(j) - \hat{\rho}(j))$, and $\hat{\rho} = \rho(f) = f_1 + f_{-1}$ is the numerical solution for the concentration field. The time step is related to the grid size according to

$$\frac{\Delta t_h}{\Delta x_h^2} = \text{const.} \quad \forall h, \tag{3.17}$$

and the parameter τ is chosen according to the diffusion constant in (3.15) (see [39, 1])

$$\tau = \frac{1}{2} + d\frac{\Delta t_h}{\Delta x_h^2}. \tag{3.18}$$

Observe that τ is independent from h in virtue of (3.17). It can be shown that the above described algorithm leads a second order accurate approximation of the solution of (3.15) [39]. Equivalently, we can rewrite (3.16) in the form [40]

$$f_h^{t_{n+1}} = P_h(I_h + \Omega_{D_h}(\tau))(I_h + \Omega_{R_h}) f_h^{t_n} = \Phi_h f_h^{t_n}, \tag{3.19}$$

highlighting the scale h and omitting the subscript i. The update $\Phi_h = P_h(I_h + \Omega_{D_h}(\tau))(I_h + \Omega_{R_h})$, has been decomposed into a diffusion part and a reaction part.

The space of states is the set $\mathbb{F}_h = \{\phi : \mathcal{G}_h \to \mathbb{R}^{2N_x}\}$, of the real functions defined on the grid $\mathcal{G}_h$. The subscript h for the operators denotes functions acting from $\mathbb{F}_h$ to itself. In detail, I_h is simply the identity on $\mathbb{F}_h$, P_h acts on a grid function shifting the value on the grid according to c_i

$$(P_h f_h)_i\,(j) = f_{i,h}(j - c_i),$$

while Ω_{D_h} and Ω_{R_h} are the operations defined in the right hand side of (3.16):

$$\left(\Omega_{D_h} f_h\right)_i = \frac{1}{\tau}(f_i^{\mathrm{eq}}(\rho(f_h)) - f_{i,h}), \quad \left(\Omega_{R_h} f_h\right)_i (j) = h^2 \frac{1}{2} R(\rho(f_h))$$

The SSM for this example is shown in Fig. 3.9. To define the CxA, we set $\Delta t_R = \Delta t_h = h^2$ for the reaction and $\Delta t_D = M h^2$ for the diffusion. Focusing on the case shown in Fig. 3.9b, the reaction is run up to a time T_R, then re-initialized after a diffusion time step. If $T_R = \Delta t_D$, the two processes are not completely separated. Figure 3.9c sketches the case when reaction leads very quickly to an equilibrium state in a typical time which is even smaller than the discrete time step of the diffusion.

We focus on the case of time-coarsening, i.e. choosing

$$\Delta x_D = \Delta x_R = h, \; \Delta t_D = M \Delta t_R = M h^2 . \tag{3.20}$$

Introducing reaction and diffusion operators $\mathcal{R}_s$, $\mathcal{D}_s$, where $s = R, D$ specifies the dependence of the discrete operators on the space-time discretization of reaction and (resp.) diffusion, the evolution of the system can be described with the state variable $f_H = (f_R, f_D)$, whose components are updated according to

$$\begin{array}{ll} (\mathrm{CA}_R) & (\mathrm{CA}_D) \\ f_R|_{t_0 = t_D} = f_D^{t_D}, & f_D^0 = f_D^{init}(\rho_0), \\ f_R^{t_R + \Delta t_R} = \mathcal{R}_R f_R^{t_R} & f_D^{t_D + \Delta t_D} = \mathcal{D}_D f_D^{t_D + M \Delta t_R}. \end{array} \tag{3.21}$$

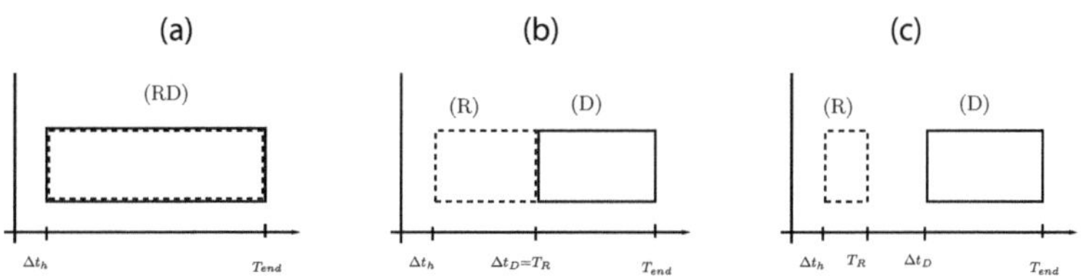

Fig. 3.9 SSM for the reaction-diffusion LBM. In **(a)** reaction (*dashed line*) and diffusion (*solid line*) are considered as a single multiscale algorithm. In **(b)** we assume to use different schemes, where the diffusion time step Δt_D is larger than the original Δt_h. **(c)** Represents the situation where the two processes are time separated, with a very fast reaction yielding an equilibrium state in a time $T_R \ll \Delta t_D$

Equation (3.21) expresses that the algorithm CA_R, which is coupled to CA_D through the initial condition (by setting at the initial time $t_0 = t_D$ (equal to a certain time of CA_D) the initial condition equal to the one obtained from CA_D, and evolves for M steps according to an update rule depending only on the reaction process. On the right, the diffusion part CA_D is coupled to the reaction through the collision operator, since the new state of f_D is locally computed starting from the output state of CA_R. With $f_D^{\text{init}}(\rho_0)$ we denoted the original initial condition, function of the initial concentration in (3.15).

In this case, the observable is represented by the concentration ρ, obtained from the numerical solution by a simple average over the particle distributions.

Following Sect. 3.2.8 we now consider the scale-splitting error $E(M)$ resulting from using a diffusion time step Δt_D M times larger than the reaction time step Δt_R. The reference solution is here the solution obtained when both reaction and diffusion act at the smallest time scale, i.e. when $M = 1$. To estimate $E(M)$ we consider M reaction steps at scale h (defined by Δt_R) followed by one diffusion step at the coarser scale h' (defined by $\Delta t_D = M \Delta t_R$) and we compare the results with M reaction-diffusion steps both at the fine scale h. In terms of the reaction and diffusion operators, $E(M)$ can be expressed as

$$\begin{aligned} E(M) &= \left\| (\mathcal{D}_h \mathcal{R}_h)^M - \mathcal{D}'_h \mathcal{R}_h^M \right\| \\ &\le \left\| (\mathcal{D}_h \mathcal{R}_h)^M - \mathcal{D}_h^M \mathcal{R}_h^M \right\| + \left\| [\mathcal{D}_h^M - \mathcal{D}'_h] \mathcal{R}_h^M \right\| \\ &= E_1(M) + E_2(M) \end{aligned} \tag{3.22}$$

Contribution E_1 can be computed from the commutator $[\mathcal{D}_h \mathcal{R}_h - \mathcal{R}_h \mathcal{D}_h]$ and E_2 follows from the time coarse-graining of the original LB model. After some calculations we obtain (see [41])

$$E(M) \le \mathcal{O}(M^2 \kappa) + \mathcal{O}(M^2 D^3) \tag{3.23}$$

3.3.1.1 Numerical Validation

We consider the problem

$$\begin{aligned} &\partial_t \rho = d \partial_{xx} \rho - \kappa(\rho - \sin(\lambda x)), \; t \in (0, T_{\text{end}}], \; x \in (0, 1] \\ &\rho(0, x) = \rho_0(x) \end{aligned} \tag{3.24}$$

with $\frac{\lambda}{2\pi} \in \mathbb{Z}$, and periodic boundary conditions in x-direction.

By selecting different values of the parameters regulating (3.24) we can tune the relevance of different time scales. Additionally, we introduce the non dimensional parameter

$$\sigma = \frac{\kappa}{\lambda^2 d}$$

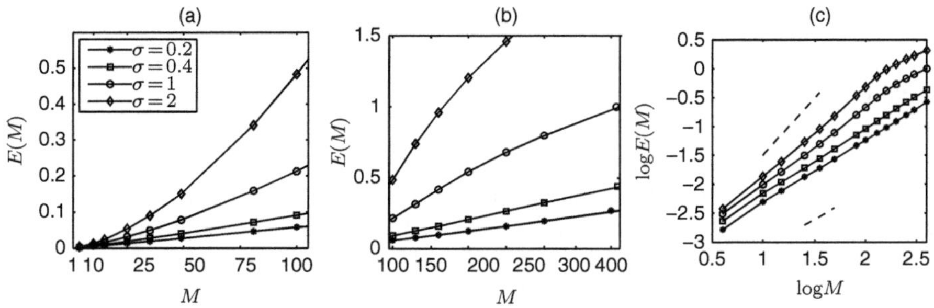

Fig. 3.10 Scale-splitting error as a function of M for a time-coarsened CxA. The different curves represent different values of σ. Simulation parameters: $h = 0.02$, $\lambda = 4\pi$, $\kappa = 10$, $d \in \{0.05, 0.1, 0.25, 0.5\}$. **(a)**: $1 < M < 100$. **(b)**: $M > 100$. The size of the scale-splitting error becomes relatively large, except for the case $\sigma = 0.2$. **(c)**: Order plot (Fig. (a)-(b) in double logarithmic scale) of maximum scale-splitting error versus M. *The dashed lines* of slope 1 (*bottom*) and 2 (*top*) indicate that $E \sim M^{\alpha}$, with $1 < \alpha < 2$

to "measure" the scale separation of the simulation. In the numerical tests, we run both the original fine scale LBM and the CxA model, measuring explicitly the scale-splitting error as the difference in the resulting concentrations. Figure 3.10 shows the results of scale-splitting error for different values of M. The order plot in Fig. 3.10c confirms estimate in Eq. (3.23).

Results of a further test to link together scale separation and scale-splitting error are shown in Fig. 3.11. Namely, for each simulation drawn in Fig. 3.10, we select the first M such that the scale splitting error lies below a certain prefixed threshold error $\bar{E}(h, H)$. These values M_{th} are plotted then as function of σ, validating the idea that better scale separation allows more efficient CxA formulations.

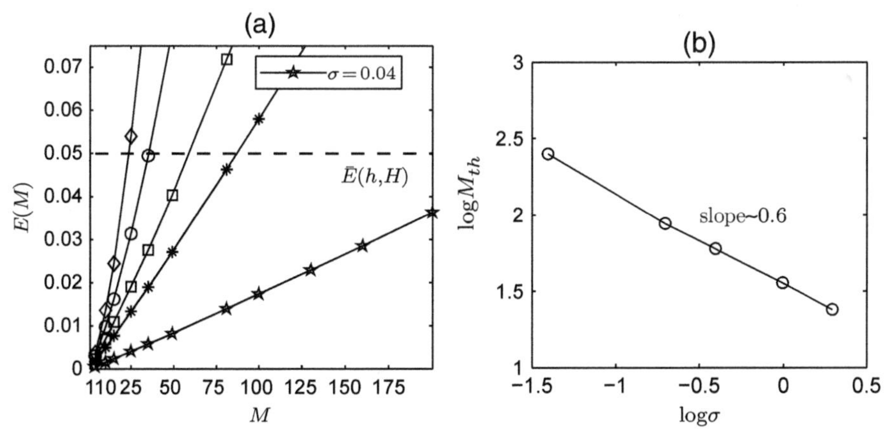

Fig. 3.11 **(a)** Zoom of the previous Fig. 3.10a, including a threshold error $\bar{E}(h, H) = 0.05$ (results with $\sigma = 0.04$ are also shown). **(b)** Values of M_{th} such that the scale-splitting error equates a threshold error $\bar{E}(h, H)$, versus the measure of scale separation $\sigma = \kappa \left(\lambda^2 D\right)^{-1}$ (in double logarithmic scale)

Detailed analysis and investigation of this example can be found in [41].

We can also compute the speedup resulting from the above time-splitting. Let us call a and b the CPU times of one iteration of respectively the reaction and the diffusion processes. If we run the full system at the finer scale Δt_R for a time T, the total CPU time will be proportional to $(a + b)(T/\delta t_R)$. With the time-splitting method, the CPU time reduces to $(Ma + b)T/(M\Delta t_R)$ and the speedup is $(a + b)/(a + b/M)$. For large M, the speedup tends to $1 + b/a$. This might not be a very big gain, unless $a << b$. However, if we would have coarse grained the spatial scale for the diffusion processes, we would get a more interesting speedup value.

3.3.2 In Stent Restenosis

A challenging application to validate the CxA methodology is represented by the *in-stent restenosis*, a coronary artery disease appearing when an arterial occlusion (stenosis), cured by deploying a small metal mesh (stent), reappears later in time, due to the maladaptive biological response of the organism. This process involves a wide range of spatial and temporal scales, spanning from micrometers to millimeters and from seconds (typical time of the cardiac cycle) to weeks (typical time of appearance of a restenosis). Details on in-stent restenosis and a formulation in terms of Complex Automata can be found in [30]. In this section, we briefly outline the methodology to construct a CxA model for the in-stent restenosis. All details of single scale models, coupling templates and simulation results will be presented elsewhere.

3.3.2.1 Single Scale Models

The SSM for a simplified in-stent restenosis model is shown in Fig. 3.12. We include the following subprocesses:

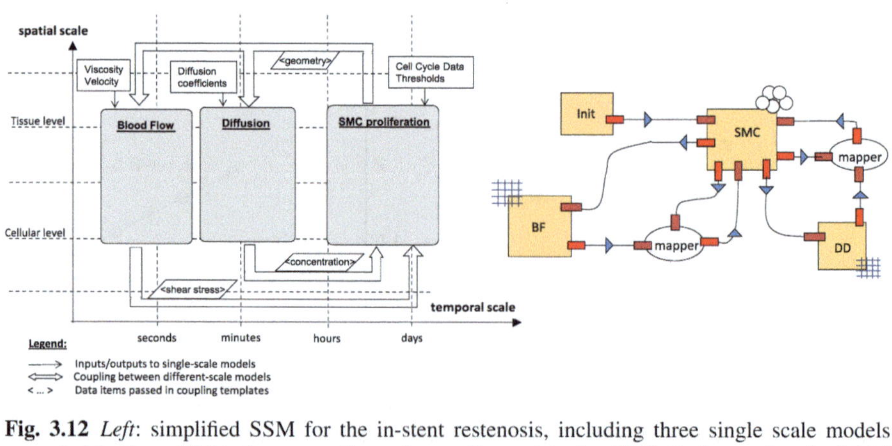

Fig. 3.12 *Left*: simplified SSM for the in-stent restenosis, including three single scale models separated in time. *Right*: the connection scheme. Respect to the SSM, it shows also the initial condition agents, and mapper agents, used when combination of multiple input or multiple output is needed

- Bulk Flow (BF): a lattice Boltzmann model for the hemodynamics, simulated on a spatial grid fine enough to resolve the flow lines near to the stent
- Smooth Muscle Cells (SMC) Hyperplasia: an Agent Based Model, where each agent represents an SMC, reacting, structurally and biologically according to the state of the neighboring cells and the flow
- Drug Diffusion (DD): a Finite Difference scheme to simulate drug eluting stents, which approximates the drug concentrations within the tissue, assuming the stent to be a source and the vessel to be a sink (since drugs are constantly flushed away by the flow).

Additionally, the computational model makes use of an *initial condition* (INIT) agent, which creates the cell configuration after the stent deployment, and two geometrical mappers, which convert the output of BF and DD (based on a lattice) into input for SMC (based on an off-grid domain). The graph driving the CxA model (*connection scheme*) is shown in Fig. 3.12.

3.3.2.2 Coupling Templates

The interaction between the single scale models can be described in the following way:

- BF to SMC: after a cardiac cycle has been completed, averaged wall shear stresses (WSS) are computed along the boundary, and distributed to the SMC in direct contact with the flow
- SMC to BF: the cells configurations (described, in case of spherical cells, by positions and radii) is *filtered*, generating the domain for the flow simulation
- SMC to DD: similarly, the space occupied by SMC is converted in domain for the DD model
- DD to SMC: after the drug concentration relaxes to steady state, the values are distributed to the cells.

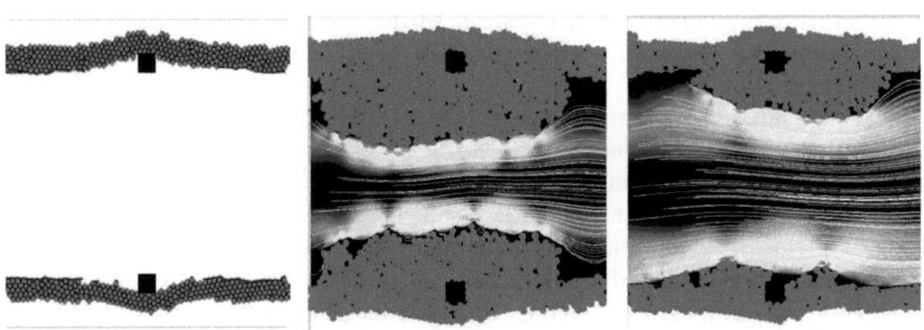

Fig. 3.13 Two-dimensional benchmark geometry (*left*), sketching a vessel of length 1.55 mm, width 1 mm, where two square struts of side 90 μm have been deployed into the cellular tissue. Smooth Muscle Cells, are depicted as *circles* with (mean) radius of 15 μm. Resulting restenosis after 16 days, with a bare metal stent (*middle*) and a drug eluting stent (*right*)

This model has been implemented using the CxA simulation software as developed in the Coast project [31] and described in detail in [37].

Figure 3.13 shows an example of a two-dimensional version of the model, showing the initial conditions after stent deployment, as well as the resulting restenosis for bare metal stents and drug eluting stents. The inhibitory effect of the drugs on the restenosis is clearly visible. Currently we are working on validating these simulations against detailed experimental data, which will be reported elsewhere [42].

3.4 Concluding Remarks

This chapter briefly described a possible approach towards multi-scale modeling and simulation using Cellular Automata. The concept of Complex Automata should allow the modeling of a large range of multi-scale systems, and the related Complex Automata simulation software provides a framework to quickly develop Complex Automata simulations. The ideas behind Complex Automata have a broader significance than Cellular Automata modeling alone and, in the near future, we will explore the possibility to enlarge the CxA idea to other modeling paradigms. Moreover, we are developing a growing set of CxA models and simulations, and we invite our readers to start doing the same.

References

1. B. Chopard, M. Droz, *Cellular Automata Modeling of Physical Systems* (Cambridge University Press, Cambridge, 1998)
2. A. Deutsch, S. Dormann, *Cellular Automaton Modeling of Biological Pattern Formation: Characterization, Applications, and Analysis* (Birkhäuser, Basel, 2005)
3. P. Sloot, A. Hoekstra, Modeling dynamic systems with cellular automata, ed. by P. Fishwick *Handbook of Dynamic System Modeling*, Chapter 21 (Chapman & Hall/CRC, London/Boca Rabin, FL, 2007)
4. S. Lloyd, Phys. Rev. Lett. **23**, 237901 (2002)
5. K. Zuse, Int. J. Theor. Phys. **21**, 580–600 (1982)
6. K. Zuse, *Rechnender Raum*, http://www.idsia.ch/~juergen/digitalphysics.html
7. S. Wolfram, *A New Kind of Science* (Wolfram Media, Inc., Champaign, IL, 2002)
8. D. Bader, *Petascale Computing: Algorithms and Applications* (Chapman & Hall/CRC, London/Boca Rabin, FL, 2008)
9. A. Hoekstra, S. Portegies Zwart, M. Bubak, P. Sloot, Towards distributed petascale computing, ed. by D. Bader, *Petascale Computing: Algorithms and Applications*, Chapter 8 (Chapman & Hall/CRC, London/Boca Rabin, FL, 2008)
10. P. Sloot, D. Frenkel, H. van der Vorst et al., White paper on computational e-science, studying complex systems in silico, a national research invitiative (2007), http://www.science.uva.nl/research/pscs/papers/archive/Sloot2007a.pdf
11. Special Issue on Multiphysics modeling, IEEE Comput. Sci. Eng. **7** 14–53, (2005)
12. SIAM Multiscale Model Simul, http://epubs.siam.org/sam-bin/dbq/toclist/MMS
13. Int J Multiscale Comput Eng, http://www.edata-center.com/journals/61fd1b191cf7e96f.html
14. A. Finkelstein, J. Hetherington, O. Margoninski, P. Saffrey, R. Seymour, A. Warner, IEEE Comput. **37**, 26–33 (2004)
15. D. Noble, Science **295**, 1678–1682 (2002)
16. B. Di Ventura, C. Lemerle, K. Michalodimitrakis, L. Serrano, Nature **443**, 527–533 (2006)

17. P. Hunter, W. Li, A. McCulloch, D. Noble, IEEE Comput. **39**, 48–54 (2006)
18. P. Sloot, A. Tirado-Ramos, I. Altintas, M. Bubak, C. Boucher, IEEE Comput. **39**, 40–46 (2006)
19. S. Smye, R. Clayton, Med. Eng. Phys. **24**, 565–574 (2002)
20. G. Ingram, I. Cameron, K. Hangos, Chem. Eng. Sci., **59**, 2171–2187 (2004)
21. E. Weinan, X. Li, W. Ren, E. Vanden-Eijnden, Commun. Comput. Phys. **2**, 367–450 (2007)
22. R. White, Modeling multi-scale processes in a cellular automata framework, ed. by J. Portugali, *Complex Artificial Environments, Simulation, Cognition and VR in the Study and Planning of Cities*, (Springer, New York, NY, 2006) pp. 165–177
23. B. Ribba, T. Alarcón, K. Marron, P. Maini, Z. Agur, The use of hybrid cellular automata models for improving cancer therapy, ed. by P. Sloot, B. Chopard, A. Hoekstra: *Cellular Automata*, 6th International Conference on Cellular Automata, ACRI 2004, LNCS, vol. 3305 (Springer, Heidelberg, 2004), pp. 444–453
24. C. Lin, Y. Lai, Phys. Rev. E. **62**, 2219–2225 (2000)
25. N. Baas, T. Helvik, Adv. Compl. Syst. **8**, 169–192 (2005)
26. N. Israeli, N. Goldenfeld, Phys. Rev. Let. **92**, 074105 (2004)
27. A. Hoekstra, E. Lorenz, J.L. Falcone, B. Chopard, Towards a complex automata framework for multi-scale modeling: Formalism and the scale separation map, ed. by Y. Shi, D. van Albada, J. Dongarra, P. Sloot, *ICCS 2007, Part I*, Lecture Notes in Computer Science, vol. 4487 (Springer, Heidelberg, 2007), pp. 922–930
28. A. Hoekstra, E. Lorenz, J. Falcone, B. Chopard, Int. J. Multiscale Comp. Eng. **5**, 491–502 (2007)
29. A.G. Hoekstra, J-L. Falcone, A. Caiazzo, B. Chopard, Multi-scale modeling with cellular automata: The complex automata approach, ed. by H. Umeo et al., ACRI 2008, Lecture Notes in Computer Science, vol. 5191, (Springer, Berlin-Heidelberg, 2008), pp. 192–199
30. D. Evans, P. Lawford, J. Gunn, D. Walker, R. Hose, R. Smallwood, B. Chopard, M. Krafczyk, J. Bernsdorf, A. Hoekstra, Phil. Trans. Roy. Soc. A **366**, 3343–3360 (2008)
31. The Coast project, http://www.complex-automata.org
32. B. Chopard, J-L. Falcone, R. Razakanirina, A.G. Hoekstra, A. Caiazzo, On the collision-propagation and gather-update formulations of a cellular automata rule, ed. by H. Umeo et al., *ACRI 2008*, Lecture Notes in Computer Science vol. 5191, (Springer, Berlin Heidelberg, 2008), pp. 144–251
33. J.A. Kaandorp, J.E. Kübler, *The Algorithmic Beauty of Seaweeds, Sponges and Corals* (Springer, Heidelberg, New York, 2001)
34. R.M.H. Merks, A.G. Hoekstra, J.A. Kaandorp, P.M.A. Sloot, J. Theor. Biol. **224**, 153–166 (2003)
35. R.M.H. Merks, A.G. Hoekstra, J.A. Kaandorp, P.M.A. Sloot, J. Theor. Biol. **228**, 559–576 (2004)
36. G. Agha *Actors: A Model of Concurrent Computation in Distributed Systems* (MIT Press, Cambridge, MA, 1986)
37. J. Hegewald, M. Krafczyk, J. Tölke, A. Hoekstra, B. Chopard, An agent-based coupling platform for complex automata ICCS 2008, Krakow. Lecture Notes in Computer Science, vol. 5102, doi:10.1007/978-3-540-69387-1 (Springer, Berlin Heidelberg, 2008), pp. 227–233
38. S. Succi, *The Lattice Boltzmann Equation for Fluid Dynamics and Beyond* (Oxford University Press, Oxford, 2001)
39. D. Alemani, B. Chopard, J. Galceran, J. Buffle, Phys. Chem. Chem. Phys. **7**, 1–11 (2005)
40. A. Caiazzo, J-L. Falcone, B. Chopard, A.G. Hoekstra, Error investigations in complex automata models for reaction-diffusion systems, ed. by H. Umeo et al., ACRI 2008, Lecture Notes in Computer Science, vol. 5191, (Springer, Berlin Heidelberg, 2008), pp. 260–267
41. A. Caiazzo, J-L. Falcone, B. Chopard, A.G. Hoekstra, Asymptotic analysis of complex automata models for reaction-diffusion systems. Appl. Num. Maths **59**, 2023–2034 (2009)
42. A. Caiazzo, D. Evans, J.L. Falcone, J. Hegewald, E. Lorenz, B. Stahl, D. Wang, J. Bernsdorff, B. Chopard, J. Gunn, R. Hose, M. Krafczyk, P. Lawford, R. Smallwood, D. Walker, A.G. Hoekstra, Towards a complex automata multiscale model of in-stent restenosis, submitted to J. Comput. Sci.

Chapter 4
Hierarchical Cellular Automata Methods

Adam Dunn

Many real-world spatial systems involve interacting processes that operate over more than scale. Whilst there has been a strong growth in knowledge about multi-scale systems in many disciplines, the advent of coupled, multiresolution, multiscale and hierarchical cellular automata has been recent in comparison. Here, the structural definition of a cellular automaton is augmented with an abstraction operator, which transforms the cellular automaton into a hierarchy of cellular spaces. Simple propagation is used as a familiar and common behavioural phenomenon in several examples of behavioural specification. The purpose of this chapter is to provide the basics of a general framework, from which hierarchical cellular automata may be constructed for specific applications. Simple examples from landscape ecology are used to elucidate the methods.

4.1 Introduction

Classical cellular automata (CA) are used as an analog of complex systems because they are fundamentally similar in operation to the self-organised systems seen in nature and the man-made world. However, there are many complex systems for which a single scale will not capture the fine-scale dynamics that influence their global behaviour, and yet others for which a single scale is intractable for practical simulation. These include complex systems in disciplines such as landscape ecology or sociology, about which we would like to make real predictions in order to guide and influence policy.

Hierarchical CA, as defined in this chapter, are capable of modelling processes that operate at different spatial scales within the same system. The approach has a lineage in coupled CA [33] and the paradigm of hierarchical patch dynam-

A. Dunn (✉)
Centre for Health Informatics, University of New South Wales UNSW,
Sydney NSW 2052, Australia; Alcoa Research Centre for Stronger Communities,
Curtin University of Technology, PO Box U1985, Perth WA 6845, Australia
e-mail: A.Dunn@unsw.edu.au; A.Dunn@curtin.edu.au

A.G. Hoekstra et al. (eds.), *Simulating Complex Systems by Cellular Automata*,
Understanding Complex Systems, DOI 10.1007/978-3-642-12203-3_4,

ics [36, 35]. The approach grew out of the need for tractable, real-time models of wildfire spread [10] and has been implemented for a model of invasive plant spread [9, 8], where humans and frugivores are responsible for the spread of invasive weeds at widely separate spatial scales. By the nature of their construction, the hierarchical CA is of practical benefit to modelling complex systems where the empirical information feeding simulations is provided at multiple scales.

The practical issues relating to hierarchical CA are discussed in this chapter. These CA fit into the categories of complex (multiscale) automata and network automata, but also use probabilistic updates. The first part of this chapter deals primarily with building the structure of a hierarchical CA and discusses the reasons behind modelling choices. In the second half of the chapter, the implementation of behaviour is discussed for both homogeneous and heterogeneous examples. The guide is by no means a complete set of potential constructions for hierarchical CA; rather, the aim is to provide enough information for any reader to replicate the approach in a range of application domains.

There are other ways in which multiresolution and multiscale processes may be modelled. A popular form using wavelets, is explored elsewhere [4]. Another example of complex CA, also presented in another chapter of this book, is discussed by Hoekstra et al. [14]. Some classes of problems are better suited to different methods, but here, examples are addressed for problems that are well-suited to the discrete and heterogeneous structure of the hierarchical CA.

4.2 Structure of Hierarchical CA

Classically, a two dimensional cellular space $\mathcal{L}$, is defined by a regular lattice of cells. Each cell's state, $\sigma = \{0, 1\}$ (for a binary-state CA), is determined by the current state of the cell and the state of a set of cells within close vicinity. A neighbourhood $\mathcal{N}$ (of size N) is defined by a function of rectilinear coordinates that produces a set of states $\{\sigma_1, \sigma_2, \ldots, \sigma_N\}$. All cells in the cellular space are updated synchronously to produce the new, collective state of the cellular space in each time step.

However, the definition is not immutable. Nearly every part of the above definition may be removed, modified or augmented. Asynchronous, heterogeneous or irregular CA may be of value for defining a model of a real system. What is fundamental to CA is the discrete structure, local connectivity and exchange of information between neighbours. As an early pioneer of CA, Burks [5] recognised the fundamental characterisation of a CA as follows:

> "Its features are a quantised time and space, a finite number of possible states for each point of space-time, and a computable local transition function or law (not necessarily deterministic or uniform over space) governing the equations of the system through time."

Burks explicitly mentions non-determinism and heterogeneity, which are tools that CA practitioners routinely use to model physical systems. Practitioners do so in order to answer questions about the real world via simulation, when it is infeasible to observe the real world directly.

4.2.1 Isotropic Propagation and CA

Before describing the structure and behaviour of a hierarchical CA, it is worth first describing how the choice of structure influences the behaviour of a model. The difference between regular and irregular structures may be explained using propagation as an example of an often-modelled behaviour, and isotropy as a measure of the model's accuracy. For one example comparing deterministic and non-deterministic update rules, it can be shown that an irregular lattice structure will eventually produce a closer approximation to isotropic propagation than a regular lattice, given the same finite number of cells. This result has consequences for modellers attempting to choose representations for the structure of a physical system.

The process of propagation is central to many complex system phenomena. One can intuitively think of many examples, but the scales and domains on which this chapter focuses includes the scales associated with the spread of bushfire, the spatial perspective of an invasive species outbreak, or the spread of infection through a population. Isotropic propagation is fundamental to many tasks where a spatially discrete model is built for a continuous space system. Imagine the spread of fire over a landscape in two dimensions, or the diffusion of a gas from a point source in three dimensions. When the environment is homogeneous, isotropic propagation is the expected behaviour. Isotropic propagation in homogeneous conditions is an important precursor to building a realistic model in heterogeneous conditions.

There are several ways in which isotropic propagation may be modelled using CA. Schönfisch [27] demonstrates that stochasticity is necessary for good estimates of isotropic propagation in CA. Similar findings are reported independently [25, 15], but none are widely cited. Schönfisch draws the conclusion that stochasticity may be implemented either in the structure of the CA via irregular grids, or in the behaviour of the CA via probabilistic updates or stochastically distributed asynchronous updates. The best estimate of isotropy in this work is achieved using an irregular grid that is constructed with a minimum distance between cell centres. In another example of modelling isotropy in a quantised space, Holland et al. [15] test both probabilistic updates and irregular spatial structures using a random walk to describe propagation through homogeneous regions.

Some examples of unusual choices for implementing isotropic propagation include a model of lava flow (Spataro et al. [28]), a model of disease spread (Pfeifer et al. [26]), and several models of wildfire spread (Sullivan and Knight [30], Trunfio [31], and Alexandridis et al. [2]). Trunfio's implementation of a CA model of wildfire uses hexagonal grids with side length approximately 20 m. It is stated explicitly that the hexagonal lattice is used to avoid the problems associated with directional bias [31]. A demonstration of the model for a homogeneous landscape is not provided. Alexandridis et al. [2] use a regular lattice and describe a set of probabilities to define a stochastic update mechanism. The probabilities are optimised to match the empirical data for one wildfire scenario and then a comparison is made to the same scenario to judge the success of the model. The homogeneous case is not presented and anisotropy is not discussed. The results of Pfeifer et al. [26] show distinct quadrilateral patterns over spatial data that do not appear to reflect such patterns.

Vicari et al. [32] more recently implemented a randomised distribution of cell centres (in the same manner as Schönfisch [27]) and a maximum radius approach, which was shown to remove the anisotropy caused by both hexagonal and square lattices. This is a simple example of an irregular lattice, since the cell shape is only used to constrain the initial placement of points rather than influence the selection of neighbourhood and evolution of the CA.

Sullivan and Knight [30] implement a two-dimensional, square-lattice CA to model wildfire. Their approach is to construct a model of convective forces into the behaviour of a wildfire model. The model's update rule allows each cell to draw information from every cell in the cellular space at each time step, breaking the local characteristic of CA. This appears to be an attempt to capture forces at multiple temporal and spatial scales – through the landscape as local heat transfer, and the much faster and larger scale forces in the atmosphere.

A regular grid is often a convenient way to quantise space because it has a pedigree that stretches at least as far back as Descartes, it corresponds to the way humans collect observations (as raster images amongst others), and it fits neatly into a computer. However, a regular grid and a probabilistic update rule may, in many cases, not be the appropriate approximation of a physical system. This is particularly true when a modeller attempts to account for structural noise (or structural heterogeneity below the level of abstraction in the model) by applying non-determinism to the behaviour.

Whilst Descartes may have been partially responsible for the grid representation of space, he may have also been responsible for the first recorded example of a Voronoi decomposition (see Okabe et al. [24]). A Voronoi decomposition is a convenient way to represent an irregular distribution of cell centres and the Delaunay triangulation associated with the spatial decomposition may be used to provide the links that define the neighbourhood. Voronoi decomposition is used to construct the spatial structure for many of the CA described in this chapter.

Each of the structures that follow comprise 4×10^4 interacting cells, either as a regular lattice (with a Moore neighbourhood, radius 1) or a Voronoi decomposition over a set of Halton points [12], with a neighbourhood defined by the Delaunay triangulation. Using Halton points ensures that the minimum distance between cell centres is maximised whilst maintaining the apparent randomness of the locations. This approach is slightly different to that of Schönfisch [27], but achieves a similar result. All examples here use a binary state set, notionally inactivate or active. All cells except for one (closest to the centre of the structure) are initially set as inactive. In the deterministic case, a cell becomes active in the first time step for which one of its neighbours is active and it remains in the active state for the duration of the simulation (Eq. (4.1)). In the non-deterministic case (Eq. (4.2)) a cell is probabilistically updated to active with probability 1 on the principal axes and probability 0.25 on the diagonal axes (as per the best approximation by Schönfisch [27] using these rules). The function $P(\mathcal{N})$ returns a cell's new state (either 0 or 1) based on a random test against the closest active cell in the neighbourhood. Here, Σ implies the sum of states for the neighbourhood (from 0 to N, the number of cells in the neighbourhood), as usual.

$$\sigma_{\alpha}^{t+1} = \begin{cases} 0\,, \text{ if } \sum_{i=0}^{N} \sigma_{\mathcal{N}(\alpha)} = 0; \\ 1\,, \text{ if } \sum_{i=0}^{N} \sigma_{\mathcal{N}(\alpha)} > 0; \\ 1\,, \text{ if } \sigma_{\alpha}^{t} = 1. \end{cases} \tag{4.1}$$

$$\sigma_{\alpha}^{t+1} = \begin{cases} 0\,, & \text{if } \sum_{i=0}^{N} \sigma_{\mathcal{N}(\alpha)} = 0; \\ P(\mathcal{N})\,, & \text{if } \sum_{i=0}^{N} \sigma_{\mathcal{N}(\alpha)} > 0; \\ 1\,, & \text{if } \sigma_{\alpha}^{t} = 1. \end{cases} \tag{4.2}$$

For the regular grid, the shape of the CA using a deterministic update rule produces a square shape, with the residuals demonstrating a characteristic shape, showing a grid-induced bias (Fig. 4.1). In the figure, residual distances are calculated by measuring the distance from the centre of the most distant activated cell to the closest point on a circle of radius equal to the number of time steps. Using a von Neumann neighbourhood produces a diamond and using a hexagonal grid produces a hexagon. The reason for the grid-induced bias is because propagation is slower in directions away from the axes of the lattice that correspond to the neighbourhood. Increasing the radius of the neighbourhood (see O'Regan et al. [25]) results in an increase in the number of sides of the polygon produced, thereby shifting the resulting propagation front closer to that of isotropic spread. For the non-deterministic example (Fig. 4.2), the rate of spread in the diagonal directions is reduced by the probabilistic update on the diagonal axes and it produces a closer approximation to isotropic propagation.

In examples of a CA whose structure is described by the Voronoi decomposition of a set of Halton points (see earlier in this section), the results show a closer

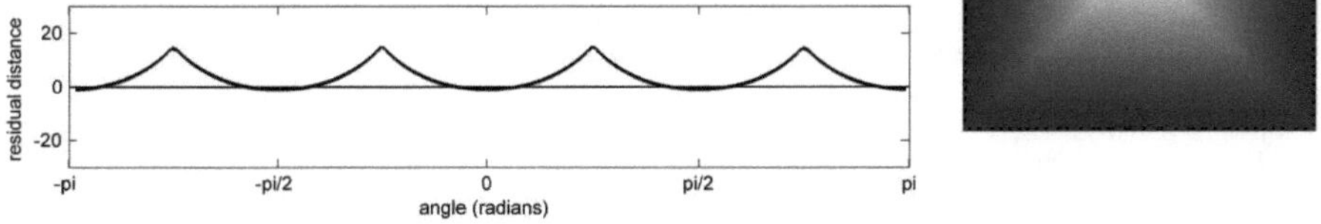

Fig. 4.1 An approximation of isotropy using a regular grid and a deterministic update rule shown as (**a**) the average distance residual during the evolution of state, (**b**) the average distance residual by angle for a given time, and (**c**) an image of the CA *shaded* by activation time (from light to dark)

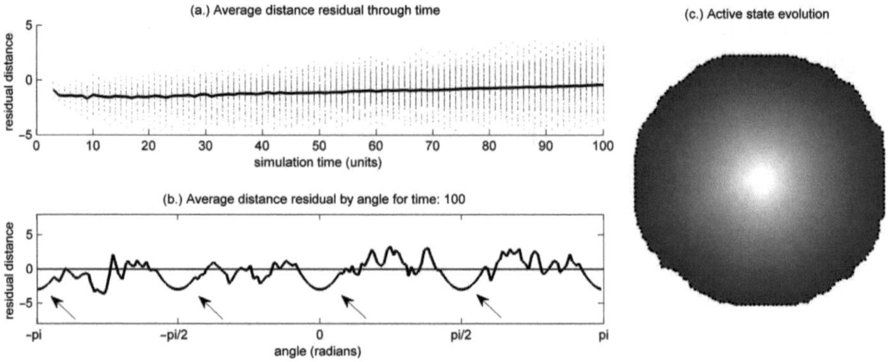

Fig. 4.2 An approximation of isotropy using a regular grid and a non-deterministic update rule (probability $\in \{0.25, 1\}$), shown as (**a**) the average distance residual during the evolution of state, (**b**) the average distance residual by angle for a given time, and (**c**) an image of the CA shaded by activation time (from light to dark). *Arrows* indicate the emerging anisotropy

approximation to isotropy as the number of cells activated increases. In the deterministic case, where cells are activated in the time step immediately following any one neighbour's activation, the rate of spread increases because some neighbours are at distances greater than 1 (Fig. 4.3). In the non-deterministic case, where the probability of activation is based on the inverse of minimum distance to an active neighbour, the rate is slower (Fig. 4.4). Since the underlying structure is determined by a stochastic process, repeated simulation yields different results. The mean propagation distance of repeated simulations produces an increasingly accurate approximation to isotropy, insofar as the residual distance curve becomes increasingly flat. This is not the case with regular grids in which the anisotropy deterministically appears in the same locations.

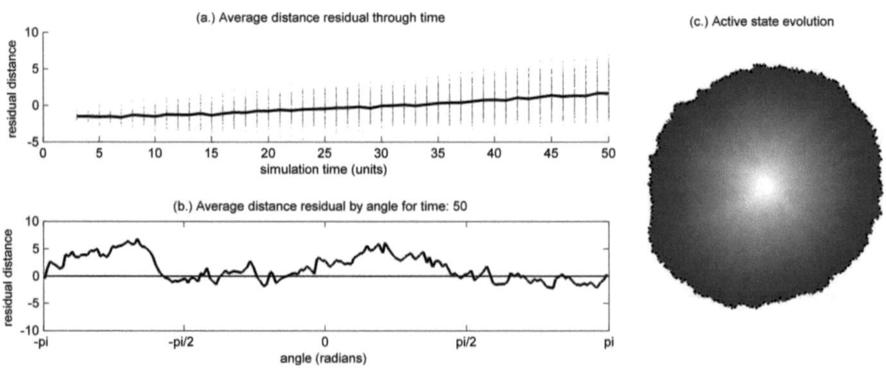

Fig. 4.3 An approximation of isotropy using an irregular structure and a deterministic update rule shown as (**a**) the average distance residual during the evolution of state, (**b**) the average distance residual by angle for a given time, and (**c**) an image of the CA shaded by activation time (from light to dark). There is no emerging pattern in the residuals

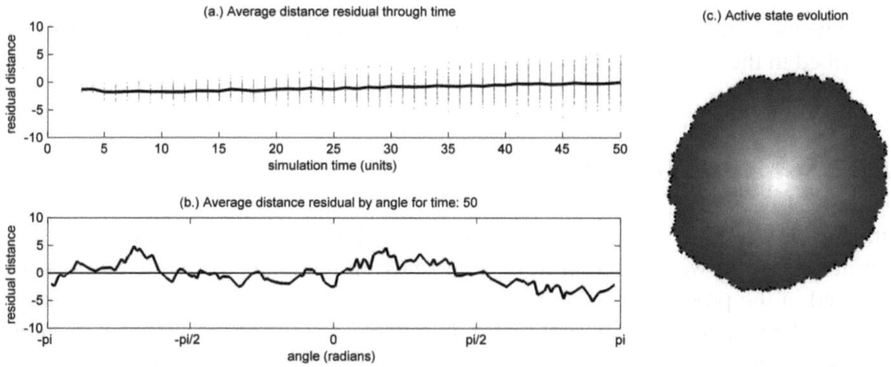

Fig. 4.4 An approximation of isotropy using an irregular structure and a non-deterministic update rule (governed by the inverse of the distance between cell centres) shown as (**a**) the average distance residual during the evolution of state, (**b**) the average distance residual by angle for a given time, and (**c**) an image of the CA *shaded* by activation time (from light to dark). There is no emerging pattern in the residuals

The results of the above demonstration produce a simple comparison between regular and irregular spatial quantisation, and between deterministic and non-deterministic update mechanisms. If a modeller is given a finite number of cells with which to represent the spatial structure of a physical system and the physical system may be modelled as a homogeneous example of propagation over a continuous two-dimensional space, then Voronoi decomposition eventually produces a closer approximation to isotropic propagation. In addition, a probabilistic update (based on distance between cell centres) produces a better approximation than a deterministic update. Note that all examples use synchronous updates.

Each of these examples above provides an approximation to isotropic propagation, but some become better than the others during the evolution of state or after averaging multiple simulation instances. In problems where we are attempting to answer questions about aggregate properties of the system or where the behaviour is homogeneous over the entire space, a simple approximation may be sufficient. Irregular grids may be more appropriate than regular grids where spatial heterogeneity is important to the behaviour, since the grid-induced bias changes the shape of propagation-diffusion models. Additionally, irregular grids may be better suited to problems in which the number of elements is relatively large, since grid-induced bias in regular grids becomes more pronounced as the evolution of state continues through a larger spatial domain, as opposed to irregular grids where the approximation improves.

The above demonstrations provide two basic structures from which a hierarchical CA may be built. Firstly, a regular lattice decomposition of a continuous physical space (the classical form of a CA) using either square or hexagonal cells. This structure may be further augmented by translating the cell centres and using minimum distances to construct irregular neighbourhoods. Secondly, an irregular structure, which sometimes provides a closer approximation to spatially explicit propagation phenomena. The second of these structures is used as a basis for the structures described in the following sections.

4.2.2 Structural Definitions

Imagine an example of a physical system in which two processes are acting at different scales and with different governing rules. In landscape ecology, this system is found in the process of seed dispersal, where multiple vectors (in the biological sense of the word) act at different scales. In models describing the spatial expansion of cities, forces relating to population growth act at different scales to forces relating to the location of resources and transport networks. Feedback exists in these examples where the propagation that results from one force influences the propagation resulting from other forces.

Weimar [33] provides what is arguably the first example of a multi-resolution CA. However, the theory of system hierarchy [1] and multi-scale phenomena outside of the CA domain have a rich theoretical history stretching through several decades of research. Others have contributed to the multiscale representation of CA models in deeply theoretical ways [17]. Yet others have described approaches for building hierarchical structures [36, 35] for specific application domains, without invoking the CA paradigm.

An abstraction operator may be defined as a function that transforms a cellular space into a new cellular space with fewer cells, and represents a physical system with fewer components. The definition is constrained such that the number of input cells is constant and all the input cells form a contiguous patch in the continuous space (i.e. they all belong to each others' neighbourhoods). The general form of the abstraction operator may be defined as follows:

$$\mathcal{L}^T = \mathcal{L} - \{\alpha_1, \alpha_2, \ldots \alpha_k\} + \{\alpha_{n+1}\}, \tag{4.3}$$

where $\mathcal{L}^T$ is the newly transformed cellular space that does not include cells α_1 through α_k, and includes a new call α_{n+1}.

The link between the original cellular space and the transformed cellular space is the interlevel neighbourhood, denoted $\mathcal{N}^*$, a neighbourhood between levels of abstraction. Compare this to the intralevel connections defined by the neighbourhood of a cell within a single level of abstraction. An interlevel connection, $\mathcal{N}^*(\alpha_{n+1})$, may be defined as a subset of $\mathcal{L} + \mathcal{L}^T$, where α_{n+1} belongs to $\mathcal{L}^T$ and takes the index $n+1$ as an indication that it is a new cell in the overall cellular space structure.

The general form of the abstraction operator alone does not provide enough information to implement a hierarchical CA, but this information may be specified for a given purpose. Further consideration must be given to the groups of cells to be abstracted, as well as the order. The following is a description of a specific implementation of the abstraction operator for a CA whose structure is defined as an irregularly distributed set of cells, and whose intralevel neighbourhoods are defined by a Delaunay triangulation [24].

Using the Delaunay triangulation to define a ternary abstraction operator is convenient for the irregular cellular space of the previous section, because it guarantees that each group of three cells forming a triangle are contiguous. The interlevel neighbourhood for this cellular space may be defined as follows:

$$\mathcal{N}^*(\alpha_{n+1}) = \{\alpha_1, \alpha_2, \alpha_3, \alpha_{n+1}\}; \quad \alpha_1, \alpha_2, \alpha_3 \in \mathcal{L}, \alpha_{n+1} \in \mathcal{L}^T \tag{4.4}$$

In this example, the abstraction operator necessarily produces larger cells from smaller ones. The transformation, $\mathcal{L} \rightarrow \mathcal{L}^T$, creates a new cellular space where three cells (whose centres form a triangle in the Delaunay triangulation) are abstracted to form a single new cell (see Fig. 4.5). The size of the cell in $\mathcal{L}^T$ that is not in $\mathcal{L}$ is always larger than each of cells in $\mathcal{L}$ that are not in $\mathcal{L}^T$. The neighbourhood of the new cell is always a subset of the aggregate of its childrens' neighbourhoods ($\mathcal{N}(\alpha_{n+1}) \subseteq \mathcal{N}(\alpha_1) \cup \mathcal{N}(\alpha_2) \cup \mathcal{N}(\alpha_3)$).

In the process of repeated transformations via the abstraction operator, the average size of cells in a cellular space increases and the hierarchy linking smaller cells to larger cells is constructed as a series of inter-level connections (see Fig. 4.6). In the example given in the figure, the irregular and hierarchical cellular space is built over a homogeneous environment. In this case, a series of abstractions are performed as follows:

1. set all cells as potentially abstractable;
2. while there are still abstractable triangles in the current Delaunay triangulation, select an abstractable triangle at random;
3. perform the single transformation from $\mathcal{L} \rightarrow \mathcal{L}^T$ over the selected set of three cells;
4. set the triangles that include the cells that are no longer in $\mathcal{L}^T$ as no longer abstractable in this iteration;

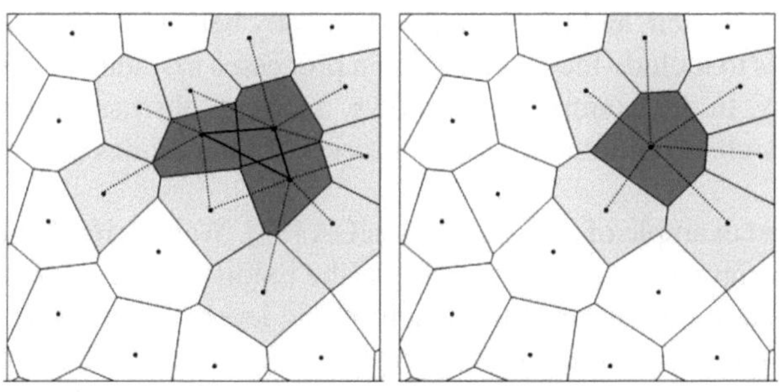

Fig. 4.5 An example of a single abstraction using a ternary abstraction operator over an irregular distribution of cells. The three darker cells on the *left* are abstracted to produce the single dark cell on the *right*. The union of the three cells' neighbourhoods are given in *light grey* and links are shown as *dotted lines* (on the *left*). The same scheme is used for the neighbourhood of the abstracted cell (on the *right*)

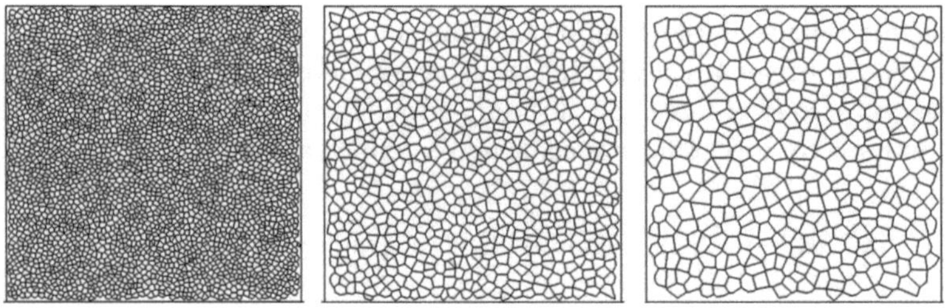

Fig. 4.6 The cellular structure for a homogeneous system using a ternary abstraction operator. The cell boundaries are delineated by *black lines*. The three cellular spaces presented here are linked through a series of abstractions that produce the cellular spaces from left to right

5. recalculate the Delaunay triangulation and repeat the abstraction process (steps 3–5) using the modified list of abstractable triangles until there are no more viable triangles; and
6. rebuild the list of abstractable cells and repeat the entire process until the average size of cells reaches some desired level (or there are less than three cells left).

In the example above, cells in the iteratively transformed cellular space become larger with relative consistency. However, a consistent increase in cell size across the cellular space may not be as useful for systems in which there is heterogeneity in the empirical information used to create the cellular space state and update rules. In the following section, the cellular space construction is modified to maintain boundaries of heterogeneous empirical data.

4.2.3 Building Structures with Heterogeneous Data

The value of implementing the hierarchical structure is evident when introducing spatial heterogeneity. In hierarchical patch dynamics [36, 35], landscape ecologists see the need to link together several scales of disparate information that is heterogeneous, spatially explicit, and may differ significantly between scales. The purpose of linking scales is to include feedback between processes that must otherwise be modelled separately. In the example that follows, a ternary abstraction operator is used to build a hierarchical cellular space for a series of three scales of binary landscape information.

In the earliest example of multiresolution CA [33], Weimar implements a coupled CA that was designed to model one process (the oxidisation of carbon monoxide on platinum) at both a microscopic and macroscopic level. The purpose for modelling both levels in the same structure is to manage a trade-off between computational costs and the need to model microscopic irregularities on a two-dimensional surface from which macroscopic patterns emerge.

In a example of CA modelling in landscape ecology, Chen and Ye [7] develop a CA that uses an irregular triangular grid and couples two specific models – the growth and development of vegetation and a hydrodynamic model. Whilst the issue of grid-based bias in regular structures is avoided by the irregular structure of the CA, the model appears to use the same scale in order to achieve the particular coupling required. In this example, there is no report of any disparity between the scale of empirical data relating to flow and the empirical data relating to the growing pattern of plant species. However, one may imagine any number of coupled processes in which the empirical data is disparate in granularity and therefore well-suited to a hierarchical CA construction.

With landscape models of ecological phenomena, it is often the boundaries between heterogeneous regions that are of specific importance. In existing studies, boundaries between heterogeneous regions of habitat are found to be significant to the behaviour of the system [3, 16, 6]. Besides affecting the dynamics of an ecological system, discrete boundaries are also formed as the result natural forces [11], not necessarily due to anthropogenic effects.

A synthesis of the general problem may be posed as follows:

> A method is required to allow for the coupling of models whose empirical inputs vary in both spatial pattern and granularity.

In the solution offered by hierarchical CA, spatially heterogeneous information is captured in the form of a hierarchy of cells. This is achieved efficiently through the use of an irregular structure where boundaries between internally homogeneous patches are captured at a finer granularity than their homogeneous counterparts. In turn, this is achieved by making specific choices about where and when to apply the abstraction operator during construction of the hierarchy. The resulting structure captures the pattern of heterogeneity more accurately than a grid of single granularity, given the same total number of cells.

Habitat fragmentation is an important problem in landscape ecology, but there are challenges in describing fragmentation. The first problem is that habitat is species specific, season specific, and not necessarily binary [21]. The second major problem is disparity of scale [19]. Conflicting approaches are used to represent species habitat [20, 8] and scale features strongly in the arguments for each approach.

Below, a fictional example of binary spatial information is used as a simple example of separate layers of spatial information contributing to the structure of a hierarchical CA. As above, the ternary abstraction operator is used, but the process is augmented with rules about where and when the operator is applied.

For each layer of empirical data, the cellular space is abstracted according to the location of boundaries between homogeneous patches, a minimum area and a maximum area. To begin with, triangles from the Delaunay triangulation [24] are chosen at random (as in the previous section). However, if the neighbourhood includes a cell with an area above the maximum area size, then the abstraction is not performed. Additionally, providing the cell areas are all greater than the minimum area, if a triangle includes cells whose neighbourhoods span separate patches (i.e. cross a boundary), then they are not abstracted. The result is that cells are bounded by a minimum and maximum area, as well as efficiently capturing the shape of boundaries between internally homogeneous regions.

The construction of a complete hierarchy involves more than one application of the above process. The method begins with a fine-scale distribution of points using a two-dimensional Halton distribution [12]. From this first cellular space, a series of abstractions are applied until the first layer of empirical data is captured (see Fig. 4.7, left). Using this second cellular space as the initial structure, another series of abstractions is applied until the third cellular space is achieved (see Fig. 4.7, middle). This process is repeated one more time to produce the final cellular space, which is the coarsest cellular space of the three results and represents the final layer of empirical data (see Fig. 4.7, right).

A grain-extent window [34] is described as the perception of an organism in terms of the processes in which it engages. For example, whilst ants and humans share the same space, ants have much smaller ranges than humans and perceive their space with a different concept of what makes a particular location habitable. In addition, ants and humans utilise resources within their habitat in different ways. This difference in grain-extent window causes problems when attempting to understand ecological processes they share, which in turn is a problem for the development of policy, the implementation of change or the valuation of specific places.

Hierarchical patch dynamics [36, 35] is an example of a method that may potentially solve the problem of varying grain-extent windows in landscape ecology. By linking separate levels of habitat information, each with a characteristic scale (or range of scales), it is possible to couple models of individual processes. In the example above (as in Fig. 4.7), the difference in scale is apparent (see Fig. 4.8). The three levels of information are captured as cellular spaces, whose cells vary in area by orders of magnitude. In this example, cell areas are measured in terms of pixels, but these pixels may represent any real-world area that is observable and recordable.

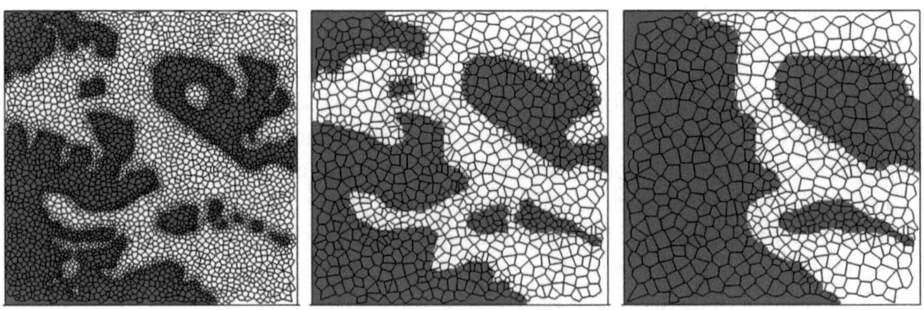

Fig. 4.7 The cellular structure for a non-homogeneous system using a ternary abstraction operator and a boundary protection rule. The cell boundaries are delineated by *black lines* and the *grey/white* areas denote different information (i.e. habitat and non-habitat). The system's spatial information is different for the three levels in the hierarchy but is linked through the series of abstractions that produces the cellular spaces iteratively, from left to right

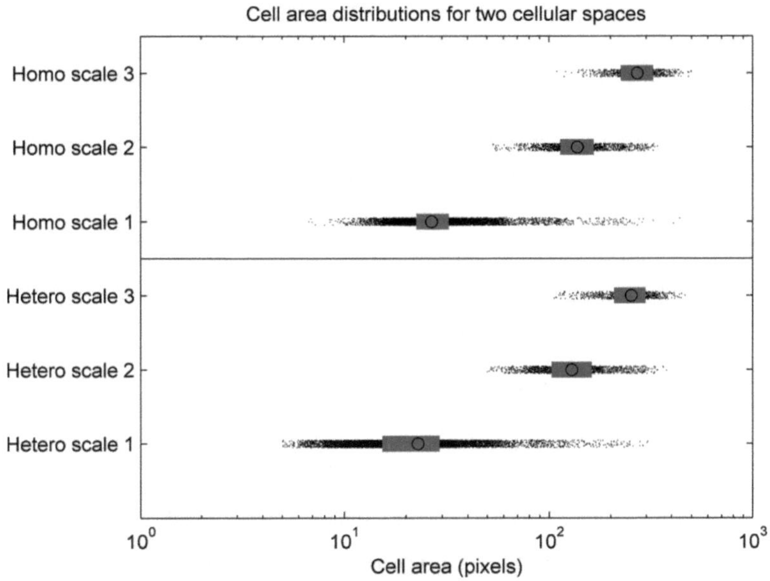

Fig. 4.8 Cell area distributions for both a homogeneous and non-homogeneous cellular space structure, using the same initial parameters. The cell areas for three levels in the hierarchy are given as *black dots*. The median is given as an *open circle* and the interquartile ranges are given as *grey rectangles*

4.3 Behaviour of Hierarchical CA

Given the structure of a hierarchical CA, as described in the previous section, the next aim is to reproduce the behaviour of a propagation phenomenon using this structure. Both homogeneous and spatially-dependent update rules are demonstrated in this section. The update rules introduced here are essentially the hierarchical version of a probabilistic CA. Before continuing to the heterogeneous examples, a homogeneous instantiation of the hierarchical CA is shown to be capable of producing an approximation to isotropic propagation. Then, a spatially heterogeneous example is shown as an analogy for the multiple-vector seed dispersal of an invasive plant.

Two types of information flow are present in a hierarchical CA model of propagation. Intralevel neighbourhoods, $\mathcal{N}(\alpha)$, are used for the information flow at specific scales in the hierarchy. For example, in a model of invasive species, the propagation of seeds via a single seed dispersal vector (such as frugivore-mediated dispersal) may be modelled using information flow via intralevel neighbourhoods. Intralevel neighbourhoods comprise a set of cells within a single cellular space, $\mathcal{L}^k$ (the kth transform of the original cellular space $\mathcal{L}^0$). Interlevel neighbourhoods, $\mathcal{N}^*(\alpha)$ are used to transfer state information up and down the hierarchy, between $\mathcal{L}^k$ and $\mathcal{L}^{k+1}$.

For example, in a model of disease spread through a social network, the presence of one additional infected person at the finest scale might modify the state at some higher level (family, school/workplace, community) from susceptible to infected.

In the general case, the update rule potentially modifies the state (σ_k) of a cell (α_k). The rule is defined over the set of states produced by the intralevel neighbourhood function ($\mathcal{N}(\alpha_k)$) or over the set of states from the cells produced by the interlevel neighbourhood function ($\mathcal{N}^*(\alpha_k)$). The corresponding update rules ϕ and ϕ^* are defined as per Eqs. (4.5) and (4.6).

$$\begin{aligned}\phi : &\overbrace{\Sigma \times \Sigma \times \cdots \times \Sigma}^{N} \rightarrow \Sigma, \\ &\text{where } \Sigma = \{0, 1\} \text{ for a binary state CA and} \\ &N \text{ is the number of cells in } \mathcal{N}\end{aligned} \tag{4.5}$$

$$\begin{aligned}\phi^* : &\overbrace{\Sigma \times \Sigma \times \cdots \times \Sigma}^{N^*} \rightarrow \overbrace{\Sigma \times \Sigma \times \cdots \times \Sigma}^{N^*} \\ &\text{where } \Sigma = \{0, 1\} \text{ for a binary state CA and} \\ &N \text{ is the number of cells in } \mathcal{N}^*\end{aligned} \tag{4.6}$$

This means that the intralevel function, ϕ, is the same as the classical definition of a CA. In ϕ^*, the states of the children cells and the parent cell are used to define the new state of both the children and the parent. In this manner, information may flow in both directions (up the hierarchy through parents and down through children) under the same state transition function.

4.3.1 A Probabilistic Update Method

The general rule may be implemented using probabilistic functions, to produce the simple propagation phenomenon detailed in previous sections. The state transition function for every cell is the same, implying a homogeneous behavioural model. Firstly, the functions are demonstrated for a spatially homogeneous system, in which cells are evenly distributed at all levels. Secondly, the functions are demonstrated for a heterogeneous system where empirical information is used to create an uneven distribution of cell centres (but the rules are still applied homogeneously). In both cases, isotropic propagation is the expected behaviour because the update rules are consistent over the whole of the cellular space at each level in the hierarchy.

The hierarchical CA is constructed as per the previous section, using three levels of data. In the first example, the absence of data produces an even distribution of cells over the space for each of the three levels. In the second example, fictional spatial data is used to create an uneven distribution of cells over the space for each of the three levels. The two structures used in the two simulations pictured below are exactly as given in Figs. 4.6 and 4.7.

Intralevel propagation is modelled using the inverse of the minimum distance as a likelihood of activation, as in the single scale example pictured in Fig. 4.4. Propagation occurs at each of the three levels in the hierarchical CA. The state of each cell, $\sigma \in \Sigma \equiv \{0, 1\}$ is 0 when unactivated and 1 when activated. Once a cell becomes activated, it remains activated indefinitely and potentially causes the further activation of cells in its neighbourhood.

Cells transmit information up and down through the hierarchy using the interlevel neighbourhoods. As per Eq. (4.7) below, ϕ^* takes the state information of the three children and the parent when determining the new state of the children and the parent. The intralevel and interlevel rules are applied in turn for each time step in the simulation.

$$\begin{gathered}\phi^* : \Sigma \times \Sigma_1 \times \Sigma_2 \times \Sigma_3 \rightarrow \Sigma \times \Sigma_1 \times \Sigma_2 \times \Sigma_3 \\ \text{where } \Sigma_1, \Sigma_2, \Sigma_3 \text{ are the state sets of the children.}\end{gathered} \tag{4.7}$$

In the simulations presented below, the transfer of information up through the hierarchy ($\sigma^{t+1} = f(\sigma^t, \sigma_1^t, \sigma_2^t, \sigma_3^t) \in \{0, 1\}$) uses a majority-based rule (perhaps the most intuitive interlevel information exchange model appropriate to propagation). The state table for both directions (up and down) is given as Table 4.1. The function P is defined here to produce either 0 or 1 depending on the outcome of a random test against the input value. In simple terms, if two cells of the three children are activated, and the parent is not, then the parent will be activated in the next step deterministically. In the reverse direction, children cells are activated with probability 0.5 if the parent is activated and the child is not.

Results of a single simulation for homogeneous data layers (Fig. 4.9), and for heterogeneous and varying data layers (Fig. 4.10) demonstrate a reasonable

Table 4.1 The update rule, ϕ^* for the interlevel transfer of information in the majority-rule hierarchical CA

Input				Output			
σ_1	σ_2	σ_3	σ	σ_1	σ_2	σ_3	σ
0	0	0	0	0	0	0	0
0	0	1	0	0	0	1	0
0	1	0	0	0	1	0	0
0	1	1	0	0	1	1	1
1	0	0	0	1	0	0	0
1	0	1	0	1	0	1	1
1	1	0	0	1	1	0	1
1	1	1	0	1	1	1	1
0	0	0	1	P(0.5)	P(0.5)	P(0.5)	1
0	0	1	1	P(0.5)	P(0.5)	1	1
0	1	0	1	P(0.5)	1	P(0.5)	1
0	1	1	1	P(0.5)	1	1	1
1	0	0	1	1	P(0.5)	P(0.5)	1
1	0	1	1	1	P(0.5)	1	1
1	1	0	1	1	1	P(0.5)	1
1	1	1	1	1	1	1	1

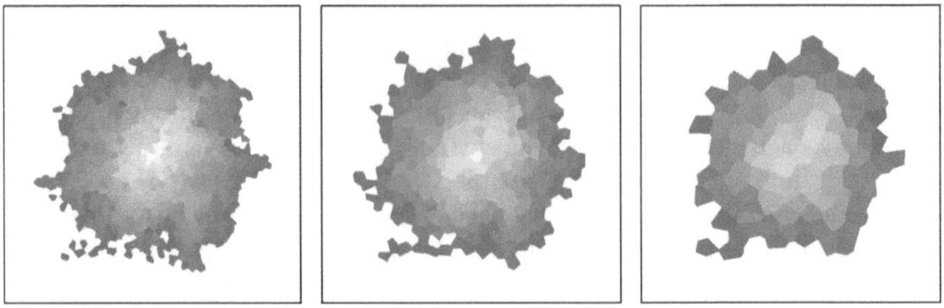

Fig. 4.9 Activation times for a single simulation of a homogeneous propagation rule over three levels of a homogeneous cellular structure. The time to activation is represented by the difference in *shading* (from light to dark). The cellular structure is as given in Fig. 4.6. Isotropic propagation is expected

approximation to isotropy. As with the simulations depicted in Figs. 2.2 and 4.4, the aggregation of repeated simulation yields closer approximations to isotropic propagation through all levels. Both types of simulation (homogeneous and heterogeneous data layers) produce similar results using the single rule across all cells. Over multiple simulations using the same two structures, the results produce identical results, showing that the difference in structure (between an absence of data and the fictional empirical data) does not influence the isotropy of propagation.

In the single simulations presented in Figs. 4.9 and 4.10, anisotropy is more apparent than in Fig. 4.4. The effect of propagation through the coarsest scale (a scale that produces the poorest approximation to isotropic propagation) filters through to the propagation at the finest scale via the interlevel mechanism. Whilst the effect might seem to be undesirable in the homogeneous case, the effect highlights the nature of coupled propagation processes – propagation at different scales contribute to the overall pattern of behaviour. The advantage of this particular feature of the hierarchical CA is demonstrated in the following section.

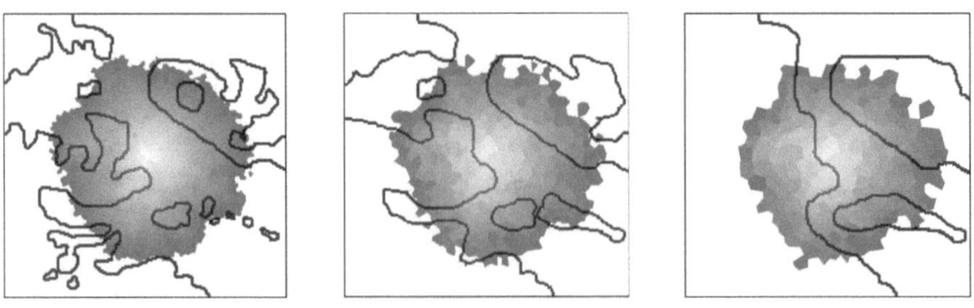

Fig. 4.10 Activation times for a single simulation of a homogeneous propagation rule over three levels of a heterogeneous cellular structure. The time to activation is represented by the difference in *shading* (from light to dark). The cellular structure is as given in Fig. 4.7. Isotropic propagation is expected

4.3.2 Processes with Heterogeneous Behaviour

The spatially dependent update rules demonstrate how forms of propagation that act at different scales (and according to different information) may be implemented within the same structure. Here, using invasive plants as an example, simulations of multiscale propagation through patchy habitats are demonstrated. The propagation rule discussed in the previous section is augmented by a second case that does not permit any propagation, to simulate the effect of disconnected patches of propagation across multiple scales. The demonstration shows that disconnections in habitat at one scale may be countered by connected habitat at a separate scale, in turn demonstrating the effect that multiple dispersal vectors may have on the spread of an invasive plant.

Humans and other fauna share their landscape in unusual ways. As described in a previous section, the manner in which different entities perceive their environment is different (see Dunn and Majer [8]). For example, in a patchy landscape as depicted in Fig. 4.11, humans exist in different densities and move through constructed transport networks. Conversely, seed-eating birds such as *Zosterops lateralis* prefer to live and move through degraded vegetation close to watercourses. In Fig. 4.11, the habitats of humans and seed-eating birds are pictured together, highlighting their differences and overlaps.

In landscape ecology, seed dispersal for many species is understood to be as a result of more than one dispersal vector and each vector may contribute a different frequency and range [23, 18]. A single vector's contribution (for example, a seed-eating bird with a specified gut-passage-time) may be defined for a homogeneous landscape pattern by a one-dimensional curve that specifies a probability of seed dispersal distance. Even if a seed-eating bird's typical range is less than 200 m, it may still disperse seeds beyond this distance with a small probability [29]. As more dispersal vectors are incorporated into a model, the volume of seeds dispersed at different distances is augmented. Traditionally, the set of dispersal "curves" are aggregated to produce a one-dimensional model of seed dispersal. These models

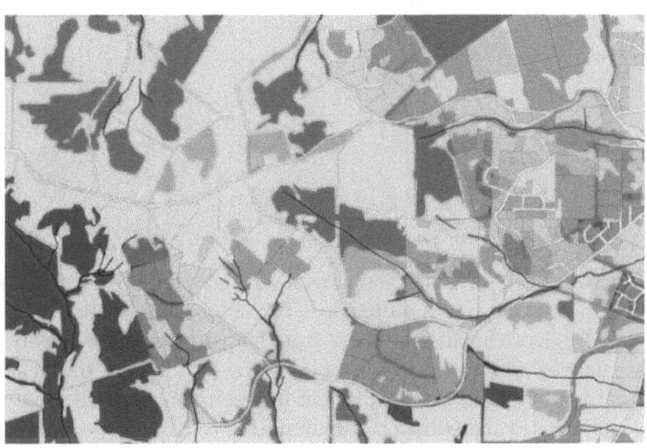

Fig. 4.11 Frugivore habitat and human population density information at two scales are superimposed. *Irregular, darker* coloured patches represent various levels of degraded vegetation (frugivore habitat) and the *transparent*, typically polygonal overlays denote human population densities greater than zero. *Dark grey lines* represent watercourses

struggle to deal with landscape heterogeneity, specifically for the different perception of habitat by the individual dispersal vectors.

A useful model of invasive species might therefore be designed to capture the individual dynamics of different seed dispersal mechanisms in the same structure. This allows a modeller to model the separate mechanisms together, allowing one mechanism's propagation to be reinforced by the other mechanism.

In the simple model depicted in Fig. 4.12, a hierarchical CA is used to simulate two mechanisms that operate in mutually exclusive spatial scales. Firstly, habitat for a seed-eating bird is associated with watercourses and remnant vegetation. The seed-eating birds distribute seeds by consumption and defecation, which means that the gut-passage time for the bird is a good estimate of potential dispersal distances. Secondly humans, who are located in both urban and exurban regions, mainly disperse seeds via transport networks within their habitat. Humans have a propensity to disperse seeds at a greater distance than seed-eating birds. For this model, humans are associated with dispersal distances that are one order of magnitude greater than the seed-eating birds.

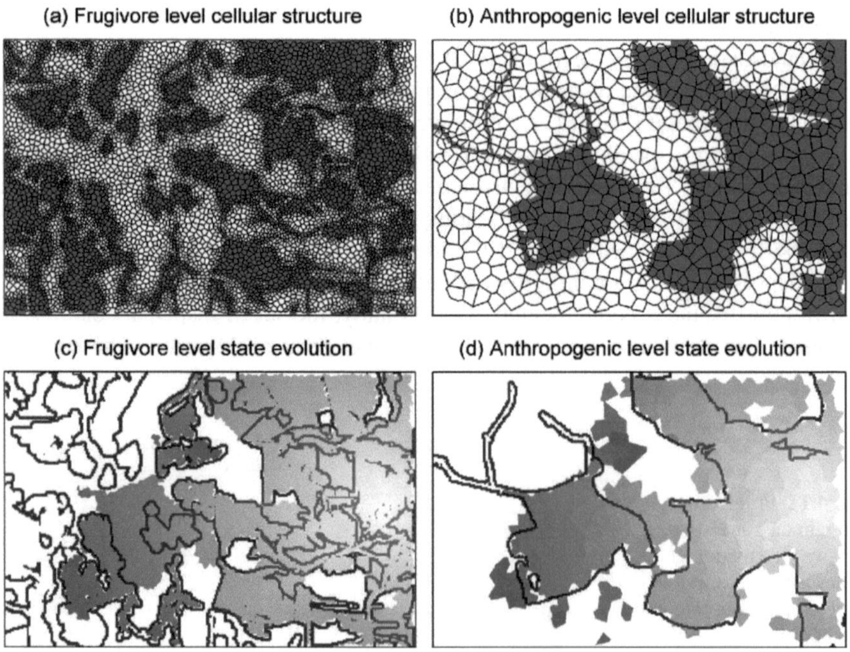

Fig. 4.12 Cell structure and activation times for two levels of structure in the hierarchy formed by a ternary abstraction operator. The spatial information is as given in Fig. 4.11, where (**a**) is from frugivore habitat and (**b**) is from human population density. In the upper sub-figures, the cells are delineated by *black lines*. In the lower sub-figures, the time to activation is given by *shading*, from light to dark, at two levels in the hierarchy with different granularities. The *lines* depict the vegetation (**c**) and human population (**d**)

The intralevel behaviour of the model is a simple constrained propagation and the interlevel behaviour is given by the majority-rule. Each cell belongs to one of two groups, either habitat or non-habitat. For cells that are considered to be habitat, the state transition function is specified as in the original definition above – activity in a cell defines the potential for seed dispersal to have reached the location at a given time. For cells that are not considered habitat, the propagation is constrained to zero – the cell is perpetually unable to propagate (see Eq. (4.8)). Recall that $P(\mathcal{N})$ is a probability function related to the inverse of the distance to the closest active cell in the neighbourhood. This means that greater the distance to the closest active cell, the lower the probability that the cell will become active in the next time step. Equation (4.8) gives the intralevel state transition rules as follows:

$$\sigma^{t+1}_{\alpha_{\text{hab}}} = \begin{cases} 0\,, & \text{if } \Sigma^{N}_{i=0}\sigma_{\mathcal{N}(\alpha)} = 0; \\ P(\mathcal{N})\,, & \text{if } \Sigma^{N}_{i=0}\sigma_{\mathcal{N}(\alpha)} > 0; \\ 1\,, & \text{if } \sigma^{t}_{\alpha} = 1. \end{cases}$$
$$\sigma^{t+1}_{\alpha_{\text{non}}} = \sigma^{t}_{\alpha_{\text{non}}} \tag{4.8}$$

In the simulation depicted in Fig. 4.12, seed dispersal begins at a point (see the centre-right of the lower sub-figures) and propagates through both the urban regions associated with humans, and the remnant vegetation regions associated with seed-eating birds. Whilst disconnections may stop dispersal at one level in the hierarchy, they may not necessarily provide a barrier to dispersal at the other level in the hierarchy. This is a simple demonstration of the manner in which multiple seed dispersal vectors may individually contribute to a larger picture of invasive plant spread.

4.4 Discussion and Summary

This chapter serves as a practical guide for the construction of hierarchical CA. The examples described here are set in the context of landscape ecology – a discipline in which the concept of hierarchy is firmly established and well understood. However, the implementation of practical hierarchical models in these areas may be considered novel because a practical construction framework has never before been detailed. The CA paradigm offers a useful setting for the construction of hierarchical models because a CA reflects the nature of real-world systems in which no central control exists and interactions are local along a series of dimensions.

Despite having been demonstrated many times in different disciplines, modellers are still modelling continuous propagation phenomena with CA that include unwanted grid-induced bias. This bias is found in deterministic CA with regular grids, including hexagonal grids. In this chapter, some specific examples of incorporating stochasticity are shown to reduce anisotropy in homogeneous cases, using the same total number of cells. Whilst none of these propagation methods are novel, their inclusion in this chapter may form a timely reminder of the necessity to first test a model under homogeneous conditions. By ensuring that the underlying propagation is consistent with expectations for homogeneous conditions, a modeller may have more confidence in constraining the model for heterogeneous conditions.

Whilst the presentation here is shallow and broad, it provides a basis from which hierarchical models may be constructed using the CA paradigm. Some examples of the basic construction of hierarchical CA are described in this chapter, including the following:

- homogeneous systems that are abstracted uniformly (using a ternary abstraction operator) to produce an approximation to isotropic propagation; and
- heterogeneous spatial systems that are abstracted to efficiently capture boundaries (using a ternary abstraction operator) and are used to demonstrate coupled propagation processes.

The examples presented here are simplistic versions of CA that one might expect of a calibrated simulation of real-world processes. However, the intention is that they serve as examples from which a modeller may implement their own hierarchical CA.

Theories of hierarchy, and associated paradigms in landscape ecology and sociology, are well-established in their respective disciplines. Theories of hierarchy in CA-based simulation methods for real-world systems are recent by comparison.

The advantages of a hierarchical CA over a classical implementation of the same system, is as follows:

- that empirical information of different granularity may be incorporated into the structure of a CA;
- processes that are best modelled at separate scales may be simulated together and within the same structure; and
- otherwise intractable simulations may be performed on more abstract structures (with fewer components), under the assumption that the abstraction is rigorous.

When building a CA for the purpose of simulating a real-world system, there is value in the process of conscious abstraction. A conscious abstraction is one in which the modeller is aware of the mapping between the structure of the real-world system and the choices that influence the structure of the CA, and the mapping between the processes of the real-world system and the choices that influence the neighbourhood and state transition functions of the CA. An especially significant component of the modelling process relates to the introduction of non-determinism in a model – the reasoning for which is often poorly discussed in existing literature. A more deliberate construction of a CA in this manner relates to the notion of structural validity [37], in which validity comes not only from a model's predictive capacity, but also from the reflection of the real-world's constituent entities in the entities that compose a CA model.

The abstraction operator discussed in detail in this chapter does not provide a failsafe mechanism for ensuring conscious abstraction. Rather, it provides one way of linking cellular structures of higher complexity (by number of components) with cellular structures of lower complexity. What is not considered in this chapter, is a rigorous method of process abstraction. Examples of process abstraction exist in the form of process algebras (see Hoare [13] and Milner [22] for original descriptions), however, this type of approach is intractable for the purpose of large-scale simulation of physical systems.

The inclusion of empirical data is fundamental to the modelling of real physical systems. Empirical data is required for statistical validation of a CA model and without explicit information about the heterogeneity of habitat, geography, demographics (amongst many others), a simulation is only an interesting game, at best. Landscape ecology and biosecurity encompass only a handful of potential applications for hierarchical CA. Hierarchical CA are designed to be used in the practical modelling of real-world systems for which predictions of the future may be used to guide policy and decision-making. For this reason, it is suggested that further investigations into the utility of hierarchical CA be centred on application-based developments for which empirical data is used to calibrate models and validate the methodology.

References

1. V. Ahl, T. F. H. Allen, *Hierarchy Theory: A Vision, Vocabulary and Epistemology* (Colombia University Press, New York, NY, 1996)
2. A. Alexandridis, D. Vakalis, C.I. Siettos, G.V. Bafas, A cellular automata model for forest fire spread prediction: The case of the wildfire that swept through Spetses Island in 1990. Appl. Math. Comput. **204**(1), 191–201 (2008)
3. H. Andrén, Effects of landscape composition on predation rates at habitat edges, *Mosaic Landscapes and Ecological Processes* (Chapman & Hall, London 1995), pp. 225–255
4. A. Brandt, Multiscale scientific computation: Review 2001. *Multiscale and Multiresolution Methods: Theory and Applications*, vol. 20 (Springer, Heidelberg, 2001), pp. 1–95
5. A.W. Burks, *Essays on Cellular Automata* (University of Illinois Press, Champaign, IL 1970)
6. M.L. Cadenasso, S.T.A. Pickett, K.C. Weathers, C.G. Jones, A framework for a theory of ecological boundaries. *BioScience* **53**(8), 750–758 (2003)
7. Q.W. Chen, F. Ye, Unstructured cellular automata and the application to model river riparian vegetation dynamics. Lecture Notes in Computer Science, ACRI 2008, vol. 5191 (Springer, Heidelberg, 2008), pp. 337–344
8. A.G. Dunn, J.D. Majer, In response to the continuum model for fauna research: A hierarchical, patch-based model of spatial landscape patterns. *Oikos* **116**(8), 1413–1418 (2007)
9. A.G. Dunn, J.D. Majer, Simulating weed propagation via hierarchical, patch-based cellular automata. Lecture Notes in Computer Science, ICCS 2007, vol. 4487 (Springer, Heidelberg 2007), pp. 762–769
10. A.G. Dunn, G.J. Milne, Modelling wildfire dynamics via interacting automata. Lecture Notes in Computer Science, ACRI 2004, vol. 3305 (Springer, Heidelberg 2004), pp. 395–404
11. D.G. Green, N. Klomp, G. Rimmington, S. Sadedin, *Complexity in Landscape Ecology, Landscape Series* (Springer, Heidelberg 2006)
12. J. Halton, G. B. Smith, Algorithm 247: Radical-inverse quasi-random point sequence. Comm ACM **7**(12), 701–702 (1964)
13. C.A.R. Hoare, Communicating sequential processes. Commun. ACM **21**(8), 666–677 (1978)
14. A. Hoekstra, E. Lorenz, J.-L. Falcone, B. Chopard, Towards a complex automata framework for multi-scale modeling: Formalism and the scale separation map. *Computational Science ICCS 2007* (Springer LNCS, Heidelberg 2007), pp. 922–930
15. E.P. Holland, J.N. Aegerter, C. Dytham, G.C. Smith, Landscape as a model: The importance of geometry. PLoS Comput Biol **3**(10), e200 (2007)

16. R.A. Ims, Movement patterns related to spatial structures. *Mosaic Landscapes and Ecological Processes* (Chapman & Hall, London, 1995), pp. 85–109
17. N. Israeli, N. Goldenfeld, Coarse-graining of cellular automata, emergence, and the predictability of complex systems. Phys. Rev. E **73**(2), 026203 (2006)
18. P. Jordano, C. Garcia, J.A. Godoy, J.L. Garcia-Castaño, Differential contribution of frugivores to complex seed dispersal patterns. Proc. Natl. Acad. Sci. USA **104**(9), 3278–3282 (2007)
19. S. Levin, The problem of pattern and scale in ecology: The Robert H. MacArthur award lecture. Ecology **73**, 1943–1967 (1992)
20. D.B. Lindenmayer, J. Fischer, R. Hobbs, The need for pluralism in landscape models: A reply to Dunn and Majer. Oikos **116**(8), 1419–1421 (2007)
21. D.B. Lindenmayer, S. McIntyre, J. Fischer, Birds in eucalypt and pine forests: landscape alteration and its implications for research models of faunal habitat use. Biol. Conserv. **110**, 45–53 (2003)
22. R. Milner, *Communication and Concurrency* (Prentice-Hall, Inc., Upper Saddle River, NJ, USA, 1989)
23. R. Nathan, Long-distance dispersal of plants. Science **313**, 786–788 (2006)
24. A. Okabe, B. Boots, K Sugihara, *Spatial Tessellations – Concepts and Applications of Voronoi Diagrams* (Wiley, New York, 2000)
25. W.G. O'Regan, P.H. Kourtz, and S. Nozaki, Bias in the contagion analog to fire spread. Forest Sci. **22**(1), 61–68 (1976)
26. B. Pfeifer, K. Kugler, M.M. Tejada, C. Baumgartner, M. Seger, M. Osl, M. Netzer, M. Handler, A. Dander, M. Wurz, A. Graber, and B. Tilg, A cellular automaton framework for infectious disease spread simulation. Open Med Inform J **2**, 70–81 (2008)
27. B. Schönfisch, Anisotropy in cellular automata. Biosystems **41**(1), 29–41 (1997)
28. W. Spataro, D. DŠAmbrosio, R. Rongo, G. A. Trunfio, An evolutionary approach for modelling lava flows through cellular automata. In Lecture Notes in Computer Science, *ACRI 2004*, vol. 3305. (Springer, Heidelberg 2004), pp. 725–734
29. C. D. Stansbury, Dispersal of the environmental weed Bridal Creeper, Asparagus asparagoides, by Silvereyes, Zosterops lateralis in south-western Australia. Emu **101**, 39–45 (2001)
30. A.L. Sullivan, I.K. Knight, A hybrid cellular automata/semi-physical model of fire growth. *Asia-Pacific Conference on Complex Systems*, Complex 09 (Cairns, Australia, 2004)
31. G.A. Trunfio, Predicting wildfire spreading through a hexagonal cellular automata model. *Cellular Automata*, LNCS (Springer, Heidelberg 2004), pp. 385–394
32. A. Vicari, H. Alexis, C. Del Negro, M. Coltelli, M. Marsella, C. Proietti, Modeling of the 2001 lava flow at Etna volcano by a cellular automata approach. Environ Model. Softw. **22**(10), 1465–1471 (2007)
33. J.R. Weimar, Coupling microscopic and macroscopic cellular automata. Parallel Comput. **27**(5), 601–611 (2001). 375183
34. J.A. Wiens, N.C. Stenseth, B. Van Horne, R.A. Ims, Ecological mechanisms and landscape ecology. *Oikos* **66**, 369–380 (1993)
35. J. Wu, J.L. David, A spatially explicit hierarchical approach to modeling complex ecological systems: Theory and applications. Ecol. Modell. **153**(1–2), 7–26 (2002)
36. J. Wu, O. Loucks, From balance-of-nature to hierarchical patch dynamics: A paradigm shift in ecology. Q. Rev. Biol. **70**, 439–466 (1995)
37. B.P. Zeigler, *Theory of Modeling and Simulation* (Krieger Publishing, Melbourne, FL, USA, 1984)

Chapter 5
Cellular Automata Composition Techniques for Spatial Dynamics Simulation

Olga Bandman

5.1 Introduction

A Cellular Automaton (CA) is nowadays an object of growing interest as a mathematical model for spatial dynamics simulation. Due to its ability to simulate nonlinear and discontinuous processes, CA is expected [1, 2] to become a complement to partial differential equations (PDE). Particularly, CA may be helpful when there is no other mathematical model of a phenomenon which is to be investigated. By now, a great variety of CA are known, whose evolution simulates certain kinds of spatial dynamics. The most known are CA-models of physical processes, such as diffusion [1, 3, 4], wave propagation [5], phase transition [2, 6], spatial self-organization [7], etc. More complicated CA called Gas–Lattice models [8, 9] are used in hydrodynamics, some of them [10, 11] dealing with a real alphabet. In chemistry and microelectronics asynchronous probabilistic CA are used, being sometimes called Kinetic Monte-Carlo methods, they are helpful for studying surface reaction on catalysts [12, 13] and processes of epitaxial growth of crystals [14]. Biology and medicine also present a wide field of phenomena to be simulated by CA-models, genetics [15], myxobacteria swarming [16], growth of tumor [17] being the examples. In solving ecological problems, CA are used more and more frequently to simulate the propagation of diseases [18] and of fire in the forests, evolution of populations, etc. Moreover, CA-simulation has now gone beyond the scope of scientific research, being used, for example, to simulate the process of cement hardening [19].

Among the above CA-models there are those, which have the PDE counterparts [4, 8], but computer implementation (especially on multiprocessors) occurs to be more efficient when based on CA-models, soliton propagation model [5] being a good illustration. However, a vast majority of natural phenomena cannot be described in continuous terms due to their inherent discreteness. For them CA are the only possible mathematical models.

O. Bandman (✉)
Supercomputer Software Department, ICM&MG,
Siberian Branch of Russian Academy of Sciences, Novosibirsk, 630090, Russia
e-mail: bandman@ssd.sscc.ru

A.G. Hoekstra et al. (eds.), *Simulating Complex Systems by Cellular Automata*,
Understanding Complex Systems, DOI 10.1007/978-3-642-12203-3_5,

The diversity of processes being simulated by CA caused the necessity to extend the cellular automaton concept by allowing it to have any kind of alphabet (Boolean, integer, real, symbolic), any kind of transition functions (deterministic, probabilistic), and any mode of functioning (synchronous, asynchronous). Although, two imperative properties of classical CA are preserved:

(1) CA consists of many identical simple processing units (cells).
(2) Interactions between cells are constrained by a small (relatively to the total amount of cells) neighborhood.

Such an extended concept of CA is sometimes referred to as *fine-grained parallelism* [20]. A wide class of CA exhibiting the properties of self-organization and emergency are considered also as models of *complex systems* [21]. Nevertheless, hereafter the term CA or CA-model is used as the most habitual one. In Fig. 5.1, CA-models of natural processes are collected and allocated according to their properties. It is worth noting that Cellular Neural Networks (CNN) [22] and explicit form of discrete representation of Partial Differential Equations (PDE) are also regarded as special cases of CA-models because two above properties are inherent in them.

A fast increase of the variety of CA-models and the growing necessity of simulating complicated processes require a general formal approach to CA composition, which is to be valid for any type of CA and any type of their interaction. It is precisely the object of the chapter, which aims to present a theoretical foundation, and based on it, the CA composition techniques in a generalized and systematic form. The necessity of such a techniques is motivated by the fact, that there is no formal procedure to construct a CA-model according to a given qualitative or quantitative specification of a space-time process. All known CA-models are the result of a trial and error work based on high level of experience in CA modeling, as well as a sophisticated understanding of the phenomenon to be simulated. However now, when a bank of CA-models is relatively large and advantages of CA simulation are

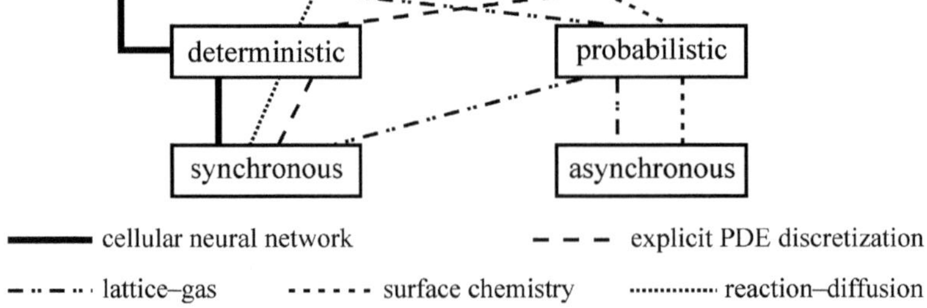

Fig. 5.1 Properties of CA simulation models: *lines* connecting the *rectangles* show the sets of properties, characterizing certain types of CA-models

known, methods for combining several simple CA into a single model for simulating complicated processes seem to be essential. The problem is similar to that in mathematical physics where a PDE is represented as a set of interacting differential operators and functions, each having its own physical meaning. But, as distinct from continuous mathematics, composition of interacting CA meets some essential problems. The solution of these problems should result in creating methods and tools for organizing common work of several CA in such a way that the evolution of a composed CA represents the required spatial process.

Complications in developing such a techniques are associated with the Boolean alphabet, because the overall impact of several Boolean CA may not be obtained by means of conventional arithmetical summation. Even more difficult is to compose CA-models having different alphabets and/or different modes of operation which is the case in reaction–diffusion and prey–predatory processes, where diffusion is given as a Boolean CA, and reaction – as a real function. For example, the snowflakes formation is usually simulated by a Boolean CA, while if it proceeds in active medium, a chemical component should be added, which may be given as a nonlinear real function. The first prototype of such a composition is proposed in [23], where a nonlinear reaction function is combined with a Boolean diffusion. A more general probabilistic variant is given in [24]. The latter is developed here in appropriate techniques for performing algebraic operations on CA configurations with all admissible numerical alphabets (Boolean, real and integer) to make compatible the CA-models with different alphabets. Based on such operations a special *CA configurations algebra* is constructed, which allows us to combine the functioning of several CA-models in a single complex process [25].

To capture all features of essential diversity of CA-models, the more general formalism for CA-algorithms representation, namely, *Parallel Substitution Algorithm* (PSA) [26], is chosen as a mathematical tool.

Recently, CA-models have aroused considerable interest in simulating the crowds behavior [27, 28]. The paper does not deal with this class of CA, concentrating on CA composition for simulation only natural phenomena.

The chapter combines author's results that are scattered about the papers. It consists of the following sections. In the next section, main concepts and formal definitions are given in terms of PSA, and operations on cellular arrays are defined. The third section presents a sequential composition (superposition) of CA-models on local and global levels. In the fourth section a parallel and a mixed composition methods are given. The fifth section concerns computational properties of composed CA, namely, accuracy, stability and complexity. All composition methods are illustrated by the original simulation results.

5.2 Main Concepts and Formal Problem Statement

For simulation of spatial dynamics, an extended concept of CA-model is further considered, whose expressive power is sufficient for simulating natural phenomena of several kinds. The concept is based on the PSA formalism [26], which seems to be the most suitable for modeling composite processes, because it is provided by an effective means (context) for interactions with external agency. Moreover, due to its flexibility, PSA allows the strict formulation of most important requirements imposed on CA composition techniques: (1) provision of behavioral correctness, and (2) compatibility of several CA with different alphabets.

5.2.1 Formal Definition of a CA-model

Simulation of a natural phenomenon comprises the determination of a suitable mathematical model, the development of an appropriate algorithm and a computer program, using the latter for computing desirable functions of time and space. If CA is chosen as a mathematical model, then time is a discrete sequence of nonnegative integers, space is a discrete set referred to as a *naming set*, function values are from an appropriate *alphabet*.

A finite naming set $M = \{m_k : k = 0, \dots, |M|\}$ is further taken for the space. Its elements $m_k \in M$ in simulation tasks are usually represented by the integer vectors of coordinates of a Cartesian space of finite size. For example, in 2D case $M = \{(i, j) : i = 0, 1, \dots, I, \ j = 0, 1, \dots, J\}$. A notation m is used instead of (i, j) for making the general expressions shorter and for indicating, that it is valid for any other kind of discrete space points.

No constraint is imposed on the alphabet A. The following cases are further used: $A_S = \{a, b, ..., n\}$ – a finite set of symbols, $A_B = \{0, 1\}$ – the Boolean alphabet, $A_R = [0, 1]$ – a set of real numbers in a closed continuous interval. Symbols from the second part of the Latin alphabet $\{v, u, x, y, \dots, z\}$ are used to denote the variables defined on A. Appealing to the above extended concept of alphabet is dictated by the aim of the study: to combine several CA of different types into a single one for simulating composite phenomena. A pair (a, m) is called a *cell*, $a \in A$ being a cell–state and $m \in M$ – a cell–name. To indicate the state of a cell named by m both notations $u(m)$ and u_m are further used.

The set of cells

$$\Omega = \{(u, m) : u \in A, m \in M\}, \tag{5.1}$$

such that there are no cells with identical names, is called a *cellular array*, or, sometimes, a *global configuration* of a CA.

On the naming set M, a mapping $\phi : M \to M$ is defined, referred to as a *naming function*. It determines a *neighboring cell* location $\phi(m)$ of a cell named m. In the naming set of Cartesian coordinates $M = \{(i, j)\}$, the naming functions are usually given in the form of shifts $\phi_k = (i+a, j+b)$, a, b being integers. The set of naming functions determines a *template*

$$T(m) = \{\phi_0(m), \phi_1(m), \dots, \phi_n(m)\}, \tag{5.2}$$

which associates a number of cell names to each name $m \in M$. The cell named as $\phi_0(m)$ is called an *active cell* of a template, where $n \ll |M|$, and $\phi_0(m) = m$ by condition.

A subset of cells

$$S(m) = \{(u_0, m), (u_1, \phi_1(m)), \ldots, (u_n, \phi_n(m))\}, \tag{5.3}$$

with the names from $T(m)$ is called a *local configuration*, $T(m)$ being its *underlying template* . The set

$$U_S(m) = (u_0, u_1, \ldots, u_n)$$

forms a *local configuration state vector* .

A cell (u_k, m) changes its state u_k to the next-state u'_k under the action of a *local operator*, which is expressed in the form of a *substitution* [26] as follows

$$\theta(m) : S(m) \star S''(m) \to S'(m), \tag{5.4}$$

where

$$\begin{aligned} S(m) &= \{(v_0, m), (v_1, \phi_1(m)), \ldots, (v_n, \phi_n(m))\}, \\ S'(m) &= \{(u'_0, m), (u'_1, \phi_1(m)), \ldots, (u'_n, \phi_n(m))\}, \\ S''(m) &= \{(v_{n+1}, \phi_{n+1}(m), \ldots, (v_{n+h}, \phi_{n+h}(m))\}. \end{aligned} \tag{5.5}$$

In (5.5) $S(m)$, $S'(m)$ and $S''(m)$ are local configurations, the first two having the same underlying template, and the third one comprises h additional cells. The next-states u'_k, $k = 0, 1, \ldots, n$, of the cells from $S'(m)$ are values of the *transition functions* f_k of the cell states from $S(m) \cup S''(m)$, i.e.

$$u'_k = f_k(v_0, \ldots, v_n, \ldots, v_{n+h}), \quad \forall k = 0, 1, \ldots n. \tag{5.6}$$

A union of the left-hand side local configurations $S(m) \cup S''(m)$ in (5.4) is called a *cell–neighborhood* where $S''(m)$ is a *context*, $S(m)$ is a *base* of $\theta(m)$. The right-hand side $S'(m)$ is the *next-state base* of the local operator.

The underlying templates $T(m)$, $T'(m)$, and $T''(m)$ of the local configuration in (5.5) are in the following relation:

$$\begin{aligned} T'(m) &= T(m), \\ T(m) \cap T''(m) &= \emptyset, \end{aligned} \tag{5.7}$$

$T(m)$ being referred to as a *basic template* of θ.

A local operator $\theta(m)$ is said to be applicable to a cell named $m \in M$ if $S(m) \cup S''(m) \subseteq \Omega$. Otherwise, it is not applicable. Application of $\theta(m)$ to a certain cell

(v, m) (*a single-shot application*) means execution of the following actions. For all $k = 0, \ldots, n$

(1) the next-states u'_k are computed according to (5.6),
(2) the cells $(v_k, \phi_k(m)) \in S(m)$ are updated by replacing the cell states u_k by u'_k.

The cells $(v_{n+l}, \phi_{n+l}(m))$, $l = 0, \ldots, h$, from the context remain unchanged. They play a role of an application condition, the states being used as variables in the transition functions (Fig. 5.2).

A subset $\hat{M} \subseteq M$, referred to as the *active naming set* is defined, such that it comprises the names of active cells, i.e., the cells to which the local operator is applied. Application of θ to all active cells $m \in \hat{M}$ comprises *an iteration* performing a *global transition*,

$$\Phi(\hat{M}) : \Omega(t) \rightarrow \Omega(t+1), \tag{5.8}$$

A sequence of global transition results

$$\Sigma(\Omega) = (\Omega, \Omega(1), \ldots, \Omega(t), \Omega(t+1), \ldots, \Omega(\hat{t})) \tag{5.9}$$

is called a *CA evolution*.

The CA evolution is the result of a simulation task, representing the process under simulation. If the process converges to a stable global state, then CA evolution has a termination, i.e., there exists such a $t = \hat{t}$, that

$$\Omega(\hat{t}) = \Omega(\hat{t}+1) = \Omega(\hat{t}+2) = \cdots = \Omega(\hat{t}+\xi), \tag{5.10}$$

where ξ is an a priori given number. If it not so, then the evolution is infinite, i.e., exhibits an oscillatory or chaotic behavior [2].

There are different modes of ordering local operator application in space and time to perform a global transition from $\Omega(t)$ to $\Omega(t+1)$. The following are the most important ones.

Synchronous mode provides for transition functions (5.6) being computed using the current state values of all their variables, i.e.

$$S(m) \cup S''(m) \in \Omega(t). \tag{5.11}$$

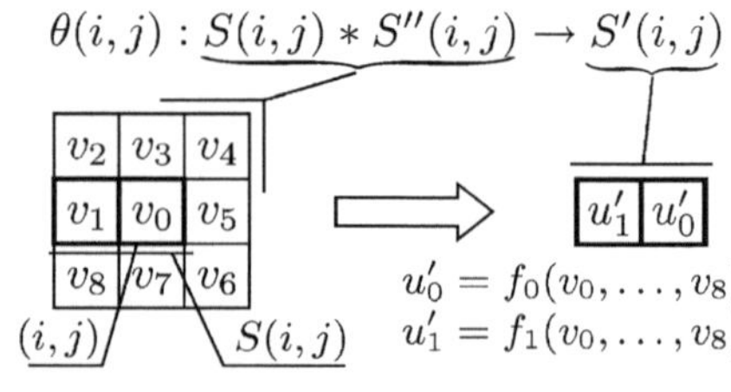

Fig. 5.2 Graphical representation of a local operator

The transition to the next cell–state values occurs after all the transition functions in cells from $S(m)$ for all $m \in \hat{M}$ are computed. Theoretically, it may be done in all cells simultaneously or in any order, which manifests the *cellular parallelism*. In fact, when a conventional sequential computer is used, such a cellular parallelism is imitated by delaying cell updating until all next states are obtained. So, the cellular parallelism is a *virtual parallelism*, which cannot be for the benefit when CA-model is run on conventional computers.

Asynchronous mode of operation suggests no simultaneous operations (neither real nor virtual). Intrinsic parallelism of CA-models is exhibited by the arbitrary order of cells to be chosen for application of $\theta(m)$, the updating of cell states of $S'(m)$ being done immediately after $\theta(m)$ is applied. The time of such an application is referred to as a *time-step* and denoted as τ. So, each global transition $\Omega(t) \rightarrow \Omega(t+1)$ consists of $|\hat{M}|$ sequential time steps, forming a sequence of cellular arrays

$$\gamma_\alpha(\Omega(t)) = \Omega(t), \Omega(t+\tau), \ldots, \Omega(t+|\hat{M}|\tau), \tag{5.12}$$

which is referred to as *global state transition sequence*. The important property of asynchronous mode of operation is that the state values used by transition functions (5.4) may belong both to $\Omega(t)$ and to $\Omega(t+1)$, i.e.,

$$S(m) \cup S''(m) \subset \Omega(t) \cup \Omega(t+1). \tag{5.13}$$

It is the reason why two CA-models with equal $\langle A, M, \hat{M}, \theta\rangle$ starting from the same Ω may have quite different evolutions when operating in different modes. Although, some exotic "very good" CA-models are known, whose evolutions and attractors are invariant whatever mode of operation is used [26].

Multi-stage synchronous mode is also frequently used. It is a mixed mode of operation, which may be regarded both as a synchronised asynchronous mode, and as an asynchronised synchronous one. The mode suggests the whole cellular array of the CA to be partitioned into nonintersecting *blocks* each containing b cells. The block partition induces a dual of partition $\{M'_1, \ldots, M'_b\}$, whose subsets are further called *stage naming subsets* . They contain representative names of all blocks (one out of each block), so that $\hat{M} = M'_1 \cup \ldots \cup M'_b$. Respectively, the iteration is divided into b stages. At each kth stage the local operator is applied to the cells of M_k synchronously, the stages being processed in asynchronous manner. Naturally, cellular parallelism is here limited by the subset cardinality.

No matter what is the mode of operation, a *global operator* is the result of application of $\theta(m)$ to all cells $m \in \hat{M}$.

From the above it follows that a *CA-model*, denoted as $\aleph$ is identified by five notions:

$$\aleph = \langle A, M, \hat{M}, \theta, \rho\rangle$$

where ρ indicates the mode of operation, $\rho = \sigma$ stands for the synchronous mode, $\rho = \beta$ – for multistage synchronous mode, and $\rho = \alpha$ – for asynchronous mode of local operator application. When the indication of operation mode is essential, the corresponding symbol is placed as an subindex, e.g., $\aleph_\alpha$ denotes an asynchronous CA.

5.2.2 Correctness of CA Simulation Process

A CA-model $\aleph = \langle A, M, \hat{M}, \theta, \rho\rangle$ is said to be correct (in computational sense) if its operation satisfies the following *correctness conditions.*

1. *Non-contradictoriness. At any moment of time, a cell is allowed to be updated by only one local operator application.* Non-contradictoriness provides absence of conflicts, which are such a situations when a local operator being applied to the cells m and $\phi_k(m)$ simultaneously is attempting to update one and the same cell by writing in it different state values. Formally, non-contradictoriness sufficient condition is formulated as follows [26]: simultaneous application of a local operator to m_k and m_l is allowed only if

$$T'(m_k) \cap T'(m_l) = \emptyset \qquad \forall (m_k, m_l) \in M. \tag{5.14}$$

It is quite clear, that the non-contradictoriness condition is always satisfied for classical synchronous CA whose local operator has a single–cell base, i.e., $|S'(m)| = 1$. It is not so if $|S'(m)| > 1$, because the local operator has to change several cells simultaneously. To avoid the above conflict situation, one has to sacrifice a bit of cellular parallelism to non-contradictoriness. It may be done either by constructing an asynchronous CA, simulating the same process, or by replacing the synchronous CA $\aleph_\sigma = \langle A, M, \hat{M}, \theta, \sigma\rangle$ by an equivalent multi-stage CA $\aleph_\beta = \langle A, M, \hat{M}_1, \dots, \hat{M}_b, \theta, \beta\rangle$. Such a sequalisation is done according to the following algorithm.

1. The naming set M is partitioned into $|M|/b$ *blocks*, a block being defined by the underlying template $B(m) = \{\psi_0(m), \psi_1(m), \dots, \psi_l(m), \dots, \psi_b(m)\}$ in such a way, that

$$B(m_j) \supseteq T'(m_j), \qquad \forall j = 1, \dots, |M|/b.$$
$$\bigcup_{j=1}^{|M|/b} B(m_j) = M, \qquad \forall j = 1, \dots, |M|/b.$$
$$B(m_h) \bigcap B(m_g) = \emptyset, \quad \forall m_h, m_g \in \hat{M}_k, \quad \forall k = 1, \dots, b, \tag{5.15}$$

   where $T'(m)$ is the basic template in θ.
2. On the active naming set $\hat{M}$ a stage partition $\{\hat{M}_1, \dots, \hat{M}_k, \dots, \hat{M}_b\}$ is defined, i.e.,

$$\hat{M}_k = \{\psi_k(m_j) : \ k = 1, \dots, b; \ j = 1, \dots, |M|/b.\} \tag{5.16}$$

   $m_j = \psi_0(m_j)$ being the active cell of a block $B(m_j) \in M$.

3. Each iteration $\Omega(t) \to \Omega(t+1)$ is divided into b sequential stages $(t_1, t_2, \ldots, t_b)$, $t_b = t+1$, the resulting arrays forming a sequence:

$$\gamma_\beta(t) = \Omega(t), \ldots, \Omega(t+t_k), \Omega(t+t_{k+1}), \ldots, \Omega(t+1), \quad t_k = \frac{\tau k}{b}, \quad (5.17)$$

referred to as a *stage transition sequence*. On the k-th stage, $k = 1, \ldots, b$, $\theta(m)$ is applied synchronously to all cells from $\hat{M}_k$.
4. The subsets $\hat{M}_k$, $k = 1, \ldots, b$, are processed sequentially in arbitrary order, hence, the total number of possible stage transition sequences is $|\{\gamma_\beta\}| = b!$

The CA-model obtained by the above algorithm satisfies the non-contradictoryness condition (5.14). Moreover, its evolution although differing from the incorrect initial one, should simulate the wanted process. As for asynchronous CA, they always satisfy non-contradictoriness conditions, because at each step only one application of $\theta(m)$ is allowed.

Fairness. At each iteration, $\theta(m)$ should be applied to all cells $m \in \hat{M}$, being applied to each cell $m \in \hat{M}$ only once. Fairness ensures that all cells have equal rights to participate in the CA operation process, therefore, it is sometimes referred to as equality in rights of cells activity [29]. Synchronous classical CA satisfy this property according to the definition of synchronicity. When multi-stage synchronous mode is used, fairness is provided by conditions (5.14) and (5.15). In asynchronous CA-models the property is the consequence of binomial probability distribution of cells chosen for local operator application.

5.2.3 Operations on Cellular Arrays

When a phenomenon under simulation consists of several interacting processes, its CA-model should be composed of a number of CA which have to interact, executing some operations on the intermediate results both on the local and global level. The problem in performing such an operation emerges when it turns to be incompatible with the alphabet of the CA-models under composition. For example, Boolean cellular arrays are incompatible with arithmetic addition. To provide such a compatibility a number of transformations on cellular arrays should be introduced, allowing to construct a kind of *algebra on CA configurations* [25]. Like in any algebraic system, unary and binary operations are defined in this algebra.

5.2.3.1 Unary Operators on Cellular Arrays

Two unary operators are defined: (1) *averaging* which transforms Boolean cellular arrays into the equivalent real ones, and (2) *state discretisation* which performs the inverse operation.

Averaging of the Boolean cellular array $\mathrm{Av}(\Omega_B)$ is a unary global operator which comprises the application of a local operator $\mathrm{Av}(m)$ to all cells of the cellular array, i.e., $\mathrm{Av}(\Omega_B) \in A_R \times M$, where $\Omega_B = \{(v, m) : v \in \{0, 1\}, m \in M\}$, $\Omega_R = \{(u, m) : u \in [0, 1], m \in M\}$.

The local operator $\mathrm{Av}(m)$ computes the average value of a cell state in the *averaging area*,

$$S_{\mathrm{Av}}(m) = \{(u_0, m), (u_1, \varphi_1(m)), \ldots, (u_q, \varphi_q(m))\}, \tag{5.18}$$

In case of 2D Cartesian cellular array, its underlying template is $T_{\mathrm{Av}}(i, j) = \{(i, j), (i+k, j+l) : k, l = -r, \ldots, r\}$, r being referred to as *averaging radius* Averaging may be regarded as a local operator

$$\mathrm{Av}(m) : (v, (i, j)) \star S_{\mathrm{Av}}(m) \to (u, m), \qquad u(m) = \langle v \rangle = \frac{1}{q} \sum_{k=0}^{q} v_k, \tag{5.19}$$

where $S_{\mathrm{Av}}(m)$ is the averaging context. Angle brackets in (5.19) and further denote the averaged state values.

Discretisation of a real cellular array $\mathrm{Dis}(\Omega_R)$ is a unary global operator $\mathrm{Dis}(\Omega_R) \in A_B \times M$, resulting from the application of a local operator $\mathrm{Dis}(m)$ to all cells of the cellular array. $\mathrm{Dis}(m)$ is a single-cell local operator which replaces a real state value $u \in [0, 1]$ by 1 with probability $p = u$.

$$\mathrm{Dis}(m) : (u, m) \to (v, m), \qquad v = \mathrm{Bool}(u) = \begin{cases} 1, & \text{if } u < \text{rand}, \\ 0 & \text{otherwise}, \end{cases} \tag{5.20}$$

where rand is a random number in the interval $[0, 1]$, $\mathrm{Bool}(u)$ means a discretised value of $u \in [0, 1]$. The above two unary operations are in the following relationship:

$$\begin{array}{ll} \mathrm{Dis}(\Omega_B) = \Omega_B, & \mathrm{Dis}(\mathrm{Av}(\Omega_B)) = \Omega_B, \\ \mathrm{Av}(\Omega_R) \ = \Omega_R, & \mathrm{Av}(\mathrm{Dis}(\Omega_R)) = \Omega_R. \end{array} \tag{5.21}$$

5.2.3.2 Binary Operators on Cellular Arrays

Binary operators are defined on cellular arrays $\Omega \in A_B \times M_1 \cup A_R \times M_2$, if between $M_1 = \{(m_i)_1\}$, and $M_2 = \{(m_i)_2\}$ there exists an one-to-one correspondence $\xi : M_1 \to M_2$,

$$\begin{array}{l} (m_i)_2 = \xi((m_i)_1), \quad \forall (m_i)_2 \in M_2, \\ (m_i)_1 = \xi^{-1}((m_i)_2), \ \forall (m_i)_1 \in M_1. \end{array} \tag{5.22}$$

The cells $(v, ((m_i)_1)) \in \Omega_1$ and $(u, ((m_i)_2)) \in \Omega_2$ are further denoted as (v_i, m_1) and (u_i, m_2), respectively, which means that v_i and u_i are states in the corresponding cells of Ω_1 and Ω_2.

Binary operations are based on the following principle: *ordinary arithmetic rules should be valid for the averaged forms of the operands*, i.e.,

$$\Omega_1 \Diamond \Omega_2 \Leftrightarrow \mathrm{Av}(\Omega_1) \diamond \mathrm{Av}(\Omega_2), \tag{5.23}$$

where $\Diamond$ stands for cellular array addition $\oplus$, cellular array subtraction $\ominus$ or cellular array multiplication $\otimes$, and $\diamond$ stands for arithmetical $+$, $-$, and $\times$, respectively.

Condition (5.23) may also be given for the cell states as follows.

$$v_i((m_i)_1) \Diamond u_i((m_i)_2) \Leftrightarrow \langle v_i((m_i)_1)\rangle \diamond \langle u_i((m_i)_2)\rangle \quad \forall i \in 1, \ldots, |M|. \tag{5.24}$$

The reason for taking averaged state values as a generalized alphabet is twofold: (1) to allow ordinary arithmetics to be used for modeling spatial functions interactions, and (2) to make the results more comprehensive from the physical point of view.

From (5.23) and (5.24) it follows that when all operands have real alphabets, the cellular array arithmetic coincides with the corresponding real cell-by-cell *arithmetical* rules. Otherwise, the rules depend on the operands alphabets.

Let $\Omega_1 = \{(v_i, m_1) : v_i \in A_1, m_1 \in M_1\}$ and $\Omega_2 = \{(u_i, m_2) : u_i \in A_2, m_2 \in M_2\}$ be the operands and $\Omega_3 = \{(w_i, m_3) : w_i \in A_3, m_3 \in M_3\}$ be a result, then binary operations are as follows.

Cellular array addition: $\Omega_1 \oplus \Omega_2 = \Omega_3$. For different alphabets of the operands the cellular addition looks somewhat different. The following cases are of main importance.

1. Both operands Ω_1 and Ω_2 are Boolean cellular arrays, and the resulting Ω_3 should have a real alphabet. Then according to (5.23) Ω_3 is computed as follows:

$$\begin{aligned} \Omega_3 &= \mathrm{Av}(\Omega_1) \bigoplus \mathrm{Av}(\Omega_2), \\ w_i &= \langle v_i\rangle + \langle u_i\rangle \qquad \forall i = 1 \ldots, |M|. \end{aligned}$$

2. Both operands are Boolean and the resulting cellular array is wanted to have Boolean alphabet. Then

$$\begin{aligned} \Omega_3 &= \mathrm{Dis}(\mathrm{Av}(\Omega_1) \oplus \mathrm{Av}(\Omega_2)), \\ w_i &= \begin{cases} 1 & \text{if rand} < (\langle u_i\rangle + \langle v_i\rangle) \\ 0 & \text{otherwise} \end{cases} \qquad \forall i = 1, \ldots, |M|. \end{aligned} \tag{5.25}$$

3. Both operands and their sum are Boolean, the latter being used as an intermediate result. Then, it is convenient to update one of the operands, say Ω_2, so, that it be equal to the resulting array, i.e.,

$$\Omega_2(t+1) = \Omega_1(t) \oplus \Omega_2(t).$$

In that case it suffices to invert a number of zero-states in the cells $(0, m_2) \in \Omega_2$. It should be done in such a way, that in every cell of Ω_2 its averaged state value be increased by $\langle v_i\rangle$. According to (5.20) the probability of such an inversion is the relation of the averaged amount of "ones" to be added to the averaged amount of "zeros" in the averaging area of each cell of Ω_2.

$$u_i' = \begin{cases} 1, & \text{if } u_i = 0 \ \& \ \text{rand} < \dfrac{\langle v_i \rangle}{1 - \langle u_i \rangle} \\ u_i & \text{otherwise,} \end{cases} \qquad \forall i = 1, \ldots, |M|. \qquad (5.26)$$

4. The operands have different alphabets. Let Ω_1 be a Boolean cellular array, Ω_2 – a real one, and Ω_3 is wanted to have the real alphabet. Then

$$\begin{aligned} \Omega_3 &= \mathrm{Av}(\Omega_1) \oplus \Omega_2, \\ w_i &= \langle v_i \rangle + u_i, \end{aligned} \qquad \forall i = 1, \ldots, |M|.$$

5. Ω_1 has Boolean alphabet, Ω_2 has a real one, and Ω_3 is wanted to be a Boolean cellular array. Two ways are possible: (1) to discretise Ω_3, obtained by (5.28), and (2) to update Ω_1 by using the following operation

$$w_i = \begin{cases} 1 & \text{if } v_i = 0 \ \& \ \text{rand} < \dfrac{u_i}{1 - \langle v_i \rangle}, \\ v_i & \text{otherwise,} \end{cases} \qquad \forall i = 1, \ldots, |M|. \qquad (5.27)$$

Cellular array subtraction $\Omega_3 = \Omega_1 \ominus \Omega_2$. The following cases are of main importance.

1. Both operands are Boolean, the result is wanted to be real or Boolean. The operations are similar to those of the cellular addition. It is merely needed to replace "+" by "−" in (5.25) or (5.26).
2. Both operands are Boolean, and Ω_2 is to be updated to obtain $\Omega_2 = \Omega_1 \ominus \Omega_2$. In that case some cell states $(1, m_2) \in \Omega_2$ should be inverted with probability equal to the relation of the amount of "ones" to be removed, to the total amount of "ones" in the averaging area.

$$u_i' = \begin{cases} 0 & \text{if } u_i = 1 \ \& \ \text{rand} < \dfrac{\langle u_i \rangle}{\langle v_i \rangle} \\ u_i & \text{otherwise} \end{cases} \qquad \forall i = 1, \ldots, |M|. \qquad (5.28)$$

3. Ω_1 has Boolean alphabet, Ω_2 has a real one, and Ω_3 is wanted to be a Boolean cellular array. Two ways are possible: (1) to discretise Ω_3, obtained by arithmetic subtraction, i.e.,

$$\Omega_3 = \mathrm{Dis}(\mathrm{Av}(\Omega_1) - \Omega_2), \qquad (5.29)$$

or (2) to update Ω_1 as follows

$$v_i' = \begin{cases} 0 & \text{if } u_i = 1 \ \& \ \text{rand} < \dfrac{u_i}{\langle v_i \rangle} \\ u_i & \text{otherwise,} \end{cases} \qquad \forall i = 1, \ldots, |M|. \qquad (5.30)$$

Cellular array multiplication $\Omega_3 = \Omega_1 \otimes \Omega_2$. The operation is defined on real cellular arrays. The cell states are computed according (5.25) with "×" instead of "+". If any or both of the operands are Boolean, they should be averaged beforehand. The operation is used in those cases when one of the two operands is a constant cellular array, i.e., such one where all cell states have the same value. This is helpful when subsets of cells have to be masked or scaled.

Since addition and subtraction are defined on cellular arrays with the alphabet restricted by the interval [0,1], the same condition should be satisfied for all cells in the resulting cellular arrays. If it is not so, the alphabet is to be renormalised.

Having the set of operation on cellular arrays in hands, it is possible to formulate CA composition techniques. General composition principles prescribe to distinguish sequential, parallel, and intermixed cases. Sequential composition represents several CA processing one and the same cellular array by alternating their application at each iteration. Parallel composition suggests each CA to process its own cellular array, albeit having neighborhoods in the others.

5.3 The Sequential Composition Techniques

Sequential composition, further referred to as *superposition*, represents a common functioning of several CA, referred to as *components*. Their local operators are applied in a certain order to one and the same cellular array. It comprises a number of techniques differing in ordering component operators application forming two groups: global and local superposition techniques.

Global superposition suggests the synchronous alternation of global operators application to the component CA. When those operators use different alphabets, their compatibility should be provided by transforming a Boolean cellular array into a real one or vice versa. Apart of the general case of global superposition, two particular cases are distinguished: (1) self-superposition, which is in fact a multistage mode of a CA operation, and (2) a so-called trivial CA superposition [20].

Local superposition is the composition when at each iteration the local operators of all components involved in the composition, are applied in any order or in random. Naturally, the components should be asynchronous CA.

5.3.1 Global Superposition

A number of CA form a global superposition $\aleph = \Psi_{Gl}(\aleph_1, \ldots, \aleph_n)$, $\aleph = \langle A_k, M, \hat{M}_k, \theta_k, \rho_k \rangle$, if its global operator $\Phi(\Omega)$ is the result of sequential application of the global operators Φ_k to $\Omega_k = \Phi_{k-1}(\Omega_{k-1})$, $k = 1, \ldots, n$, providing compatibility of A_k and A_{k-1}, i.e.,

$$\Phi(\Omega) = \Phi'_n(\Phi'_{n-1}(\ldots \Phi'_1(\Omega_1))), \tag{5.31}$$

each Φ'_k being itself a superposition of Φ_k and a unary operator, i.e.,

$$\Phi'_k = \Phi_k(\mathrm{Un}(\Omega_k)), \tag{5.32}$$

where

$$\mathrm{Un}(\Omega_k) = \begin{cases} \mathrm{Av}(\Omega_k), & \text{if} \quad A_k = [0,1]\ \&\ A_{k-1} = \{0,1\}, \\ \mathrm{Dis}(\Omega_k), & \text{if} \quad A_k = \{0,1\}\ \&\ A_{k-1} = [0,1]. \end{cases}$$

Components of the superposition may differ in alphabets, local operators and modes of operating, but the same naming set should be used.

The following particular cases of global superposition are of especial importance: self-superposition, trivial superposition, and the general type of superposition of CA with different types of alphabets.

5.3.1.1 Global Self-Superposition

This type of composition is the most simple one, being defined only for synchronous CA. A CA $\aleph = \langle A, M, \hat{M}, \theta, \sigma \rangle$ is a self-superposition $\aleph = \Psi_{SS}(\aleph_1, \ldots, \aleph_n)$, $\aleph_k = \langle A, M, \hat{M}_k, \theta, \sigma \rangle$, if its components differ only in active subsets $\hat{M}_k$. Since the same local operator is applied at all stages of the superposition, there is no need to take care about their compatibility, so, $\Phi(\Omega) = \Phi_n(\Phi_{n-1}(\ldots(\Phi_1(\Omega))))$.

Self-superposition is usually obtained by modifying a synchronous CA-model of a process which requires several neighboring cells to be updated at once. In that case non-contradictoryness condition (5.14) may be violated, hence, conflicts and data loss are possible. To avoid such a situations some amount of cellular parallelism should be sacrificed by performing the global transition in several stages. It is done as follows.

1. Each tth iteration is divided into n stages $t_1(t), \ldots, t_n(t)$, the results of kth stage being $\Omega(t_k)$.
2. At the $t_k(t)$th stage, $\theta(m)$ is applied to all cells named $m_k \in \hat{M}_k$ of $\Omega(t_k(t))$.

Example 1 Diffusion is a random wandering of particles aiming to even distribution. The process may be simulated by the exchange of cell states in any pair of adjacent cells. Since synchronous simulation of such a process is contradictory, as it is shown in Sect. 5.2.2, self–superposition of two CA [1, 3, 4]: $\aleph_1 = \langle A, M, \hat{M}_1, \theta, \sigma \rangle$ and $\aleph_2 = \langle A, M, \hat{M}_2, \theta, \sigma \rangle$, is used, where $A = \{0, 1\}$, $M = \{(i, j) : i, j = 0, 1, \ldots, N\}$,

$$\begin{aligned} \hat{M}_1 &= \{(i, j) : i_{\mathrm{mod}2} = 0, j_{\mathrm{mod}2} = 0\}, \\ \hat{M}_2 &= \{(i, j) : i_{\mathrm{mod}2} = 1, j_{\mathrm{mod}2} = 1\}, \end{aligned} \tag{5.33}$$

$\hat{M}_1$ and $\hat{M}_2$ being referred to as *even active subset* and *odd active subset*, respectively. The local operator is as follows:

$$\begin{aligned} \theta(i, j) : \{&(v_0, (i, j)), (v_1, (i, j+1)), (v_2, (i, j+1)), (v_3, (i, j+1))\} \\ &\to \{(u_0, (i, j)), (u_1, (i, j+1)), (u_2, (i, j+1)), (u_3, (i, j+1))\}, \end{aligned} \tag{5.34}$$

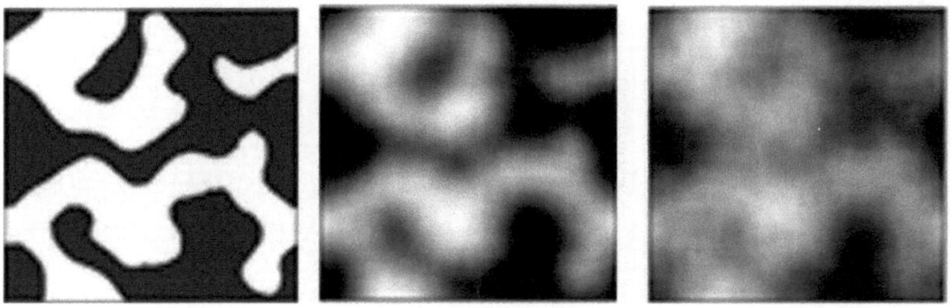

Fig. 5.3 Three snapshots of diffusion process, simulated by the CA-model with a local operator (5.34). *Black pixels* stand for $\langle v \rangle = 1$, *white pixels* – for $\langle u \rangle = 0$

$$u_k = \begin{cases} v_{(k+1)(\mathrm{mod}4)} & \text{if rand} < p, \\ v_{(k-1)(\mathrm{mod}4)} & \text{if rand} > (1-p), \end{cases} \qquad k = 0, 1, 2, 3,$$

the probability p depending on the diffusion coefficient.

Each iteration of a composed CA is divided into two stages: even stage and odd stage. At the odd stage $\theta(m)$ is applied to all cells from $\hat{M}_1$, at the even stage $\theta(m)$ is applied to all cells from $\hat{M}_2$.

In Fig. 5.3 three snapshots are shown of the CA evolution simulating the diffusion of a black dye slopped onto the water surface.

5.3.1.2 Global Trivial Superposition

Trivial superposition $\aleph = \Psi_{Tr}(\aleph_1, \ldots, \aleph_n)$, where $\aleph_k = \langle A_k, M, \hat{M}_k, \theta_k, \rho_k \rangle$, suggests the evolution of $\aleph$ be a sequential composition of the evolutions $\Sigma_{\aleph_k}(\Omega'_k(\hat{t}_k))$ of its components, $\Omega'_k(\hat{t}_k)$ being a result of a unary operator (5.34) application to $\Omega_k(\hat{t}_k)$, if A_k and A_{k+1} are incompatible. The alphabets, local operators, modes of operation, and active naming subsets in the components may be different. But the order of component application is essential.

Example 2 Pattern formation process starts in the cellular array which has been obtained by a short-time application of a diffusion CA to a cellular array with two areas of high concentration (black bands along vertical borders) and empty (white) background (Fig. 5.4a). Diffusion is simulated by an asynchronous probabilistic CA, called in [1] a naive diffusion $\aleph_1 = \langle A, M, \hat{M}, \theta_1, \alpha \rangle$. Pattern formation is simulated by synchronous CA $\aleph_2 = \langle A, M, \hat{M}, \theta_2, \sigma \rangle$. Both CA have a Boolean alphabet $A = \{0, 1\}$, their naming sets are identical as well as active naming subsets, $M = \hat{M} = \{(i, j) : i, j = 0, \ldots, 300\}$. The local operators $\theta_1(m)$ and $\theta_2(m)$ are as follows:

$$\begin{aligned} \theta_1(m) : \{&(v_0, (i, j)), (v_1, (i-1, j)), (v_2, (i, j+1)), (v_3, (i+1, j)), (v_4, (i-1, j))\} \rightarrow \\ \{&(u_0, (i, j)), (u_1, (i-1, j)), (u_2, (i, j+1)), (u_3, (i+1, j)), (u_4, (i-1, j))\} \end{aligned} \tag{5.35}$$

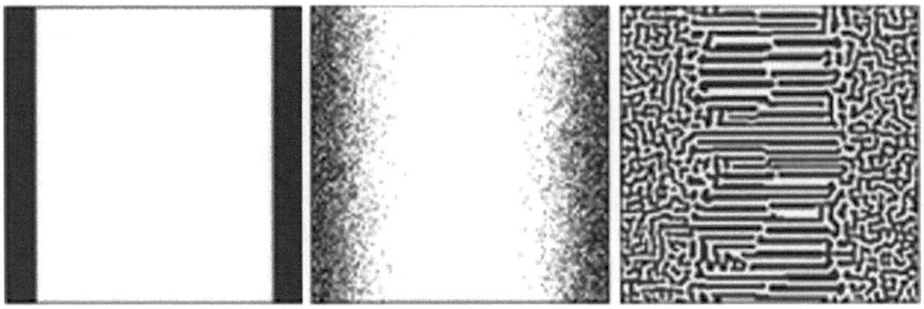

Fig. 5.4 Three snapshots of the process, simulated by trivial superposition of an asynchronous diffusion CA and a synchronous pattern formation CA: (**a**) initial array, (**b**) $\hat{t}_1 = 10$, (**c**) $\hat{t}_2 = 12$

with the transition functions

$$\begin{array}{ll} u_0 = & v_k, \quad \text{if} \quad 0.25k < \text{rand} < 0.25(k+1), \\ u_k = & \begin{cases} v_0 & \text{if} \quad 0.25k < \text{rand} < 0.25(k+1), \\ v_k & \text{otherwise.} \end{cases} \end{array} \quad k = 1, \ldots, 4. \tag{5.36}$$

$$\theta_2(m) : (u_0, (i,j)) \star \{(u_{gh}, \phi_{gh}(i,j)) : g, h = -3, -2, -1, 1, 2, 3\} \rightarrow (v_0, (i,j)), \tag{5.37}$$

has a transition function

$$v_0 = \begin{cases} 1, \text{ if} \quad S_w > 0, \\ 0, \text{ otherwise,} \end{cases} \tag{5.38}$$

where the weighted sum

$$S_w = \sum_{g=-3}^{3} \sum_{h=-3}^{3} (w_{gh} \cdot v_{(i+g,j+h)}) \quad \text{with} \quad w_{gh} = \begin{cases} 1, \text{ if} \quad g \leq 1 \;\&\; h \leq 1 \\ -0.2, \text{ otherwise.} \end{cases}$$

In Fig. 5.4 three snapshots of trivial composition of two CA ($\aleph_1$ simulating diffusion and $\aleph_2$ simulating pattern formation) are shown. Cellular array size is 300×300, $\hat{t}_1 = 10$, $\hat{t}_2 = 12$. The obtained pattern is a stable one, further application of θ_2 to $\Omega_2(\hat{t}_2)$ implies no change in it.

5.3.1.3 Global Superposition of Arbitrary CA

Global superposition of arbitrary CA is a technique for obtaining a CA $\aleph_{Gl} = \Psi_{Gl}$ $(\aleph_1, \ldots, \aleph_n)$, which combines operation of several CA $\aleph_k = \langle A_k, M, \hat{M}_k, \theta_k, \rho \rangle$, $k = 1, .., n$, whose alphabets and local operators are allowed to be incompatible, and modes of operation may be different. The operation of the composed CA is as follows.

1. Each tth iteration of the composed CA consists of n stages $t_1(t), \ldots, t_n(t)$, the results of t_kth stage being $\Omega(t_k(t)) = \Phi_{k-1}(t_{k-1}(t))$.
2. At the $t_k(t)$th stage θ_k is applied to all cells $m \in \hat{M}_k$ of $\Omega'(t_k(t))$. The latter should be obtained by transforming $\Omega(t_k(t))$ according to (5.34), if needed.

Example 3 Simulation of the alga spreading over the water is considered to combine three elementary processes: (1) agglomeration of randomly distributed alga, (2) diffusion of alga into water, and (3) procreation of alga.

The first process is represented by a Boolean CA $\aleph_1$ sometimes called a *phase-separation CA* [20, 30], the second – by the two-stage diffusion CA $\aleph_2$ given in Sect. 5.3.1(Example 1), the third – by $\aleph_3$ computing a nonlinear logistic function [31]. Accordingly, each tth iteration of the composed CA has three stages. At the first stage t_1, the transition $\Omega(t) \to \Omega(t_1)$ is performed by a synchronous CA $\aleph_1 = \langle A_1, M, \hat{M}_1, \theta_1, \sigma\rangle$ with $A_1 = \{0, 1\}$, $M = \{(i, j) : i, j = 0, \ldots, N\}$, $\hat{M} = M$, and a single-cell updating local operator

$$\theta_1(i, j)) : (v, (i, j)) \star S''(i, j) \to \{(v', (i, j))\} \qquad \forall (i, j) \in M, \tag{5.39}$$

where

$$S''(i, j) = \{(v_k, \phi_k(i, j)) : \phi_k(i, j) = (i + g, j + h)\},\\ g, h \in \{-3, -2, -1, 1, 2, 3\},$$

$$v' = \begin{cases} 1, & \text{if} \quad s < 24 \text{ or } s = 25, \\ 0, & \text{if} \quad s > 25 \text{ or } s = 24. \end{cases} \qquad \text{where} \quad s = \sum_{g=-2}^{2} \sum_{h=-2}^{2} v_{i+g,j+h}.$$

At the second stage $\aleph_2$ given in Sect. 5.3.1(Example 1) performs a transition $\Omega(t_1) \to \Omega(t_2)$ by application θ_2 (5.36) to all cells of $\Omega(t_1)$), the value of the probability in (5.36) being $p = 0.5$. As the alphabet A_2 is compatible with A_1, θ_2 is applied directly to the cells of Ω_1 resulting in a Boolean array $\Omega(t_2) = \{(u, (i, j))\}$.

At the third stage alga procreation CA $\aleph_3 = \langle A_3, M, \hat{M}_k, \theta_3, \sigma\rangle$ is applied to $\Omega(t_2)$. But since $A_3 = [0, 1]$ and, hence, θ_3 is incompatible with Ω_2, the latter is transformed into $\Omega(t_2)'$ by averaging, i.e. the operator Av(i, j) is applied to $\Omega(t_2)$ replacing each cell state $u(i, j)$ by $\langle u(i, j)\rangle$. The latter is computed according to (5.19) with the averaging template $T_{\mathrm{Av}}(i, j) = \{(i + k, j + l) : k, l = -8, \ldots, 8\}$. The local operator

$$\theta_3(i, j) : (\langle u(i, j)\rangle, (i, j)) \to (F(\langle u(i, j)\rangle), (i, j)) \tag{5.40}$$

is applied to $\Omega(t_2)'$ replacing a cell state $\langle u(i, j)\rangle$ by the value of a nonlinear logistic function $F(\langle u(i, j)\rangle) = 0.5\langle u(i, j)\rangle(1 - \langle u(i, j)\rangle)$. The resulting cellular array having real states should be discretized according to (20) to obtain $\Omega(t_3)' = \Omega(t + 1)$.

The composition has been applied to an initial Boolean cellular array $\Omega(0)$ with $v = 1$ randomly distributed with probability $p = 0.5$, so that $\langle v(i, j)\rangle \approx 0.5$ for all $(i, j) \in M$, the border conditions being periodic.

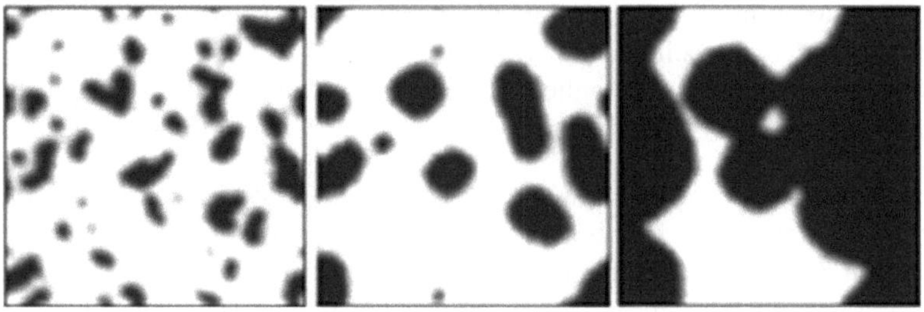

Fig. 5.5 Three snapshots of alga spreading in water, simulated by synchronous global superposition of $\aleph_1$ with θ_1 (5.39), $\aleph_2$ with θ_2 (5.34) and $\aleph_3$ with θ_3 (5.40). *Black pixels* stand for maximal concentration of alga, *white pixels* – for clear water

In Fig. 5.5, three snapshots of the simulation process are shown, cellular arrays being averaged for making the observation more comprehensive. Black pixels stand for maximum concentration of alga, white ones represent clear water. It is seen that on the first iterations, the total amount of alga decreases, but if some compact spots remain large enough, the procreation activeness enhances their growth up to the saturation.

5.3.2 *Local Superposition*

Asynchronous local superposition is mainly used in simulating biological processes and nano-kinetics, i.e., the processes on micro- or nano-levels, which are considered to be completely stochastic by nature [13]. This technique aims at obtaining a CA-model $\aleph = \Psi_{Loc}(\aleph_1, \ldots, \aleph_n\}$ composed of n asynchronous CA $\aleph_k = \langle A, M, \hat{M}_k, \theta_k, \alpha \rangle$, $k = 1, \ldots, n$, which differ only in local operators and (perhaps) in active subsets. The way of their common functioning is as follows. An iteration $\Omega(t) \rightarrow \Omega(t+1)$ consists of $|M|$ cycles, a cycle being a sequence of single-shot applications of $\theta_k(m)$, $k = 1, \ldots, n$, to a randomly chosen cell from $\Omega(t)$. Each $\theta_k(m)$ is executed immediately after the application. There is no constraints neither on the order of choosing a cell during an iteration, nor on the order of choosing $\theta_k(m)$ for application during a cycle.

Example 4 A chemical reaction of CO oxidation over platinum catalysts, well known in surface chemistry as Ziff-Guilari-Barshod model [32], is represented by a local superposition of four simple local operators, mimicking elementary actions of adsorption, reaction, oxidation, and diffusion. The cellular array Ω corresponds to a catalysts plate, each site on it being named as $(i, j) \in M$, $|M| = N \times N$, $\hat{M} = M$. The alphabet contains three symbols $A = \{a, b, 0\}$, so that $(a, (i, j))$, $(b, (i, j))$, and $(0, (i, j))$ are cells corresponding to the sites occupied by the molecules of CO, O, or being empty, respectively. In the initial array, all cells are empty. The CO oxidation process consists of the following four elementary molecular actions in any cell named (i, j) (Fig. 5.6).

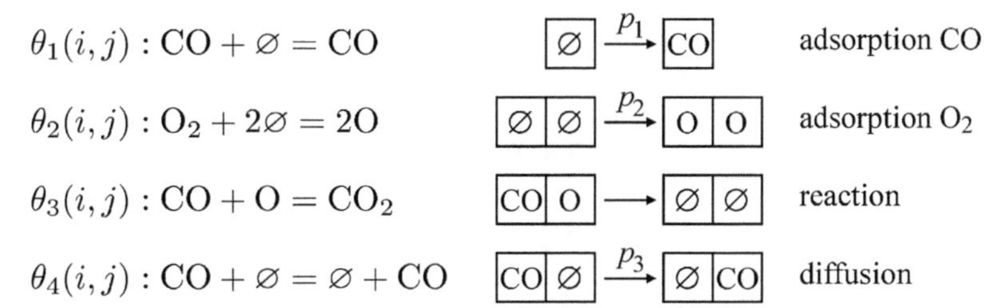

Fig. 5.6 Graphical representation of local operators involved in an asynchronous local superposition simulating chemical oxidation of CO on platinum

(1) Adsorption of CO from the gas: if the cell (i, j) is empty, it becomes occupied by a CO molecule with probability p_1.
(2) Adsorption of the oxygen O_2 from the gas: if the cell (i, j) is empty and has an empty adjacent cell, both become occupied by an atom of oxygen with probability p_2. One out of $h < 4$ adjacent cells of the cell (i, j) is chosen with probability $p_n = 1/h$.
(3) Reaction of oxidation of CO (CO+O $\rightarrow$ CO_2): if the cell (i, j) occurs to be in a CO state and its adjacent cell is in O state, then the molecule CO_2, formed by the reaction, transits to the gas and both cells become empty. One out of $h < 4$ adjacent cells occupied by oxygen is chosen with probability $p_n = 1/h$.
(4) Diffusion of CO over the plate: if the cell (i, j) occurs to be in a CO state when one of its adjacent cells is empty, the cell (i, j) becomes empty, and the empty cell gets the state CO. This occurs with probability p_3. One out of $h < 4$ adjacent cells of the cell (i, j) is chosen with probability $p_n = 1/h$.

Formally, local operators of the above actions are represented as follows.

$$
\begin{aligned}
&\theta_1(i,j) : \{(0,(i,j))\} \rightarrow \{(a,(i,j))\}, \quad \text{if} \quad p_1 > \text{rand},\\
&\theta_2(i,j) : \{(0,(i,j))(0,\phi_k(i,j))\} \rightarrow \{(b,(i,j)),(b,\phi_k(i,j))\},\\
&\qquad\qquad \text{if} \quad (k-1)p_n < \text{rand} < kp_n \ \&\ p_2 > \text{rand}\\
&\theta_3(i,j) : \{(a,(i,j))(b,\phi_k(i,j))\} \rightarrow \{(0,(i,j)),(0,\phi_k(i,j))\},\\
&\qquad\qquad \text{if} \quad (k-1)p_n < \text{rand} < kp_n\\
&\theta_4(i,j) : \{(a,(i,j))(0,\phi_k(i,j))\} \rightarrow \{(0,(i,j)),(a,\phi_k(i,j))\},\\
&\qquad\qquad \text{if} \quad (k-1)p_n < \text{rand} < kp_n) \ \&\ p_3 > \text{rand},
\end{aligned}
$$

for $k = 1, \ldots, 4$.

In Fig. 5.7 three snapshots of the simulation process are shown, the initial cellular array $\Omega(0) = \{(0,(i,j)) : \forall(i,j) \in M\}$, $|M| = 200 \times 200$.

In the general case local superposition is not a commutative operation, i.e., if $\theta_1 \neq \theta_2$, then

$$\theta_1(\theta_2(m)) \neq \theta_2(\theta_1(m)). \tag{5.41}$$

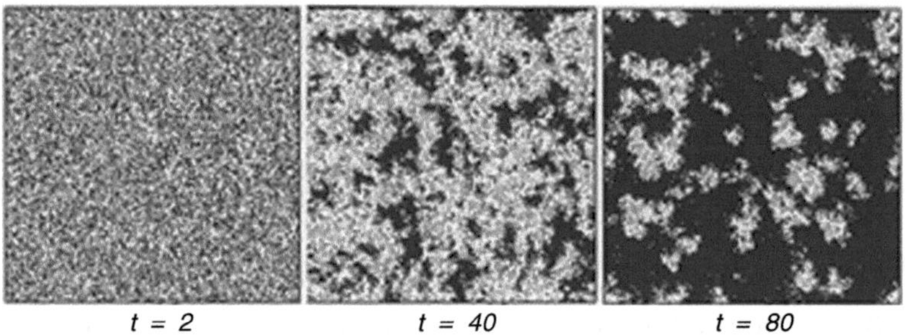

Fig. 5.7 Three snapshots of the oxidation reaction simulation by an asynchronous superposition of local operators shown in Fig. 5.6. *Black pixels* stand for CO, *gray pixels* – for O, and *white pixels* – for empty sites

The above property is very important, because the results of the simulation may differ essentially if the order of superpositions is changed. Although in case of long evolution, the repetitive sequence of superpositions, for example, such as $\theta_1(\theta_2(\theta_1(\theta_2(m)\ldots)))$, makes the composition insensitive of the substitution being the first. If it is not the case, the only way to make the result independent of the order of substitutions in the composition is their random choice at any step of application (the Monte-Carlo method).

5.4 The Parallel Composition Techniques

Parallel composition suggests functioning of n interacting CA, each processing its own cellular array. Taking into account that the number of possible interactions in the composition exponentially increases with n, and for clearness of presentation, the composition $\aleph = \Upsilon(\aleph_1, \aleph_2)$ of not more than two CA is further considered. The components $\aleph_k = \langle A_k, M_k, \hat{M}_k, \theta_k, \rho_k \rangle, k = 1, 2$, are allowed to have different alphabets, different modes of operation, different local operators, and between $M_1 = \{(m_i)_1\}$, and $M_2 = \{(m_i)_2\}$, $i = 1, 2 \ldots, |M|$, the condition (5.22) is satisfied.

Since $\theta_1((m)_1)$ and $\theta_2((m)_2)$ are to be executed simultaneously, the computation is dangerous from the point of view of non-contradictoryness condition (5.14), and at the same time the transition function in $\theta_1((m)_1)$ and $\theta_2((m)_2)$ should interact. Hence, with respect to (5.11) and (5.13), the left-hand sides of the operators should have nonempty intersection, i.e.

$$(S_1((m_i)_1) \cup S_1''((m_i)_1)) \cap (S_2((m_i)_2) \cup S_2''((m_i)_2)) \neq \emptyset.$$

Combining this statement with (5.14) the correctness condition yields:

$$T_k((m_i)_k) \subseteq M_k, \tag{5.42}$$

$$T_k''((m_i)_k) \subseteq (M_1 \cup M_2) \quad \forall k \in \{1, 2\}. \tag{5.43}$$

From (5.45) it follows that $\theta_1((m_i)_1)$ and $\theta_2((m_i)_2)$ may update cells only from their own cellular arrays, whereas from (5.46) they are allowed to use cell states of the both. It means, that the neighborhoods of cells $(m_i)_1$ and $(m_i)_2$, may intersect only by their contexts.

The above conditions are valid both for local and global composition techniques, as well as both for CA with synchronous and asynchronous modes of operation.

5.4.1 Global Parallel Composition

5.4.1.1 Trivial Parallel Composition

Trivial parallel composition $\aleph = \Upsilon_{Tr}(\aleph_1, \aleph_2)$, $\aleph_k = \langle A_k, M_k, \hat{M}_k, \theta_k, \rho_k\rangle$, $k = 1, 2$, is a degenerate particular case of parallel composition, when the components are completely independent, i.e.,

$$(S_1((m_i)_1) \cup S_1''((m_i)_1)) \cap (S_2((m_i)_2) \cup S_2''((m_i)_2)) = \emptyset. \tag{5.44}$$

Nonetheless, after both components have terminated, a binary operation on the resulting cellular arrays may be performed. So,

$$\Omega(\hat{t}) = \Omega_1(\hat{t}_1) \diamond \Omega_2(\hat{t}_2), \tag{5.45}$$

where $\diamond$ is any binary operator given in Sect. 5.2.3.2.

Example 5 Two phase separation models are to be compared by computing the difference of two resulting cellular arrays:

(1) $\Omega_1(\hat{t}_1)$ obtained by the evolution of a totalistic CA $\aleph_1 = \langle A_1, M_1, \hat{M}_1, \sigma\rangle$, which is described in Sect. 5.3.1(Example 1) with $\theta_1(i, j)$, given as (5.41), and
(2) $\Omega_2(\hat{t}_2)$ obtained by solving a PDE proposed in [33] which describes the same process,

$$u_{t'} = 0.2(u_{xx} + u_{yy} - 0.2(u - 1)(u - 0.5)(u - 0.9). \tag{5.46}$$

Let us consider the finite-difference representation of (5.49) as a synchronous CA $\aleph_2 = \langle A_2, M_2, \hat{M}_2, \theta_2, \sigma\rangle$, where $A_2 = [0, 1]$, $M_2 = \hat{M}_2 = \{(i, j)_2 : i = x/h; j = y/h; i, j = 0, \dots, N\}$, h being a space step, $t = t'/(\Delta t)$,

$$\begin{aligned}\theta_2(i, j) : (u_0, (i, j)_2) \star \{&(u_1, (i - 1, j)_2), (u_2, (i, j + 1)_2), (u_3, (i + 1, j)_2),\\ &(u_4, i, j - 1)_2)\} \to (u_0', (i, j)_2),\\ &u_0' = (u_1 + u_2 + u_3 + u_4 - 4u_0)/h^2.\end{aligned} \tag{5.47}$$

The initial cellular arrays $\Omega_1(0)$ and $\Omega_2(0)$ for $\aleph_1$ and $\aleph_2$ are identical, so that $\langle v(i, j)_1\rangle = u(i, j)_2 = 0.5$ for all $(i, j)_1 \in M_1$ and all $(i, j)_2 \in M_2$.

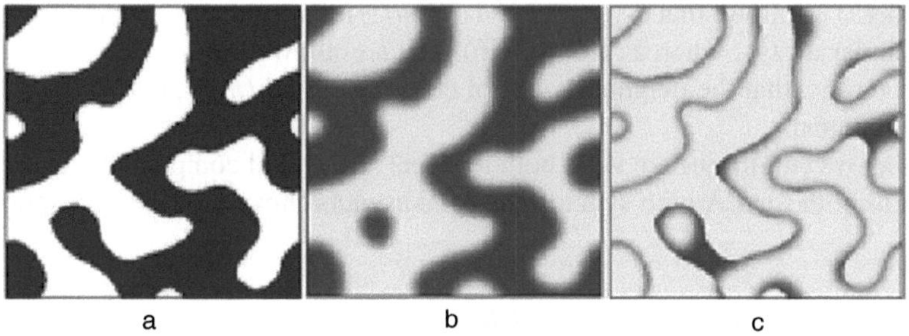

Fig. 5.8 Three snapshots of parallel trivial composition of two CA simulating phase separation: (**a**) resulting cellular array obtained by a totalistic CA (5.39), (**b**) resulting cellular array obtained by a CA based on PDE (5.46), and (**c**) their difference. *Black pixels* stand for 1, *white* for 0, *gray scale* intensity corresponds to values from [0, 1]

The comparison of the results of both components evolutions $\Omega_1(\hat{t}_1)$ and $\Omega_2(\hat{t}_2)$, is done by computing the absolute value of their cellular arrays subtraction (Sect. 5.2.3). Since $\Omega_1(\hat{t}_1)$ and $\Omega_2(\hat{t}_2)$ are incompatible the first is to be averaged according to (5.19). The final result is obtained as $\Omega_2'(\hat{t}) = \{(u_2', (i,j)_2)\}$, where

$$u_2'((i,j)_2) = |\langle v_1((i,j)_1)\rangle - u((i,j)_2)|. \tag{5.48}$$

The three resulting cellular arrays: $\Omega_1(\hat{t}_1)$, $\Omega_2(\hat{t}_2)$, and $\Omega_2'(\hat{t})$ are shown in Fig. 5.8.

5.4.1.2 Nontrivial Parallel Composition

Nontrivial parallel composition $\aleph = \Psi(\aleph_1, \aleph_2)$ suggests that both components $\aleph_1$ and $\aleph_2$ interact at each iteration. Two types of interaction between them determine two types of parallel composition techniques: unidirectional parallel composition and bidirectional parallel composition [20].

In *unidirectional parallel composition*, one of the components, say $\aleph_1$, functions independently. But, the transition functions of $\aleph_2$ depend on states of both cellular arrays. Hence, condition (5.43) takes the following form.

$$T_1''((m_i)_1) \subseteq M_1, \quad \forall (m_i)_1 \in M_1, \tag{5.49}$$

$$T_2''((m_i)_2) \subseteq (M_1 \cup M_2) \quad \forall (m_i)_2 \in M_2. \tag{5.50}$$

Such a kind of composition is frequently used when simulating a certain process by $\aleph_1$, and using auxiliary CA $\aleph_2$ for transforming simulation results of $\aleph_1$ into a proper form for analyzing or visualizing its evolution. For example, $\aleph_1$ is a Boolean CA, and observation of its evolution requires it to be real numbers. Then, $\aleph_1$ works independently, and $\aleph_2$ performs the averaging of $\Omega_1(t)$ at each iteration using cell states of Ω_1 in its transition functions.

In *bidirectional parallel composition* transition functions of both components depend on states of cells from both cellular arrays, i.e.

$$\begin{aligned} T_1''((m_i)_1) &\subseteq (M_1 \cup M_2) \quad \forall (m_i)_1 \in M_1, \\ T_2''((m_i)_2) &\subseteq (M_1 \cup M_2) \quad \forall (m_i)_2 \in M_2, \end{aligned} \tag{5.51}$$

(5.42) being preserved as well. If the alphabets of $\aleph_1$ and $\aleph_2$ are incompatible, then a suitable unary transformations of $\Omega_1(t)$ or $\Omega_2(t)$ should be done after each iteration.

Example 6 A 2D reaction–diffusion process of autocatalytic reaction propagation in a domain with obstacles is simulated by bidirectional parallel composition of two CA: (1) a two-stage synchronous diffusion CA $\aleph_1 = \langle A_1, M_1, \hat{M}_1, \theta_1, \sigma \rangle$ given in Sect. 5.3.1(Example 1), and (2) a single cell synchronous CA $\aleph_2 = \langle A_2, M_2, \hat{M}_2, \theta_2, \sigma \rangle$ which computes a real nonlinear function of the cell state.

Since A_1 and A_2 are incompatible, unary operators Dis(i, j) and Av(i, j) should be added, which is done by means of incorporating them into the local operators $\theta_1((i, j)_1)$ and $\theta_2((i, j)_2)$, respectively. In $\theta_1((i, j)_1)$ the operator Dis$(u(i, j)_1)$ is included in the transition function as follows:

$$\begin{aligned} \theta_1((i, j)_1) : \; & \{(v_0, (i, j)_1), (v_1, (i, j + 1)_1), (v_2, (i + 1, j)_1), (v_3, (i, j - 1)_1)\} \\ & \star \{(u_0, (i, j)_2), (u_1, (i, j + 1)_2), (u_2, (i + 1, j)_2), (u_3, (i, j - 1)_2)\} \\ & \to \{(v_0', (i, j)_1), (v_1', (i, j + 1)_1), (v_2', (i + 1, j)_1), (v_3', (i, j - 1)_1)\}, \\ v_k' = \; & \begin{cases} \mathrm{Bool}(u_{(k+1)\mathrm{mod}4}) \text{ if rand} < p, \\ \mathrm{Bool}(u_{(k-1)\mathrm{mod}4}) \text{ if rand} > (1 - p), \end{cases} \end{aligned} \tag{5.52}$$

The local operator $\theta_2((i, j)_2)$ is combined with Av$((i, j)_1)$ which results in the following.

$$\begin{aligned} \theta_2((i, j)_2) : (u, (i, j)_2) \star \{S_{\mathrm{Av}}((i, j)_1)\} &\to f(\langle v((i, j)_1)\rangle, (i, j)_2), \\ f(\langle v((i, j)_1)\rangle) &= 0.5\langle v((i, j)_1)\rangle(1 - \langle v((i, j)_1)\rangle), \end{aligned} \tag{5.53}$$

$\langle v((i, j)_1)\rangle$ being obtained according to (5.19).

The process is simulated on a square area 300×300 cells with a number of rectangular obstacles (Fig. 5.9).

5.4.2 Local Parallel Composition

Like in sequential case this type of composition aims at obtaining an asynchronous CA-model $\aleph_\alpha = \Upsilon_{\mathrm{Loc}}(\aleph_1, \aleph_2)$ composed of two asynchronous CA $\aleph_k = \{A_k, M_k, \hat{M}_k, \theta_k, \alpha\}$, $k = 1, 2$. The components may differ in alphabets and in local operators, naming sets M_1 and M_2 being in the relation (5.22). The way of the composed CA functioning is as follows. Both components operate in parallel in asynchronous mode: at each tth iteration the local operator $\theta_k(m)$ is applied to all cells of $\hat{M}_k$, the cells being selected in any order and updated immediately after selection.

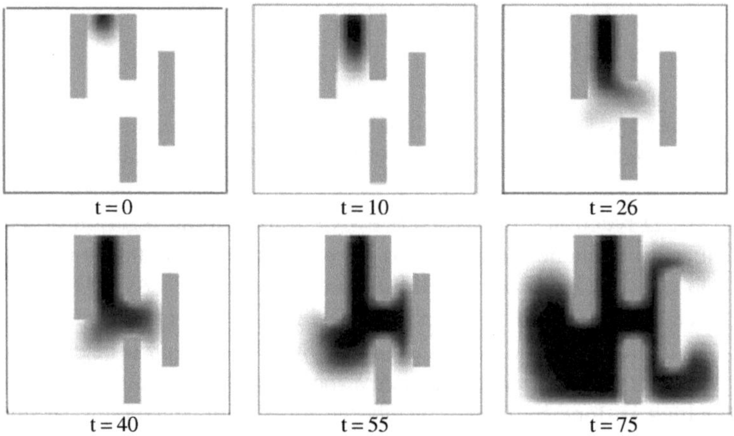

Fig. 5.9 Six snapshots of $\aleph_2$ evolution of a parallel bidirectional composition simulating the front propagation of autocatalytic reaction. *Black pixels* stand for obstacles, *grey pixels* – for maximal concentration of the reactant, *white* – for reactant absence

Example 7 A soliton-like 1D process is simulated by a parity totalistic CA [5] $\aleph_1 = \{A_1, M_1, \hat{M}_1, \theta_1, \alpha\}$. Since A_1 is a Boolean alphabet the process is difficult to recognize as two moving waves passing one through the other. So, to make the process observable in a habitual form, $\aleph_1$ is combined with another CA $\aleph_2 = \{A_2, M_2, \hat{M}_2, \theta_2, \alpha\}$ which performs averaging of any cell state in Ω_1 just after its updating. The naming sets $M_1 = \hat{M}_1 = \{i_1 : i = 0, \ldots, N\}$, and $M_2 = \hat{M}_2 = \{i_2 : i = 0, \ldots, N\}$, are in one-to one correspondence (5.22).

$$\theta_1(i_1) : (v_0, i_1) \star \{(v_j, i_1 + j) : \; j = -r, \ldots, -1, 1, \ldots, r\} \rightarrow (v_0', i_1), \quad (5.54)$$

$$v_0'(i_1) = \begin{cases} 1, \text{ if } \;\; w \neq 0 \;\&\; w = 0_{\text{mod2}} \\ 0, \text{ otherwise}, \end{cases}, \qquad w = \sum_{j=-r}^{r} v_j. \quad (5.55)$$

The mode of $\aleph_1$ operation is an ordered asynchronous one: $\theta_1(i_1)$ is applied sequentially according to the cell numbers $i_1 = 0, 1, \ldots N$, each cell (v, i_1) being immediately updated, so, that the cell states situated leftwards of i_1, are already in the next state, while the rightward cells are yet in the current state. Border conditions are periodic. The initial global cellular state $\Omega_1(0)$ has certain patterns referred to as "particles" [5]. Here, the two following particles are used: $P_1 = 1101$, and $P_2 = 10001001$ with $r = 4$. All others cells are in zero states. The evolution of $\aleph_1$ shows that P_1 appears in $\Omega_1(t)$ any 2 iterations being displaced by $d_1 = 7$ cells to the left. And P_2 appears in $\Omega_1(t)$ any 6 iteration being displaced by $d_2 = 12$ cells also to the left. So, each 6 iterations the distance between the particles diminishes by 9 cells. After the start ($t = 0$) during the period from $t = 12$ till $t = 24$ the particles are superimposed, and after $t = 30$ the first particle is ahead, as it is shown in the following global states.

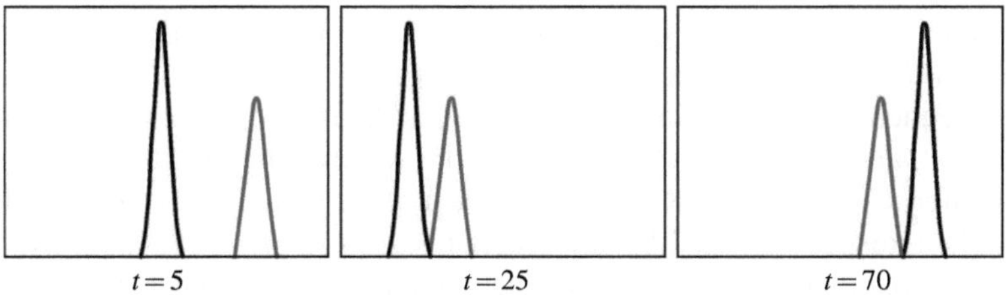

Fig. 5.10 Three snapshots of the soliton propagation obtained by simulating the process using local parallel composition of two CA with local operators given by (5.54) and (5.55)

$t = 0$: 0000...00000000000000**10001001**0000000000000000000000000**11011**00
$t = 6$: 0000...00**10001001**000000000000000**11011**000000000000000000000000
$t = 30$: 000000000000000000000000**11011**00001**10001001**00...0000000000000
$t = 36$: 00**11011**000000000000000000**10001001**00000...0000000000000000000

The second CA $\aleph_2$ performs an asynchronous averaging of Ω_1, in order to transform patterns displacement into waves propagation. The steps of $\aleph_2$ are synchronized with those of $\aleph_1$ and the order of cell selection is the same (Fig. 5.10).

$$\theta_2(i_2) : (v_0, i_2) \star \{(v_j, i_1 + j) : \ j = -r, \dots, -1, 1, \dots, r\} \to (\langle v_0 \rangle, i_2), \quad (5.56)$$

$$\langle v_0 \rangle = \frac{1}{(2r+1)} \sum_{j=-r}^{r} v_j.$$

5.4.3 Mixed Composition

In practice, complex phenomena simulation requires a number of CA-models to be included in a composition forming a complicated scheme of different composition techniques. The main principle for constructing such a mixed composition is that any component may be itself a composed CA. Hence, mixed composition is a hierarchical structure, any level of hierarchy being a composed CA.

Example 8 A simplified process of vapor nucleation in binary system (vapor, gas–carrier) is simulated using a mixed CA composition. The process has been studied in a number of investigations on self-organizing reaction–diffusion systems. For example, in [34] an attempt is made to solve the PDE system which describes the process as follows.

$$\begin{aligned} v_t &= 0.025(v_{xx} + v_{yy}) + 0.2v - v^3 - 1.5u, \\ u_t &= 0.0025(v_{xx} + v_{yy}) + v - u. \end{aligned} \quad (5.57)$$

Since two species are involved in the process, a bidirectional parallel composition should be used. The resulting CA $\aleph = \Upsilon(\aleph_1, \aleph_2)$ has two components, each simulating a reaction–diffusion process in $\Omega_1 = \{(v, (ij)_1)\}$ (vapor) and in $\Omega_2 = \{(u, (ij)_2)\}$ (gas), respectively. Each component $\aleph_k = \Psi_{Gl}(\aleph_{Dk}, \aleph_{Rk})$, in its turn, is a sequential composition of $\aleph_{Dk} = \langle A_D, M_k, \hat{M}_k, \theta_{Dk}, \beta\rangle$ which represents the diffusion, and $\aleph_{Rk} = \langle A_R, M_k, \hat{M}_k, \theta_{Rk}, \sigma\rangle$, which represents the reaction. The two diffusion CA, $\aleph_{D1}$ and $\aleph_{D2}$, operate each in its own cellular array independently. Their results are used by the reaction CA $\aleph_{R1}$ or $\aleph_{R2}$, which are in the bidirectional parallel composition with each other. Since the alphabets A_D and A_R are incompatible, the diffusion global operator result $\Phi_{Dk}(\Omega_{D_k}(t))$ is averaged, and that of the reaction $\Phi_{Rk}(\Omega_{R_k}(t))$ is discretized, which yields the following superposition of global operations.

$$\Phi_k(\Omega_k(t)) = \mathrm{Dis}(\Phi_{Rk}(\mathrm{Av}(\Phi_{Dk}(\Omega_k(t-1))))), \quad k = 1, 2. \tag{5.58}$$

Diffusion is simulated by the two-stage synchronous CA given in Sect. 5.3.1 (Example 1) with $\theta_{Dk}(ij)_1$ given as (5.34). The difference between $\aleph_{D1}$ and $\aleph_{D2}$ is in the values of probabilities used in the transition function. They are: $p_v = 0.5$, $p_u = 0.05$, which corresponds to the diffusion coefficients in (5.57), provided the time step $\Delta t = 0.6$ s and space step $h = 0.1$ cm.

Reaction is simulated by a single cell context-free CA with the following local operators.

$$\begin{aligned}
&\theta_{R1}(i,j)_1) : (\langle v\rangle, (i,j)_1) \to (f_v(\langle v((i,j)_1)\rangle, \langle u((i,j)_2)\rangle), (i,j)_1),\\
&\theta_{R2}(i,j)_2) : (\langle u\rangle, (i,j)_2) \to (f_u(\langle v((i,j)_1)\rangle, \langle u((i,j)_2)\rangle), (i,j)_2),
\end{aligned} \tag{5.59}$$

where

$$\begin{aligned}
f_v &= 0.2\langle v((i,j)_1)\rangle - \langle v((i,j)_1)\rangle^3 - 1.5\langle u(i,j)_2)\rangle,\\
f_u &= \langle v((i,j)_1)\rangle - \langle u((i,j)_2)\rangle,
\end{aligned}$$

The size of both cellular arrays is 300×300 cells with periodic border conditions. The initial conditions are Boolean cellular arrays with the following evenly distributed concentrations of vapor and gas: $\langle v(i,j)_1\rangle = 0.1$, $\langle v(i,j)_2\rangle = 0.9$.

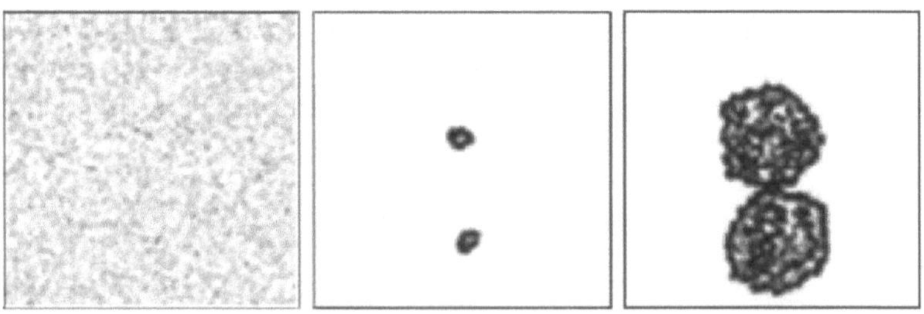

$\Omega_1(0)$ $\Omega_1(12)$ $\Omega_1(40)$

Fig. 5.11 Three snapshots of vapor nucleation process obtained by simulating it as a parallel composition of two superpositions or diffusion and reaction. *Black pixels* stand for vapor particles

The evolutions of $\aleph_1$ and $\aleph_2$ show the processes of vapor and gas space-time distribution, respectively. In Fig. 5.11 three snapshots are shown for vapor nucleation process. Emergency of small vapor bubbles from a fog is observed. The bubbles grow in size exhibiting oscillations of vapor density inside them.

5.5 Computational Properties of Composed CA

In real simulation tasks when dealing with large CA size and large amount of iterations, the computational properties, such as accuracy, stability, and complexity are of main importance. Hence, the impact of above composition techniques on these properties should be assessed. As for the accuracy, the study of this property is focused on the procedures which are beyond the conventional cellular automata theory, namely, cellular array transformations for providing compatibility, since it is precisely these operations that may contribute some errors. Stability assessment of the composition directly depends on the stability of its components, which may exhibit different kind of behavior [2], their evolutions tending to a stable state or never reaching it, or being chaotic. So, the attention is focused on stability conservation, provided the components of CA composition are stable. The property of complexity is concerned with the additional operations inserted for eliminating incompatibility between the interacting components.

It should be noticed that contemporary mathematics has no well-established concepts of CA computational properties, as well as no methods for their quantitative assessment. So, the subsections below may be regarded as some considerations for the problem, indicating the points for further investigation.

5.5.1 *Accuracy of the Composed CA*

One of Boolean CA advantages is that they are absolutely accurate from the computational standpoint, i.e. no errors are incorporated by rounding off. But, once averaging $\text{Av}(\Omega)$ or discretisation $\text{Dis}(\Omega)$ is used and, hence, real numbers are processed, the errors may be brought in.

In trivial compositions, both sequential and parallel ones, the two above operations are performed only once at the start and at the end of the simulation process, bringing in inessential approximation error. But in nontrivial compositions, when $\text{Av}(\Omega)$ and $\text{Dis}(\Omega)$ are used at each iteration, their impact on the result may be significant. So, just this pair of operations are further considered from the point of view of the accuracy problem.

Let Ω_B be the tth iteration result of a composed CA, and $\Omega_R = \text{Av}(\Omega_B)$ should be obtained to make next operation compatible. Then according to (5.19) Boolean states $(v, m) \in \Omega_B$ are replaced by real ones from the finite set of numbers $Q = \{0, 1/q, \ldots, 1\}$, where $q = |\text{Av}(m)|$. Hence, the error $E_{\text{Av}}(m)$ incorporated by approximating a Boolean representation of a spatial function by discrete values from a finite set Q is constrained by

$$E_{\mathrm{Av}} \leq \frac{1}{|\mathrm{Av}(m)|} = \frac{1}{q}. \tag{5.60}$$

Boolean discretisation of $\Omega_R = \{(u, m)\}$ performed according to (5.20) and resulting in $\Omega_B = \{(v, m)\}$ also brings in some errors. Probabilistic formula (5.20) provides that the obtained Ω_B in its averaged form is equal to the averaged state value $\langle v(m) \rangle \in \mathrm{Av}(\Omega_B)$, which yields the following condition of the discretisation accuracy.

$$\Omega_R = \mathrm{Av}(\Omega_B), \quad u(m) = \langle v(m) \rangle \quad \forall m \in M, \tag{5.61}$$

discretisation error $E_{\mathrm{Dis}}(m)$ being the difference

$$E_{\mathrm{Dis}}(m) = |u(m) - \langle v(m) \rangle|. \tag{5.62}$$

The error vanishes in those cells where

$$u(m) = \langle v(m) \rangle = \frac{1}{q} \sum_{k=0}^{q-1} v(\phi_k(m)), \tag{5.63}$$

which happens very rarely, for example, when a fragments of a linear function or a parabola of odd degree is discretised. The error is most serious at the cells where $u(m)$ has extremes.

The most correct representation of discretisation error is a function $E_{\mathrm{Dis}}(m, t)$, which shows possible deviations of $u(m, t)$ in all cells during the evolution. But, sometimes in the particular cases error values in a certain part of Ω, or maximal error in extremes of the spatial function at a certain time is of interest. For a general assessment of CA composition the mean discretisation error at a given $t = \hat{t}$

$$E_{\mathrm{Dis}}(\hat{t}) = \frac{1}{|M|} \sum_{m \in M} |u(m, \hat{t}) - \langle v(m, \hat{t}, \rangle|, \tag{5.64}$$

is also used.

From (5.65) and (5.64) it follows that discretisation errors depend on the averaging area size $q = |\mathrm{Av}(m)|$ and on the smoothness of $u(m)$ on $T_{\mathrm{Av}}(m)$. Both these parameters are conditioned by the discretisation step h, which should be taken small, allowing q to be chosen large enough to smooth the extremes. The following experiment gives a quantitative insight to the accuracy problem.

Example 9 A half-wave of a sinusoid $u = \sin x,\ 0 < x < \pi$, is chosen for experimental assessment of discretisation error dependence of $E_{\mathrm{Dis}}(\hat{t})$ on $|M|$ and on $\mathrm{Av}(m)$. The cellular array representation of a given continuous function is as follows

$$\Omega = \{(u(m), m)\}, \quad u(m) = \sin\left(\frac{\pi}{|M|} m\right), \quad m = 0, 1, 2, \ldots, |M|. \tag{5.65}$$

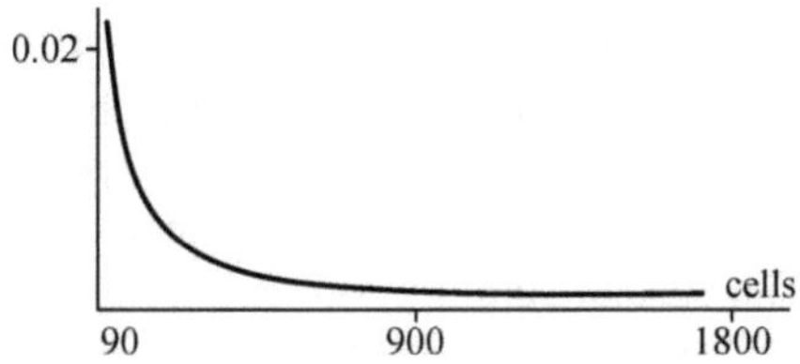

Fig. 5.12 Mean discretisation error dependence on the naming set size $|M|$ with $|Av(m)| = 0.2|M|$ for cellular array (5.64)

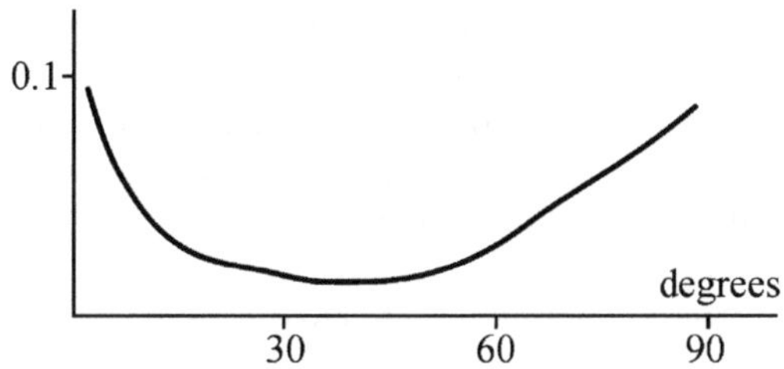

Fig. 5.13 Mean discretisation error dependence on the naming set size of averaging area $|\mathrm{Av}(m)|$ with $|M| = 360$ for cellular array (5.64)

To obtain the dependence $E_{\mathrm{Dis}}(|M|)$, 30 discretisations $\{\mathrm{Dis}_k(\Omega) : k = 1, 2, \ldots, 30\}$ of the function given as (5.65) have been obtained with $|M_k| = 60 \times k$, that corresponds to the argument domain of the cellular array equal to $60 < |M_k| < 1800$, or to the sinus' argument domain in angular form equal to $2° > h > 0.1°$. Each $\mathrm{Dis}_k(\Omega)$ has been averaged with $|\mathrm{Av}_k(m)| = 0.2|M_k|$, and the mean errors $E_{\mathrm{Dis}}(|M_k|)$ have been computed according to (5.67) (Fig. 5.12).

To obtain the dependence $E_{\mathrm{Dis}}(q)$, for the same Ω given as (5.65) 30 discretisations $\{\mathrm{Dis}_j(\Omega) : j = 1, 2, \ldots, 30\}$ have been obtained with fixed $|M| = 360$ but different $q_j = |\mathrm{Av}_j|$, where $q_j = 5 \times j$. Each $\mathrm{Dis}_j(\Omega)$ has been averaged with $\mathrm{Av}_j(m)$, and the mean errors $E_{\mathrm{Dis}}(q_j)$ have been computed according to (5.64) (Fig. 5.13).

From Figs. 5.12 and 5.13 it may be concluded that

(1) the mean error $E_{\mathrm{Dis}}(|m|)$ decreases with the increase of $|M|$ and does not exceed 1% with $|M| > 360$ which correspond to $h < 0.5°$;
(2) the mean error $E_2(|\mathrm{Av}(m)|)$ has a minimum when $|\mathrm{Av}(m)| \approx 36°$, i.e. $|\mathrm{Av}(m)|_{E_{\mathrm{Dis}}=\min} = 0.2|M|$.

From this example it follows, that regulating the smoothness of the extremes by appropriate choice of the CA size, the needed accuracy of CA composition may be achieved. Of course, the complexity of simulation increases linearly with the increase of $|M|$.

5.5.2 CA Composition Stability

There are two aspects of stability regarding CA composition. The first aspect concerns *behavioral stability,* which determines whether the CA evolution tends to a stable state or to a periodic cycling. The property is studied for simple Boolean

CA-models in [2], but no method is known to check behavioral stability for an arbitrary CA. As for CA with real alphabets and nonlinear functions, their behavioral stability is a subject of nonlinear dynamic system theory and may be checked using its methods, as it is usually done in continuous mathematics (see for example, [35]). The second stability aspect is *computational stability*. This property is associated with the round-off errors, which are inevitable when float point arithmetics is used. This aspect of stability is more effectual for study because there are at least two particular cases of composition methods for which computational stability may be quantitatively assessed.

The first case comprises local and global sequential composition of Boolean CA-models. Since all alphabets are Boolean, there is no round-off errors, and since cellular arrays under processing have finite size, the resulting averaged values are bounded and stable.

The second case includes sequential or parallel global composition techniques of Boolean and real CA-models, where cellular array transformations $\mathrm{Av}(\Omega_R)$ and $\mathrm{Dis}(\Omega_B)$ are used at each iteration. In this case the following assertion is true: if $\aleph_R$ is stable, and, hence, its state values may be made bounded by the real closed interval [0, 1], then the composition is computationally stable. This assertion is evident, at the same time it is of considerable importance for widely used diffusion–reaction processes, because it asserts, that composition of Boolean diffusion and real reaction is free of the so called Courant constraint imposed on the PDE counterpart of the process. The Courant constraint in PDE explicit solution is associated with second order partial derivatives of spatial coordinates (Laplace operator), representing a diffusive part in the PDE. For example, for 2D case, it forbids the value $C_{\mathrm{PDE}} = \frac{\tau d}{h^2}$ to exceed 0.25, where h is a space step, d is a diffusion coefficient. From the above it follows that the time step $\tau < \frac{1}{2}\frac{h^2}{d}$, should be small enough, which results in significant increase of computation time. Meanwhile, simulating the diffusion part by a CA, no care should be taken about the stability, the constraint being imposed by the dimensionless diffusion coefficient, which is the characteristic of a CA-model.

Example 10 A diffusion–reaction process called a propagating front is simulated by two models: (1) explicit finite-difference method of PDE solution and (2) composition of a Boolean diffusion CA and a real reaction CA. The PDE is as follows. The process is initiated by a dense square 80 × 80 of propagating substance in the center of an area 639 × 639. The border conditions are periodic.

$$u_t = d(u_{xx} + u_{yy}) + 0.5u(1 - u), \tag{5.66}$$

where $d = 0.33\,\mathrm{cm}^2/\mathrm{s}$. The discretized 2D space is $M = \{(i, j) : i, j = 0, 1, \ldots, i, \ldots, 639\}$. Initial state for PDE solution is

$$u^{(0)}(i, j) = \begin{cases} 1, & \text{if } 280 < i, j < 360, \\ 0, & \text{otherwise,} \end{cases}$$

The finite difference representation of the diffusion part of (5.69) is as follows

$$u^{(t+1)}(i,j) = u^{(t)}(i,j) + 0.33(u^{(t)}(i-1,j) + u^{(t)}(i+1,j) + u^{(t)}(i,j+1) \\ + u^{(t)}(i,j-1) - 4u^{(t)}(i,j)). \tag{5.67}$$

With the time-step $\tau = 1\,\text{s}$ and the space step $h = 1\,\text{cm}$, the Courant value $C_{\text{PDE}} = td/h^2 = 0.33$, which is out of Courant constraint. So, the function $u^{(t)}(i,j)$ obtained by (5.70) is not stable.

The same process may be simulated by a superposition $\aleph = \Psi(\aleph_{\text{diff}}, \aleph_{\text{reac}})$. The first component $\aleph_{\text{diff}} = \langle A, M, \hat{M}, \theta_{\text{diff}}, \sigma\rangle$ is in its turn a superposition of the two-stage synchronous CA from Sect. 5.3.1(Example 1) with $\theta_{\text{diff}}(i,j)$ given as (5.34), and the operator of averaging, i.e.

$$\theta_{\text{diff}}(i,j) = \text{Av}(\theta(i,j)), \tag{5.68}$$

resulting in a global configuration $\Omega_{\text{diff}}(t) = \{(u,(i,j)) : u \in [0,1]\}$

The second component $\aleph_{\text{reac}} = \langle A, M, \hat{M}, \theta_{\text{reac}}, \sigma\rangle$ is a context-free CA computing in each cell a reaction function $f(u) = 0.5u(1-u)$ with subsequent discretisation, where $A, M, \hat{M}, \sigma$ are equal to those of $\aleph_{\text{diff}}$, the local operator being

$$\theta_{\text{reac}} : (u,(i,j)) \to (v,(i,j)), \quad v = \text{Dis}(f(u),(i,j)). \tag{5.69}$$

Snapshots of both processes (PDE solution) and (CA superposition) after 20 iterations are shown in Fig. 5.14. It is seen, that evolution of CA superposition is absolutely stable, while finite-difference solution of (5.67) exhibits a divergence.

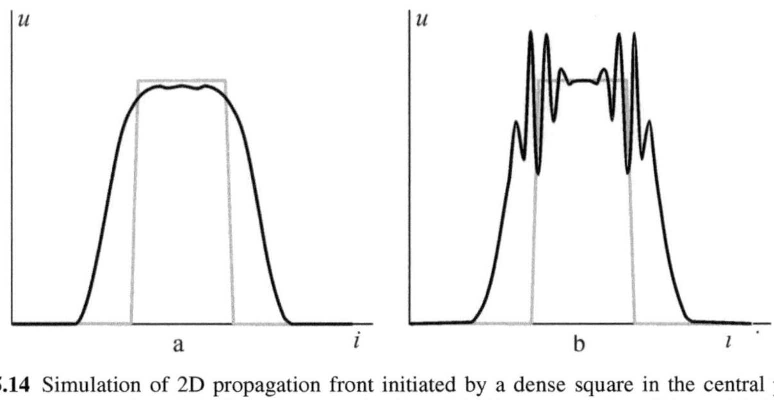

a b

Fig. 5.14 Simulation of 2D propagation front initiated by a dense square in the central part of cellular space, a profile $u(i, 319)$ is given obtained by : (**a**) CA superposition of $\aleph_{\text{diff}}$ with θ_1 (5.71) and $\aleph_{\text{reac}}$ with θ_2 (5.72), (**b**) solution of finite-difference equation (5.70)

5.5.3 Composition Complexity

Here, an attempt is made to assess how much of additional work a composed CA has to do, as compared with the total complexity of the components. Such an assessment cannot be precisely done. There are many reasons for that. The most significant are the following: (1) complexity relations between different arithmetic operations strongly depend on hardware architecture; (2) the same is true for comparing Boolean and real operations; (3) complexity of performing CA transition functions range from $O(n)$ to $O(2^n)$, n being the cardinality of the neighborhood in the local operator. Nonetheless, an insight may be given on the relation between the complexity of transition function computation and that of transformations needed for providing compatibility.

In case when sequential local asynchronous composition and global synchronous composition techniques contain no averaging and discretisation operations, no additional time is needed and the total number of elementary operations is equal to the sum of those of the component CA.

When global synchronous composition, no matter sequential or parallel, is used, transformations of Boolean cellular array into a real one and vice versa are to be performed at each iteration. In such a case, iteration time is increased by t_{add} additional elementary operations.

$$t_{\text{add}} = |M| \times (t_{\text{Av}} + t_{\text{Dis}}), \tag{5.70}$$

where t_{Av} and t_{Dis} are numbers of elementary operations which have to be executed by a cell while performing averaging according to (5.19), or discretisation according to (5.20), respectively. As for t_{Av}, it is clearly seen from (5.19), that the time needed to compute $\text{Av}(v_m)$ may be assessed as $t_{\text{Av}} = C_{\text{Av}} \times |\text{Av}(m)| \times \tau$ where $C_{\text{Av}} \approx 1$ is a constant, τ – the time of elementary function execution. The discrertisation time $t_{\text{Dis}} = C_{\text{rand}}$, so, according to (5.20) it depends only on the random number generator time, which may be taken $C_{\text{rand}} < 5$. Since the transformation is used in the composition techniques where both Boolean and real components are included, the time t_{add} should be compared with Boolean and real transition functions computation time $t_{\text{comp}} = t_B + t_R$, where $t_B = C_B \times \tau$ and $t_R = C_R \times \tau$. The coefficients C_B and C_R essentially depend on the character of the transition functions but, usually, both functions require to execute not more than 100 elementary operations.

Comparison of t_{add} with t_{comp} yields:

$$\frac{t_{\text{add}}}{t_{\text{comp}}} = \frac{C_{\text{Av}} + C_{\text{Dis}}}{C_B + C_R},$$

which enables us to conclude that t_{add} and t_{comp} have identical order of complexity, hence, Boolean–real transformations increase the computation time about twice.

5.6 Conclusion

Till now, no mathematical method and no promising approach is known to CA synthesis from a given description of its evolution. Nevertheless, some way out should be found. A simple one is to follow a well known approach used in PDE theory which implies composing PDE systems out of a set of differential operators and functions. Such an approach seems to be expedient when considering the following similarities between CA composition and PDE system construction. For example, first order and second order differential operators in PDEs over the space have their CA counterparts in the form of shift and diffusion local operators, respectively. And in both cases for obtaining a mathematical model of reaction–diffusion process those operators are composed with nonlinear reaction functions. Unfortunately, the above similarities are only particular cases. In general, there is no formal procedure to obtain a CA simulating space-time nonlinear behavior. It is just the fact that has provoked the development of compatible algebraic operations on cellular arrays, allowing to integrate continuous functions into a CA composition techniques.

But the most important destination of CA composition is not in presenting another way of simulating processes which may be described in terms of PDE, but in obtaining capability of constructing mathematical models for those phenomena, for whom no other mathematical description is known. Such a processes are mostly associated with the fields of science which are in the initial stage of development. For example, plant growth mechanisms, embryo fetation, cellular division, morphogenesis – from biology; surface oxidation, chemical reaction on catalyst, dissociation, adsorption – from chemistry; epitaxial growth, crack formation, rubber deformation, robotics – from engineering; tumor growth – from medicine, etc. Of course, the available experience in science and engineering is not sufficient to forecast the future of CA simulation methodology. Anyway, now it is clear that only a small part of the huge amount of CA-models have evolutions which resemble natural phenomena, and, hence, may be used for simulation. Moreover, those, which occur to be helpful, ought to be enriched by some additional properties, such as probability in transition functions, complicated modes of operations, composite alphabet, non-homogeneous cellular space, etc. All these, being oriented to obtain CA-models of complex phenomena, require a unique formalism for composing complex CA-models from a number of more simple ones. The above considerations allow to hope that the presented attempt to construct a systematic approach to CA composition is not futile.

References

1. T. Toffolli, N. Margolus, *Cellular Automata Machines* (MIT Press, Cambridge, MA, 1987)
2. S. Wolfram, *A New Kind of Science* (Wolfram Media Inc., Champaign, IL, 2002)
3. O. Bandman, Comparative study of cellular automata diffusion models, ed. by V. Malyshkin, PaCT-1999, LNCS vol. 1662 (Springer, Berlin, 1999), pp. 395–404

4. G.G. Malinetski, M.E. Stepantsov, Modeling diffusive processes by cellular automata with Margolus neighborhood. Zhurnal Vychislitelnoy Matematiki i Mathematicheskoy Phisiki **36**(6), 1017–1021 (1998)
5. J.K. Park, K. Steiglitz, W.P. Thurston, Soliton-like behavior in automata. Physica D **19**, 423–432 (1986)
6. C. Vannozzi, D. Fiorentino, M. D'Amore et al., Cellular automata model of phase transition in binary mixtures. Ind. Eng. Chem. Res. **45**(4), 2892–2896 (2006)
7. M. Creutz, Celllular automata and self-organized criticality, ed. by G. Bannot, P. Seiden, *Some New Directions in Science on Computers* (World Scientific, Singapore), pp. 147–169
8. U. Frish, D. d'Humieres, B. Hasslacher et al., Lattice-gas hydrodynamics in two and three dimensions. Compl. Syst. **1**, 649–707 (1987)
9. D.H. Rothman, S. Zalesky, *Lattice-Gas Cellular Automata. Simple Model of Complex Hydrodynamics* (Cambridge University Press, Cambridge, UK, 1997)
10. S. Succi, *The Lattice Boltzmann Equation for Fluid Dynamics and Beyond* (Oxford University Press, New York, NY, 2001)
11. L. Axner, A.G. Hoekstra, P.M.A. Sloot, Simulating time harmonic flows with the lattice Boltzmann method. Phys. Rev. E **75**, 036709 (2007)
12. V.I. Elokhin, E.I. Latkin, A.V. Matveev, V.V. Gorodetskii, Application of statistical lattice models to the analysis of oscillatory and autowave processes in the reaction of carbon monoxide oxidation over platinum and palladium surfaces. Kinet. Catal. **44**(5), 672–700 (2003)
13. A.P.J. Jansen, *An Introduction to Monte-Carlo Simulation of Surface Reactions*. ArXiv: cond-mat/0303028 v1 (2003)
14. I.G. Neizvestny, N.L. Shwartz, Z.Sh. Yanovitskaya, A.V. Zverev, 3D-model of epitaxial growth on porous 111 and 100 Si Surfacex. Comput. Phys. Commun. 147, 272–275 (2002)
15. M.A. Saum, S. Gavrilets, CA simulation of biological evolution in genetic hyperspace, ed. by S. El Yacoubi, B. Chopard, S. Bandini, *ACRI-2006*. LNCS vol. 4176 (Springer, Berlin, 2006), pp. 3–13
16. Y. Wu, N. Chen, M. Rissler, Y. Jiang et al., CA models of myxobacteria sworming ed. by S. El Yacoubi, B. Chopard, S. Bandini, ACRI-2006, LNSC vol. 4176 (Springer, Berlin, 2006), pp. 192–203
17. M. Ghaemi, A. Shahrokhi, Combination of the cellular potts model and lattice gas cellular automata for simulating the avascular cancer growth, ed. by S. El Yacoubi, B. Chopard, S. Bandini, *ACRI-2006*, LNSC vol. 4176 (Springer, Berlin, 2006), pp. 297–303
18. R. Slimi, S. El Yacoubi, Spreadable probabilistic cellular automata models, ed. by S. El Yacoubi, B. Chopard, S. Bandini, *ACRI-2006*, LNSC vol. 4176 (Springer, Berlin, 2006), pp. 330–336
19. F. Biondini, F. Bontempi, D.M. Frangopol, P.G. Malerba, Cellular automata approach to durability analysis of concrete structures in aggressive environments. J. Struct. Eng. **130**(11), 1724–1737
20. O. Bandman, Composing fine-grained parallel algorithms for spatial dynamics simulation, ed. by V. Malyshkin, *PaCT-2005*, LNCS Vol. 3606 (Springer, Berlin, 2005), pp. 99–113
21. S. Wolfram, Universality and complexity in cellular automata. Physica D **10**, 1–35 (1984)
22. L.O. Chua, *CNN: A Paradigm for Complexity* (World Scientific, Singapore, 2002)
23. L.R. Weimar, J.P. Boon, Class of cellular automata for reaction-diffusion systems. Phys Rev E **49**, 1749–1752 (1994)
24. O. Bandman, Simulating spatial dynamics by probabilistic cellular automata, ed. by S. Bandini, B. Chopard, M. Tomassini, *ACRI-2002*, LNCS vol. 2493 (Springer, Berlin, 2002), pp. 10–20
25. O. Bandman, Spatial functions approximation by boolean arrays. Bulletin of Novosibirsk Computer Center, series Computer Science 19. ICMMG, Novosibirsk:10–19 (2003)
26. S. Achasova, O. Bandman, V. Markova, S. Piskunov, *Parallel Substitution Algorithm. Theory and Application* (World Scientific, Singapore, 1994)

27. S. Bandini, S. Manzoni, G. Vizzari, SCA: A model to simulate crowding dynamics. IEICE Trans. Inf. Syst. **E87-D**, 669–676 (2004)
28. A. Adamatsky, *Dynamics of Crowd-Minds*. in Series on Nonlinear Science, vol. 54 (World Scientific, Singapore, 2005)
29. O. Bandman, Coarse-grained parallelisation of cellular-automata simulation algorithms, ed. by V. Malyshkin, *PaCT-2007* LNCS vol. 4671 (Springer, Berlin, 2007), pp. 370–384
30. G. Vichniac, Simulating physics by cellular automata. Physica D **10**, 86–112 (1984)
31. Y. Svirezhev, *Nonlinear Waves, Dissipative Structures and Catastrophes in Ecology* (Nauka, Moscow, 1987)
32. R.M. Ziff, E. Gulari, Y. Barshad, Kinetic phase transitions in an irreversible surface-reaction model. Phys. Rev. Lett. **56**, 2553 (1986)
33. F. Schlogl, Chemical reaction models for non-equilibrium phase transitions. Z. Physik **253** 147–161 (1972)
34. C.P. Schrenk, P. Schutz, M. Bode, H.-G. Purwins, Interaction of selforganised quaziparticles in two-dimensional reaction diffusion system: the formation of molecules. Phys. Rev. E **5** (6), 6481–5486 (1918)
35. A.N. Michel, K. Wang, B. Hu, *Qualitative Theory of Dynamics Systems: The Role of Stability Preserving*. (CRC Press, New York, NY, 2001).

Chapter 6
Problem Solving on One-Bit-Communication Cellular Automata

Hiroshi Umeo

6.1 Introduction

In recent years, interest in cellular automata (CA) has been increasing in the field of modeling real phenomena that occur in biology, chemistry, ecology, economy, geology, mechanical engineering, medicine, physics, sociology, and public transportation. Cellular automata are considered to provide a good model of complex systems in which an infinite array of finite state machines (cells) updates itself in a synchronous manner according to a uniform local rule. In the present paper, we study a problem solving on a special subclass of cellular automata: one-bit inter-cell communication cellular automaton. The problems dealt with are a firing squad synchronization problem, an integer sequence generation problem, an early bird problem, and a connectivity recognition problem for two-dimensional binary images, all of which are classical, fundamental problems that have been studied extensively on O(1)-bit communication models of cellular automata. The O(1)-bit communication model is a conventional CA in which the number of communication bits exchanged in one step between neighboring cells is assumed to be O(1) bits. However, such bit information exchanged between inter-cells is hidden behind the definition of conventional automata-theoretic finite state descriptions. On the other hand, the 1-bit inter-cell communication model studied in the present paper is a new subclass of CAs, in which inter-cell communication is restricted to 1-bit communication. We refer to this model as the 1-bit CA and denote the model as $CA_{1\text{-bit}}$. The number of internal states of the $CA_{1\text{-bit}}$ is assumed to be finite as in a usual sense. The next state of each cell is determined based on the present state of the cell and two binary 1-bit inputs from its left and right neighbor cells. Thus, the $CA_{1\text{-bit}}$ is one of the weakest and simplest models among the variants of the CAs. A main question in this paper is whether the $CA_{1\text{-bit}}$ can solve problems solved by conventional cellular automata without any overhead in time complexities.

H. Umeo (✉)
University of Osaka Electro-Communication, Neyagawa-shi, Hatsu-cho, 18-8, Osaka, 572-8530, Japan
e-mail: umeo@cyt.osakac.ac.jp

A.G. Hoekstra et al. (eds.), *Simulating Complex Systems by Cellular Automata*, Understanding Complex Systems, DOI 10.1007/978-3-642-12203-3_6,

In Sect. 6.2, we define the 1-bit communication cellular automaton and review a computational relation between the conventional O(1)-bit-communication CA and the $CA_{1\text{-bit}}$. In Sect. 6.3, a firing squad synchronization problem is studied and several state-efficient 1-bit implementations of synchronization algorithms for one-dimensional cellular arrays are presented. In Sect. 6.4 we consider an integer sequence generation problem on the $CA_{1\text{-bit}}$ and present a real-time prime generator with 34 states. In Sect. 6.5, we study an early bird problem and its 37-state implementation operating in twice real-time will be given. In Sects. 6.6 and 6.7, a two-dimensional (2-D) version of the $CA_{1\text{-bit}}$ is introduced and the firing squad synchronization problem is studied again on the 2-D $CA_{1\text{-bit}}$. In Sect. 6.7, a connectivity recognition algorithm for two-dimensional binary images will be presented.

6.2 One-Bit-Communication Cellular Automata

A one-dimensional 1-bit inter-cell communication cellular automaton ($CA_{1\text{-bit}}$) consists of a finite array of identical finite state automata, each located at a positive integer point. Each automaton is referred to as a cell. The cell at point i is denoted by C_i where $i \geq 1$. Each C_i, except for C_1 and C_n, is connected with its left and right neighbor cells via a left or right one-way communication link, where those communication links are indicated by right- and left-going arrows, respectively, as shown in Fig. 6.1. Each one-way communication link can transmit only one bit at each step in each direction.

A cellular automaton with 1-bit inter-cell communication (abbreviated as $CA_{1\text{-bit}}$) consists of a finite array of finite state automaton $A = (Q, \delta)$, where

1. Q is a finite set of internal states.
2. δ is a function that defines the next state of any cell and its binary outputs to its left and right neighbor cells such that δ: $Q \times \{0, 1\} \times \{0, 1\} \rightarrow Q \times \{0, 1\} \times \{0, 1\}$ where $\delta(p, x, y) = (q, x', y')$, $p, q \in Q$, $x, x', y, y' \in \{0, 1\}$, has the following meaning: We assume that, at step t, the cell C_i is in state p and receives binary inputs x and y from its left and right communication links, respectively. Then, at the next step t+1, C_i takes a state q and outputs x' and y' to its left and right communication links, respectively. Note that binary inputs to C_i at step t are also outputs of C_{i-1} and C_{i+1} at step t. A quiescent state $q \in Q$ has a property such that $\delta(q, 0, 0) = (q, 0, 0)$.

Thus, the $CA_{1\text{-bit}}$ is a special subclass of *normal* (i.e., *conventional*) cellular automata. Let N be any normal cellular automaton with a set of states Q and a transition function $\delta : Q^3 \rightarrow Q$. The state of each cell on N depends on the

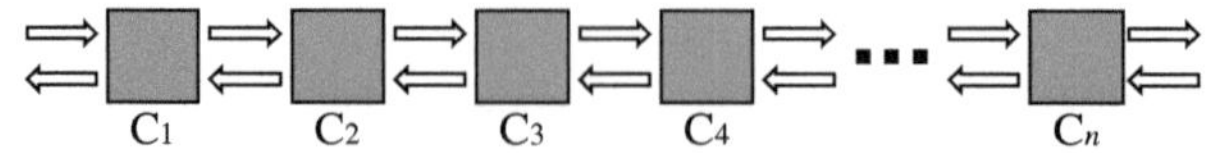

Fig. 6.1 One-dimensional cellular automaton connected with 1-bit inter-cell communication links

cell's previous state and states on its nearest neighbor cells. This means that the total information exchanged per step between neighboring cells is O(1) bits. Each state in Q can be encoded with a binary sequence of length $\lceil \log_2 |Q| \rceil$ and then sending the binary sequences sequentially bit-by-bit in each direction via each one-way communication link. The sequences are then received bit-by-bit and decoded into their corresponding states in Q. Thus, the $\mathrm{CA}_{1\text{-bit}}$ can simulate one step of N in $\lceil \log_2 |Q| \rceil$ steps. This observation gives the following computational relation between the normal CA and $\mathrm{CA}_{1\text{-bit}}$.

Theorem 1 (Mazoyer [21], Umeo and Kamikawa [37]) *Let N be any normal cellular automaton operating in $T(n)$ steps with internal state set Q. Then, there exists a $CA_{1\text{-bit}}$ that can simulate N in $kT(n)$ steps, where k is a positive constant integer such that $k = \lceil \log_2 |Q| \rceil$.*

A question is whether the $\mathrm{CA}_{1\text{-bit}}$ can solve problems solved by conventional cellular automata without any overhead in time complexities. In some cases, the answer is *yes*.

6.3 Firing Squad Synchronization Problem

Section 6.3 studies the firing squad synchronization problem (FSSP) on $\mathrm{CA}_{1\text{-bit}}$, the solution of which yields a finite-state protocol for large-scale synchronization of cellular automata. This problem was originally proposed by J. Myhill in Moore [23] to synchronize all parts of self-reproducing cellular automata. The firing squad synchronization problem has been studied extensively for more than 50 years. Recent developments in the FSSP algorithms are given in Umeo [33] and Umeo et al. [35]. An optimum-time (i.e., $(2n-2)$-step for n cells) synchronization algorithm for one-dimensional array was devised first by Goto [10]. The algorithm needed many thousands of internal states for its realization. Afterwards, Waksman [45], Balzer [4], Gerken [9] and Mazoyer [19] developed an optimum-time algorithm and reduced the number of states realizing the algorithm, each with 16, 8, 7 and 6 states on the conventional O(1)-bit communication model.

The FSSP is defined as follows: At time $t = 0$, the left end cell C_1 is in the *fire-when-ready* state, which is the initiation signal for the array. The FSSP is to determine a description (state set and next-state function) for cells that ensures all cells enter the *fire* state at exactly the same time and for the first time. The set of states and the next-state function must be independent of n.

6.3.1 FSSP with a General at One End

Here we briefly sketch the design scheme for the firing squad synchronization algorithm according to Waksman [45] in which the first transition rule set was presented. It is quoted from Waksman [45].

> The code book of the state transitions of machines is so arranged to cause the array to progressively divide itself into 2^k equal parts, where k is an integer and an increasing function of time. The end machines in each partition assume a special state so that when the last partition occurs, all the machines have for both neighbors machines at this state. This is made the *only* condition for any machine to assume terminal state.

Figure 6.2 (left) is a space-time diagram for the Waksman's optimum-step firing squad synchronization algorithm. The general at time $t = 0$ emits an infinite number of signals which propagate at $1/(2^{k+1} - 1)$ speed, where k is positive integer. These signals meet with a reflected signal at half point, quarter points, ..., etc., denoted by $\odot$ in Fig. 6.2 (left). It is noted that these cells indicated by $\odot$ are synchronized. By increasing the number of synchronized cells exponentially, eventually all of the cells are synchronized.

Most of the implementations for the optimum-time synchronization algorithms developed so far on $\mathrm{CA}_{1\text{-bit}}$ are based on the space-time diagram shown in Fig. 6.2 (left). Mazoyer [21] developed an optimum-time synchronization algorithm for the $\mathrm{CA}_{1\text{-bit}}$ based on Balzer [4]. Each cell of the constructed $\mathrm{CA}_{1\text{-bit}}$ had 58 internal states. The original set of transition rules constructed in Mazoyer [21] included a small error. Here we show a reconstructed version in Table 6.1. Figure 6.3 shows some snapshots of the synchronization processes on 21 cells, each for Balzer's

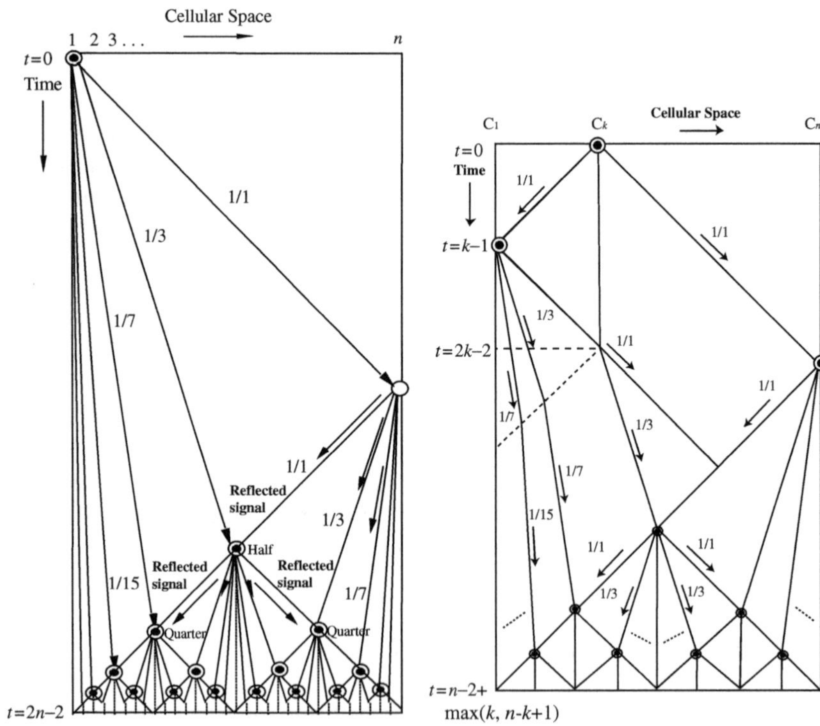

Fig. 6.2 Space-time diagram for optimum-time synchronization algorithms with a general at the left end (*left*) and a generalized case where a general at an arbitrary point (*right*)

Table 6.1 Reconstructed transition table for Mazoyer's 1-bit implementation (Mazoyer [21])

1

0	R = 0	R = 1
L = 0	(0,0,0)	--
L = 1	(ir,0,1)	--

2

ir	R = 0	R = 1
L = 0	(kr,1,0)	(kr,1,0)
L = 1	(2ar,0,0)	(2ar,0,0)

3

F	R = 0	R = 1
L = 0	--	--
L = 1	--	--

4

kr	R = 0	R = 1
L = 0	(Exl,0,1)	(il,1,0)
L = 1	(Oddr,0,0)	(Ex!r,1,0)

5

pF	R = 0	R = 1
L = 0	(pF,0,0)	(F,0,0)
L = 1	(F,0,0)	(F,0,0)

6

2ar	R = 0	R = 1
L = 0	(2b,0,1)	(pF,1,1)
L = 1	--	--

7

2b	R = 0	R = 1
L = 0	--	(2c,0,1)
L = 1	(2c,1,0)	--

8

2c	R = 0	R = 1
L = 0	(2c,0,0)	(pF,1,1)
L = 1	(pF,1,1)	--

9

Oddr	R = 0	R = 1
L = 0	--	(Oor,1,0)
L = 1	--	(3m,0,1)

10

Oor	R = 0	R = 1
L = 0	(Oo*r,0,0)	(il,1,0)
L = 1	--	--

11

Oo*r	R = 0	R = 1
L = 0	(Oor,0,0)	(Oe,1,0)
L = 1	(Op,0,0)	(Orl,1,0)

12

Oe	R = 0	R = 1
L = 0	(Oe*,0,0)	(Exr,1,0)
L = 1	(Exl,0,1)	--

13

Oe*	R = 0	R = 1
L = 0	(Oe,0,0)	(Oor,1,0)
L = 1	(Ool,0,1)	--

14

Op	R = 0	R = 1
L = 0	(Op*,0,0)	(il,1,0)
L = 1	(ir,0,1)	--

15

Op*	R = 0	R = 1
L = 0	(Op,0,0)	(Orl,1,0)
L = 1	(Orr,0,1)	--

16

Orr	R = 0	R = 1
L = 0	(Or*r,0,0)	(Rr,1,0)
L = 1	(pLl,0,1)	--

17

Or*r	R = 0	R = 1
L = 0	(Orr,0,0)	(Oe,1,0)
L = 1	(Om,1,1)	--

18

Om	R = 0	R = 1
L = 0	(Om*,0,0)	(M,1,1)
L = 1	(M,1,1)	--

19

Om*	R = 0	R = 1
L = 0	(Om,0,0)	(Orr,0,0)
L = 1	(Orl,0,0)	--

20

Exr	R = 0	R = 1
L = 0	(il,0,0)	(Ex!l,0,0)
L = 1	(Ee,0,1)	(Err,0,1)

21

Ex!r	R = 0	R = 1
L = 0	(2ar,0,0)	(Ex!l,0,0)
L = 1	--	(pF,1,1)

22

3m	R = 0	R = 1
L = 0	(3m*,0,0)	(M,1,1)
L = 1	(M,1,1)	--

23

3m*	R = 0	R = 1
L = 0	(3m,0,0)	(3r,0,0)
L = 1	(3r,0,0)	--

24

3l	R = 0	R = 1
L = 0	(3l*,0,0)	(pLr,1,0)
L = 1	(pLl,0,1)	--

25

3l*	R=0	R=1
L=0	(3l,0,0)	(3m,0,1)
L=1	(3m,1,0)	--

26

Err	R=0	R=1
L=0	(Er*r,0,0)	(Rr,1,0)
L=1	(pLl,0,1)	--

27

Er*r	R=0	R=1
L=0	(Err,0,0)	(Eor,1,0)
L=1	(Em,1,1)	--

28

Ep	R=0	R=1
L=0	(Ep*,0,0)	(il,1,0)
L=1	(ir,0,1)	--

29

Ep*	R=0	R=1
L=0	(Ep,0,0)	(Erl,1,0)
L=1	(Err,0,1)	--

30

Eor	R=0	R=1
L=0	(Eo*r,0,0)	(il,1,0)
L=1	--	--

31

Eo*r	R=0	R=1
L=0	(Eor,0,0)	(Ee,1,0)
L=1	(Ep,0,0)	(Erl,1,0)

32

Em	R=0	R=1
L=0	(Em*,0,0)	(M,1,1)
L=1	(M,1,1)	--

33

Em*	R=0	R=1
L=0	(Em,0,0)	(Err,0,0)
L=1	(Erl,0,0)	--

34

Rr	R=0	R=1
L=0	--	(pF,1,0)
L=1	--	--

35

M	R=0	R=1
L=0	(pF,1,1)	--
L=1	--	--

36

pLr	R=0	R=1
L=0	(Lr,1,1)	(Lr,1,1)
L=1	--	--

37

Lr	R=0	R=1
L=0	(pF,0,1)	--
L=1	(F,0,0)	--

38

irW	R=0	R=1
L=0	(pF,0,1)	--
L=1	--	--

39

0W	R=0	R=1
L=0	(0W,0,0)	--
L=1	(ilW,1,0)	--

40

3r	R=0	R=1
L=0	(3r*,0,0)	(Rr,1,0)
L=1	(Rl,0,1)	--

41

3r*	R=0	R=1
L=0	(3r,0,0)	(3l,0,0)
L=1	(3l,0,0)	--

42

Ee	R=0	R=1
L=0	(Ee*,0,0)	(Exr,1,0)
L=1	(Exl,0,1)	--

43

Ee*	R=0	R=1
L=0	(Ee,0,0)	(Eor,1,0)
L=1	(Eol,0,1)	--

44

ilW	R=0	R=1
L=0	(pF,1,0)	--
L=1	(F,0,0)	--

45

2al	R=0	R=1
L=0	(2b,1,0)	--
L=1	(pF,1,1)	--

46

Oddl	R=0	R=1
L=0	--	--
L=1	(Ool,0,1)	(3m,1,0)

47

Ool	R=0	R=1
L=0	(Oo*l,0,0)	--
L=1	(ir,0,1)	--

48

Oo*l	R=0	R=1
L=0	(Ool,0,0)	(Op,0,0)
L=1	(Oe,0,1)	(Orr,0,1)

49

Orl	R=0	R=1
L=0	(Or*l,0,0)	(pLr,1,0)
L=1	(Rl,0,1)	--

50

Or*l	R=0	R=1
L=0	(Orl,0,0)	(Om,1,1)
L=1	(Oe,0,1)	--

51

Exl	R=0	R=1
L=0	(ir,0,0)	(Ee,1,0)
L=1	(Ex!r,0,0)	(Erl,1,0)

52

Ex!l	R=0	R=1
L=0	(2al,0,0)	--
L=1	(Ex!r,0,0)	(pF,1,1)

53

Erl	R=0	R=1
L=0	(Er*l,0,0)	(pLr,1,0)
L=1	(Rl,0,1)	--

54

Er*l	R=0	R=1
L=0	(Erl,0,0)	(Em,1,1)
L=1	(Eol,0,1)	--

55

Eol	R=0	R=1
L=0	(Eo*l,0,0)	--
L=1	(ir,0,1)	--

56

Eo*l	R=0	R=1
L=0	(Eol,0,0)	(Ep,0,0)
L=1	(Ee,0,1)	(Err,0,1)

57

Rl	R=0	R=1
L=0	--	--
L=1	(pF,0,1)	--

58

pLl	R=0	R=1
L=0	(Ll,1,1)	--
L=1	(Ll,1,1)	--

59

Ll	R=0	R=1
L=0	(pF,1,0)	(F,0,0)
L=1	--	--

60

il	R=0	R=1
L=0	(kl,0,1)	(2al,0,0)
L=1	(kl,0,1)	(2al,0,0)

61

kl	R=0	R=1
L=0	(Exr,1,0)	(Oddl,0,0)
L=1	(ir,0,1)	(Ex!l,0,1)

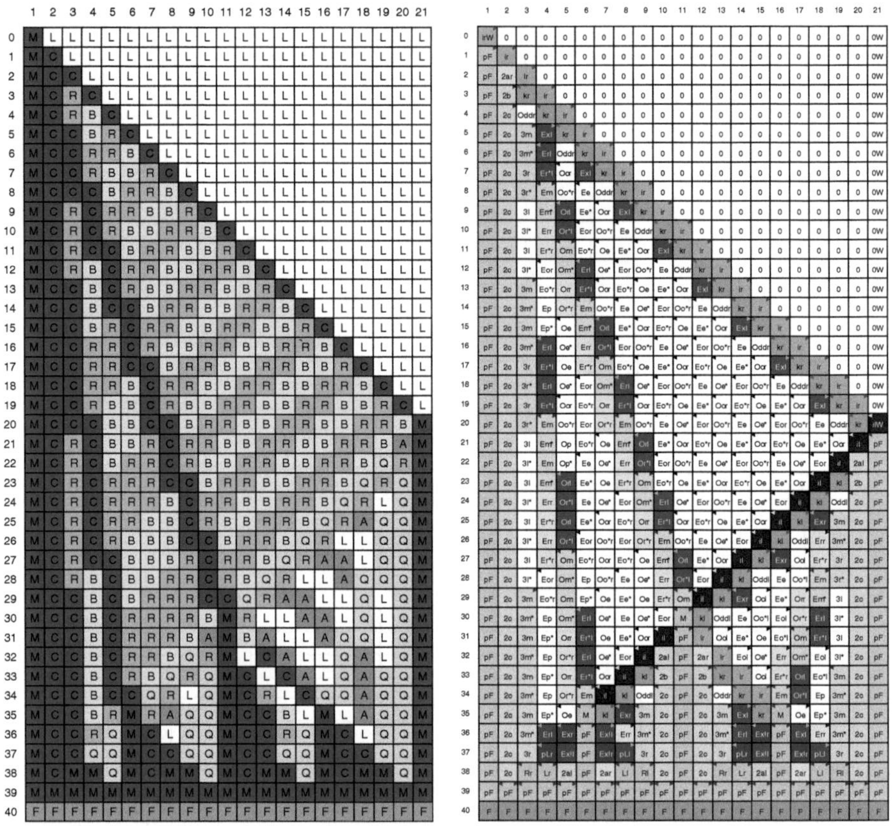

Fig. 6.3 Snapshots for synchronization processes on 21 cells, each for Balzer's algorithm [4] on the O(1)-bit-communication model (*left*) and the reconstructed 1-bit implementation on the $CA_{1\text{-bit}}$ (*right*)

algorithm [4] on the O(1)-bit-communication model (left) and the reconstructed 1-bit implementation on the $CA_{1\text{-bit}}$ (right). The small right- and left-facing black triangles, ► and ◄, in the figure, indicate a 1-bit signal transfer in the right or left direction between neighbor cells. The symbol in each cell shows its internal state. Nishimura et al. [25] also constructed an optimum-time synchronization algorithm (NSU algorithm for short) based on Waksman's algorithm [45]. Each cell had 78 internal states and 208 transition rules. Figure 6.4 shows snapshots for synchronization processes on 21 cells, each for Waksman's algorithm [45] on O(1)-bit-communication model (left) and NSU algorithm [25] on $CA_{1\text{-bit}}$ (right).

Theorem 2 (Mazoyer [21], Nishimura, Sogabe and Umeo [25]) *There exists a $CA_{1\text{-bit}}$ that can synchronize n cells with the general at a left end in $2n-2$ steps.*

Umeo et al. [42] developed a non-optimum-step synchronization algorithm for $CA_{1\text{-bit}}$ based on Mazoyer's 6-state algorithm [19] for the O(1)-bit model. The

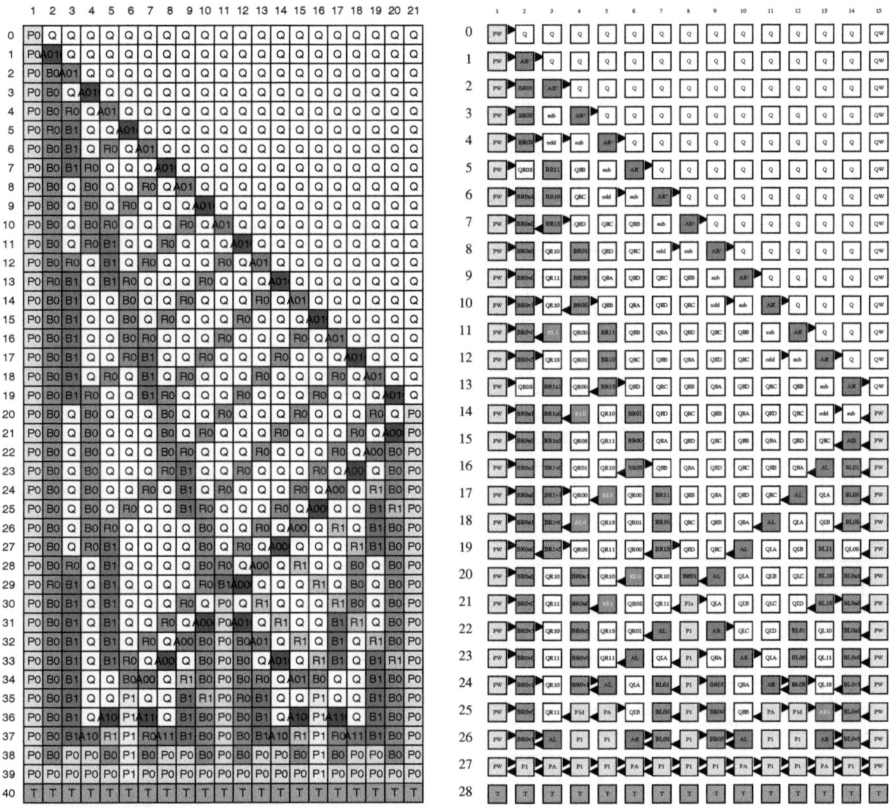

Fig. 6.4 Snapshots for synchronization processes on 21 cells, each for Waksman's algorithm [45] on O(1)-bit-communication model (*left*) and NSU implementation [25] on $CA_{1\text{-bit}}$ (*right*)

constructed $CA_{1\text{-bit}}$ synchronizes n cell in $2n-1$ steps and each cell has 54 states and 207 transition rules.

Theorem 3 (Umeo, Yanagihara and Kanazawa [42]) *There exists a 54-state $CA_{1\text{-bit}}$ that can synchronize any n cells in $2n-1$ non-optimum-step.*

Umeo and Yanagihara [41] also constructed a smaller optimum-time implementation based on Gerken's synchronization algorithm [9] on the O(1)-bit-communication model. The constructed $CA_{1\text{-bit}}$ has 35 internal states and 114 transition rules. Table 6.2 presents its transition rule set for the 35-state synchronization protocol and Figure 6.5 shows snapshots for synchronization processes on 17 cells, each for Gerken's algorithm [9] on O(1)-bit-communication model (left) and our 35-state algorithm [41] on the $CA_{1\text{-bit}}$ (right).

Theorem 4 (Umeo and Yanagihara [41]) *There exists a 35-state $CA_{1\text{-bit}}$ that can synchronize n cells with the general on the left end in $2n-2$ steps.*

Table 6.2 Transition table for a 35-state implementation of the optimum-time synchronization algorithm (Umeo and Yanagihara [41])

1

Q	R = 0	R = 1
L = 0	(Q,0,0)	(Q,0,0)
L = 1	(RA,0,1)	(LGW,1,1)

2

RGW	R = 0	R = 1
L = 0	(RGW,0,1)	(F,0,0)
L = 1	(RGW,1,1)	(F,0,0)

3

RPW	R = 0	R = 1
L = 0	(RPW,0,1)	(LGW,1,1)
L = 1	--	--

4

RA	R = 0	R = 1
L = 0	(RQoS,0,0)	--
L = 1	(RP,0,0)	--

5

RQoS	R = 0	R = 1
L = 0	(RQ0A,0,1)	(LP',1,0)
L = 1	(RQeS,0,0)	(LP,1,0)

6

RQeS	R = 0	R = 1
L = 0	(RQ1B,1,0)	--
L = 1	(RG1,0,1)	--

7

RQ1A	R = 0	R = 1
L = 0	(RQ0A,0,0)	(LG,1,0)
L = 1	--	--

8

RQ0A	R = 0	R = 1
L = 0	(RQ1A,0,0)	(RQ1A,1,0)
L = 1	(RQ1A,0,0)	(RP1,1,1)

9

RQ1B	R = 0	R = 1
L = 0	(RQ0B,0,0)	(LP,1,0)
L = 1	--	--

10

RQ0B	R = 0	R = 1
L = 0	(RQ1B,0,0)	(RQ1B,1,0)
L = 1	(RQ1B,0,0)	(RG1,1,1)

11

RQ1C	R = 0	R = 1
L = 0	(LQ1B,0,0)	(LQ1C,0,0)
L = 1	(RQ0C,0,0)	(LP,1,0)

12

RQ0C	R = 0	R = 1
L = 0	(LQ1A,0,1)	--
L = 1	(RQ1C,0,0)	(RG1,0,1)

13

RG1	R = 0	R = 1
L = 0	(RG0,0,0)	(LPW,1,0)
L = 1	(RG0,0,0)	(LPW,1,0)

14

RG0	R = 0	R = 1
L = 0	(RG1,0,1)	(RQ1B,0,0)
L = 1	(RG1,0,1)	(RQ1C,0,0)

15

RP1	R = 0	R = 1
L = 0	(RP0,0,0)	(LGW,1,1)
L = 1	--	--

16

RP0	R = 0	R = 1
L = 0	(RP1,0,1)	(RQ1A,1,0)
L = 1	--	--

17

RG	R = 0	R = 1
L = 0	(LQ1C,0,0)	(LPW,1,0)
L = 1	(LQ1C,0,0)	(LPW,1,0)

18

RP	R = 0	R = 1
L = 0	(LQ0C,0,0)	(LGW,1,1)
L = 1	(RP,0,1)	(LGW,1,1)

19

LGW	R = 0	R = 1
L = 0	(LGW,1,0)	(LGW,1,1)
L = 1	(F,0,0)	(F,0,0)

20

LPW	R = 0	R = 1
L = 0	(LPW,1,0)	--
L = 1	(RGW,1,1)	-

21

LQ1A	R = 0	R = 1
L = 0	(LQ0A,0,0)	(LQ1B,0,0)
L = 1	(RG,0,1)	(LP1,1,0)

22

LQ1B	R = 0	R = 1
L = 0	(LQ0B,0,0)	(LQ1A,0,0)
L = 1	(RP,0,1)	(RP,0,0)

23

LQ0A	R = 0	R = 1
L = 0	(LQ1A,0,0)	(LQ1A,0,0)
L = 1	(LQ1A,0,1)	(LP1,1,1)

24

LQ0B	R = 0	R = 1
L = 0	(LQ1B,0,0)	(LQ1B,0,0)
L = 1	(LQ1B,0,1)	(LG1,1,1)

25

LQ1C	R = 0	R = 1
L = 0	(RQ1B,0,0)	(LQ0C,0,0)
L = 1	(RQ1C,0,0)	(RP,0,1)

26

LQ0C	R = 0	R = 1
L = 0	(RQ1A,1,0)	(LQ1C,0,0)
L = 1	--	(LG1,1,0)

27

LG1	R = 0	R = 1
L = 0	(LG0,0,0)	(LG0,0,0)
L = 1	(RPW,0,1)	(RPW,0,1)

28

LG0	R = 0	R = 1
L = 0	(LG1,1,0)	(LG1,1,0)
L = 1	(LQ1B,0,0)	(LQ1C,0,0)

29

LP1	R = 0	R = 1
L = 0	(LP0,0,0)	(LP0,0,0)
L = 1	(RGW,1,1)	(RPW,0,0)

30

LP0	R = 0	R = 1
L = 0	(LP1,1,0)	(LP1,1,0)
L = 1	(LQ1A,0,1)	(LQ1B,0,0)

31

LG	R = 0	R = 1
L = 0	(RQ1C,0,0)	(RQ1C,0,0)
L = 1	(RPW,0,1)	(RPW,0,1)

32

LP	R = 0	R = 1
L = 0	(RQ0C,0,0)	(LP,1,0)
L = 1	(RGW,1,1)	(RGW,1,1)

33

LP'	R = 0	R = 1
L = 0	--	(LQ1A,0,0)
L = 1	--	(RPW,0,0)

34

F	R = 0	R = 1
L = 0	--	--
L = 1	--	--

35

QW	R = 0	R = 1
L = 0	(QW,0,0)	--
L = 1	(LGW,1,0)	--

6.3.2 *Generalized FSSP with a General at an Arbitrary Point*

Section 6.3.2 considers a *generalized* firing squad synchronization problem which allows the initial general to be located anywhere on the array. It has been shown to be impossible to synchronize any array of length n less than $n - 2 + \max(k, n - k + 1)$ steps, where the general is located on C_k, $1 \leq k \leq n$. Moore and Langdon [24], Szwerinski [30] and Varshavsky et al. [43] developed a generalized optimum-time synchronization algorithm for O(1)-bit cellular automaton each with 17, 10 and 10 internal states, respectively, that can synchronize any array of length n at exactly $n - 2 + \max(k, n - k + 1)$ steps. Recently, Settle and Simon [28] and Umeo et al. [34] have also proposed a 9-state generalized synchronization algorithm operating in optimum-step for the O(1)-bit model.

Umeo et al. [34] developed a generalized synchronization algorithm on the $CA_{1\text{-bit}}$ model operating in non-optimum steps. The implementation for the $CA_{1\text{-bit}}$

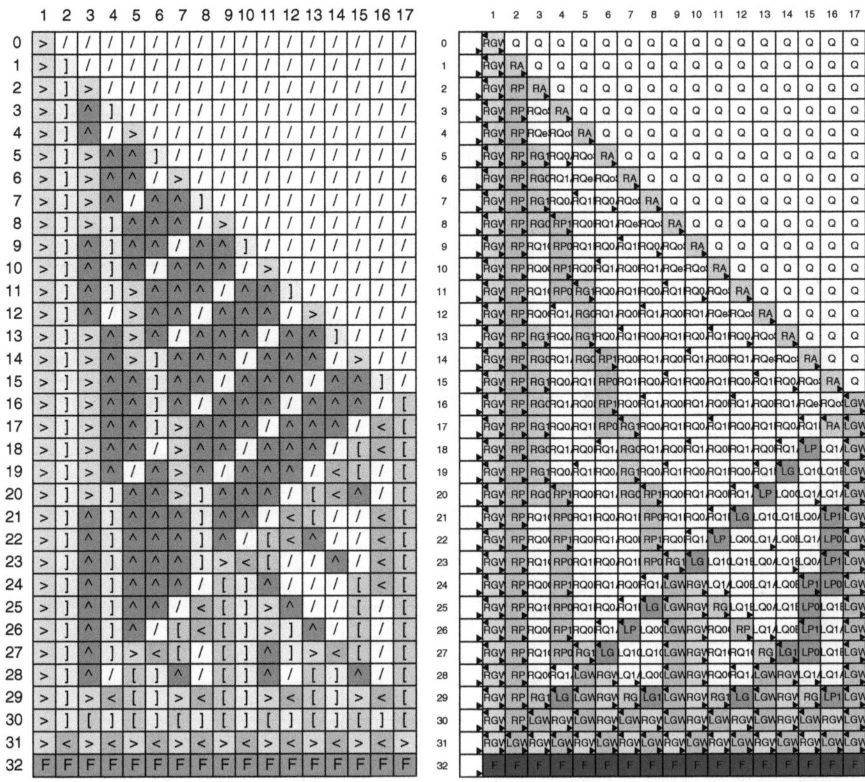

Fig. 6.5 Snapshots for synchronization processes on 17 cells, each for Gerken's algorithm [9] on O(1)-bit-communication model (*left*) and the 35-state implementation (Umeo and Yanagihara [41]) on $CA_{1\text{-bit}}$ (*right*)

has 282-state and 721 transition rules. Kamikawa and Umeo [12] also developed a generalized synchronization algorithm on the $CA_{1\text{-bit}}$ model operating in $n+\max(k, n-k+1)$ steps, which is one-step larger than optimum-step. The total numbers of internal states and transition rules of the constructed $CA_{1\text{-bit}}$ are 219 and 488, respectively. Figure 6.2 (right) shows a space-time diagram for the optimum-time generalized firing squad synchronization algorithm. We also show some snapshots for the synchronization processes on 21 cells with a general at C_7 on $CA_{1\text{-bit}}$ in Fig. 6.6. We present Table 6.3 that shows a quantitative comparison of synchronization algorithms and their implementations proposed so far with respect to the number of internal states of each finite state automaton, the number of transition rules realizing the synchronization and time complexity.

Theorem 5 (Kamikawa and Umeo [12]) *There exists a 219-state, 488-transition-rule $CA_{1\text{-bit}}$ that can synchronize n cells in $n-1+max(k, n-k+1)$ steps, where k is any integer such that $1 \leq k \leq n$ and a general is located on the kth cell from the left end of the array.*

	1	2	3	4	5	6	7	8	9	10	11	12	13	14	15	16	17	18	19	20	21
0	SW	S	S	S	S	S	M	S	S	S	S	S	S	S	S	S	S	S	S	S	SW
1	SW	S	S	S	S	AL	Mp1	AR	S	S	S	S	S	S	S	S	S	S	S	S	SW
2	SW	S	S	S	AL	RL	Mp2	RR	AR	S	S	S	S	S	S	S	S	S	S	S	SW
3	SW	S	S	AL	RL	QLC	Mp3	QRC	RR	AR	S	S	S	S	S	S	S	S	S	S	SW
4	SW	S	AL	RL	QLa	BLo	Ms1	BRo	QRa	RR	AR	S	S	S	S	S	S	S	S	S	SW
5	SW	AL	RL	QLC	QLb	QLA	Ms2	QRA	QRb	QRC	RR	AR	S	S	S	S	S	S	S	S	SW
6	MW	RL	QLa	BLo	QLc	QLB	Ms3	QRB	QRc	BRo	QRa	RR	AR	S	S	S	S	S	S	S	SW
7	MW	ECo	QLb	QLA	BLe	QLC	Ms1	QRC	BRe	QRA	QRb	QRC	RR	AR	S	S	S	S	S	S	SW
8	MW	Cp0	ECe	QLB	QLa	BLo	Ms1	BRo	QRa	QRB	QRc	BRo	QRa	RR	AR	S	S	S	S	S	SW
9	MW	Cp1	pRE0	ECo	QLb	QLA	Ms2	QRA	QRb	QRC	BRe	QRA	QRb	QRC	RR	AR	S	S	S	S	SW
10	MW	Cp2	pRE1	pBO0	ECe	QLB	Ms3	QRB	QRc	BRo	QRa	QRB	QRc	BRo	QRa	RR	AR	S	S	S	SW
11	MW	Cp1	pRE2	pBO1	pRE0	ECo	Ms1	QRC	BRe	QRA	QRb	QRC	BRe	QRA	QRb	QRC	RR	AR	S	S	SW
12	MW	Cp2	Ct1	pBO2	pRE1	pBO0	sFL1	BRo	QRa	QRB	QRc	BRo	QRa	QRB	QRc	BRo	QRa	RR	AR	S	SW
13	MW	Cp1	Ct2	RO1	pRE2	BO1	sFL2	QRAx	QRb	QRC	BRe	QRA	QRb	QRC	BRe	QRA	QRb	QRC	RR	AR	SW
14	MW	Cp2	Ct1	RO2	BE1	BO2	sFLx	Fl1	QRc	BRo	QRa	QRB	QRc	BRo	QRa	QRB	QRc	BRo	QRa	RR	MW
15	MW	Cp1	Ct2	CO1	BE2	RO1	sFLx	Fl2	BRe	QRA	QRb	QRC	BRe	QRA	QRb	QRC	BRe	QRA	QRb	EQo	MW
16	MW	Cp2	Rt1	CO2	RE1	RO2	RE1	Fl3	QRa	QRB	QRc	BRo	QRa	QRB	QRc	BRo	QRa	QRB	EQe	Qp0	MW
17	MW	Cp1	Rt2	CO1	RE2	RO1	RE2	RO1	FL1	QRC	BRe	QRA	QRb	QRC	BRe	QRA	QRb	EQo	pLE0	Qp1	MW
18	MW	Cp2	Rt1	CO2	RE1	RO2	RE1	RO2	FL2	BRo	QRa	QRB	QRc	BRo	QRa	QRB	EQe	pAO0	pLE1	Qp2	MW
19	MW	Cp1	Rt2	CO1	RE2	RO1	RE2	BO1	FL3	QRAx	QRb	QRC	BRe	QRA	QRb	EQo	pLE0	pAO1	pLE2	Qp1	MW
20	MW	Cp2	Rt1	CO2	RE1	RO2	BE1	BO2	BE1	Fl1	QRc	BRo	QRa	QRB	EQe	pAO0	pLE1	pAO2	Qt1	Qp2	MW
21	MW	Cp1	Rt2	CO1	RE2	BO1	BE2	BO1	BE2	Fl2	BRe	QRA	QRb	EQo	pLE0	pAO1	pLE2	LO1	Qt2	Qp1	MW
22	MW	Cp2	Rt1	CO2	CE1	BO2	BE1	BO2	RE1	Fl3	QRa	QRB	EQe	pAO0	pLE1	pAO2	AE1	LO2	Qt1	Qp2	MW
23	MW	Cp1	Rt2	BO1	CE2	BO1	BE2	RO1	RE2	RO1	FL1	EQo	pLE0	pAO1	pLE2	LO1	AE2	QO1	Qt2	Qp1	MW
24	MW	Cp2	Ct1	BO2	CE1	BO2	RE1	RO2	RE1	RO2	MC	pAO0	pLE1	pAO2	AE1	LO2	LE1	QO2	Lt1	Qp2	MW
25	MW	Cp1	Ct2	BO1	CE2	RO1	RE2	RO1	RE2	BO1	MC	pAO1	pLE2	LO1	AE2	AO1	LE2	QO1	Lt2	Qp1	MW
26	MW	Cp2	Ct1	BO2	CE1	RO2	RE1	RO2	BE1	eQo	MC	eCo	AE1	LO2	LE1	AO2	QE1	QO2	Lt1	Qp2	MW
27	MW	Cp1	Ct2	BO1	CE2	RO1	RE2	BO1	eQe	Qp1	MC	Cp1	eCe	AO1	LE2	LO1	QE2	AO1	Lt2	Qp1	MW
28	MW	Cp2	Ct1	BO2	CE1	RO2	BE1	eQo	pLE1	Qp2	MC	Cp2	pRE1	eCo	AE1	LO2	QE1	AO2	Qt1	Qp2	MW
29	MW	Cp1	Ct2	BO1	CE2	CO1	eQe	pAO1	pLE2	Qp1	MC	Cp1	pRE2	pBO1	eCe	QO1	QE2	AO1	Qt2	Qp1	MW
30	MW	Cp2	Ct1	BO2	RE1	MC	pLE1	pAO2	Qt1	Qp2	MC	Cp2	Ct1	pBO2	pRE1	MC	LE1	AO2	Qt1	Qp2	MW
31	MW	Cp1	Ct2	RO1	OQe	MC	OCe	LO1	Qt2	Qp1	MC	Cp1	Ct2	RO1	OQe	MC	OCe	LO1	Qt2	Qp1	MW
32	MW	Cp2	Ct1	OQo	Qp0	MC	Cp0	OCo	Qt1	Qp2	MC	Cp2	Ct1	OQo	Qp0	MC	Cp0	OCo	Qt1	Qp2	MW
33	MW	Cp1	pMC	MC	Qp1	MC	Cp1	MC	pMC	Qp1	MC	Cp1	pMC	MC	Qp1	MC	Cp1	MC	pMC	Qp1	MW
34	MW	MC	MC	MC	MC	MC	MC	MC	MC	MC	MC	MC	MC	MC	MC	MC	MC	MC	MC	MC	MW
35	F	F	F	F	F	F	F	F	F	F	F	F	F	F	F	F	F	F	F	F	F

Fig. 6.6 Snapshots for the generalized synchronization processes on 21 cells with a general at C_7 on $CA_{1\text{-bit}}$

Table 6.3 A list of firing squad synchronization algorithms for $\mathrm{CA}_{1\text{-bit}}$. A symbol "$*$" indicates the reconstructed rule set given in Table 6.1

Implementations	No. of states	No. of rules	Time complexity	Prototype algorithms
Mazoyer [21]	61*(58)	167*	$2n-2$	Balzer [4]
Nishimura et al. [25]	78	208	$2n-2$	Waksmann [45]
Umeo et al. [42]	54	207	$2n-1$	Mazoyer [19]
Umeo and Yanagihara [41]	35	114	$2n-2$	Gerken [9]
Umeo et al. [34]	282	721	$n+\max(k, n-k+1)$	–
Kamikawa and Umeo [12]	219	488	$n-1+\max(k, n-k+1)$	–

6.4 Prime Sequence Generation Problem

Sequence generation is an important, fundamental problem in cellular automata. Arisawa [3], Fischer [8], Korec [4] and Mazoyer and Terrier [20] have considered the sequence generation problem on the conventional O(1)-bit cellular automata model. Fischer [8] showed that the prime sequence can be generated in real-time on the O(1)-bit cellular automata with 11 states for C_1 and 37 states for $C_i (i \geq 2)$. Arisawa [3] also developed a real-time prime generator and decreased the number of states of each cell to 22. Korec [14] reported a real-time prime generator having 11 states on the same model.

Here we study a real-time prime generator on $\mathrm{CA}_{1\text{-bit}}$. The sequence generation problem on $\mathrm{CA}_{1\text{-bit}}$ can be defined as follows: Let M be a $\mathrm{CA}_{1\text{-bit}}$, and $\{t_n \mid n = 1, 2, 3, \ldots\}$ be an infinite monotonically increasing positive integer sequence defined on natural numbers such that $t_n \geq n$ for any $n \geq 1$. We then have a semi-infinite array of cells, and all cells, except for C_1, are in the quiescent state at time $t = 0$. The communication cell C_1 assumes a special state r in Q and outputs 1 to its right communication link at time $t = 0$ for initiation of the sequence generator. We say that M generates a sequence $\{t_n \mid n = 1, 2, 3, \ldots\}$ in k *linear-time* if and only if the leftmost end cell of M falls into a special state in $F \subseteq Q$ and outputs 1 to its leftmost communication link at time $t = kt_n$, where k is a positive integer. We call M a *real-time generator* when $k = 1$.

In this section, we present a real-time prime generation algorithm on $\mathrm{CA}_{1\text{-bit}}$. The algorithm is implemented on a $\mathrm{CA}_{1\text{-bit}}$ using 34 internal states and 71 transition rules. Our real-time prime generation algorithm is based on the well-known sieve of Eratosthenes. Details can be found in Umeo and Kamikawa [37]. Figure 6.7 is a space-time diagram for the real-time prime generation algorithm. We have implemented the algorithm on a computer. Each cell has 34 internal states and 71 transition rules. The transition rule set is given in Table 6.4. We have tested the validity of the rule set from $t = 0$ to $t = 20000$ steps. In Fig. 6.8, we show a number of snapshots of the configuration from $t = 0$ to 40. The readers can see that the first 11 primes can be generated in real-time by the left end cell. Now we have:

Theorem 6 (Umeo and Kamikawa [37]) *Prime sequence can be generated by a* $\mathrm{CA}_{1\text{-bit}}$ *in real-time.*

Table 6.5 is a list of typical non-regular sequences generated by $\mathrm{CA}_{1\text{-bit}}$ in real-time.

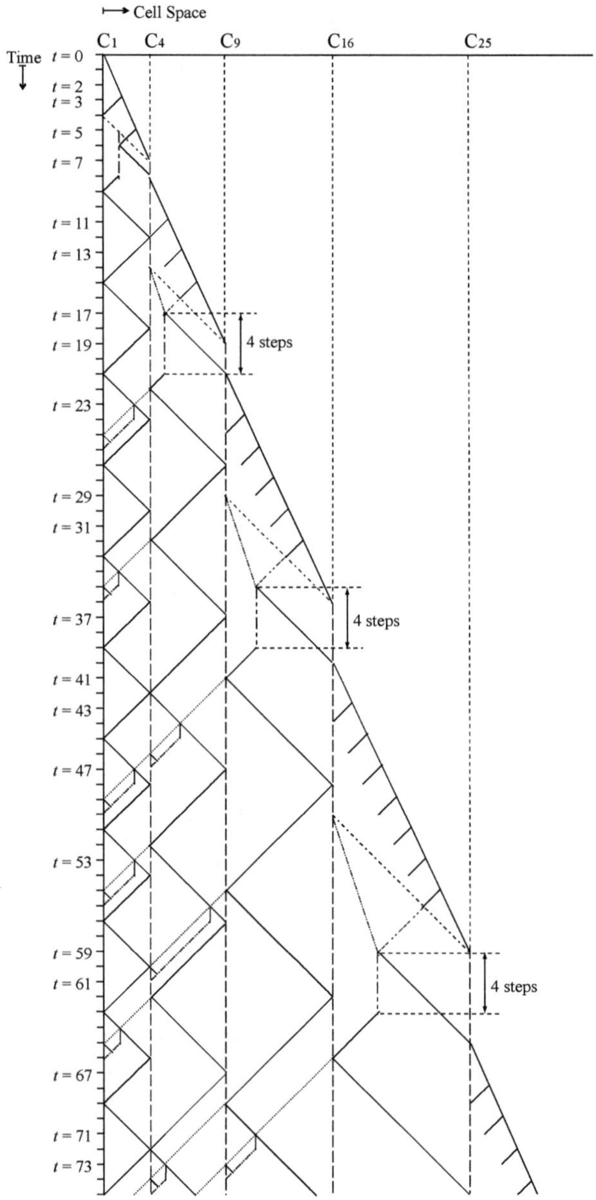

Fig. 6.7 Space-time diagram for real-time prime generation

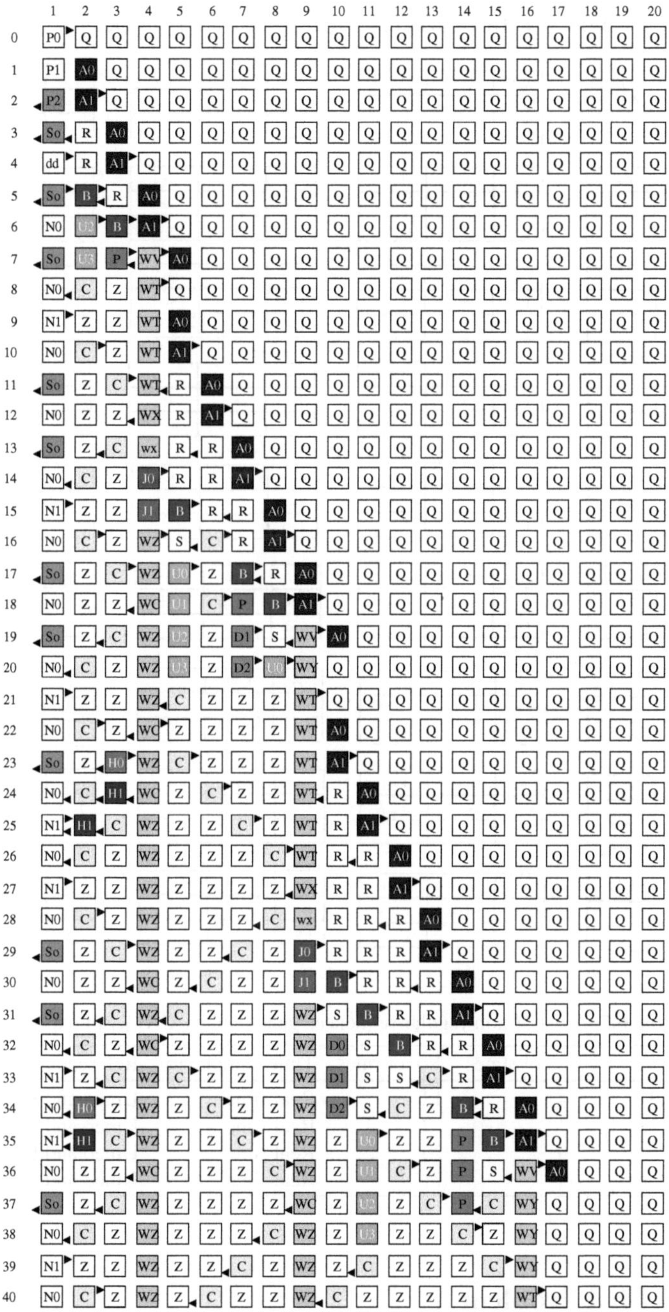

Fig. 6.8 A configuration of real-time generation of prime sequences on the $CA_{1\text{-bit}}$ with 34 states

Table 6.4 Transition rule set for real-time prime generator

Internal states : {Q, P0, P1, P2, R, S, Z, So, N0, N1, A0, A1, B, C, D0, D1, D2, U0, U1, U2, U3, WV, WY, WX, wx, WT, WZ, WC, dd, P, J0, J1, H0, H1}

Current state	Input from right link
Input from left link	(next state, left output, right output)

1

Q	R =0	R= 1
L =0	(Q,0,0)	--
L =1	(A0,0,0)	--

2

P0	R =0	R= 1
L =0	(P1,0,0)	--
L =1	--	--

3

P1	R =0	R= 1
L =0	(P2,1,0)	--
L =1	--	--

4

P2	R =0	R= 1
L =0	(So,1,0)	--
L =1	--	--

5

R	R =0	R= 1
L =0	(R,0,0)	(R,0,0)
L =1	(B,0,1)	(C,1,1)

6

S	R =0	R= 1
L =0	(S,0,0)	(C,1,0)
L =1	(D0,0,0)	(U0,0,1)

7

Z	R =0	R= 1
L =0	(Z,0,0)	(C,1,0)
L =1	(C,0,1)	(H0,1,1)

8

So	R =0	R= 1
L =0	(N0,0,0)	(dd,0,1)
L =1	--	--

9

N0	R =0	R= 1
L =0	(So,1,0)	(N1,0,1)
L =1	--	--

10

N1	R =0	R= 1
L =0	(N0,0,0)	(N0,0,0)
L =1	--	--

11

A0	R =0	R= 1
L =0	(A1,0,1)	--
L =1	(Q,0,0)	--

12

A1	R =0	R= 1
L =0	(R,1,0)	--
L =1	(WV,1,1)	--

13

B	R =0	R= 1
L =0	(S,0,0)	(P,0,0)
L =1	(P,0,1)	(U2,0,1)

14

C	R =0	R= 1
L =0	(Z,0,0)	(H1,1,0)
L =1	--	--

15

D0	R =0	R= 1
L =0	(D1,0,0)	--
L =1	--	--

16

D1	R =0	R= 1
L =0	(D2,0,1)	--
L =1	--	--

17

D2	R =0	R= 1
L =0	(Z,0,0)	--
L =1	--	--

18

U0	R =0	R= 1
L =0	(U1,0,0)	--
L =1	(Z,0,0)	--

19

U1	R =0	R= 1
L =0	(U2,0,0)	--
L =1	--	--

20

U2	R =0	R= 1
L =0	(U3,0,0)	--
L =1	--	--

21

U3	R =0	R= 1
L =0	(C,1,0)	--
L =1	--	--

22

WV	R =0	R= 1
L =0	(WY,0,0)	--
L =1	(WT,0,1)	--

23

WY	R =0	R= 1
L =0	(WY,0,0)	(WY,0,0)
L =1	(WT,0,1)	(WX,0,0)

24

WX	R =0	R= 1
L =0	(wx,0,0)	--
L =1	--	--

25

wx	R =0	R= 1
L =0	(J0,0,1)	--
L =1	--	--

26

WT	R =0	R= 1
L =0	(WT,0,0)	(WT,0,0)
L =1	(WX,1,0)	(WX,1,0)

27

WZ	R =0	R= 1
L =0	(WZ,0,0)	(WC,1,1)
L =1	(WC,1,0)	(WC,1,1)

28

WC	R =0	R= 1
L =0	(WZ,0,0)	(WZ,0,0)
L =1	--	--

29

dd	R =0	R= 1
L =0	(So,1,1)	--
L =1	--	--

30

P	R =0	R= 1
L =0	(P,0,0)	(Z,0,0)
L =1	(D1,0,1)	(C,0,1)

31

J0	R =0	R= 1
L =0	(J1,0,0)	--
L =1	--	--

32

J1	R =0	R= 1
L =0	(WZ,0,1)	--
L =1	--	--

33

H0	R =0	R= 1
L =0	(H1,1,0)	--
L =1	--	--

34

H1	R =0	R= 1
L =0	(Z,0,0)	(C,1,0)
L =1	(Z,0,0)	(C,1,0)

Table 6.5 A list of non-regular sequences generated by $CA_{1\text{-bit}}$ in real-time

Sequences	No. of states	No. of rules	Time complexity	References
$\{2^n \mid n = 1, 2, 3, \ldots\}$	4	12	Real-time	Umeo and Kamikawa [36]
$\{n^2 \mid n = 1, 2, 3, \ldots\}$	3	7	Real-time	Umeo and Kamikawa [36]
Fibonacci	9	26	Real-time	Umeo and Kamikawa [36]
Prime	34	71	Real-time	Umeo and Kamikawa [37]

6.5 Early Bird Problem

In this section, we study an early bird problem on $CA_{1\text{-bit}}$. Consider a one-dimensional $CA_{1\text{-bit}}$ consisting of n cells in which any cell initially in a quiescent state may be excited from outside world. The problem is to describe the automata (state set and next state function) so that the first excitation(s) can be distinguished from the later excitations. This problem was originally devised by Rosenstiehl et al. [27] to design some graph-theoretic algorithms operating on networks of finite state automata with O(1)-bit communication. Rosenstiehl et al. [27] presented a $2n$-step solution on a condition that at most one excitation occurs at each step. Vollmar [44] extended the problem allowing more than one cell to be excited at a given step. Legendi and Katona [16] gave a 5-state solution with multiple excitations operating in $3n+O(1)$ steps on a conventional CA of length n. Kleine-Büning [13] showed that the 5-state solution developed by Legendi and Katona [16] is the optimal solution with regard to the number of internal states of each cell on the O(1)-bit communication model.

Based on the 5-state solution given by Legendi and Katona [16], Umeo et al. [40] have given a 37-state implementation on $CA_{1\text{-bit}}$ of size n operating in $6n+O(1)$ steps. In our implementation, multiple birds are allowed to appear at only even steps. Two steps are required for the simulation of each one step of Legendi and Katona's solution. Thus the time complexity for the implemented algorithm is twice. An improvement in the time complexity seems to be difficult. Figure 6.9 shows some snapshots of the 37-state implementation. An appearance of the early bird is repre-

	1	2	3	4	5	6	7	8	9	10	11	12	13	14	15	16	17	18	19	20
0	QW	Q	B	Q	Q	Q	Q	Q	Q	Q	Q	B	Q	Q	Q	Q	B	Q	Q	QW
1	QW	q	B_	q	Q	Q	Q	Q	Q	Q	q	B_	q	Q	Q	q	B_	q	Q	QW
2	QW	La	B_	Ra	Q	Q	Q	Q	B	Q	La	B_	Ra	Q	Q	La	B_	Ra	Q	QW
3	QW_	Lb	B_	Rb	q	Q	Q	q	B_	q	Lb	B_	Rb	q	q	Lb	B_	Rb	q	QW
4	QW_	L3a	B_	R3a	Ra	B	Q	La	B_	q	L3a	B_	R3a	Ra	La	L3a	B_	R3a	Ra	QW
5	LRW	L3b	B_	R3b	Rb'	B10	q	Lb	B_	N	L3b	B_	R3b	Rb'	Lb'	L3b	B_	R3b	Rb	QW_
6	Nl	L3a	B_	R3a	Nr	R4a	q	L3a	B01	L2a	Nl	B_	R3a	Nr	Nl	L3a	B_	R3a	R2a	QW_
7	Nlb	L3b	B_	R3b	Nrb	R3b	N	L3b	B01	L2b	Nlb	B_	R3b	Nrb	Nlb	L3b	B_	R3b	R2b	LRW
8	L2a	Nl	B_	Nr	R4a	Nr	N	Nl	L4a	Nla	N	B10	Nr	R2a	L2a	Nl	B_	R3a	R2a	Nr
9	L2b	Nlb	B_	Nrb	R3b	Nrb	N	Nlb	L3b	Nlb	N	B10	Nrb	R2b	L2b	Nlb	B_	R3b	R2b	Nrb
10	Nla	N	B11	Nr2	Nr	R2a	N	L2a	Nl	Nl2	N	B11	N	Nra	Nla	N	B10	R3a	Nr	R2a
11	Nlb	N	B11	Nr2	Nrb	R2b	N	L2b	Nlb	Nl2	N	B11	N	Nrb	Nlb	N	B10	R3b	Nrb	R2b
12	N	N	B11	N	N	Nra	N	Nla	N	N	N	B11	N	N	N	N	B10	Nr	R2a	Nra
13	N	N	B11	N	N	Nrb	N	Nlb	N	N	N	B11	N	N	N	N	B10	Nrb	R2b	Nrb
14	N	N	B11	N	N	N	N	N	N	N	N	B11	N	N	N	N	B11	N	Nra	R2a
15	N	N	B11	N	N	N	N	N	N	N	N	B11	N	N	N	N	B11	N	Nrb	R2b
16	N	N	B11	N	N	N	N	N	N	N	N	B11	N	N	N	N	B11	N	N	Nra
17	N	N	B11	N	N	N	N	N	N	N	N	B11	N	N	N	N	B11	N	N	Nrb
18	N	N	B11	N	N	N	N	N	N	N	N	B11	N	N	N	N	B11	N	N	N

Fig. 6.9 Snapshots of a 37-state implementation of the early bird problem on $CA_{1\text{-bit}}$

sented by an internal state "B" with output signal "1" to both neighbors. In Fig. 6.9, three birds appear on C_3, C_{12} and C_{17} at time $t = 0$ and one bird appear on C_9 at time $t = 2$ and on C_6 at time $t = 4$, respectively. The first three birds can survive and the last two birds will be *killed*. The transition rule set is given in Table 6.6.

Theorem 7 (Umeo, Michisaka, Kamikawa, and Kanazawa [40]) *There exists a 37-state $CA_{1\text{-bit}}$ that can solve the early bird problem in $6n+O(1)$ steps under an assumption such that multiple excitations are allowed to appear at only even steps.*

Table 6.6 Transition rule set for early bird generator

Internal state : {Q, q, QW, QW, B, B_, BW, BW, B01, B10, B11, Ra, Rb, R2a, R2b, R3a, R3b, R4a, La, Lb, L2a, L2b, L3a, L3b, L4a, LRW, Nr, Nra, Nrb, Nr2, Nl, Nla, Nlb, Nl2, N, Lb', Rb'}

Current state	Input from right link
Input from left link	(next state, left output, right output)

1

Q	R=0	R=1
L= 0	(Q,0,0)	(q,0,0)
L= 1	(q,0,0)	(q,0,0)

2

q	R=0	R=1
L= 0	(N,1,1)	(La,1,0)
L= 1	(Ra,0,1)	(q,0,0)

3

QW	R=0	R=1
L= 0	(QW,0,0)	(QW_,0,0)
L= 1	(QW_,0,0)	--

4

QW_	R=0	R=1
L= 0	(LRW,0,0)	(QW_,0,0)
L= 1	(QW_,0,0)	--

5

B	R=0	R=1
L= 0	(B_,1,1)	(B01,1,0)
L= 1	(B10,0,1)	(B11,0,0)

6

B_	R=0	R=1
L= 0	(B_,0,0)	(B01,0,0)
L= 1	(B10,0,0)	(B11,0,0)

7

BW	R=0	R=1
L= 0	(BW_,1,1)	(B11,0,0)
L= 1	(B11,0,0)	--

8

BW_	R=0	R=1
L= 0	(BW_,0,0)	(B11,0,0)
L= 1	(B11,0,0)	--

9

B01	R=0	R=1
L= 0	(B01,0,0)	(L4a,0,0)
L= 1	(B11,0,0)	(L2a,0,0)

10

B10	R=0	R=1
L= 0	(B10,0,0)	(B11,0,0)
L= 1	(R4a,0,0)	(R2a,0,0)

11

B11	R=0	R=1
L= 0	(B11,0,0)	(L2a,0,0)
L= 1	(R2a,0,0)	(N,0,0)

12

Ra	R=0	R=1
L= 0	(Rb,0,1)	(Rb',0,1)
L= 1	--	--

13

Rb	R=0	R=1
L= 0	(R3a,0,0)	(Nr,0,0)
L= 1	(R2a,0,0)	(Nr,0,0)

14

R2a	R=0	R=1
L= 0	(R2b,0,1)	--
L= 1	--	--

15

R2b	R=0	R=1
L= 0	(Nra,0,0)	(Nra,0,0)
L= 1	(R2a,0,0)	(Nr,0,0)

16

R3a	R=0	R=1
L= 0	(R3b,0,1)	--
L= 1	--	--

17

R3b	R=0	R=1
L= 0	(R3a,0,0)	(Nr,0,0)
L= 1	(R2a,0,0)	(Nr,0,0)

18

R4a	R=0	R=1
L= 0	(R3b,1,1)	--
L= 1	--	--

19

La	R=0	R=1
L= 0	(Lb,1,0)	--
L= 1	(Lb',1,0)	--

20

Lb	R=0	R=1
L= 0	(L3a,0,0)	(L2a,0,0)
L= 1	(Nl,0,0)	(Nl,0,0)

21

L2a	R=0	R=1
L= 0	(L2b,1,0)	--
L= 1	--	--

22

L2b	R=0	R=1
L= 0	(Nla,0,0)	(L2a,0,0)
L= 1	(Nla,0,0)	(Nl,0,0)

23

L3a	R=0	R=1
L= 0	(L3b,1,0)	--
L= 1	--	--

24

L3b	R=0	R=1
L= 0	(L3a,0,0)	(L3a,0,0)
L= 1	(Nl,0,0)	(Nl,0,0)

25

L4a	R=0	R=1
L=0	(L3b,1,1)	--
L=1	--	--

26

LRW	R=0	R=1
L=0	(q,0,0)	(Nl,0,0)
L=1	(Nr,0,0)	--

27

Nr	R=0	R=1
L=0	(Nrb,1,0)	--
L=1	--	--

28

Nra	R=0	R=1
L=0	(Nrb,0,0)	--
L=1	--	--

29

Nrb	R=0	R=1
L=0	(N,0,0)	(Nr2,0,0)
L=1	(R2a,0,0)	(R4a,0,0)

30

Nr2	R=0	R=1
L=0	(Nr2,0,0)	(N,0,0)
L=1	(R4a,0,0)	(R2a,0,0)

31

Nl	R=0	R=1
L=0	(Nlb,0,1)	--
L=1	--	--

32

Nla	R=0	R=1
L=0	(Nlb,0,0)	--
L=1	--	--

33

Nlb	R=0	R=1
L=0	(N,0,0)	(L2a,0,0)
L=1	(Nl2,0,0)	(L4a,0,0)

34

Nl2	R=0	R=1
L=0	(Nl2,0,0)	(L4a,0,0)
L=1	(N,0,0)	(L2a,0,0)

35

N	R=0	R=1
L=0	(N,0,0)	(L2a,0,0)
L=1	(R2a,0,0)	(N,0,0)

36

Lb'	R=0	R=1
L=0	(Nl,0,0)	(Nl,0,0)
L=1	(Nl,0,0)	(Nl,0,0)

37

Rb'	R=0	R=1
L=0	(Nr,0,0)	(Nr,0,0)
L=1	(Nr,0,0)	(Nr,0,0)

6.6 Firing Squad Synchronization Problem on 2-D $CA_{1\text{-bit}}$

Here we consider the FSSP again on two-dimensional arrays. The FSSP on 2-D arrays for O(1)-bit communication model is studied in Umeo et al. [38]. Figure 6.10 shows a finite two-dimensional (2-D) cellular array consisting of $m \times n$ cells. A cell on (i, j) is denoted by $C_{i,j}$. Each cell is an identical (except the border cells) finite state automaton. The array operates in lock-step mode in such a way that the next state of each cell (except border cells) is determined by both its own present state and the present binary inputs from its north, south, east and west neighbors. The cell also outputs four binary values to its north, west, south and east neighbors, depending on both its own present state and the present binary inputs from its north, south, east and west neighbors. Thus we assume a von Neumann-like neighborhood with the 1-bit communication. All cells except for the general cell are initially in the quiescent state and have a property such that the next state of a quiescent cell with four 0 inputs is the quiescent state and outputs 0 to its four neighbors.

The FSSP on 2-D $CA_{1\text{-bit}}$ is defined as follows: Given an array of $m \times n$ identical cells, including a *General* on $C_{1,1}$ cell that is activated at time $t = 0$, we want to describe (state set and next-state function) the automata such that, *at some future time*, all of the cells will *simultaneously* and *for the first time* enter a special *firing* state. The set of states and transition rules must be independent of m and n. The difficult part of this problem is that the same types of cells with a fixed number of states must be synchronized, regardless of the size m and n of the array. The firing squad synchronization problem on 2-D 1-bit communication cellular automata has been studied by Torre et al. [31], Gruska et al. [11], and Umeo et al. [40]. This section presents two 1-bit implementations for square and rectangular arrays.

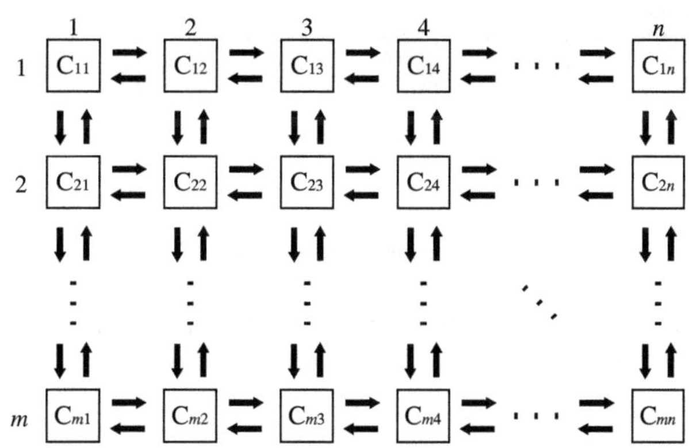

Fig. 6.10 Two-dimensional cellular automaton

6.6.1 Synchronization Algorithm on Square Arrays

The first one is for square arrays given in Umeo et al. [40]. It runs in $(2n - 1)$ steps on $n \times n$ square arrays. The proposed implementation is one step slower than

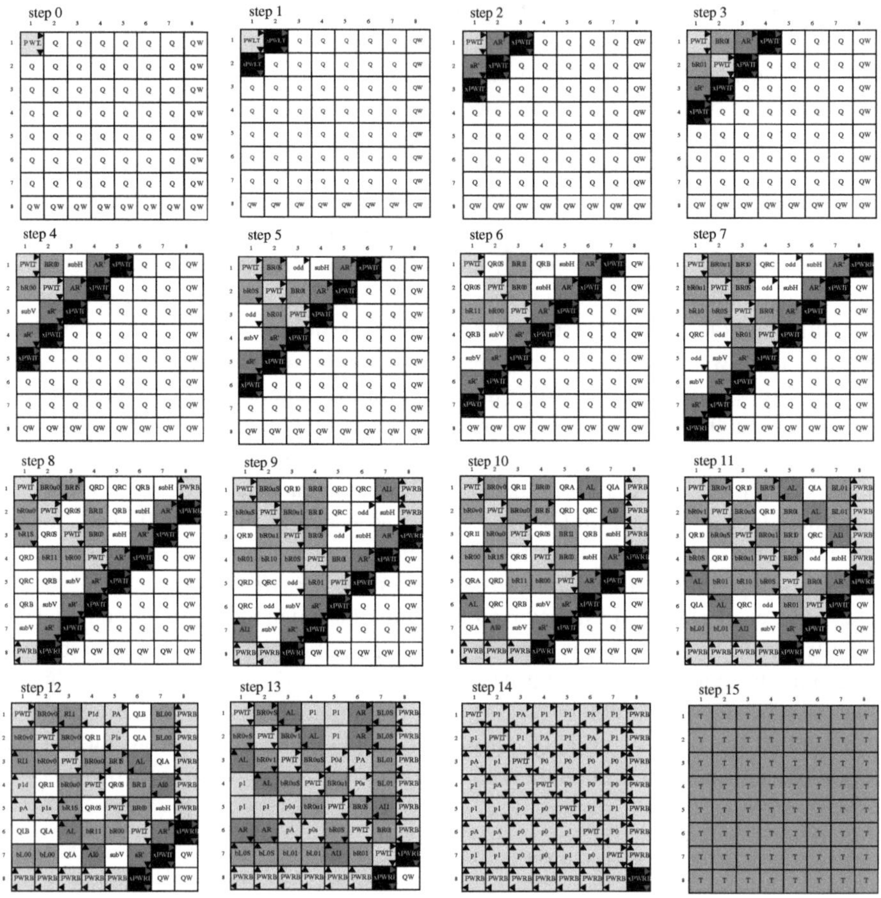

Fig. 6.11 Snapshots of the $(2n - 1)$-step square synchronization algorithm with the general on the northwest corner

optimum-time for the O(1)-bit communication model. The total numbers of internal states and transition rules of the $CA_{1\text{-bit}}$ are 127 and 405, respectively. Figure 6.11 shows snapshots of configurations of the 127-state implementation running on a square of size 8×8. Gruska, Torre, and Parente [11] presented an optimum-time algorithm.

Theorem 8 (Gruska, Torre, and Parente [11]) *There exists a 2-D* $CA_{1\text{-bit}}$ *that can synchronize any* $n \times n$ *square arrays in* $2n - 2$ *steps.*

6.6.2 Synchronization Algorithm on Rectangle Arrays

The generalized firing squad synchronization algorithm for 1-D arrays presented in Sect. 6.3.2 can be applied to the problem of synchronizing rectangular arrays with the general at the northwest corner. The rectangular array is regarded as $\min(m, n)$ L-shaped 1-D arrays that are synchronized independently using the generalized firing squad synchronization algorithm. Configurations of the generalized synchronization on 1-D $CA_{1\text{-bit}}$ can be embedded on 2-D array. The original embedding scheme for O(1)-bit communication model was presented in Beyer [5] and Shinahr [29] in order to synchronize any $m \times n$ arrays in optimum $m + n + max(m, n) - 3$ steps. Umeo et al. [40] have implemented the rectangular synchronization algorithm for 2-D $CA_{1\text{-bit}}$. The total numbers of internal states and transition rules of the $CA_{1\text{-bit}}$ are 862 and 2217, respectively. Figure 6.12 shows snapshots of the synchronization process on a 5×8 rectangular array. Thus we have:

Theorem 9 (Umeo, Michisaka, Kamikawa, and Kanazawa [40]) *There exists a 2-D $CA_{1\text{-bit}}$ that can synchronize any $m \times n$ rectangular arrays in $m + n + max(m, n)$ steps.*

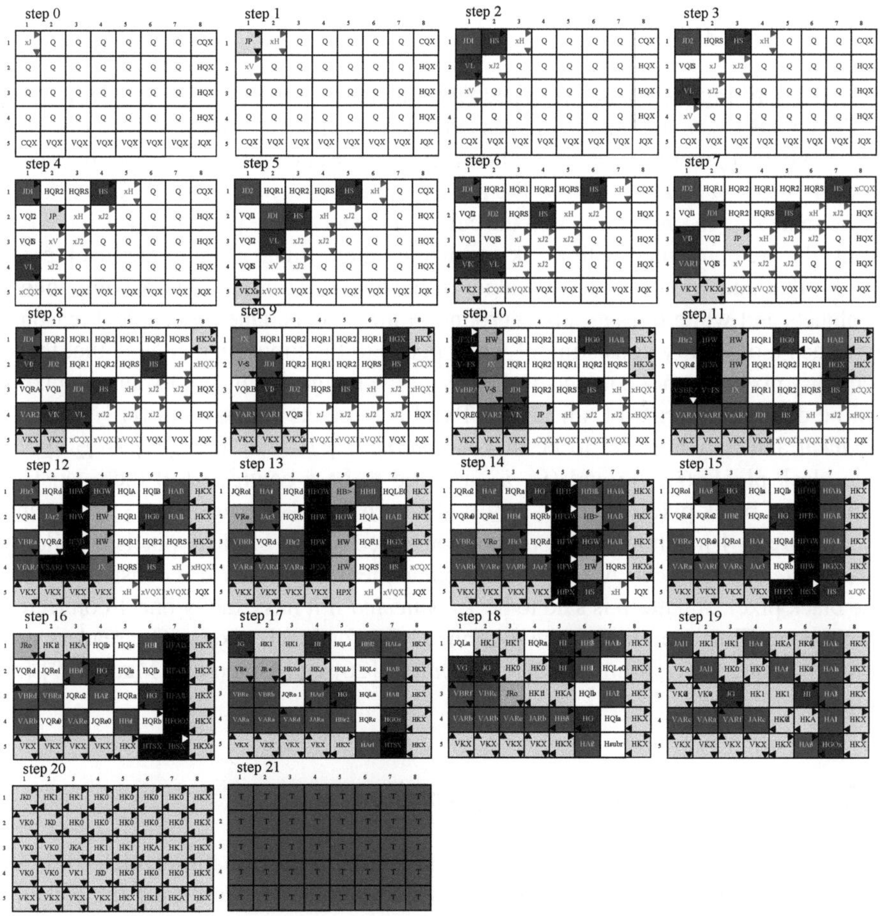

Fig. 6.12 Snapshots of the proposed rectangular firing squad synchronization algorithm with the general at the northwest corner

6.7 Connectivity Recognition Problem

Recognizing and labeling connected regions of images are important problems in image processing and machine vision, and many parallel algorithms for them have been developed on a rich variety of parallel architectures. See Alnuweiri and Prasanna [1, 2], Cypher et al. [6], Cypher and Sanz [7], Leighton [15], Manohar and Ramapriyan [18], and Miller and Stout [22]. In this section, we consider a connectivity recognition problem on a 2-D $CA_{1\text{-bit}}$. The connectivity recognition problem of binary images on cellular automata has been investigated by Beyer [5] and Levialdi [17]. We present a linear-time connectivity recognition algorithm for two-dimensional binary images. Precisely, it is shown that a set of two-dimensional connected binary images of size $m \times n$ can be recognized in $2(m+n)+O(1)$ steps by a 2-D $CA_{1\text{-bit}}$.

6.7.1 Connectivity

Before describing the connectivity recognition algorithm, we need some definitions of the connectivity for binary images. We assume that the given image is of size $m \times n$ where a pixel (i, j) denotes the pixel in row i and column j of the image for every $1 \leq i \leq m$ and $1 \leq j \leq n$. We put an input of size $m \times n$ on the 2-D $CA_{1\text{-bit}}$ of the same size in such a way that the cell (i, j) receives the pixel (i, j) as its initial input. We are concerned with black and white binary images where the black pixel has 1-value and white one has 0-value, respectively. We regard black components as objects and white ones as a background of the objects. Due to technical reasons, we attach a boundary consisting of white pixels to the input image. Note that those boundary pixels are not counted as the size of the image. Connectivity among pixels can be defined in terms of adjacency. Two black pixels (i_1, j_1) and (i_2, j_2) are *4-adjacent* and they are said to be in *4-neighbor*, if $\mid i_1 - i_2 \mid + \mid j_1 - j_2 \mid \leq 1$. Two black pixels (i_1, j_1) and (i_k, j_k) are said to be *4-connected*, if there exists a sequence of black pixels (i_p, j_p), $2 \leq p \leq k$ such that each pair of (i_{p-1}, j_{p-1}) and (i_p, j_p) are in 4-neighbor. A maximum connected region of black pixels is called a *4-connected component*. A 4-connected component is *isolated* if it consists of only one black pixel. A pattern is said to be *4-connected* if it has exactly one 4-connected component. Thus we employ the 4-connectivity for black pixels. The readers can define 8-connectivity, similarly. See Rosenfeld [26] and Umeo and Mauri [39] for details.

6.7.2 Parallel Shrinking Transformation

Beyer [5] proposed an interesting parallel shrinking transformation which trims all 4-connected components of binary images simultaneously, preserving the connectivity of binary images. The transformation was implemented on a conventional O(1)-bit communication model of cellular automaton. The recognition algorithm we develop here is based on the parallel shrinking algorithm proposed by Beyer [5]. We first review the Beyer's algorithm. The algorithm is based on a connectivity-preserving operation which trims all connected components simultaneously in the diagonal (from south-east to north-west) direction of each component. Figure 6.13 shows the Beyer's connectivity-preserving operation consisting of two rules R_1 for black pixels and R_2 for white pixels. If a black pixel has a white pixel in its south and east neighbors, then the rule R_1 is applied to the black pixel and the pixel becomes white at the next step. If a white pixel has three black pixels in its south, east, and south-east (diagonal) neighbors, respectively, then the rule R_2 is applied to the white pixel and it becomes black at the next step. The symbols x and y denote any value in {white, black}. When we apply the above operations repeatedly to all pixels of an image simultaneously, we observe that each connected component of the image is reduced to one isolated black pixel and then vanishes after one application of the rule R_1. What is important is that, all the while, every distinct connected component remains distinct and either vanishes at each different position or in the same position at different time. This is the reason why the Beyer's shrinking rule is called connectivity preserving operation.

Precisely, the above statement is described as follows: Let c be a non-isolated connected component and $n(c)$, $w(c)$, and $se(c)$ be positive integers defined as follows:

$$
\begin{aligned}
n(c) &= \min\{i \mid \text{cell } C_{i,k} \text{ is in } c \text{ for some } k\},\\
w(c) &= \min\{k \mid \text{cell } C_{i,k} \text{ is in } c \text{ for some } i\},\\
se(c) &= \max\{i+k \mid \text{cell } C_{i,k} \text{ is in } c\}.
\end{aligned}
$$

The connected component c is within the triangle consisting of the row $n(c)$, column $w(c)$, and the 45° diagonal line containing the farthest cells from the cell $C_{n(c),w(c)}$, shown in Fig. 6.14. Let ψ denote the Beyer's transformation and c be an

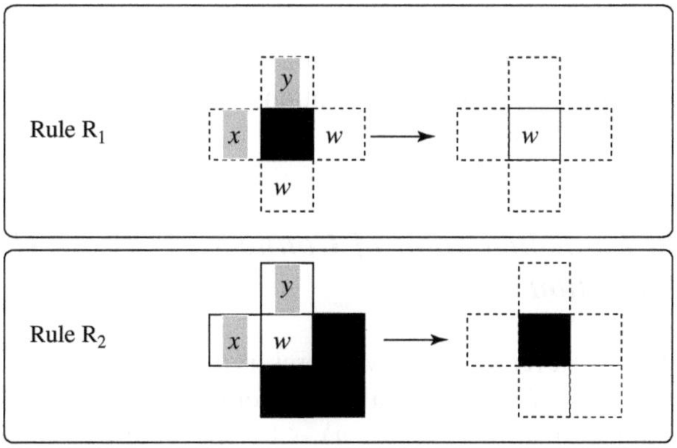

Fig. 6.13 Beyer's 4-connectivity preserving operation

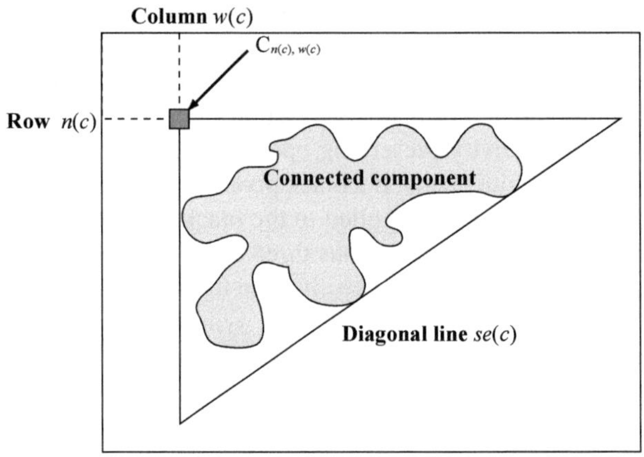

Fig. 6.14 Beyer's parallel shrinking transformation

image. A function $\psi^{t+1}(c)$ is defined as follows:

$$\psi^0(c) = c,$$
$$\psi^{t+1}(c) = \psi(\psi^t(c)), t \geq 0.$$

Then, the following lemmas are given in Beyer [5].

Lemma 10 (Beyer [5]) *For any non-isolated component c, we have* $n(\psi(c)) = N(c)$, $w(\psi(c)) = w(c)$, *and* $se(\psi(c)) = se(c) - 1$.

Lemma 11 (Beyer [5]) *For any non-isolated component c, let k be an integer such that* $k = se(c) - n(c) - w(c)$. *Then,* $\psi^k(c)$ *is an isolated component located at* $C_{n(c),w(c)}$.

Lemma 10 assures the exact shrinking to the north-west corner of the connected component. From Lemma 11, we can know the number of applications of the ψ operation necessary to shrink the original connected component to an isolated black pixel. It is shown that, for any image of size $m \times n$, all of the connected components will vanish within $(m+n-1)$-time applications of ψ. In Fig. 6.15, we show several snapshots obtained after consecutive applications of ψ to a binary image. Note that T means the application times.

6.7.3 One-Bit Implementation of Connectivity-Preserving Transformation

Here we show that the Beyer's connectivity-preserving transformation can be implemented on 2-D $CA_{1\text{-bit}}$. We construct a two-dimensional $CA_{1\text{-bit}}$ M that can simulate the Beyer's transformation in $2(m+n)-1$ steps for any given binary image x of size $m \times n$. Each cell has two auxiliary registers X and Y. The register X holds a binary pixel value during the transformation and Y acts as a temporary register for storing a pixel value in the south neighbor cell. Any operation of each cell at step

Fig. 6.15 Beyer's connectivity-preserving transformation on O(1)-bit communication model

t (≥ 1) is classified into two categories according to the parity of global clock step t such that $t \equiv 1 \pmod 2$ or $t \equiv 0 \pmod 2$. We refer to the former operation as A-phase operation and the latter B-phase operation, respectively. Each cell repeats an A- and B-phase operation alternatively, that is, it repeats two operations, one in A-phase followed by the other in B-phase.

The 1-bit implementation is as follows: At time $t = 1$, each register X in $C_{i,j}$ holds an initial pixel value (i, j) of x and the register Y has been set empty. At the beginning of the A-phase, each cell outputs a 0 or 1 signal to its north and west output communication links depending on the pixel value 0 or 1 in the X register, and the signals are received at that step by its north and west neighbor cells through their input communication links. In the shrinking transformation shown in Fig. 6.13, one step is sufficient for the execution of the Rule R_1, however, it takes two steps for the execution of the Rule R_2. To get the pixel value in the east and south neighbor cells, each cell uses its east and south input communication links in the A-phase operation. In order to get south-east (diagonal) pixel value, which is necessary for the execution of the rule R_2, each cell uses its east input communication link in the B-phase operation. For this purpose, the diagonal pixel value has been stored in the Y register in its east neighbor cell in the latest A-phase. Thus, with those two steps of the A- and B-phases, each cell can get all pixel values that are necessary to perform one application of the transformation. At odd step t of M where $t = 2k + 1$, for any k such that $0 \le k \le m + n - 1$, we can see the values of $\psi^0(x), \psi^1(x), \ldots, \psi^k(x), \ldots, \psi^{m+n-1}(x)$ in the X registers of each cell on the array. It is observed that, in the construction above, both east-to-west horizontal

Table 6.7 Transition rule set for 1-bit shrinking transformation

Internal State : {B, W, B/B, R, BR, W/B-B}

Current State	Input from Right and Left Link
Input from Upper and Lower Link	(Next State, Left Output, Right Output, Upper Output, Lower Output)

1

B	R=0,L=0	R=1,L=0	R=0,L=1	R=1,L=1
U=0,D=0	(R,1,0,1,0)	(BR,1,1,0,0)	--	--
U=1,D=0	(R,1,0,1,0)	(BR,1,1,0,0)	--	--
U=0,D=1	(B/B,0,1,1,1)	(B/B,0,1,1,1)	--	--
U=1,D=1	(B/B,0,1,1,1)	(B/B,0,1,0,1)	--	--

2

W	R=0,L=0	R=1,L=0	R=0,L=1	R=1,L=1
U=0,D=0	(W,0,0,0,0)	(W,0,0,0,0)	(W,0,0,0,0)	(W,0,0,0,0)
U=1,D=0	(W,0,0,0,0)	- -	--	--
U=0,D=1	(W,0,0,0,0)	(W/B-B,0,0,0,0)	(W,0,0,0,0)	--
U=1,D=1	--	(W/B-B,0,0,0,0)	--	--

3

B/B	R=0,L=0	R=1,L=0	R=0,L=1	R=1,L=1
U=0,D=0	(B,1,0,1,0)	--	(B,1,0,1,0)	--
U=1,D=0	(B,1,0,1,0)	--	(B,1,0,1,0)	--
U=0,D=1	(B,1,0,1,0)	(B,1,0,1,0)	(B,1,0,1,0)	(B,1,0,1,0)
U=1,D=1	(B,1,0,1,0)	(B,1,0,1,0)	(B,1,0,1,0)	(B,1,0,1,0)

4

R	R=0,L=0	R=1,L=0	R=0,L=1	R=1,L=1
U=0,D=0	(W,0,0,0,0)	--	(W,0,0,0,0)	--
U=1,D=0	(W,0,0,0,0)	--	(W,0,0,0,0)	--
U=0,D=1	--	--	--	--
U=1,D=1	--	--	--	--

5

BR	R=0,L=0	R=1,L=0	R=0,L=1	R=1,L=1
U=0,D=0	(B,1,0,1,0)	(B,1,0,1,0)	(B,1,0,1,0)	(B,1,0,1,0)
U=1,D=0	(B,1,0,1,0)	(B,1,0,1,0)	(B,1,0,1,0)	(B,1,0,1,0)
U=0,D=1	--	--	--	--
U=1,D=1	--	--	--	--

6

W/B-B	R=0,L=0	R=1,L=0	R=0,L=1	R=1,L=1
U=0,D=0	(B,1,0,1,0)	--	(B,1,0,1,0)	--
U=1,D=0	--	--	--	--
U=0,D=1	(B,1,0,1,0)	(W,0,0,0,0)	(B,1,0,1,0)	(W,0,0,0,0)
U=1,D=1	--	--	--	--

and south-to-north vertical one-way communication links are utilized in the A-phase operation, however, in the B-phase operation, only an east-to-west horizontal link is used in each cell. Thus we have:

Theorem 12 (Umeo [32]) *For any binary image of size $m \times n$, the Beyer's connectivity-preserving transformation can be performed on a 2-D $CA_{1\text{-bit}}$ in $2(m+n)-1$ steps.*

We have implemented the 1-bit shrinking transformation algorithm on a 2-D $CA_{1\text{-bit}}$ with 6 internal states. Table 6.7 gives the transition rule set for the shrinking operation and Fig. 6.16 shows some snapshots of the shrinking process on a binary image of size 14×14.

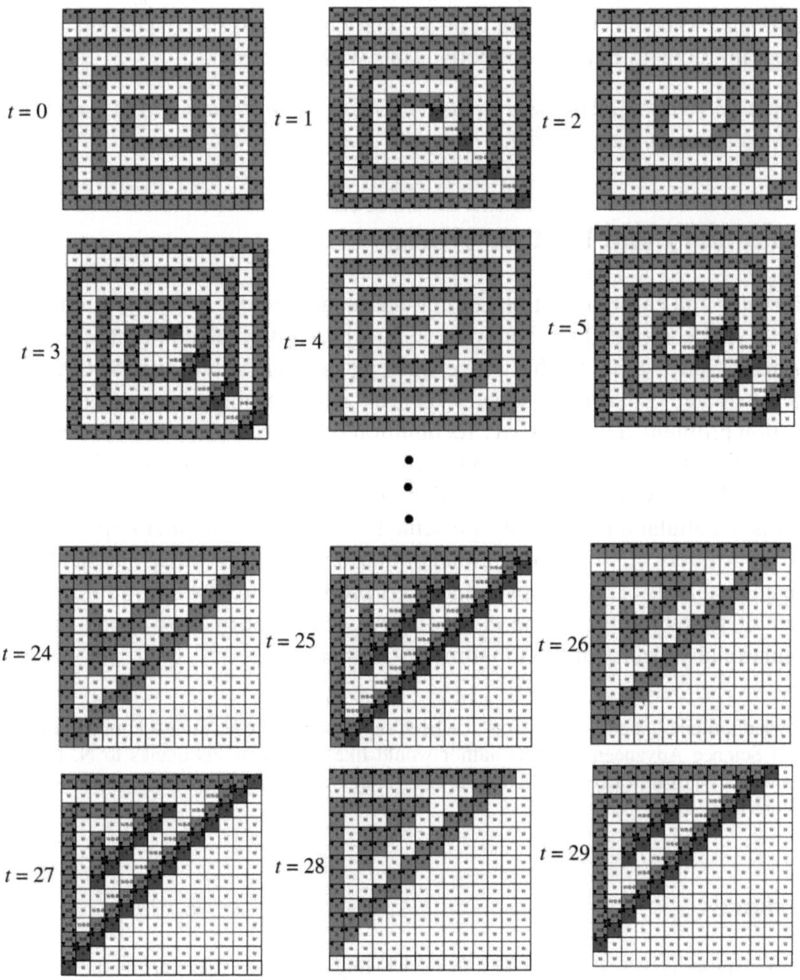

Fig. 6.16 Snapshots for connectivity-preserving shrinking transformation on 1-bit communication model

It is shown that the connectivity of any binary images can be also detected in linear-time by a 2-D $CA_{1\text{-bit}}$. The algorithm is based on our previous 1-bit implementation of the shrinking transformation. The detection of the connectivity of binary images can be done by counting up the number of vanished isolated black pixels by the accept cell located in the north-west corner of the array. The array accepts the input if and only if the count is exactly one. Every cell works not only for the transformation of images into isolated black pixels but also for the transmission of the vanished isolated black pixels toward the accept cell. Both of the operations can be simultaneously implemented on a $CA_{1\text{-bit}}$. See Umeo [32] for details.

Thus it has been shown that the $CA_{1\text{-bit}}$ can recognize the connectivity of any binary images of size $m \times n$ in $2(m+n) + O(1)$ steps. We have implemented our algorithm on a computer program, which simulates a $CA_{1\text{-bit}}$ with 61 states, recognizing 2-D connectivity. For typical binary images of size from 4×4 to 45×47, the program recognizes them correctly. Thus we have:

Theorem 13 (Umeo [32]) *There exists a 2-D $CA_{1\text{-bit}}$ that can recognize a set of 2-D 4-connected binary images of size $m \times n$ in $2(m+n) + O(1)$ steps.*

6.8 Summary and Further Works

A 1-bit inter-cell communication cellular automaton model ($CA_{1\text{-bit}}$) studied in this paper is a subclass of cellular automata (CA) whose inter-cell communication at one step is restricted to 1-bit. We have investigated a problem solving on the $CA_{1\text{-bit}}$. The problems treated are a firing squad synchronization problem, an integer sequence generation problem, a connectivity recognition problem for two-dimensional binary images, an early bird problem, and a connectivity recognition problem for two-dimensional binary images, all of which are known as the classical, fundamental problems in cellular automata. We presented several state-efficient implementations on the 1-bit inter-cell communication cellular automata for those classical cellular automata problems. Those implementations presented are not optimum ones in the number of states required. The class of $CA_{1\text{-bit}}$ is confirmed to be an interesting computational subclass of CAs that merits further study.

Acknowledgements A part of this work has been supported by the Kayamori Foundation of Informational Science Advancement. The author would like to express his thanks to N. Kamikawa, T. Yanagihara, M. Kanazawa, K. Michisaka, and T. Fujiwara who helped to develop several implementations on $CA_{1\text{-bit}}$.

References

1. H.M. Alnuweiri, V.K. Prasanna, Fast image labeling using local operators on mesh-connected computers. IEEE Trans. PAMI. **13**(2), 202–207 (1991)
2. H.M. Alnuweiri, V.K. Prasanna, Parallel architectures and algorithms for image component labeling. IEEE Trans. PAMI. **14**(10), 1014–1034 (1992)

3. M. Arisawa, On the generation of integer series by the one-dimensional iterative arrays of finite state machines (in Japanese). The Trans. IECE. 71/8 **54-C**(8), 759–766 (1971)
4. R. Balzer, An 8-state minimal time solution to the firing squad synchronization problem. Inf. Control, **10**, 22–42 (1967)
5. W.T. Beyer, *Recognition of Topological Invariants by Iterative Arrays*. Ph.D. Thesis, (MIT, Massachusetts, 1969)
6. R.E. Cypher, J.L.C. Sanz, L. Snyder, Algorithms for image component labeling on SIMD mesh-connected computers. IEEE Trans. Comput. **39**(2), 276–281 (1990)
7. R.E. Cypher, J.L.C. Sanz, *The SIMD Models of Parallel Computation*. (Springer-Verlag, Heidelberg, 1994)
8. P.C. Fischer, Generation of primes by a one-dimensional real-time iterative array. J. ACM. **12**(3), 388–394 (1965)
9. H.D. Gerken, Über Synchronisations – Probleme bei Zellularautomaten. *Diplomarbeit* (Institut für Theoretische Informatik, Technische Universität Braunschweig, Braunschweig, 1987)
10. E. Goto, A minimal time solution of the firing squad problem. Dittoed course notes for Applied Mathematics 298, Harvard University (1982)
11. J. Gruska, S.L. Torre, M. Parente, The firing squad synchronization problem on squares, toruses and rings. Intern. J. Found. Comput. Sci. **18**(3), 637–654 (2007)
12. N. Kamikawa, H. Umeo, A generalized FSSP algorithm on one-bit communication cellular automata. (draft version) (2008)
13. H. Kleine-Büning, The early bird problem is unsolvable in a one-dimensional cellular space with 4 states. Acta Cybernetica, **6**, 23–31 (1983)
14. I. Korec, Real-time generation of primes by a one-dimensional cellular automaton with 11 states. Proc. 22nd Int. Symp. MFCS '97. **LNCS 1295**, 358–367 (1997)
15. F.T. Leighton, Introduction to parallel algorithms and architectures: arrays, trees, hypercubes. (Morgan Kaufmann, San Fransisco, CA, 1992)
16. T. Legendi, E. Katona, A 5-state solution of the early bird problem in a one-dimensional cellular space. Acta Cybernetica. **5**(2), 173–179 (1981)
17. S. Levialdi, On shrinking binary picture patterns. Commun. ACM. **15**(1), 7–10 (1972)
18. M. Manohar, H.K. Ramapriyan, Connected component labeling of binary images on a mesh connected massively parallel processor. Comput. Visi. Graph. Image Process. **45**, 133–149 (1989)
19. J. Mazoyer, A six-state minimal time solution to the firing squad synchronization problem. Theore. Comput. Sci. **50**, 183–238 (1987)
20. J. Mazoyer, V. Terrier, Signals in one-dimensional cellular automata. Theor. Comput. Sci., **217**, 53–80 (1999)
21. J. Mazoyer, On optimal solutions to the firing squad synchronization problem. *Theor. Comput. Sci.* **168**, 367–404 (1996)
22. R. Miller, Q.F. Stout, Parallel algorithms for regular architectures: meshes and pyramids. (The MIT Press, Massachusetts, 1996)
23. E.F. Moore, The firing squad synchronization problem, ed. by E.F. Moore, *Sequential Machines, Selected Papers* (Addison-Wesley, Reading MA, 1964)
24. F.R. Moore, G.G. Langdon, A generalized firing squad problem. Information and Control. **12**, 212–220 (1968)
25. J. Nishimura, T. Sogabe, H. Umeo, A design of optimum-time firing squad synchronization algorithm on 1-bit cellular automaton. *Proceedings of The 8th International Symposium on Artificial Life and Robotics*, 381–386 (2003)
26. A. Rosenfeld, Connectivity in digital pictures. J. ACM. **17**(1), 146–160 (1970)
27. P. Rosenstiehl, J.R. Fiksel, A. Holliger, Intelligent graphs: Networks of finite automata capable of solving graph problems, ed. by R.C. Reed. *Graph Theory and Computing* (Academic, New York, NY, 1973)
28. A. Settle, J. Simon, Smaller solutions for the firing squad. Theoretical Computer Science. **276**, 83–109 (2002)

29. I. Shinahr, Two- and three-dimensional firing squad synchronization problems. Inf. Control. **24**, 163–180 (1974)
30. H. Szwerinski, Time-optimum solution of the firing-squad-synchronization-problem for *n*-dimensional rectangles with the general at an arbitrary position. Theor. Comput. Sci. **19**, 305–320 (1982)
31. S.L. Torre, M. Napoli, M. Parente, Firing squad synchronization problem on bidimensional cellular automata with communication constraints. Proc. MCU 2001. LNCS **2055**, 264–275 (2001)
32. H. Umeo, Linear-time recognition of connectivity of binary images on 1-bit inter-cell communication cellular automaton. Parallel Comput. **27**, 587–599 (2001)
33. H. Umeo, Firing squad synchronization problem in cellular automata. ed. by R.A. Meyers, *Encyclopedia of Complexity and Systems Science* (Springer, Heidelberg, 2009)
34. H. Umeo, M. Hisaoka, K. Michisaka, N. Nishioka, M. Maeda, Some new generalized synchronization algorithms and their implementations for large scale cellular automata. LNCS **2509**, 276–286 (2002)
35. H. Umeo, M. Hisaoka, T. Sogabe, A survey on optimum-time firing squad synchronization algorithms for one-dimensional cellular automata. Int. J. Unconventional Comput. **1**, 403–426 (2005)
36. H. Umeo, N. Kamikawa, A design of real-time non-regular sequence generation algorithms and their implementations on cellular automata with 1-bit inter-cell communications. Fundamenta Inf. **52**, 255–275 (2002)
37. H. Umeo, N. Kamikawa, Real-time generation of primes by a 1-bit-communication cellular automaton. Fundamenta Inf. **58**(3, 4), 421–435 (2003)
38. H. Umeo, M. Maeda, M. Hisaoka, M. Teraoka, A state-efficient mapping scheme for designing two-dimensional firing squad synchronization algorithms. Fundamenta Inf. **74**(4), 603–623 (2006)
39. H. Umeo, G. Mauri, A duality in two topology-preserving parallel shrinking algorithms - Between Beyer's and Levialdi's algorithms –. Future Generation Comput. Syst. **18** 931–937 (2001)
40. H. Umeo, K. Michisaka, N. Kamikawa, M. Kanazawa, State-efficient one-bit communication solutions for some classical cellular automata problems. Fundamenta Inf. **78**(3), 449–465 (2007)
41. H. Umeo, T. Yanagihara, State-efficient optimum-time implementations of synchronization algorithms on $CA_{1\text{-bit}}$. (draft version) (2008)
42. H. Umeo, T. Yanagihara, M. Kanazawa, State-efficient firing squad synchronization protocols for communication-restricted cellular automata. LNCS 4173, 169–181 (2006)
43. V.I. Varshavsky, V.B. Marakhovsky, V.A. Peschansky, Synchronization of interacting automata. Mathematical Systems Theory. **4**(3), 212–230 (1970)
44. R. Vollmar, On two modified problems of synchronization in cellular automata. Acta Cybernetica. **3**(4), 293–300 (1978)
45. A. Waksman, An optimum solution to the firing squad synchronization problem. Inf. Control. **9**, 66–78 (1996)

Chapter 7
Minimal Cellular Automaton Model of Inter-species Interactions: Phenomenology, Complexity and Interpretations

Andrew Adamatzky and Martin Grube

7.1 Introduction

Cellular automata have been used to simulate population dynamics for over 20 years. A "mass-usage" of automata to imitate space-time dynamics of species has started with the popular article by Dewdney [20], supported by scientific publications on lattice-gas automata [9] and automata models of host-parasite interaction [29, 17], see a brief overview in [24] and discussion on advantages of cellular automata models of population dynamics in [18].

Quickly automata models became uncontested models of pattern formation in population dynamics [19], pattern-oriented ecological modeling [28], spatial ecology [7, 51, 50, 21], and the geomorphology-ecology interface [12].

Cellular automata "substrates" are proved to be successful in imitating and simulating propagation of species [10], developments of plant populations [6], prediction of epidemics dynamics in spatially heterogeneous environments [23], competitive interactions [15], hierarchical ecological systems [54], stochastic species invasion [11, 35], predation chains [38] and competition in complex landscapes [14], and prey–predator systems [16].

Most cellular automata models of population dynamics aimed to simulate real-world phenomena. They are tied therefore to some particular species, landscapes or development scenarios, which leads to increasing number of parameters and characteristics. Any attempt to bring a model closer to reality blurs skeletal features of the model, and prevents, due to complicated designs, complete classification of space-time dynamics. In the present paper we decided to strip automata models of inter-species interactions to bare bones and consider the most primitive model of two-species population without resources. We utilize basic ideas previously developed in the context of using automaton-based population dynamics to generate complex patterns [3] and our designs of cell-state transition rules covering all types of inter-species interactions [2].

A. Adamatzky (✉)
University of the West of England, Bristol, UK
e-mail: andrew.adamatzky@uwe.ac.uk

A.G. Hoekstra et al. (eds.), *Simulating Complex Systems by Cellular Automata*, Understanding Complex Systems, DOI 10.1007/978-3-642-12203-3_7,

Given two species a and b we can depict their relationships by tuples (γ_a, γ_b), where γ_a (γ_b) shows how much species a (b) benefits or suffers from interaction with species b (a) [45]. The following interactions are considered in the paper:

- Mutualism (++): Both species benefit from inter-species interaction. Previously indirect mutualism was discussed in [1], in models where increase of one species in numbers leads to saturation of predators and thus indirectly allows another species (also preyed by the same predator) to prosper.
- Commensalism (+0): One species benefits while another is not affected, e.g. mixed cultures of milk-fermenting bacteria [27, 5], relationships between hermit crab and its associated species [53], commensalims between predators [30].
- Parasitism (predation, herbivory) (+−): One species benefits while another species suffers. Positive and negative features of parasitism have been studied via cellular automaton simulation in [36], cellular automaton models of starfish predation in corral reef developed in [37], cellular automata equivalents of Lotka-Volterra model discussed in [55, 22], automaton models reflecting distribution of individual characteristics of predating species and influence of predation on mimicry are studied in [31] and [34].
- Amensalim (0−): One species is not affected while the other species suffers. Examples include mixed cultures of easts [47], relationships between surface and subsurface feeding polechaeta, where ragworms took advantage from the absence of lugworms [52], and "apparent" amensalism: wildcats in nature parks of central Spain are negatively affected by red deers and wild boards (due to direct competition between these hoofed animals and rodents which are preys of the wildcats) [41].
- Neutralism (00): None of the species is affected by interaction. In two-state cellular automaton neutralism-rule means that cells never change their states and any initial configuration becomes the fixed one immediately. Therefore we will not discuss neutralism further.
- Competition (−−): Both species are badly affected by the interaction. Competition was simulated in cellular automata in the context of spatially explicit models of competition [13], analysis of propagation in population of competing species (as a function of the species' competing abilities) [4], emergence of rare species [26], and competition between plants [42, 10].

These six types of interactions are clearly idealistic and hardly realized in this clarity in nature. Naturally occurring interaction rather locate a continuum between beneficial (mutualistic) and pathogenic (parasitic) poles. In certain cases, the context of ecological conditions, such as habitat conditions can modify the behaviour of interacting species [33]. Moreover, some classic cases are hard to place in these anthropocentrical categories. The lichen symbiosis, for example, is widely considered as a mutualism of fungi and algae, but the original view of lichens as fungi that enslaved algae cannot be rejected [49].

The paper is structured as follows. In Sect. 7.2 we introduce cellular automata models of two-species interactions and define global characteristics of their space-time dynamics. We present and analyze space-time dynamics of the cellular automata models in Sects. 7.3, 7.4, 7.5 and 7.6, and discuss the results in Sect. 7.7.

7.2 Cellular Automaton Model

We study a two-dimensional hexagonal cellular automaton, every cell of which takes two states: species a and species b, and updates its state in discrete time depending on its own state and just the numbers of a and b cell-states of its six neighbors.

Let $\sigma_a^t(x)$ and $\sigma_b^t(x)$ be sums of cell x's neighbors in state a and b, respectively, at time step t. Then the cell-state transition function will be as follows:

$$x^{t+1} = f(x^t, \sigma_a^t(x), \sigma_b^t(x)). \tag{7.1}$$

A cell in state a updates its state depending on a number of neighbors in state b, a cell in state b updates its state depending on a number of neighbors in state a. We represent the rule (7.1) by threshold conditions for a cell to keep its state, i.e. we consider only conditions of the transitions $x^t = a \rightarrow x^{t+1} = a$ and $x^t = b \rightarrow x^{t+1} = b$. The conditions of the cell-state transitions $a \rightarrow a$ and $b \rightarrow b$ are as follows:

	Transition	
Interaction	$a \rightarrow a$	$b \rightarrow b$
Mutualism	$\sigma_b > \theta$	$\sigma_a > \delta$
Commensalism	$\sigma_b > \theta$	
Parasitism	$\sigma_b < \theta$	$\sigma_a > \delta$
Amensalism	$\sigma_b < \theta$	
Neutralism		
Competition	$\sigma_b < \theta$	$\sigma_a < \delta$

where $0 \leq \theta, \delta \leq 6$. If no conditions are attached to a transition, the transition takes place by default (unconditionally).

Values of parameters θ and δ indicate a strength of an influence of one species to another. Smaller values correspond to higher strength because they imply that less amount of one species is required to influence another species. Higher values of the parameters mean weaker strength of inter-species interactions.

For example, consider a parasitic interaction with parameters $\theta = 3$ and $\delta = 1$. A cell in state a at time step t takes the same state a at time step $t + 1$ if it has at least three neighbors in state b, otherwise the cell takes state b. A cell in state b at time step t takes the same state b if it has at least one neighbor in state a.

In case of amensalism, a cell in state a at time step t will remain in state a at time step $t + 1$ if not more than θ of its neighbors are in state b; otherwise the cell takes state b. A cell in state b will remain in the state b indefinitely.

We employ the two following integral parameters to characterize space-time dynamics of studied cellular automata: neighborhood frequency and cell-update frequency.

Let $\nu(u)$ be a frequency of occurrence of neighborhood configuration $u \in \{0, 1\}^7$ (the neighborhood includes central cell) in configuration of a cellular automata, and η be an average distance of each frequency from the frequency of neighborhood configurations in a random configuration, $\eta = \sqrt{(}\sum_{u \in \{0,1\}^7} \nu(u) - \frac{1}{128})$. For a random configuration, where each cell-state occurs with probability $\frac{1}{2}$, the value of η is 0; configuration where cells are in the same state is characterized by $\eta = 1$. The parameter η characterizes a completeness of local representation in the system. When $\eta = 0$ all possible configurations of neighborhood are presented in the single configuration, and thus the configuration is locally complete. A configuration becomes more and more incomplete when η tends to 1.

The parameter ϕ is an average cell-state update frequency, the value measured in t steps indicates a probability for a cell to change its state during any given step of evolution. The parameter ϕ may be thought of as a degree of local visible activity: the higher ϕ the more often cells change their states. We stress that this only applies to the visible activity because each cell of a cellular automaton updates its state at every step of the automaton development, the cell might not change its state however.

To compare morphological characteristics of configurations emerged in development of the cellular automata models, we initiated the systems with random pattern of species a and b and then recorded configurations after a transient period. The transient period is assumed to be completed when the automaton reaches its global fixed point of evolution (still configuration) or a cycle (when succession of configuration is looped). Exemplar configurations, recorded after the transient period are shown in Figs. 7.1 – 7.16.

7.3 Mutualism

Spatial dynamics of cellular automata imitating mutualistic interactions, in scenarios when both species are equally represented in initial configurations, is characterized by still or switching patters (Fig. 7.1). The size of the same species domains increases when dependency of one species increases relatively to dependency of other species, see e.g. increase of θ while δ is fixed in Fig. 7.1. When dependencies of both species increase simultaneously (e.g. the same increases in values of θ and δ, Fig. 7.1) then spatial configurations of the automata change their morphological characteristics from fine-granular (see $\theta = 1$ and $\delta = 1$, Fig. 7.1) to pronounced switching clusters (see $\theta = 3$ and $\delta = 3$, Fig. 7.1) and then return back to fine-granular configurations (see $\theta = 5$ and $\delta = 5$, Fig. 7.1).

Typical examples of switching configuration for parameters $\delta = 2, \theta = 3$ are shown in Fig. 7.2a and b. One can observe compact clusters of one species on the lattice homogeneously filled with other species (Fig. 7.2a), and strips of elongated

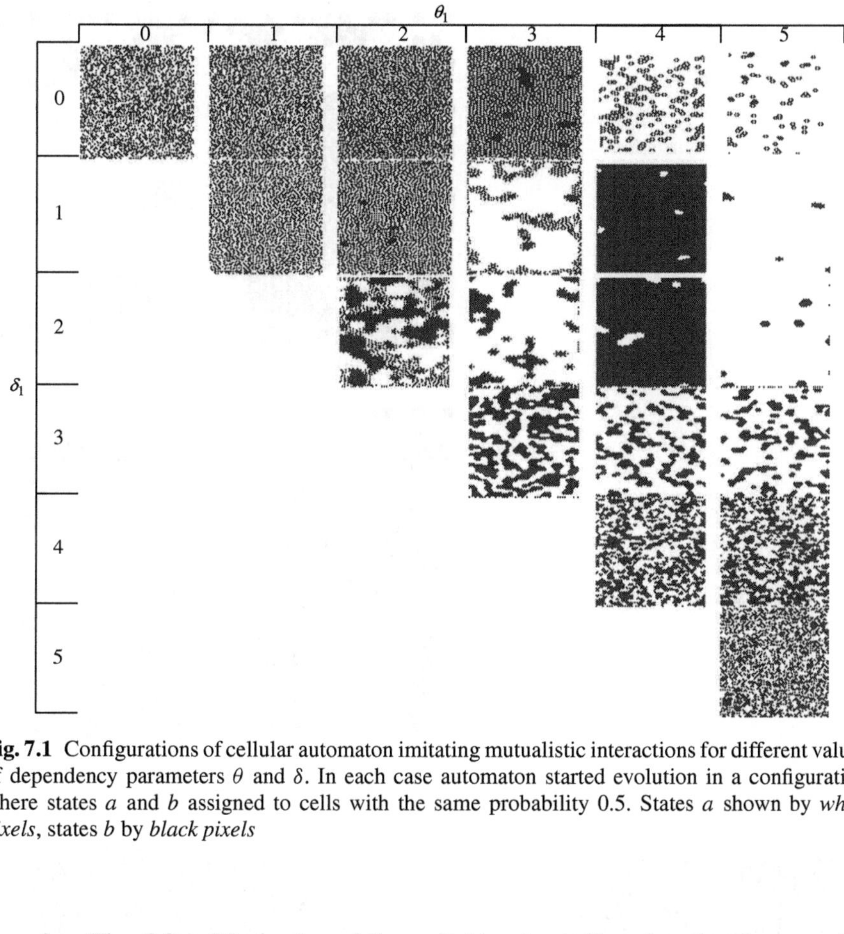

Fig. 7.1 Configurations of cellular automaton imitating mutualistic interactions for different values of dependency parameters θ and δ. In each case automaton started evolution in a configuration where states a and b assigned to cells with the same probability 0.5. States a shown by *white pixels*, states b by *black pixels*

domains (Fig. 7.2a). Mechanics of the switching is attributed to the absence of a no-species states: when cell in state a lacks enough neighbors in state b, the cell switches to state b.

In a cellular automaton imitating mutualism, a small concentration of one species, e.g. a, gives rise to traveling target waves on the lattice filled with another species b, see example in Fig. 7.3a–e. Cells not occupied by propagating wave patterns switch synchronously between b and a states. When wave-fronts of propagating patterns collide they stop their propagation and a quasi-stationary structure is formed (Fig. 7.3f).

As to integral characteristics of cellular automaton dynamics we can see (Fig. 7.4a) that local representation becomes incomplete and less and less rich when influence of one species to another decreases while influence of other species remains unchanged. There is just slight decrease in completeness when influences of both species decrease simultaneously, see Fig. 7.4aη. Activity of the cellular automata simulating the interacting systems increases with decrease of influences between species, see Fig. 7.4aϕ.

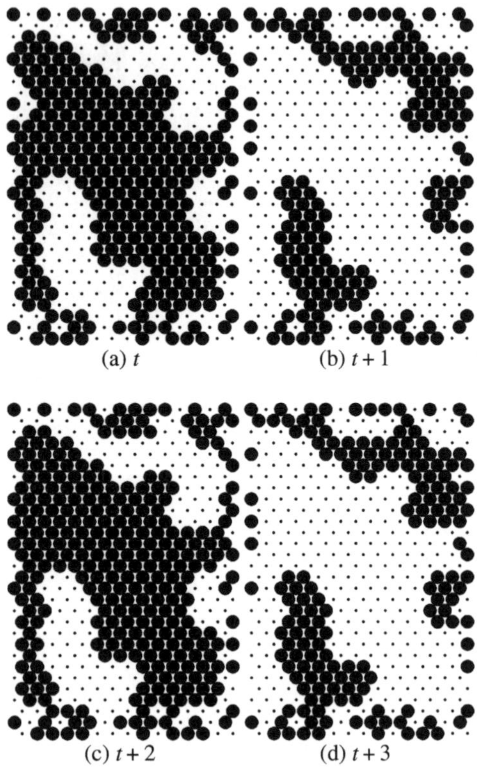

Fig. 7.2 Examples of switching patterns in mutualistic model, $\delta = 2, \theta = 3$

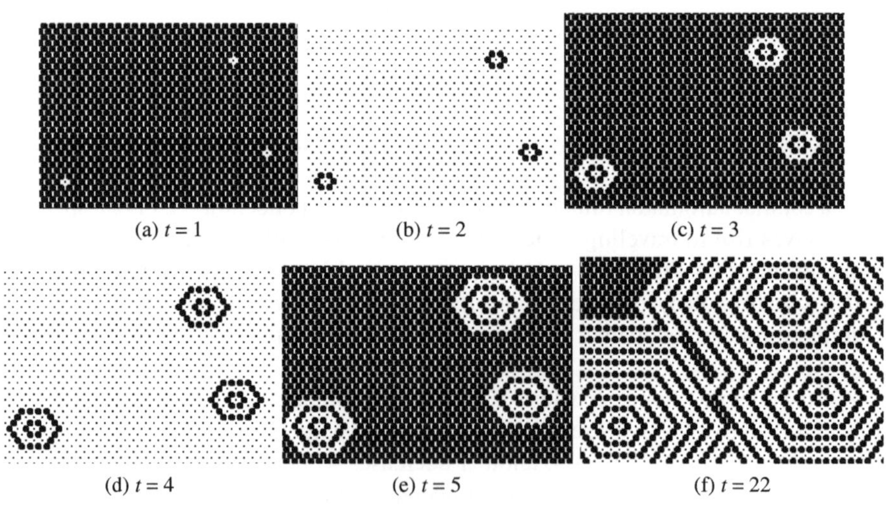

Fig. 7.3 Propagating patterns in cellular automaton imitating mutualistic interactions: (**a**)–(**f**) $\delta = 0$, $\theta = 0$, filled with a, with few, probability 0.001, cells in state b at initial stage of wave propagation

$\eta_{mutualism}$	θ = 0	1	2	3	4	5
δ = 0	0.02	0.04	0.1	0.2	0.45	0.85
1		0.06	0.11	0.7	0.93	0.95
2			0.27	0.69	0.88	0.95
3				0.37	0.48	0.54
4					0.16	0.16
5						0.03

$\phi_{mutualism}$	θ = 0	1	2	3	4	5
δ = 0	0	0	0	0.03	0.62	0.9
1		0	0.02	0.82	0.98	0.99
2			0.67	0.95	0.98	0.99
3				0.99	0.99	0.99
4					0.99	0.99
5						0.99

(a) mutualism

$\eta_{commensalism}$	θ = 0	1	2	3	4	5
	0.01	0.04	0.09	0.29	0.66	0.93

$\phi_{commensalism}$	θ = 0	1	2	3	4	5
	0	0	0	0	0	0

(b) commensalism

$\eta_{parasitism}$	θ = 1	2	3	4	5	6
δ = 0	0.26	0.24	0.20	0.14	0.03	0.01
1	0.18	0.15	0.11	0.09	0.05	0.04
2	0.13	0.11	0.07	0.15	0.1	0.09
3	0.1	0.07	0.48	0.21	0.2	0.26
4	0.04	0.03	0.59	0.3	0.43	0.62
5	0.016	0.07	0.73	0.49	0.64	0.87

$\phi_{parasitism}$	θ = 1	2	3	4	5	6
δ = 0	0.47	0.45	0.37	0.07	0.03	0.01
1	0.66	0.57	0.47	0.05	0.02	0.002
2	0.77	0.66	0.42	0.01	0.005	0.001
3	0.85	0.74	0.004	0.008	0.002	0.004
4	0.97	0.89	0.008	0.002	0.003	0.001
5	0.99	1	0.008	0.002	0.001	0.001

(c) parasitism

$\eta_{amensalism}$	θ = 1	2	3	4	5	6
	1	1	1	0.84	0.13	0.02

$\phi_{amensalism}$	θ = 1	2	3	4	5	6
	0	0	0	0	0.0	0

(d) amensalisms

$\eta_{competition}$	θ = 1	2	3	4	5	6
δ = 1	0.02	0.06	1	1	1	1
2		0.06	1	1	1	1
3			0.3	1	1	1
4				0.38	0.55	0.78
5					0.16	0.13
6						0.03

$\phi_{competition}$	θ = 1	2	3	4	5	6
δ = 1	1	1	0	0	0	0
2		1	0	0	0	0
3			0.3	0	0	0
4				0	0	0
5					0	0
6						0

(e) competition

Fig. 7.4 Integral characteristics of cellular automata imitating two-species interactions η is a completeness of local representation in configurations and ϕ is a cell-update frequency

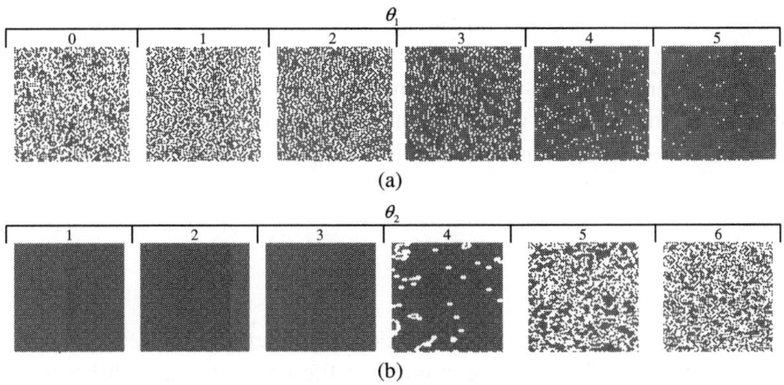

Fig. 7.5 Configurations of cellular automaton imitating (**a**) commensalism, different values of dependency parameter θ, (**b**) amensalism, different values of dependency parameter θ. In each case automaton starts its development in a random configuration where states a and b assigned to cells with the same probability 0.5. States a are shown by *white pixels*, states b by *black pixels*

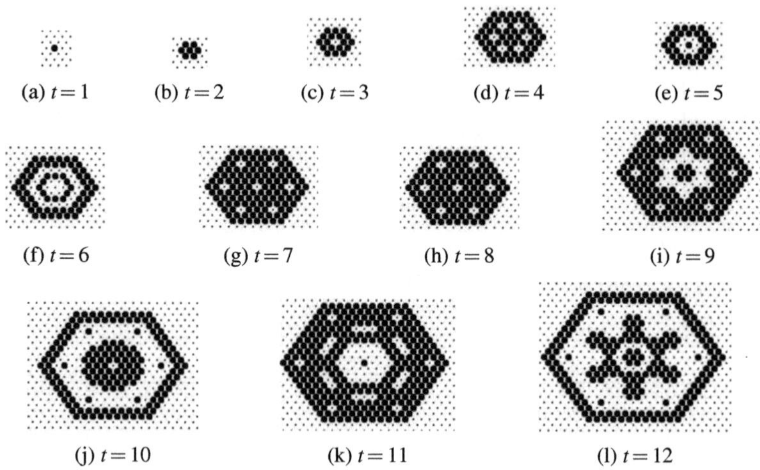

Fig. 7.6 Propagation of parasites in a lattice filled with hosts, parasitism $\delta = 1$ and $\theta = 1$, initial density of parasites is 0.001

7.4 Commensalism and Amensalism

Cell-state transition functions related to commensalism – we consider the case when species a benefits from presence of species b, but species b are not affected by species a – do not support any propagating patterns. This is because once a cell takes a state b it stays in the state b forever. As a consequence of this, an amount of cells in state a can either decrease or remain unchanged in the development of commensalistic cellular automaton. With increase of parameter θ the number of cells in state a decreases, as illustrated in Fig. 7.5.

In amensalistic interactions species a suffers, up to a degree θ, from the presence of species b. When value θ is small (Fig. 7.5b) then species a quickly extinct and species b propagate on the lattice till the wholes space becomes filled with b. With increasing θ localized domains of a emerge and then dominate the space (Fig. 7.5b).

By weakening links between species one can control morphology. Thus, a decrease of influence of one species to another leads to increase of incompleteness of configuration morphology in models of commensalism but increase of completeness of morphology in cellular automata imitating amensalism Fig. 7.4b and d.

7.5 Parasitism

In parasitic interaction, particularly for strong dependencies between host species a and parasite species b, parasites propagate on the lattice filled with hosts, see e.g. Fig. 7.6.

When dependencies between species become weaker propagation of parasites gets less trivial. Thus for $\delta = 2$ and $\theta = 2$ and an initial configuration of lattice filled with hosts a and just few parasites b, small domains of parasites stay still

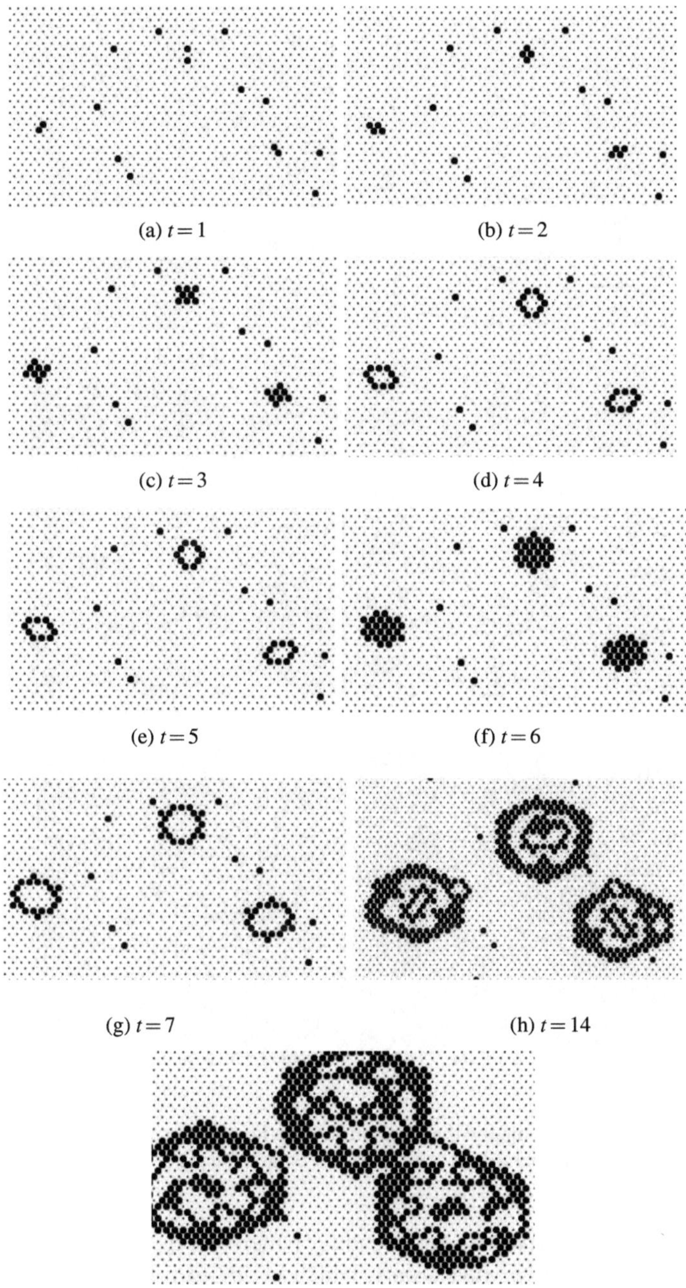

(a) $t=1$ (b) $t=2$

(c) $t=3$ (d) $t=4$

(e) $t=5$ (f) $t=6$

(g) $t=7$ (h) $t=14$

(i) $t=18$

Fig. 7.7 Propagation of parasites in a space filled with hosts for minimal dependencies between hosts and parasites: $\delta = 2$ and $\theta = 2$, initially cells were assigned parasite states with probability 0.01, the rest are hosts

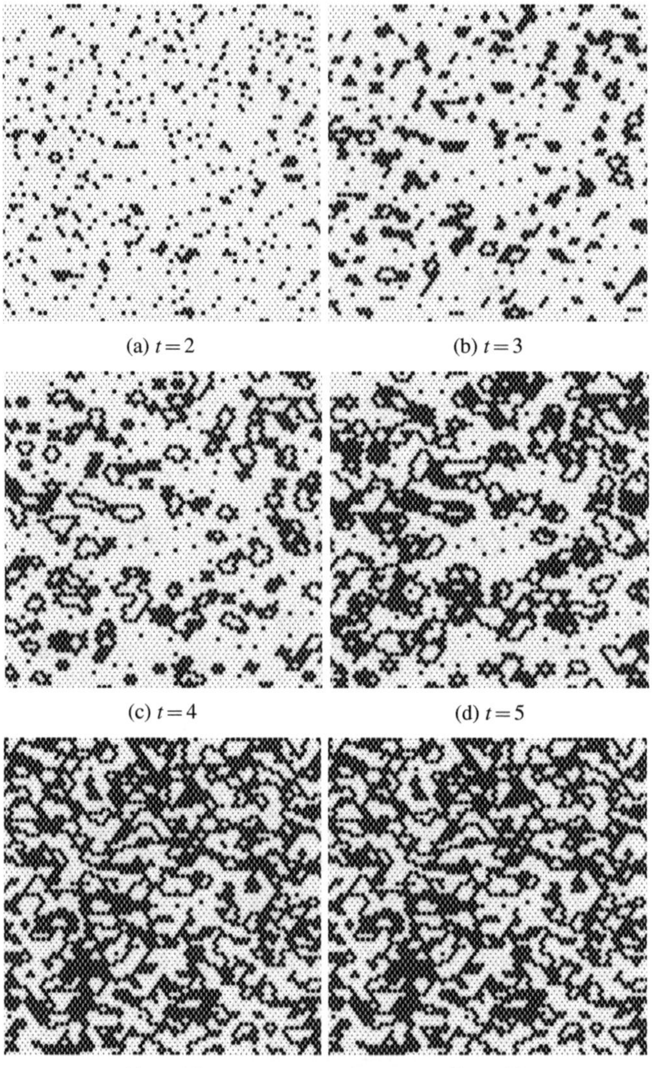

(a) $t = 2$ (b) $t = 3$

(c) $t = 4$ (d) $t = 5$

(e) $t = 29$ (f) $t = 30$

Fig. 7.8 Example of spatio-temporal dynamics in parasite-host cellular automaton for weak dependencies between host a and parasite b, $\delta = 2$ and $\theta = 3$; initial density of parasites is 0.1

while bigger domains give rise to spreading patterns (Fig. 7.7). The propagating patterns exhibit distinctive wave fronts, which is due to parasites' need to have a "comfortable" concentration of hosts around them.

By weakening further dependency parameters δ and θ we shift character of a spatio-temporal dynamic towards formation of slowly expanding irregular clusters of parasites with short-living mobile localizations of parasitic activity propagating between them (Fig. 7.8).

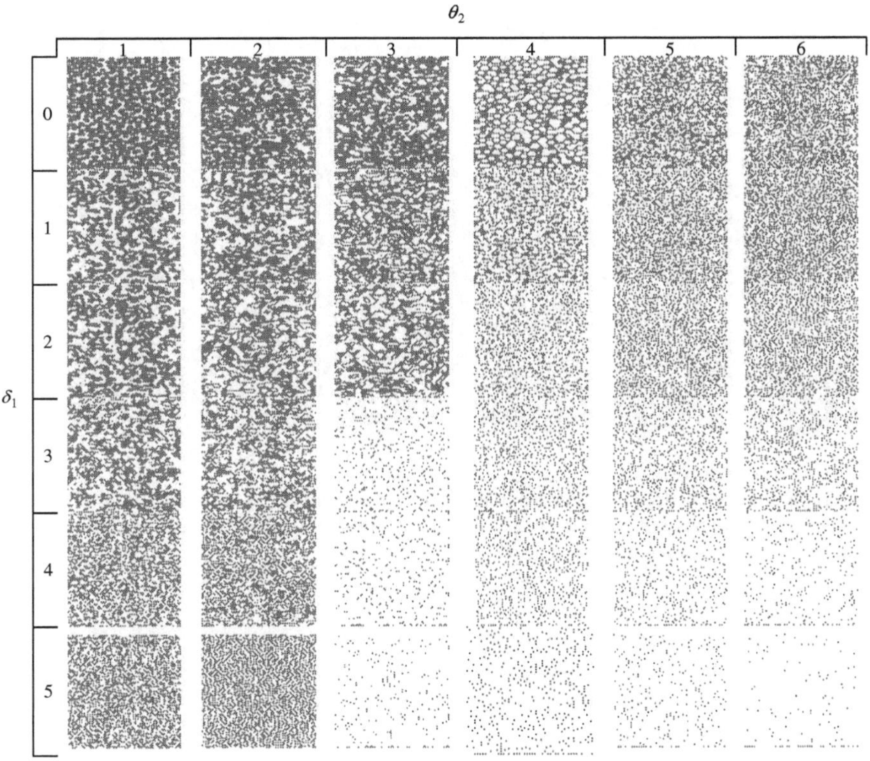

Fig. 7.9 Configurations of cellular automaton imitating parasitic interactions for different values of dependency parameters θ and δ. In each case the automaton starts its development in a configuration where states a and b assigned to cells with the same probability 0.5. States a shown by *white pixels*, states b by *black pixels*

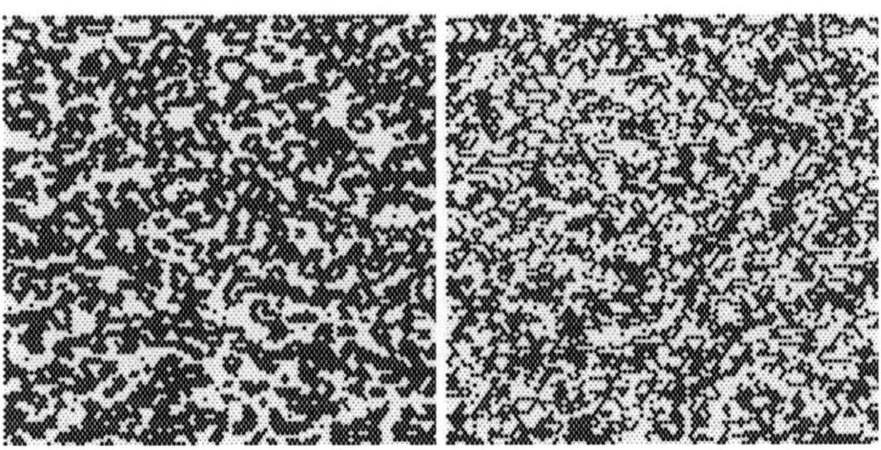

(a) $t = 2$ (b) $t = 2$

Fig. 7.10 Examples of structures formed in parasite-host systems (**a**) $\delta = 1$ and $\theta = 1$; (**b**) $\delta = 2$ and $\theta = 3$; initial density of parasites is 0.5. Left configuration represent automaton configuration at time step t, right configuration at time step $t + 1$

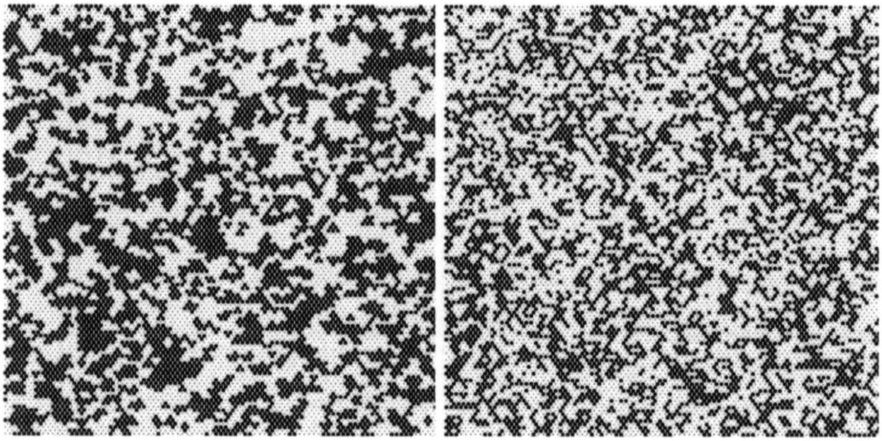

(a) $t = 2$ (b) $t = 2$

Fig. 7.11 Examples of structures formed in parasite-host systems (**a**) $\delta = 0$ and $\theta = 4$; (**b**) $\delta = 3$ and $\theta = 2$. initial density of parasites is 0.5. *Left* configuration represent automaton configuration at time step t, *right* configuration at time step $t + 1$

In parametric portrait Fig. 7.9 we observe that the increase in dependency parameters is reflected by the following transitions in spatio-temporal dynamics: from waves of parasites to still patterns of parasites, exchanging with each other mobile localizations, to still patterns.

In situations of equal densities of hosts and parasites in initial configuration one can observe the following:

- small domains switching between few-site groups of parasites to domains of parasites spanning significant part of the lattice, see Fig. 7.10a;
- branching clusters of parasites, "generating" singletons of parasites, Fig. 7.10b;
- networks of parasites penetrating hosts (each network consists of relatively ordered domains of parasites, nodes of the network exhibits dynamical links with their closest neighbors), Fig. 7.11a;
- combination of disordered clusters of parasites and loci of highly ordered groups of parasites, Fig. 7.11b.

In parasitic interactions, when the influence of one species stay fixed while the influence of the other species decreases, we observe increase of local completeness of morphological representation (Fig. 7.4cη). Simultaneous decrease in species influences first leads to increase of local completeness, followed by decrease of the completeness. The turning point is at mid-range influence, degrees of influence between two and three.

The activity ϕ decreases with decreasing influence of the parasite species on host species. The activity increasing with decreasing of the benefits the parasites species receive from the host species (Fig. 7.4cϕ). Simultaneous decrease of influences of both host and parasite species leads to decrease in the activity.

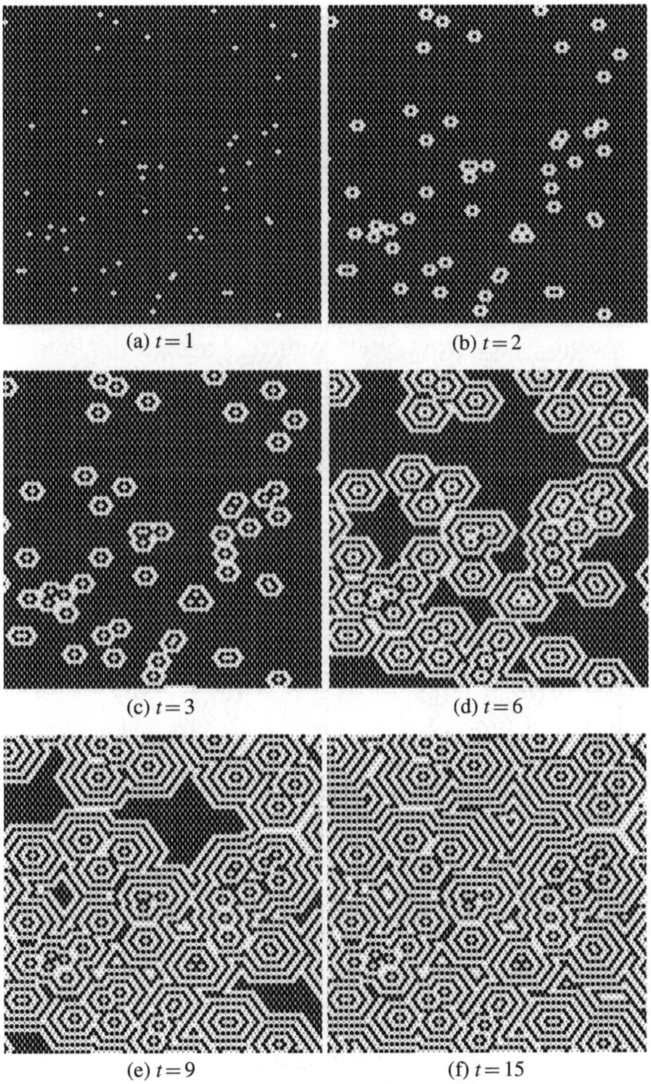

(a) $t = 1$ (b) $t = 2$ (c) $t = 3$ (d) $t = 6$ (e) $t = 9$ (f) $t = 15$

Fig. 7.12 Propagation and interaction of wave-patterns on the lattice with competing species a and b, $\theta = 1$ and $\delta = 1$, initially almost every cell takes state b and few (probability 0.001) cells take state a

7.6 Competition

Cellular automata imitating competition interaction exhibit wave patterns, similar to that observed in excitable media. If competition between species is strong, e.g. $\theta = 1$ and $\delta = 1$ (Fig. 7.12), and one of the species is sparsely distributed in the automaton initial configuration then target waves are formed and propagate (Fig. 7.12).

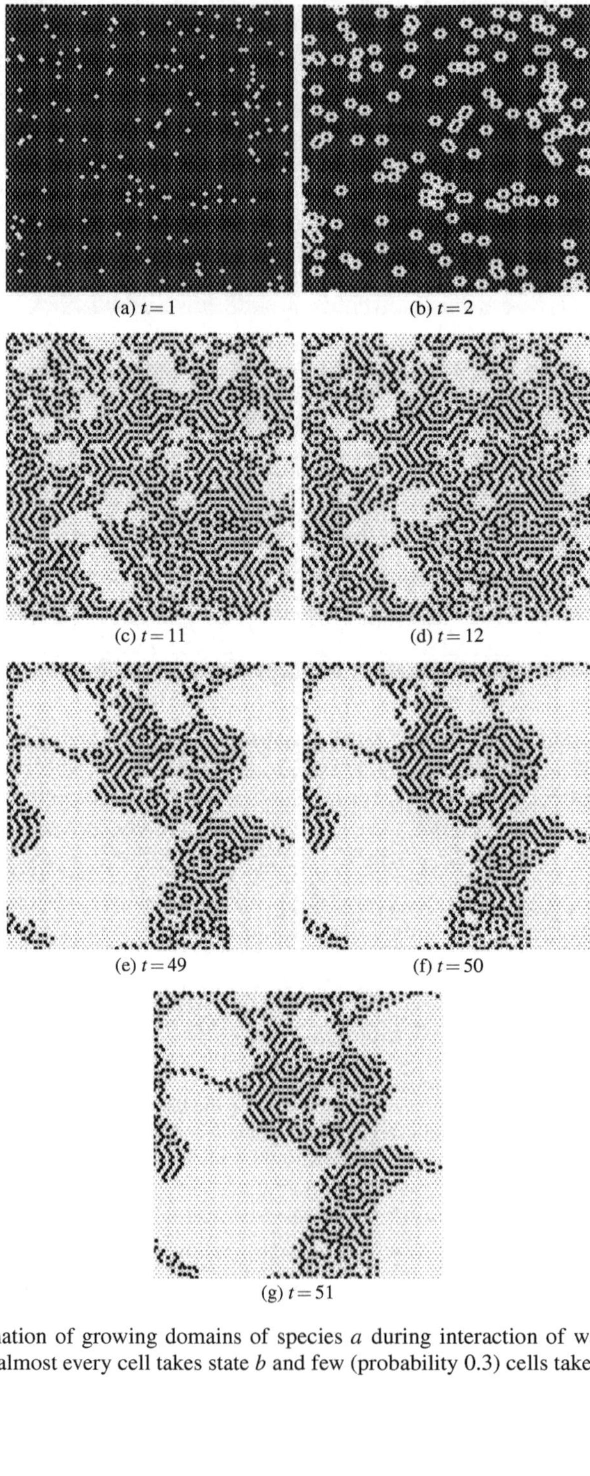

(a) $t=1$ (b) $t=2$

(c) $t=11$ (d) $t=12$

(e) $t=49$ (f) $t=50$

(g) $t=51$

Fig. 7.13 Formation of growing domains of species a during interaction of waves, $\theta = 1$ and $\delta = 3$, initially almost every cell takes state b and few (probability 0.3) cells take state a

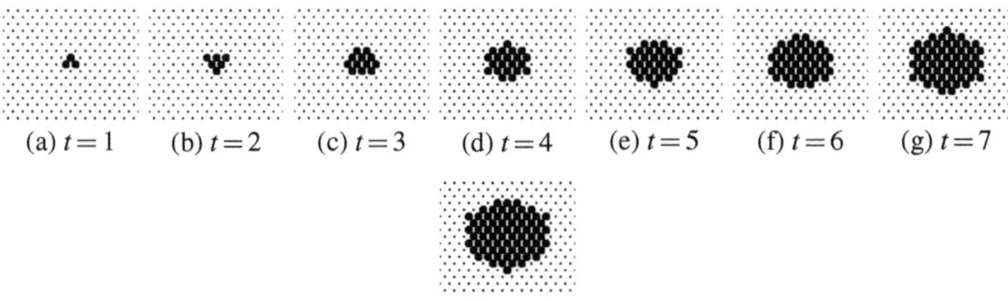

(a) $t=1$ (b) $t=2$ (c) $t=3$ (d) $t=4$ (e) $t=5$ (f) $t=6$ (g) $t=7$ (h) $t=8$

Fig. 7.14 Another way for stationary domains of species a to form: the still and "solid" domain of a states grows from a tiny cluster of cells in the state a, $\theta = 5$ and $\delta = 2$

In scenario of dissimilar dependencies between species, e.g. $\delta = 1$ and $\theta = 2$, propagating target waves give rise to stationary domains when waves collide with each other (Fig. 7.13). The still domains are filled with the species which is lesser affected by other species, and thus eventually wins competition.

Propagating "solid" (filled with one species) domains is yet another feature of the cellular automata imitating competition. This is typical for scenarios where there is a significant difference between dependencies of each species, e.g. $\theta = 5$ and $\delta = 2$ (Fig. 7.14). There we can see that certain configurations of one species in the lattice filled with another species form domains which increase in size and ultimately fill the whole lattice.

When almost all cells of automaton lattice take the same state apart of few cells which take different state, then breathing, oscillating and switching patterns can be observed (Fig. 7.15). This is particularly visible for the following values of dependencies: $\delta = 2$ and $\theta > 1$ or $\delta > 1$ and $\theta = 2$. The structures formed are similar to switching patters in Conway's Game of Life and to stationary waves in cellular automaton models of reaction-diffusion systems.

Momentary configurations, recorded after a transient period, of cellular automata imitating competing species are shown in Fig. 7.16. The automata developed from initial random configurations, where both species have been represented equally. Multiplicity of wave sources makes target waves almost unrecognizable. Therefore we can only concentrate on general characteristics of the configurations morphology.

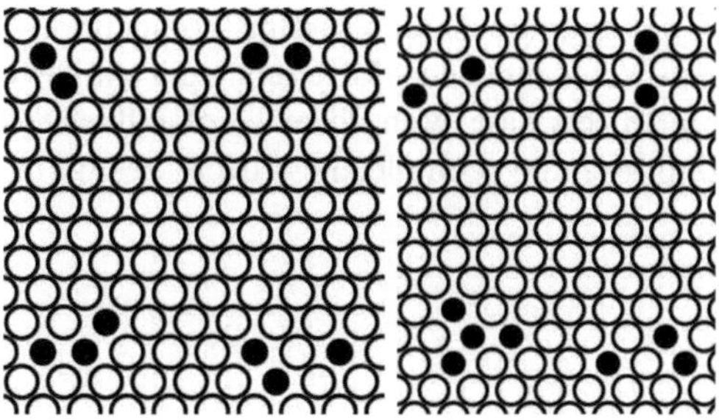

(a) $t = 1$ (b) $t = 2$

Fig. 7.15 Typical oscillators in competing systems for $\theta = 2$ and $\delta = 5$. Similar oscillators can be for the parameters $\theta = 3$ and $\delta = 2$; $\theta = 3$ and $\delta = 2$; $\theta = 5$ and $\delta = 2$

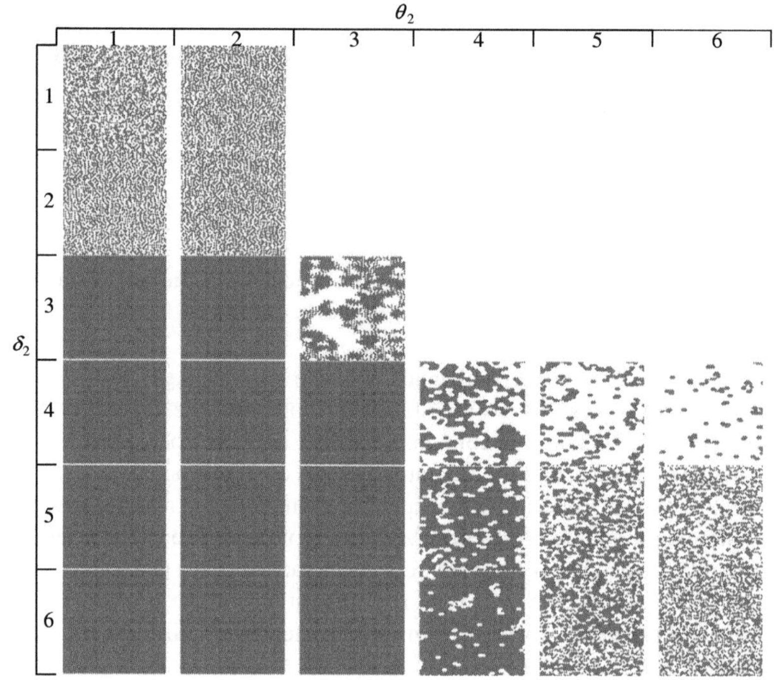

Fig. 7.16 Configurations of cellular automaton imitating competition interactions for different values of dependency parameters θ and δ. In each case automaton started evolution in a configuration where states a and b assigned to cells with the same probability 0.5. States a shown by *white pixels*, states b by *black pixels*

We see that the increase in competing dependencies lead to multiplication of large one-species domains into many smaller domains, which further multiply to singletons (one or two cells in one state in the "ocean" of other states).

In cellular automata imitating competition activity level is high (almost every cell changes its state every time step) only when degrees of influence between species are high, namely for $\theta, \delta \in \{1, 2\}$ and also when $\theta = \delta = 3$ (Fig. 7.4e). The activity deceases for weak influences between species.

Morphologies of configurations representing competition are only complete when degrees of inter-species influences have relatively similar values, see diagonal $\theta = \delta$ in Fig. 7.4eη. In this case with decrease of influences (corresponding to simultaneous growth of θ and δ) completeness decreases till $\theta = \delta = 4$ and then start to increase.

7.7 Discussion

We aimed to study a minimal model of inter-species interactions, therefore we did not include an empty space or a substrate (which would be a third cell-state) in the cellular automaton model. Because the model is reduced to the minimal number of parameters, it is actually difficult to find similar configurations in natural interactions. For example, the consequence of the lack of an "empty" space in the model implies instantaneous occupation of space by species a (b) if species b (a) "dies". This may lead to somewhat paradoxical interpretation of interaction rules, e.g. in case of mutualism we have the following: species a and b benefit from presence of each other in their vicinities, however they also benefit from extinction of other species. Such a shift in interpretation of results is not entirely nonsense. In real-world the relationships can be combined, channeled via different routes, and transformed when applied in a complex networks of interaction species. Even the most simple cases of bacterial interactions are certainly more complex, as they do not restrict on two dimensions and frequently involve additional mechanisms, such as diffusive quorum-sensing signals. Symbioses among organisms with different level of organisation can be much more complex. Ant gardens are among the best examples: there one can find commensalism (relationships between ants and bacteria), amensalim (between bacteria and parasitic fungi) and one more commensalism (between bacteria and cultivated fungi) [48]. The leaf-cutter ant symbiotis is similarly complex with five recognized members [39], including a parasitism between fungi, amensalism between bacteria and parasitic fungi, and mutualism between ants and fungi. The strength of one interaction may here also vary with the abundance of a partner influencing another interaction [40]. Another example – a transformation of preying to mutualism is provided in [25, 8]. They discuss that preys of American Alligator do actually benefit not suffer from the alligators' predation because for any particular species A, alligators not only consume organisms of the species A but also other species, which involved in competitive relationships with the species A.

The complexity of these systems show that a full biological understanding of symbiotic associations requires examining the direct and indirect interactions of symbionts in their ecological community context. In other cases, however, such as the human gut microbial community (with a long-tailed rarefaction curve of involved species), it is difficult to assess the number of symbiotic partners and their interaction strengths. Basic spatial processes of symbiotic interaction, as assessed by our model, are necessarily blurred in excessively complex natural systems, which usually involve more than two partners. Yet, the simple model of interactive excitation could serve as a primer to investigate more complex systems. The knowledge of basic configurations of symbiotic excitation might also help to design nature-inspired and massively parallel computing approaches based on (variable numbers of) symbiotic/interacting agents.

What is a complexity hierarchy of inter-species interactions? Neutralism is out of question in our setup because cells do not change their states from the beginning of the systems' development. Amensalism and commensalism have lowest complexities: cell-state update frequencies are minimal, transient periods are short, no non-trivial propagating patterns observed. Competition occupies the next above bottom level of complexity. Cellular automata models of competing species do exhibit target waves and growing domains however they are characterized by low values of local completeness and minimal frequency of cell-state updates.

Parasitism and mutualism are top candidates for the most complex inter-species interactions. Cell-state update frequencies and local completeness of spatial configurations are in the whole higher in automaton models of mutualistic interactions than in parasitic ones. Both types of interaction support formation and propagation of wave patterns, domains and switching patterns. Cellular automata imitating mutualism exhibit wider spectrum of breathing domains than automata imitating parasitism. These findings persuade us to conclude that mutualistic interactions are more complex than parasitic interactions.

The proposed hierarchy of complexity of two-species interactions, as inferred from analysis of cellular automata models, will be as follows:

$$\{\text{commensalism, amensalism}\} \prec \text{competition} \prec \text{parasitism} \prec \text{mutualism}.$$

It will now be interesting to proceed with our model of simple symbiotic excitation to trace the consequences when interaction strength or the transition function are allowed to evolve. Facing the ubiquity of mutualistic (or cooperative) interactions in nature, the question could be raised whether evolution will also promote the more complex mutualism even in simple models. Evolution has favored more complex networks of interactions via supporting the aggregation of positively interacting organisms. In a long run this apparently also evolved towards mutual dependence, e.g. when metabolic processes are sorted among the community members. The mutual dependence on metabolic capacities is quite common in microbial systems (such as the phototrophic consortium of *Chlorochromatium aggregatum* [46]) or in bacteria-insect symbioses, where insects rely on their gut bacteria for production of essential nutrients and where bacteria with their reduced genomes can only survive in the host [43]. More generally, the low number of culturable bacteria detected by microbial ecologists in most habitats is a clear sign for the ubiquity of obligate symbiotic interactions in nature and it's evolutionary success. But how did the evolutionary success of cooperation [44] evolve actually. Often the biological interactions resemble more a bargaining relationship, where one organism exchanges easily produced good for others that require more own effort. In this context, fast-exploiting parasites would fail in a long run, as they would just overexploit their victims (it could be the same in economy). Natural selection towards slowing down the exploitation rate is perhaps the first step towards mutualistic behaviour. That a parasitic relationship can evolve into a mutualistic one by minimizing damage to the host, was shown some time ago by a long term experiment. In 1966, K. W. Jeon discovered a culture of amoebas that had become infected by large numbers of bacteria (up to 150,000 per cell). The infection slowed their rate of amoebal growth and made the amoeba more fragile. But 5 years later, the amoebas still were infected but without pathogenic effects and the amoeba had apparently become dependent on the bacteria [32]. It appears that synchronization with the host biology and low exploitation (weak interaction strength) are key factors for transforming from parasitism towards mutualism.

References

1. P.A. Abrams, H. Matsuda, Positive indirect effects between prey species that share predators. Ecology **77**, 610–616 (1996)
2. A. Adamatzky, *Identification of Cellular Automata* (Taylor & Francis, London 1994)
3. A. Adamatzky, Cellular automaton labyrinths and solution finding. Comput. Graph. **21**, 519–522 (1997)
4. K. Arii, L. Parrott, Examining the colonization process of exotic species varying in competitive abilities using a cellular automaton model. Ecolo. Model. **199**, 219–228 (2006)
5. M. Aziza, A. Amrane, Commensalism during submerged mixed culture of *Geotrichum candidum* and *Penicillium camembertii* on glutamate and lactate. Process Biochem. **41**, 2452–2457 (2006)
6. H. Balzter, P.W. Braun, P. Köhler, Cellular automata models for vegetation dynamics. Ecol. Model. **107**, 113–125 (1998)
7. N. Boccara, Automata network models of interacting population, ed. by E. Goles, S. Martinez, *Cellular Automata, Dynamical Systems and Neural Networks* (Springer, Heidelberg, 1994)
8. C. Bondavalli, R.E. Ulanowicz, Unexpected effects of predators upon their prey: The case of the American Alligator. Ecosystems **2**, 49–63 (1999)
9. S. Camazine, Self-organizing pattern formation on the combs of honey bee colonies. Beh. Ecol. Sociobiol. **28**, 61–76 (1991)
10. S.A. Cannas, S.A. Páez, D.E. Marco, Modeling plant spread in forest ecology using cellular automata. Comput. Phys Commun. **121/122**, 131–135 (1999)
11. S.A. Cannas, D.E. Marco, S.A. Páez, Modelling biological invasions: species traits, species interactions, and habitat heterogeneity. Mathe. Biosci. **183**, 93–110 (2003)
12. Q. Chen, A.E. Mynett, Modelling algal blooms in the Dutch coastal waters by integrated numerical and fuzzy cellular automata approaches. Ecol. Model. **199**, 73–81 (2006)
13. G. de Cardozo, D. de Silvestre, A. Colato, Periodical cicadas: A minimal automaton model. Physica A **382**, 439–444 (2007)
14. H. Caswell, R. Etter, Cellular automaton models for competition in patchy environments: Facilitation, inhibition, and tolerance. Bull Math Biol **61**, 625–649 (1999)
15. Q. Chen, A. E. Mynett, A.W. Minns, Application of cellular automata to modelling competitive growths of two underwater species *Chara aspera* and *Potamogeton pectinatus* in Lake Veluwe. Ecol Modell **147**, 253–265 (2002)
16. Q. Chen, A.E. Mynett, Effects of cell size and configuration in cellular automata based prey-predator modelling. Simulation Model. Pract. Theory **11**, 609–625 (2003)
17. H.N. Comins, M.P. Hassell, R.M. May, The spatial dynamics of host-parasitoid systems. J. Anim. Ecol. **61**, 735–748 (1992)
18. P.J. Darwen, D.G. Green, Viability of populations in a landscape. Ecol. Model. **85**, 165–171 (1996)
19. A. Deutsch, S. Dormann, *Cellular Automaton Modeling of Biological Pattern Formation* (Birkhäuser, Basel, 2006)
20. A.K. Dewdney, *Armchair Universe: An Exploration of Computer Worlds* (Freeman and Co, New York, 1988).
21. U. Dieckmann, R. Law, J.A.J. Metz, *The Geometry of Ecological Interactions: Simplifying Spatial Complexity* (Cambridge University Press, Cambridge 2000)
22. M. Droz, A. Pekalski, Dynamics of populations in extended systems, ed. by S. Bandini, B. Chopard, M. Tomassini, *Cellular Automata*: 5th International Conference on Cellular Automata for Research and Industry, Springer, Berlin 2002
23. M. Duryea, T. Caraco, G. Gardner, W. Maniatty, B.K. Szymanski, Population dispersion and equilibrium infection frequency in a spatial epidemic. Physica D **132**, 511–519 (1999)
24. G.B. Ermentrout, L. Edelstein-Keshet, Cellular automata approaches to biological modeling. J. Theor. Biol. **160**, 97–133 (1993)

25. B.D. Fath, Network mutualism: Positive community-level relations in ecosystems. Ecol. Model. **208**, 56–67 (2007)
26. S. Galam, B. Chopard, M. Droz, Killer geometries in competing species dynamics. Physica A **314**, 256–263 (2002)
27. F. Grattepanche, P. Audet, C. Lacroi, Milk fermentation by functional mixed culture producing nisin Z and exopolysaccharides in a fresh cheese model. Int. Dairy J. **17**, 123–132 (2007)
28. V. Grimm, K. Frank, F. Jeltsch, R. Brandl, J. Uchmanski, C. Wissel, Pattern-oriented modelling in population ecology. Sci Total Environ **183**, 151–166 (1996)
29. M.P. Hassell, S.W. Pacala, R.M. May, P.L. Chesson, The persistence of host-parasitoid associations in patchy environments. I. A general criterion. Ameri Nat. **138**, 568–583 (1991)
30. S.B. Heard, Pitcher-plant midges and mosquitoes: A processing chain commensalism. Ecology **75**, 1647–1660 (1994)
31. C. Hui, M.A. McGeoch, Evolution of body size, range size, and food composition in a predator-prey metapopulation. Ecol. Complex. **3**, 148–159 (2006)
32. K.W. Jeon, Development of cellular dependence on infective organisms: Micrurgical studies in amoebas. Science **176**, 1122–1123 (1972)
33. J. Karst, L. Marczak, M.D. Jones, R. Turkington, The mutualism-parasitism continuum in ectomycorrhizas: A quantitative assessment using meta-analysis. Ecology **89**, 1032–1042 (2008)
34. I. Kawaguchi, A. Sasaki, The wave speed of intergradation zone in two-species lattice Müllerian mimicry model. J. Theor. Biol. **243**, 594–603 (2006)
35. S. Kizaki, M. Katori, A stochastic lattice model for locust outbreak. Physica A **266**, 339–342 (1999)
36. C.-H. Kuo, V. Corby-Harris, D.E.L. Promislow, The unavoidable costs and unexpected benefits of parasitism: Population and metapopulation models of parasite-mediated competition. J. Theor. Biol. **250**, 244–256 (2008)
37. J.D. van der Laan, R.H. Bradbury, Futures for the Great Barrier Reef ecosystem. Mathe. Comput. Model. **14**, 705–709 (1990)
38. J.D. van der Laan, L. Lhotka, P. Hogeweg, Sequential predation: A multi-model study. J. Theor. Biol. **174**, 149–167 (1995)
39. A.E.M. Little, C.R. Currie, Symbiotic complexity: Discovery of a fifth symbiont in the attine ant-microbe symbiosis. Biol. Lett. **3**, 501–504 (2007)
40. A.E.M. Little, C.R. Currie, Black yeast symbionts compromise the efficiency of antibiotic defenses in fungus-growing ants. Ecology **89**, 1216–1222 (2008)
41. J. Lozano, E. Virgós, S. Cabezas-Díaz, J.G. Mangas, Increase of large game species in Mediterranean areas: Is the European wildcat (*Felis silvestris*) facing a new threat? Biol. Conser. **138**, 321–329 (2007)
42. Y.G. Matsinos, A.Y. Troumbis, Modeling competition, dispersal and effects of disturbance in the dynamics of a grassland community using a cellular automaton model. Ecol. Model. **149**, 71–83 (2002)
43. N. Moran, Symbiosis. Curr. Biol. **16**, R866–R871 (2006)
44. M.A. Nowak, Five rules for the evolution of cooperation. Science **314**, 1560–1563 (2006)
45. E.P. Odum, *Basic Ecology* (Harcourt Brace College Publisher, Fortworth, TX 1983)
46. J. Overmann, Phototrophic consortia. A tight cooperation between non-related eubacteria, ed. by J. Seckbach, *Symbiosis. Mechanisms and Model Systems*. (Kluwer, Dordrecht, 2001), 239–255
47. S. Pommier, P. Strehaiano, M.L. Délia, Modelling the growth dynamics of interacting mixed cultures: A case of amensalism. Int. J. Food Microbiol. **100**, 131–139 (2005)
48. P. Salles, B. Bredeweg, N. Bensusan, The ants garden: Qualitative models of complex interactions between populations. Ecol. Model. **194**, 90–101 (2006)
49. S. Schwendener, Die Algentypen der Flechtengonidien. Programm für die Rectorsfeier der Universität Basel **4** (1869) 1–42.

50. T. Szaran, *Spatiotemporal Models of Population and Community Dynamics* (Springer, Heidelberg 1997)
51. D. Tilman, P. Kareiva, *Spatial Ecology* (Princeton University Press, Princeton, NJ, 1997)
52. N. Volkenborn, K. Reise, Lugworm exclusion experiment: Responses by deposit feeding worms to biogenic habitat transformations. J. Exp. Marine Biol. Ecol. **330**, 169–179 (2006)
53. J.D. Williams, J.J. McDermott, Hermit crab biocoenoses: A worldwide review of the diversity and natural history of hermit crab associates. J. Exp. Marine Biol. Ecol. **305**, 1–128 (2004)
54. J. Wu, J.L. David, A spatially explicit hierarchical approach to modeling complex ecological systems: Theory and applications. Ecol. Model. **153**, 7–26 (2002)
55. M. He, J. Lin, H. Jiang, X. Liu, The two populations cellular automata model with predation based on the Penna model. Physica A **312**, 243–250 (2002)

Chapter 8
Cellular Evolutionary Algorithms

Marco Tomassini

8.1 What Are Evolutionary Algorithms?

Evolutionary algorithms (EAs) are a family of heuristic search methods that are often used nowadays to find satisfactory solutions to difficult optimization and machine learning problems. EAs are loosely based on a few fundamental evolutionary ideas introduced by Darwin in the nineteenth century. These concepts revolve around the notion of populations of organisms adapting to their environment through genetic inheritance and survival of the fittest. Innovation is provided by various biological recombination and mutation mechanisms. EAs make use of a metaphor whereby an optimization problem takes the place of the environment; feasible solutions are viewed as individuals living in that environment and an individual's degree of adaptation to its surrounding environment is the counterpart of the objective function evaluated on a feasible solution. In the same way, a set of feasible solutions takes the place of a population of organisms.

Each individual may be viewed as a representation, according to an appropriate encoding, of a particular solution to an algorithmic problem, of a strategy to play a game, or even of a simple computer program that solves a given problem. To implement an EA for solving a given problem, at least approximately, the user must provide the following pieces of information:

8.1.1 Representation

Individuals in the population represent solutions to a problem and must be encoded in some way to be manipulated by the EA processes. They may be just strings of binary digits, which is a widespread and universal representation, or they may be any other data structure that is suitable for the problem at hand such as strings of

M. Tomassini (✉)
Information Systems Department, HEC, University of Lausanne, Lausanne, Switzerland
e-mail: marco.tomassini@unil.ch

A.G. Hoekstra et al. (eds.), *Simulating Complex Systems by Cellular Automata*, Understanding Complex Systems, DOI 10.1007/978-3-642-12203-3_8,

integers, alphabetic characters, strings of real numbers, and even trees or graphs. In this chapter individuals will be represented either by strings of binary digits or of real numbers.

8.1.2 Genetic Operators

The first operator to be applied to the population is *selection*. Its aim is to simulate the Darwinian law of "survival of the fittest". One often used selection method is the so-called fitness proportionate selection. In order to create a new intermediate population of n "parents", n independent extractions of an individual from the old population are performed, where the probability of each individual being extracted is linearly proportional to its fitness. Therefore, above average individuals will expectedly have more copies in the new population, while below average individuals will risk extinction.

Once the population of parents, that is of individuals that have been selected for reproduction, has been extracted, the individuals for the next generation will be produced through the application of a number of reproduction operators, which can involve just one parent (thus simulating asexual reproduction) in which case we speak of mutation, or more parents (thus simulating sexual) reproduction in which case we speak of recombination.

Crossover is a standard recombination in which two parent individuals recombine to form one or two offspring. To apply crossover, couples are formed with all parent individuals; then, with a certain probability, called crossover rate p_c, each couple actually undergoes crossover: if the individuals are represented by bit strings, the two bit strings are cut at the same random position and the second halves are swapped between the two individuals, thus yielding two novel individuals, each containing characters from both parents.

After crossover, all individuals undergo mutation. The purpose of mutation is to simulate the effect of transcription errors that can happen with low probability (p_m) when a chromosome is duplicated. This is accomplished by flipping each bit in every individual with a very small probability, called mutation rate. This is the typical mutation found in *Genetic Algorithms*; the mutation operator is more complicated, and more important, in an EA family called *Evolution Strategies*.

8.1.3 The Evolutionary Cycle

An evolutionary algorithm starts with a population of randomly generated individuals, although it is also possible to use a previously saved population, or a population of individuals encoding for solutions provided by a human expert or by another heuristic algorithm. Once an initial population has been created, an evolutionary algorithm enters a loop. At the end of each iteration a new population will have been created by applying the previously described genetic stochastic operators to the previous population. One such iteration is referred to as a *generation*. The evolutionary cycle can be summarized by the following pseudo-code:

```
generation = 0
Seed Population
while not termination condition do
    generation = generation + 1
    Evaluate fitness of individuals
    Selection
    Crossover(p_c)
    Mutation(p_m)
end while
```

The stopping condition is given either by finding a globally optimal individual (if known), a satisfactory solution, or simply after a pre-determined number of generations. This points out that EAs are just heuristics and not exact algorithms. They cannot give any guarantee of finding the optimum, although they usually find good enough solutions quickly. Indeed it can be shown that, under mild conditions, EAs do converge to the globally optimal solution but only do so with probability 1 when time increases without bound [1], which can hardly be considered a useful result for practitioners.

This introductory section has been necessarily brief. Readers wishing to know more about EAs can consult good standard textbooks such as [1].

8.2 Cellular Evolutionary Algorithms

The previous section provided the basics of EAs but the reader should be aware that there exist many variations on this common theme. For example, EAs may differ by the kind of representation they are using, or by the kind of genetic operators, or both, and there may be other differences as well. One possible source of difference is the assumed population structure and we shall focus on this particular aspect in this chapter. Usually EAs assume that the structure of the population is *panmictic*, which means that any individual may interact with any other individual in the population. However, this need not be always the case: we often see populations in the biological and social world in which individuals only interact with a subset of the rest of the population. This situation can usefully be depicted by using the concept of a *population graph*. In this undirected graph, vertices correspond to individuals, while edges correspond to interactions between pairs of individuals. From this point of view, the standard panmictic, also called *mixing*, population would be represented by a complete graph in which there are edges between any individual and all the others. Not all conceivable graph structures make sense to describe real populations but for EAs, which are just computer algorithms, any suitable graph structure can be used in principle. Among the many possibilities, *regular graphs* in which any vertex

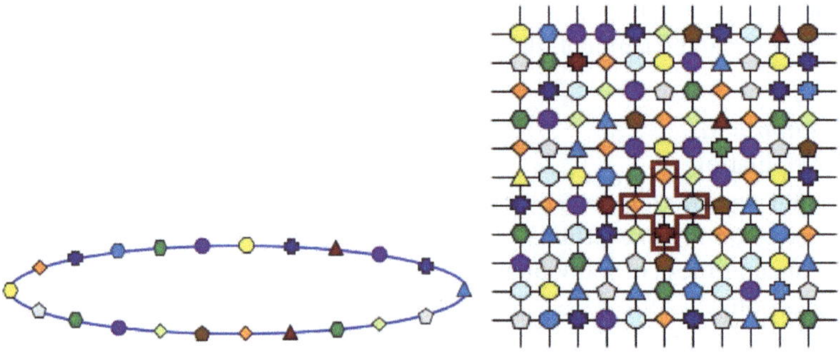

(a) (b)

Fig. 8.1 A ring cellular structure (**a**), and a grid cellular structure with a von Neumann neighborhood highlighted (**b**)

(individual) has the same number of edges (links to other individuals) have emerged early as an easy and useful graph topology, examples of which are given in Fig. 8.1.

Comparing these regular grids of low dimension with those described in Chap. 1, Sect. 1.5, it is clear that they are isomorphic with cellular automata (CAs). This also holds for the commonly used neighborhoods, which are the same as those in CAs namely, essentially regular radius 1 or radius 2 von Neumann or Moore neighborhoods (see Chap. 1, Sect. 1.5). We shall pursue the analogy further in a moment. For the time being, the important thing to note is that individuals now interact *locally*, instead of globally as in the customary mixing population. We shall call the EAs based on these particular population structures *Cellular Evolutionary Algorithms* (CEAs). A CEA starts with the cells in a random state and proceeds by successively updating them using evolutionary operators, until a termination condition is satisfied. Updating a cell in a CEA means selecting parents in the individual's neighborhood, applying genetic operators to them, and finally replacing the individual if the offspring obtained has a better fitness (other replacement policies can be used). Note that each individual step in the algorithm takes place in lockstep for a synchronous CEA. As a result, the general EA pseudo-code of the previous section takes the following form for a synchronous CEA:

```
for each cell i in the grid do
    generate a random individual i
end for
while not termination condition do
    for each cell i in the grid do
        Evaluate individual i
        Select individual(s) in the neighborhood of i
        Produce offspring
        Evaluate offspring
        assign one of the offspring to cell i according to a given criterion
    end for
end while
```

8.2.1 CEAs and CAs

While the analogy between CAs and CEAs is quite clear as far as the agents' population structure is concerned, the relationship between a CA transition function and the local evolution mechanism of a CEA needs some explanation. First of all, what is the finite set of states Σ in the case of a CEA? A simple example should be useful to introduce the general idea. Suppose that the individuals that compose the evolutionary population are coded as binary strings. We have seen that this does not imply a loss of generality as any other finite data structure can be encoded in this way. Further assume that the length of the strings is, for instance, equal to 8. Thus there are $2^8 = 256$ possible strings. Remembering that an individual in this context represents a possible solution in the configuration space of the problem, we can conclude that the cardinality of the set Σ is equal to the number of configurations in the search space. This number can be quite large in practice, but it is always finite.

Now let's see how CEAs evolve in time. Equation (1.3) in Chap. 1 is repeated here for the sake of convenience. It describes the calculation of the state $\sigma_{i,j}(t+1)$ of individual at position $\{i, j\}$ in a two-dimensional grid at the next time step from the states of the same individual and the state of the neighbors at the current time step:

$$\sigma_{i,j}(t+1) = \phi(\sigma_{k,l}(t) \mid \sigma_{k,l}(t) \in \mathcal{N}).$$

The function $\phi(.)$ is the transition rule. For a CEA, ϕ is in general the composition of three functions: the selection function $s(.)$ that selects two individuals in the neighborhood $\mathcal{N}$ of individual $\{i, j\}$ including the latter, the crossover function $c(.)$ which recombines the selected individuals, and the mutation function $m(.)$ which makes random variations to the individual that will replace the original one. Now, knowing that at least mutation and crossover, and most often also selection, are stochastic operators, it follows that the transition function $\phi(.)$ is a stochastic one in a CEA and thus CEAs are *probabilistic cellular automata*. If we now call $C(t) = \{\sigma_1(t), \sigma_2(t), \ldots, \sigma_n(t)\}$ the ensemble of states of all the cells in the grid at time t (this is also called a *global configuration* of the corresponding CA), the global evolution of the automaton will be given by the following symbolic equation:

$$C(t+1) = m(\, c\,(s\,(C(t)\,)\,)\,).$$

This description assumes that cells change their state simultaneously, i.e. in a synchronous manner. Other update policies in which cells change state in some, perhaps random, order can also be used and will be described later.

8.2.2 Brief Historical Background

Influenced by biological and ecological approaches in which the structure of the population plays an important role, evolutionary computation researchers have been quick in adapting structured populations ideas and, in particular, the decentralized grid population structure model seems to have occurred to several people in the mid-eighties. Among those early papers, we mention here Gorges-Schleuter's [2] and Manderick's works [3]. Other early contributors were [4–6]. The relationship between the CEAs and CA models was explicitly recognized independently by Tomassini [7] and Whitley [8]. The field has been a little bit "dormant" for several years but today there is a regain of interest as witnessed by two new books which are totally or partly devoted to CEAs [9, 10].

8.3 Selection Pressure

Selection is the driving force behind any evolutionary algorithm. It is the filter that ensures that better than average individuals will have more chances to reproduce with respect to the worse ones. In other words, the purpose of selection in evolutionary algorithms is to concentrate the use of the available computational resources in promising regions of the search space. There is a relationship of reciprocity between the aspects of *exploration* and *exploitation* of the search space and, clearly, the stronger the pressure exerted by selection toward a concentration of the computational effort, the smaller the fraction of resources utilized to explore other possibilities. Selection pressure is thus a key parameter in the operation of an EA: with high selection pressure, diversity in the population is quickly lost, and the search stagnates, unless a large population is used or a lot of disruption is caused by the variation operators. On the other hand, if the selection pressure is weak, convergence slows down and the search may wander in the problem space without focusing on very good solutions. An effective search thus requires a careful trade-off between the selection method, the variation operators, and other EA parameters such as the population size. Thus, as a first fundamental step toward an understanding of the workings of a CEA, in this section I shall present a theoretical and empirical study of the effects of selection alone, without the use of variation operators.

In standard EAs the selection pool, i.e. the set of individuals that undergo selection is the whole population. However, we have seen that CEAs introduce locality into the population structure and the selection pool in CEAs is formed only by the individuals belonging to the neighborhood. In this case two selection methods are particularly useful and easy to implement: *linear ranking* selection and *binary tournament* selection. In linear ranking selection the individuals in the neighborhood of a given cell are ranked according to their fitness: each individual then has a probability $2(s-i)/(s(s-1))$ of being selected for the replacement phase, where s is the number of cells in the neighborhood and i is its rank in the neighborhood. In binary tournament selection two individuals are randomly chosen with replacement in the neighborhood of a given cell, and the one with the better fitness is selected for the replacement phase. In what follows, the selected individual replaces the original individual only if it has better fitness.

8.3.1 Takeover Time

Selection methods are characterized by their *takeover time*. The takeover time is the time it takes for a single, best individual to take over the entire population. In other words, it represents the speed at which the best solution in the initial population propagates and conquers the whole population under the application of the selection operator alone. It can also be seen as a simplified epidemic process, in which infection means being replaced by the best individual and in which infected individuals remain infected forever. These simple epidemic models are well known in the CA community. The takeover time can be estimated experimentally by measuring the proportion of the best individual as a function of time, under the effect of selection only, without any variation operator. A shorter takeover time indicates a higher selection pressure, and thus a more exploitative algorithm. If the selection intensity is lowered, the algorithm becomes more explorative.

In the study described here, we consider cEAs defined on a one-dimensional lattice of size n or a square lattice of size $n = m \times m$. Both the linear cEA and the two-dimensional case have periodic boundary conditions, i.e. the structures are a ring and a torus respectively.

The main neighborhoods that we consider are the radius-1 neighborhood in the one-dimensional case, which comprises the cell itself and its first right and left neighbors and, in the two-dimensional case, the radius-1 von Neumann neighborhood, which is constituted by the central cell and the four first-neighbor cells in the directions north, east, south, and west (see also Chap. 1, Sect. 1.5).

8.3.2 Asynchronous Updating

In addition to the customary synchronous updating, we shall also use a few sequential update schemes. Synchronous update, with its idealization of a global clock, is customary in cellular automata. However, perfect synchronicity is only an abstraction. In fact, in any spatially extended system, signals require a finite time to propagate. Of course, this "unphysicality" is not a problem in artificial evolutionary algorithms, where we are free to use any solution that makes sense computationally; but there are other reasons that make asynchronous cEAs potentially useful as problem solvers, as we shall see.

In the asynchronous case, cells are updated one at a time in some order. There are thus many ways for sequentially updating the cells of a cEA, including "mixed" ones in which whole blocks of cells are updated asynchronously with respect to each other, while cells belonging to the block are updated in parallel. Here I consider four commonly used asynchronous update methods for cellular automata in which cells are updated one by one [11]:

- In *fixed line sweep* (LS), the n cells are updated sequentially from left to right for rings, and line by line, starting from the upper left corner cell, for grids.

- In *fixed random sweep* (FRS), the next cell to be updated is chosen with uniform probability without replacement; this will produce a certain update sequence $(c_1^j, c_2^k, \ldots, c_n^m)$, where c_q^p means that cell number p is updated at time q and $(j, k, \ldots, m)$ is a permutation of the n cells. The same permutation is then used for all update cycles.
- The method of *new random sweep* (NRS) works like FRS, except that a new random cell permutation is used for each sweep through the array.
- In *uniform choice* (UC), the next cell to be updated is chosen at random with uniform probability and with replacement.

A *time step* is defined as the process of updating n times sequentially, which corresponds to updating *all* the n cells in the grid for LS, FRS, and NRS, and possibly fewer than n different cells in the uniform-choice method, since some cells might be updated more than once. Fixed line sweep is a rather degenerate updating policy, always imposing a one-by-one sequential scan of the array. In spite of this, it can be useful at times, and is also an interesting bounding asynchronous case to consider.

8.3.3 Mathematical Models

In this section I shall present a synthesis of the mathematical considerations that allow one to set up evolution equations in recurrent form for the expected growth of the number of best individuals in the population as a function of (discrete) time. For more details, the reader is referred to [12] and references therein.

Let us consider the random variables $V_i(t) \in \{0, 1\}$ indicating the presence in cell i $(1 \le i \le n)$ of a copy of the current best individual $(V_i(t) = 1)$ or of a worse one $(V_i(t) = 0)$ at time step t, where n is the population size. The random variable

$$N(t) = \sum_{i=1}^{n} V_i(t) \tag{8.1}$$

denotes the number of copies of the best individual in the population at time step t. Initially $V_i(1) = 1$ for some individual i, and $V_j(1) = 0$ for all $j \neq i$.

If the extinction probability is 0, which is guaranteed with the selection methods used here, then the expectation $E[T]$, where $T = \min\{t \ge 1 : N(t) = n\}$, is called the takeover time of the selection method. In the case of spatially structured populations the quantity $E_i[T]$, denoting the takeover time if cell i contains the best individual at time step 1, is termed the takeover time with initial cell i. Assuming a uniformly distributed initial position of the best individual over all cells, the takeover time is therefore given by

$$E[T] = \frac{1}{n}\sum_{i=1}^{n} E_i[T]. \tag{8.2}$$

The above expression is valid for arbitrary undirected connected graphs. For rings and toruses, which are vertex-transitive (i.e. any vertex has a successor in the graph) the takeover time does not depend on the initial vertex.

In what follows, recurrences are given that describe the growth of the random variable $N(t)$ in CEAs with different regular lattice topologies for synchronous and asynchronous update policies.

8.3.3.1 Upper Bounds on the Takeover Time for One- and Two-Dimensional Systems

It can easily be shown [13] that in finite panmictic populations the speed of growth of the best individual follows the well known logistic behavior: there is an initial exponential increase, followed by an exponential decrease after an inflexion point, followed by saturation when the finite prefixed population size is reached.

However, in the artificial evolution of locally interacting, spatially structured populations, the assumption of logistic growth does not hold anymore. Instead, in these locally interacting structures, although the curves have the familiar "S shape" denoting growth followed by saturation, they are not exponential but rather polynomial, with a time dependence $\propto t^d$, where d is the lattice dimension. In fact, in the case of a ring or a torus structure we have a linear or a subquadratic growth, respectively.

To see this, let us consider the limiting case for a structured population, which represents an upper bound on the growth rate, in which the selection mechanism is deterministic, i.e. a cell always chooses its best neighbor for updating with probability 1. If we consider a population of size n with a ring structure, and consider a neighborhood radius of r (i.e. the neighborhood of a cell contains $2r + 1$ cells), the following recurrence describes the growth of the number of copies of the best individual:

$$\begin{cases} N(0) = 1, \\ N(t) = N(t-1) + 2r. \end{cases}$$

This recurrence can be described by the closed equation $N(t) = 1 + 2rt$, which clearly shows the linear character of the growth rate.

In the case of a population of size n on a toroidal grid of size $\sqrt{n} \times \sqrt{n}$ (assuming $\sqrt{n}$ odd) and a von Neumann generalized neighborhood structure of radius r, the growth of the number of copies of the best individual can be described by the following recurrence:

$$\begin{cases} N(0) = 1, \\ N(t) = N(t-1) + 4\sum_{i=0}^{r-1}(rt - i) \qquad , \ 0 \le t \le (\sqrt{n} - 1)/2, \\ N(t) = N(t-1) + 4\sum_{i=0}^{r-1}(\sqrt{n} - rt - i) \ , \ t \ge (\sqrt{n} - 1)/2. \end{cases}$$

which reduces to the following:

$$\begin{cases} N(0) = 1, \\ N(t) = N(t-1) + 4r^2t - 2r(r+1) & , \ 0 \le t \le (\sqrt{n}-1)/2, \\ N(t) = N(t-1) - 4r^2t + 4r\sqrt{n} - 2r(r+1) & , \ t \ge (\sqrt{n}-1)/2. \end{cases}$$

This growth is described by a convex quadratic equation followed by a concave one, as the two closed forms of the recurrence clearly show:

$$\begin{cases} N(t) = 2r^2t^2 + 2r(2r+1)t + 1 & , \ 0 \le t \le (\sqrt{n}-1)/2, \\ N(t) = -2r^2t^2 + 2r(2\sqrt{n} - 3r - 1)t + 1 & , \ t \ge (\sqrt{n}-1)/2. \end{cases}$$

Figure 8.2 depicts graphically the growth described by the above equations for a population of 81 individuals on a 9×9 torus structure using a radius-1 von Neumann neighborhood.

Thus, we conclude that a more accurate fit should take into account the nonexponential growth followed by saturation.

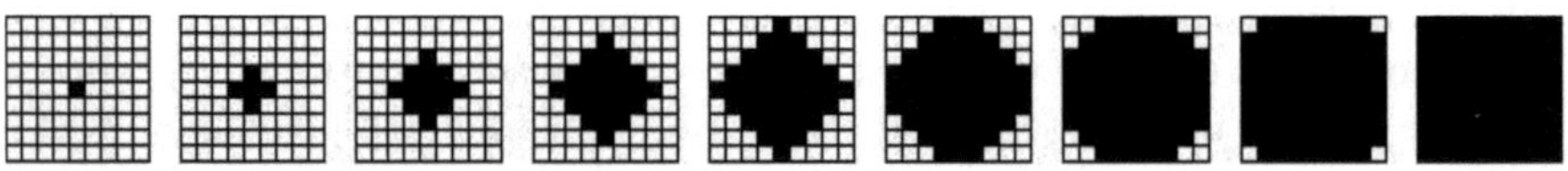

Fig. 8.2 Example of deterministic growth of $N(t)$ for a population structured as a torus with a von Neumann neighborhood

8.3.3.2 Takeover Times in Rings

Here we calculate the theoretical takeover times for rings. For simplicity, only the synchronous case is shown, the interested reader will find more details of the asynchronous cases in [12].

Since we are assuming neighborhoods of radius 1 and $N(0) = 1$, the set of cells containing a copy of the best individual will always be a connected region of the ring. Therefore, at each time step, only two more cells (the two adjacent to the connected region of the ring) will contain a copy of the best individual, with probability p. The growth of the quantity $N(t)$ can be described by the following recurrence:

$$\begin{cases} N(0) = 1, \\ E[N(t)] = \sum_{j=1}^{n} P[N(t-1) = j](j + 2p), \end{cases}$$

where $P[N(t-1) = j]$ is the probability that the random variable N takes the value j at time step $t-1$. Since $\sum_{j=1}^{n} P[N(t-1) = j] = 1$, and the expected number $E[N(t-1)]$ of copies of the best individual at time step $t-1$ is by definition $\sum_{j=1}^{n} P[N(t-1)=j]j$, the above recurrence is equivalent to

$$\begin{cases} N(0) = 1, \\ E[N(t)] = E[N(t-1)] + 2p. \end{cases}$$

The closed form of this recurrence is trivially $E[N(t)] = 2pt + 1$, and therefore the expected takeover time $E[T]$ for a synchronous ring cEA with n cells is

$$E[T] = \frac{1}{2p}\,(n-1).$$

8.3.3.3 Takeover Times in Two-Dimensional Grids

We consider CEAs defined on a square lattice of finite size $\sqrt{n} \times \sqrt{n}$ and radius-1 von Neumann neighborhood. Because of the wrapping properties of the torus, at each time step t the expected number of copies $N(t)$ of the best individual is independent of its initial position. Therefore, the expected takeover time is $E[T] = E_i[T], \forall i$.

We have seen above the limiting case of growth with deterministic selection (i.e. a mechanism that selects the best individual in the neighborhood with probability $p = 1$). In that case, the time variable t in the equations determines the half-diagonal of a square rotated by 45° (see Fig. 8.2). When a probabilistic selection method is modeled, the exact recurrences, corresponding to those derived for the ring topology in the previous subsection, become very complicated. In fact, as can be seen in Fig. 8.3, the phenomenon that has to be modeled implies different selection probabilities at different locations in the grid.

To keep the models simple and easily interpretable, the geometry of the propagation is approximated as the growth of a rotated square in the torus (see Fig. 8.4).

Fig. 8.3 Example of growth of $N(t)$ with probabilistic selection for a population of 81 individuals on a 9×9 torus structure

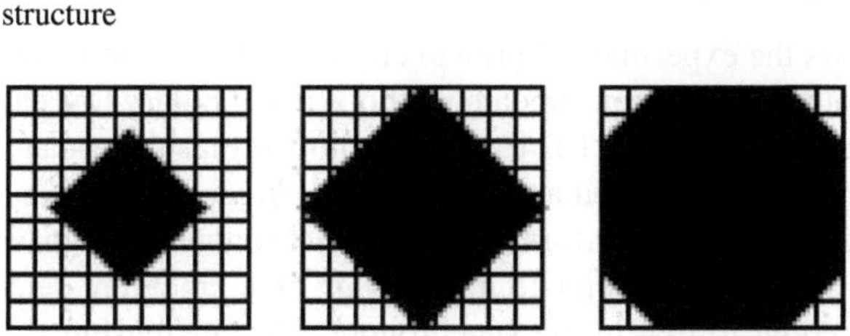

Fig. 8.4 Geometric approximation of growth with probabilistic selection in a torus-structured population: a rotated square grows as a function of time; there is unrestricted growth until the square reaches the edges of the grid, and then the population saturates

Using this geometric growth, the side length s and the half-diagonal d of the rotated square can be approximated by

$$s = \sqrt{N(t)}, \quad d = \frac{\sqrt{N(t)}}{\sqrt{2}}.$$

With these quantities, we shall now focus on synchronous takeover times, using the relevant probabilities in each case. The asynchronous cases are more difficult and the analytical expressions can be found in [12].

Let us consider the growth of such a region with a selection mechanism that has probabilities p_1, p_2, p_3, p_4, and p_5 of selecting the best individual when there are respectively 1, 2, 3, 4 and 5 copies of it in the neighborhood. Assuming that the region containing the copies of the best individual expands such that it maintains the shape of a square rotated by 45°, we can model the growth of $N(t)$ with the following recurrence

$$\begin{cases} N(0) = 1, \\ N(t) = N(t-1) + 4p_2\sqrt{N(t-1)}/\sqrt{2}\ , \ N(t) \leq n/2, \\ N(t) = N(t-1) + 4p_2\sqrt{n - N(t-1)}\ , \ N(t) > n/2. \end{cases}$$

As it is extremely difficult to find a closed analytic form of this recurrence, as well of those corresponding to the asynchronous cases, we shall made use of the explicit recurrences.

8.3.4 Experimental Validation

In this section we see how extensive numerical simulation confirms that the previously described models are sufficiently accurate ones. Furthermore, it will become clear that the cell update policy has a rather marked influence on the global induced selection pressure in the population, and thus on the exploitation/exploration characteristic of the CEA.

8.3.4.1 The Ring Structure

Figure 8.5a shows the experimental growth curves of the best individual for the synchronous and four asynchronous update methods using binary tournament selection and a population of size $n = 1024$. We may notice that the mean curves for the two asynchronous methods fixed and new random sweep show a very similar behavior while synchronous and uniform choice asynchronous overlap. The graph also shows that the asynchronous update methods give an emergent selection pressure greater than or equal to that in the synchronous case, increasing from the case of uniform choice to that of line sweep, with fixed and new random sweep in between. Figure 8.5b shows the predicted and experimental curves for the synchronous case and the mean square error between them. It is clear that the model faithfully predicts

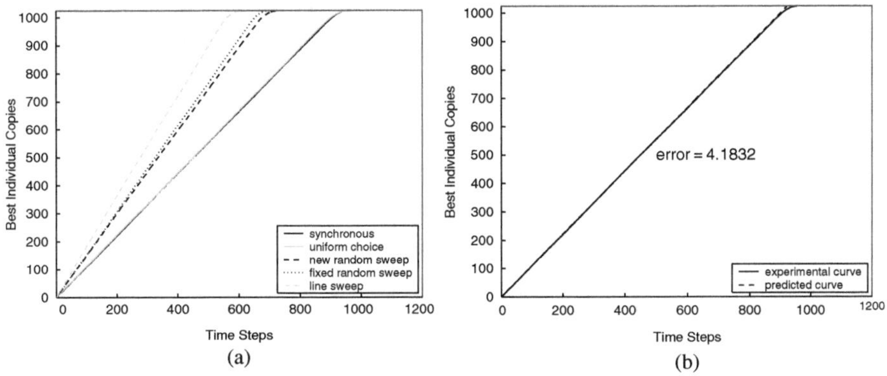

Fig. 8.5 (**a**) Takeover times with binary tournament selection: mean values over 100 runs. The vertical axis represents the number of copies $N(t)$ of the best individual in each population as a function of the time step t. (**b**) Comparison between calculated and theoretical curves for the synchronous case

the observed takeover time. Similar comparisons for the other update schemes are similarly good and can be found in [12].

Numerical values of the mean takeover times for the five update methods, along with their standard deviations, are shown in Table 8.1, where it can be seen that the fixed-random-sweep and new-random-sweep methods give results that are statistically indistinguishable, and can therefore be described by a single model. The same can be said for the synchronous and uniform-choice methods.

Similar results are obtained when using the linear ranking selection methods. To save space they are not reported here but the interested reader can consult [12].

Table 8.1 Mean takeover time and standard deviation for binary tournament selection and the five update methods in rings. Mean values over 100 independent runs

	Synchro	LS	FRS	NRS	UC
Mean takeover time	925.03	569.82	666.18	689.29	920.04
Standard deviation	20.36	24.85	17.38	20.27	26.68

8.3.4.2 The Grid Structure

Figure 8.6a shows the growth curves of the best individual for the panmictic, synchronous, and three asynchronous update methods for binary tournament selection. The mean curves for the two asynchronous methods, i. e. fixed and new random sweep, show a very similar behavior, and thus only the results for new random sweep are plotted. The graphics shows that the asynchronous update methods give an emergent selection pressure greater than in the synchronous case, increasing from the case of uniform choice to that of line sweep, with fixed and new random sweep in between (similarly to our findings for the ring topology). The logistic curve

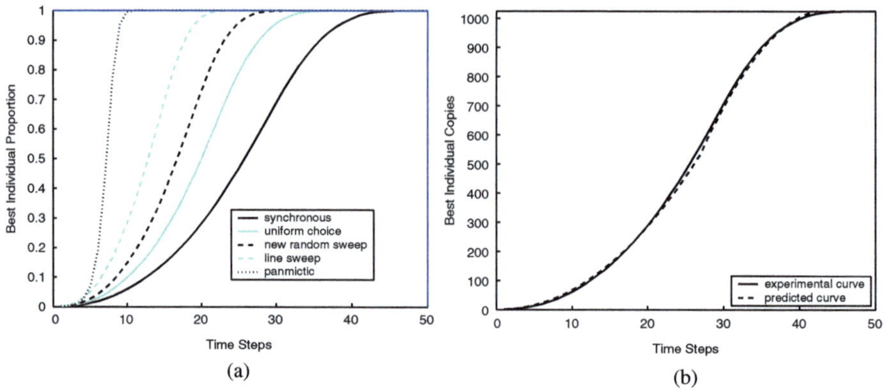

Fig. 8.6 (**a**) Takeover times with binary tournament selection. Mean values over 100 runs. The vertical axis represents the fraction of the best individual in each population as a function of the time step t. (**b**) Comparison of experimental takeover time curves (*full line*) with the model (*dashed*) for the synchronous update case

corresponding to a mixing (panmictic) population is also shown for comparison. Figure 8.6b shows the predicted and experimental curves for the synchronous update method. It can be observed that the agreement between theory and experiment is very good, in spite of the approximations made in the models, and this is also true for the asynchronous update methods [12].

The numerical values of the mean takeover times for the five update methods, together with their standard deviations, are shown in Table 8.2, where it can be seen that the fixed-random-sweep, and new-random-sweep methods give results that are statistically indistinguishable. However, this time the differences between the uniform-choice and synchronous update are meaningful in the case of torus. Results for binary tournament selection are analogous [12].

Table 8.2 Mean takeover time and standard deviation for the binary tournament selection and the five update methods in grids. Mean values over 100 independent runs

	Synchro	LS	FRS	NRS	UC
Mean takeover time	44.06	21.8	27.21	28.26	35.73
Standard deviation	1.6746	1.7581	1.5654	1.8996	2.4489

8.3.4.3 The Influence of Neighborhood Size and Grid Shape

Sarma and De Jong have empirically shown that the neighborhood's size and shape have an important influence on the induced global selection pressure in grid-structured populations [14, 15]. Theoretical models similar to those of Sect. 8.3.3 confirm that this is indeed the case [12]. Figure 8.7 depicts the behavior of takeover time for increasing values of the radius for generalized von Neumann neighborhoods. It is clear that both in rings and in two-dimensional grids selection intensity

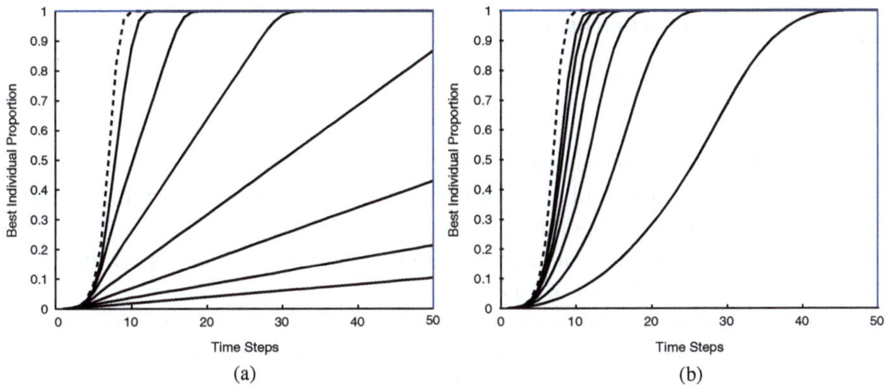

Fig. 8.7 (**a**) Growth curves for rings with neighborhoods of increasing radius with binary tournament selection. From right to left the radii are 2, 4, 8, 16, 32, 64, and 128. (**b**) Growth curves for tori with neighborhoods of radius 1, 2, 3, 4, 5, 6, and 7, increasing from right to left. The *dashed curves* in (a) and (b) represents the case of a panmictic population

grows with increasing neighborhood size. In the limit of a neighborhood size equal to the population size one obviously recovers the panmictic case, which is shown as a dashed line in the figures.

It is also possible to influence the takeover times, and thus the selection pressure, by changing the shape of a two-dimensional grid [16]. Starting from a square grid, selection pressure tends to decrease when the grid becomes rectangular. In the limit when the linear dimension becomes equal to the number of cells one recovers the ring case. Figure 8.8 is an example of the numerical behavior, which is in agreement with the model described in [12].

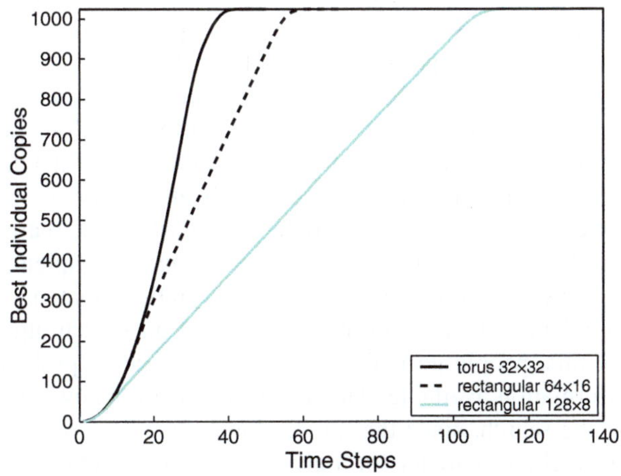

Fig. 8.8 Growth curves for synchronous evolution on toroidal structures with different ratios of axes using linear ranking selection. Mean values over 100 independent runs

8.3.4.4 Concluding Remarks on Selection Pressure

To wrap-up the study of the effects of selection in CEAs, the following observations can be made.

- Takeover times are larger in rings and grids than in mixing populations. Speeds vary by a large amount and go from linear in rings, to quadratic at most in grids, to exponential in mixing populations (see Figs. 8.5 and 8.6).
- In lattices (rings and grids) all asynchronous update methods are faster than synchronous update or at least of equal speed.
- In lattices there is a hierarchy of takeover times among the asynchronous update methods: line sweep is the faster while uniform choice is the slower.
- Selection pressure can also be controlled by using different neighborhood sizes and different grid shapes.

In standard EAs there are several ways for influencing the selection pressure; however, they all require either changing the selection method or parameterizing it in an ad hoc manner. We have just seen that in CEAs this is easier to achieve, for example, by just using a grid or a ring-structured population or by changing the cell update scheme. The selection pressure in a given structured population can also be varied at will by using flatter grids or by working with larger neighborhoods, even dynamically, i.e. during the CEA's execution. From this point of view, lattice-structured populations appear to offer a high degree of flexibility in problem-solving.

8.4 Benchmarking CEAs

Now that we understand how lattice structured populations provide us with new degrees of freedom to work with, we add variation operators in order to probe the workings of CEAs on actual optimization problems. Before describing some real-world problems to which CEAs have been successfully applied, we first study their behavior on a number of benchmark problems. However, before jumping to the empirical results, a few cautionary comments are in order. In the first place, choosing a set of test problems for benchmarking purposes is always a risky exercise for no test suite can represent all the kinds of problems that are likely to arise in practice. Thus, any test problem suite will be necessarily biased and incomplete. Furthermore, a general result, known as *no free lunch theorem* [17] tells us that, averaged over all problems and problem instances, the performance of all search algorithms is the same. In spite of this, people normally want to solve specific problems for which there is extra information available. This problem-specific knowledge can thus be used to improve the search, since the no free lunch theorem does not prevent one from finding an excellent searcher for a given problem or problem family; all that it says is that this same algorithm will necessarily perform badly on some other problems. Therefore, although no algorithm can be said to be superior in all cases, there is still scope for algorithms that behave well on particular problems or problem classes. With these limitations in mind, we now describe the standard CEA and the test suite used for performance evaluation purposes. This section draws from [18].

8.4.1 The Algorithm

The CEA is actually a *cellular genetic algorithm* (CGA) since it uses binary representation of solutions, double point crossover, and standard single bit mutation, mutation rates being much lower than crossover rates. However, it would be easy to transform the CGA into a CEA that uses, for instance, strings of reals and different recombination and mutation operators; the changes would be limited to these aspects while the global evolutionary process would remain the same, as described in the pseudo-code of Sect. 8.2. Another important consideration is that the tests have only an illustrative purpose. The CEA used, a straightforward genetic cellular algorithm, has not been tuned, nor does it include local search capabilities or problem knowledge other than what is contained in the individual representation and the fitness function. Of course, such improvements would be needed if the algorithms were intended to compete with the best solvers for a given problem or problem class. Here we are mainly interested in comparing CEAs among themselves as a function of two important parameters that have been dealt with at length before: operation timing and grid shape.

The configuration of the algorithm is detailed in Table 8.3. Besides the synchronous CGA, the four asynchronous models were also used. In the case of rectangular grids, the shapes of the grids used are shown in Table 8.4. The CGAs based on the rectangular grids were synchronous.

Table 8.3 Parameterization used in the algorithm

Population size	400 individuals
Selection of parents	Binary tournament
Recombination	Double-point crossover, $p_c = 1.0$
Bit mutation	Bit-flip, $p_m = 1/L$
Length of individual	L
Replacement	Replace if better

Table 8.4 Grid shapes studied

Name	(Shape of population)
Square	(20 × 20 individuals)
Rectangular	(10 × 40 individuals)
Narrow	(4 × 100 individuals)

8.4.2 Test Suite: Discrete Optimization Problems

The full test suite comprises combinatorial optimization problems and continuous optimization problems. Here I only show the results on the combinatorial problems. Full results and a more detailed description can be found in [18, 9] and references therein.
The problems used are:

1. the massively multimodal deceptive problem (MMDP),
2. the multimodal problem generator (P-PEAKS)
3. the error-correcting-code (ECC) design problem,
4. the problem of maximum cut of a graph (MAXCUT).

Although there cannot be an optimal choice, as explained above, this set of problems seems at least to be rather representative of different degrees of difficulty and of various important application domains. Given the computational limitations that any experiment must face, this should be enough for us to obtain a good level of confidence in the results. What follows is a brief description of each problem, original references are in [18].

Massively Multimodal Deceptive Problem (MMDP)

The MMDP is a problem that has been specifically designed to be difficult for an EA. It is made up of k deceptive subproblems (s_i) of 6 bits each, whose values depend on the number of ones (*unitation*) in a binary string (see Fig. 8.9). It is easy to see that these subfunctions have two global maxima and a deceptive attractor at the midpoint.

In the MMDP, each subproblem s_i contributes to the fitness value according to its unitation (Fig. 8.9). The global optimum has a value of k and is attained when every subproblem is composed of zeros or six ones. The number of local optima is quite large (22^k), while there are only 2^k global solutions. Therefore, the degree of multimodality is regulated by the parameter k. We use here a considerably large problem instance with $k = 40$ subproblems. The instance we try to maximize for

Unitation	Subfunction value
0	1.000000
1	0.000000
2	0.360384
3	0.640576
4	0.360384
5	0.000000
6	1.000000

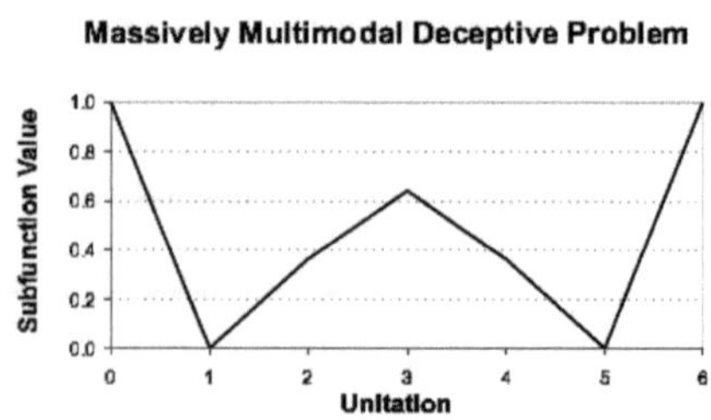

Fig. 8.9 Basic deceptive bipolar function (s_i) for MMDP

solving the problem is shown in the following equation, and its maximum value is equal to k:

$$f_{\text{MMDP}}(\mathbf{s}) = \sum_{i=1}^{k} \text{fitness}_{s_i}.$$

Multimodal Problem Generator (P-PEAKS)

The P-PEAKS problem is a multimodal problem generator. A problem generator is an easily parameterizable task which has a tunable degree of epistasis, thus allowing one to derive instances of increasing difficulty at will. Also, using a problem generator removes the opportunity to hand-tune algorithms to a particular problem, therefore allowing more fairness when comparing algorithms. With a problem generator, the algorithms are run on a high number of random problem instances, since a different instance is solved each time the algorithm runs, the predictive power of the results for the problem class as a whole is increased.

The idea of P-PEAKS is to generate P random N-bit strings that represent the location of P peaks in the search space. The fitness value of a string is the number of bits that the string has in common with the nearest peak in that space, divided by N (as shown in 8.3). By using a small/large number of peaks we can obtain weakly/strongly epistatic problems. In the work described here we have used an instance of $P = 100$ peaks of length $N = 100$ bits each, which represents a medium to high epistasis level. The maximum fitness value for this problem is 1.0. The fitness value is given by

$$f_{\text{P-PEAKS}}(\mathbf{x}) = \frac{1}{N} \max_{1 \le i \le p} \{N - \text{Hamming}D(\mathbf{x}, \text{Peak}_i)\}. \tag{8.3}$$

Error-Correcting-Code Design Problem (ECC)

We shall consider a three-tuple (n, M, d), where n is the length of each codeword (number of bits), M is the number of codewords, and d is the minimum Hamming distance between any pair of codewords. The objective is to find a code which has a value of d as large as possible (reflecting greater tolerance to noise and errors), given previously fixed values of n and M. Here we search half of the codewords ($M/2$) that will make up the code, and the other half is made up by the complement of the codewords computed by the algorithm.

The fitness function to be maximized is

$$f_{\text{ECC}} = \frac{1}{\sum_{i=1}^{M} \sum_{j=1, i \ne j}^{M} d_{ij}^{-2}},$$

where d_{ij} represents the Hamming distance between codewords i and j in the code C (made up of M codewords, each of length n). We consider here an instance where $M = 24$ and $n = 12$. The search space is of size $\binom{4096}{24}$, which is approximately 10^{87}. The optimum solution for $M = 24$ and $n = 12$ has a fitness value of 0.0674.

Maximum Cut of a Graph (MAXCUT)

The MAXCUT problem looks for a partition of the set of vertices (V) of a weighted graph $G = (V, E)$ into two disjoint subsets V_0 and V_1 such that the sum of the weights of the edges with one endpoint in V_0 and the other one in V_1 is maximized. Individuals are encoded as binary strings $(x_1, x_2, \ldots, x_n)$ of length n, where each digit corresponds to a vertex. If a digit is 1 then the corresponding vertex is in the set V_1; if it is 0 then the corresponding vertex is in the set V_0. The function to be maximized is

$$f_{\mathrm{MAXCUT}}(\mathbf{x}) = \sum_{i=1}^{n-1} \sum_{j=i+1}^{n} w_{ij} \cdot \left[x_i \cdot (1 - x_j) + x_j \cdot (1 - x_i) \right]$$

Note that w_{ij} contributes to the sum only if nodes i and j are in different partitions. While one can generate random instances of a graph to test the algorithm, here we have used the case "cut20.09", with 20 vertices and a probability 0.9 of having an edge between any two randomly chosen vertices. The maximum fitness value for this instance is 56.740064.

8.4.2.1 Experimental Results

The following tables show the results for the problem suite: MMDP, Table 8.5; P-PEAKS, Table 8.6; ECC, Table 8.7; and MAXCUT, Table 8.8. One hundred independent runs were performed for each algorithm and for every problem in the test suite. Note that only the synchronous version of the CGA was run on square, narrow, and rectangular grids, the asynchronous versions results are all for the square grid.

The tables report the average of the final best fitness over all runs, the average number of time steps needed to obtain the optimum value (if obtained), and the hit rate (percentage of successful runs). Therefore, the final distance from the optimum

Table 8.5 MMDP problem with a maximum of 1000 generations

Algorithm	Avg. solution (best=20)	Avg. generations	Hit rate (%)
Square	19.813	214.18	57
Rectangular	19.824	236.10	58
Narrow	19.842	299.67	61
LS	19.518	343.52	23
FRS	19.601	209.94	31
NRS	19.536	152.93	28
UC	19.615	295.72	36

Table 8.6 P-PEAKS problem with a maximum of 100 generations

Algorithm	Avg. solution (best=1)	Avg. generations	Hit rate (%)
Square	1.0	51.84	100
Rectangular	1.0	50.43	100
Narrow	1.0	53.94	100
LS	1.0	34.75	100
FRS	1.0	38.39	100
NRS	1.0	38.78	100
UC	1.0	40.14	100

Table 8.7 ECC problem with a maximum of 500 generations

Algorithm	Avg. solution (best=0.0674)	Avg. generations	Hit rate (%)
Square	0.0670	93.92	85
Rectangular	0.0671	93.35	88
Narrow	0.0673	104.16	94
LS	0.0672	79.66	89
FRS	0.0672	82.38	90
NRS	0.0672	79.46	89
UC	0.0671	87.27	86

Table 8.8 MAXCUT problem with a maximum of 100 generations

Algorithm	Avg. solution (best=56.74)	Avg. generations	Hit rate (%)
Square	56.74	11.26	100
Rectangular	56.74	11.03	100
Narrow	56.74	11.88	100
LS	56.74	9.46	100
FRS	56.74	9.69	100
NRS	56.74	9.55	100
UC	56.74	9.58	100

(especially interesting when the optimum is not found), the effort expended by the algorithm, and its expected efficacy, respectively, are reported.

From inspection of these tables some conclusions can be drawn. First, the asynchronous algorithms tend to need a smaller number of generations to locate an optimum than do the synchronous ones, except in the case of the MMDP problem. Statistical tests not shown here (see [18]) confirm that the differences between the asynchronous and synchronous algorithms are significant. This indicates that the asynchronous versions perform more efficiently with respect to CGAs with different grid shapes, a result that confirms the influence of the stronger selection of asynchronous CGAs.

With regard to the grid shape influence on synchronous CEAs, one can see that the flatter the grid, the slower the algorithm, which is in line with takeover time results, taking into account that here recombination and mutation are active as well. On the other hand, the grid shape does not seem to have an influence on the solution quality.

Conversely, if we pay attention to the success (hit) rate, it can be concluded that the synchronous policies with various rectangular shapes outperform the asynchronous algorithms (except for the ECC problems): slightly in terms of the average final fitness, and clearly in terms of the probability of finding a solution (i.e. the frequency of location of the optimum).

Another interesting result is the fact that we can define two classes of problems: those solved by all methods to optimality (100% hit rate) and those in which no 100% rate is achieved at all. The former seem to be suitable for straight CGAs, while the latter would need some help, for example by including local search.

In order to summarize the large set of results and draw some useful conclusions, a final ranking of the algorithms following three different metrics is presented: average best final solution, average number of generations for success, and hit rate. Table 8.9 shows the three rankings, which go from 1 (best) to 7 (worst) according to the three criteria.

As one would expect after the previous comments, synchronous algorithms with "narrow" and "rectangular" shapes are in general more accurate than all the asynchronous algorithms, according to the criteria of average best final fitness and of hit ratio, at least for the test problems used here, with a special leading position for narrow population grids. On the other hand, the asynchronous versions clearly outperform any of the synchronous algorithms in terms of the average number of generations, with a trend towards NRS as the best-ranked flavor of CGA for the test suite.

In conclusion, asynchronous algorithms seem to be numerically faster than synchronous ones for the P-PEAKS, ECC, and MAXCUT problems, but not for the MMDP. On the other hand, synchronous algorithms outperform asynchronous ones in terms of the hit rate for these benchmarks, which could be an important issue for many applications. In particular, the more explorative character of the narrow population structure seems to allow a more accurate search in most cases. Again, it has to be pointed out that the results cannot be immediately generalized to other problems or problem types. However, the picture that emerges from this empirical investigation is a coherent one, and it essentially confirms the importance of selection intensity considerations.

Table 8.9 Ranking of the algorithms

Avg. solution		Avg. generations		Hit rate	
1 Narrow	4	1 NRS	8	1 Narrow	4
2 Rectangular	9	2 LS	10	2 Rectangular	9
2 FRS	9	3 FRS	11	2 FRS	9
4 NRS	10	4 UC	16	4 NRS	11
5 UC	11	5 Rectangular	19	5 Square	12
5 LS	11	6 Square	21	5 UC	12
7 Square	12	7 Narrow	27	5 LS	12

8.5 CEAs and Real-World Problem Solving

Although less popular than other kinds of EAs, CEAs have been successfully used for solving difficult real-life optimization problems. However, the bare-bones CEAs presented in the previous Sects. would not be efficient enough to compete with well established problem solving techniques. Plain CEAs have the advantage of being easy to apply to almost any problem once a representation for individuals has been found. However, one needs to put more knowledge into the algorithm to hope to be able to successfully tackle hard engineering problems. There are several ways in which a CEA, or a standard panmictic EA for that matter, can be enhanced. In order to convey the main ideas, here I list some modifications that would make a CEA perform better on a given problem or problem class (the list is not exhaustive, as several other possibilities have been left out):

- Memetic and Hybrid algorithms
- Metaheuristics
- Multi-Objective optimization

Memetic and hybrid algorithms are heuristic techniques that combine, for instance, population based methods with local search approaches or, in general, add problem-specific knowledge to the heuristic search. In the case of CEAs, the evolutionary scenario would typically be supplemented with a technique that searches for locally optimal solutions among the best members of the population.

Metaheuristics are generic problem-solving techniques in which a higher-level heuristic guides or controls an underlying problem-specific heuristic method. These techniques overlap to some extent with memetic algorithms and there isn't a generally accepted definition.

In multi-objective optimization one is interested in optimizing under the constraint that several objectives must be taken care of, not only one such as maximization or minimization of a specific fitness function. As these objectives are usually incompatible, one is led to consider tradeoffs in the way in which resources are allocated. Actually, many real-world problems are of this kind.

Evidently, a description of the techniques and achievements in this vast interdisciplinary field would be impossible in a short space. I thus briefly describe a couple of problems for which these enhanced CEAs have been successful to give the flavor of present-day research and applications. For more information, the reader is referred to the original articles and to the book [10] which presents the state of the art in the field.

8.5.1 Vehicle Routing Problem

The Vehicle Routing Problem (VRP) includes a large set of problems that can be seen as multiple traveling salesperson problems. The problems are obviously NP-hard and they are very important in practice as they represent the framework that is found in many industrial applications in which finding better solutions means large savings of resources. This kind of problems has been studied for years and good solutions have been found. Recently, a CEA-based approach has been able to outperform previous results [19]. In combination with the standard CEA framework, the authors have used a specialized representation for the routes and the deliveries, together with crossover and mutation operators that were tailored to this same representation. Furthermore, a local search phase was added after the application of variation operators in which special tour operators were applied (2-opt and 1-exchange) in order to further improve good solutions. The method was applied to many instances of the VRP and a number of best-so-far solutions were found. On the other hand, the solutions found in the other cases were equal or very close to the best solutions known.

8.5.2 Diffusion in Mobile Ad-Hoc Networks

A mobile ad-hoc network (MANET) is a kind of wireless network, and is a self-configuring network of mobile routers (and associated hosts) connected by wireless links, the union of which form an arbitrary topology since there is no need of a previously existing infrastructure. The routers are free to move randomly and organize themselves arbitrarily; thus, the network's wireless topology may change rapidly and unpredictably. In [20] the authors apply a multi-objective CEA to the problem of finding good strategies for information transmission and diffusion to MANETS that are situated in metropolitan areas. The objectives typically are: minimizing the duration of the diffusion process, maximizing the network coverage, and minimizing the network usage. Results were very good and competitive with other established multi-objective optimization methods.

8.6 Conclusions

I have described a particular class of probabilistic cellular automata called cellular evolutionary algorithms. They are formally CAs but they also belong to the large family of evolutionary algorithms. From that point of view, I have presented mathematical models that correctly describe the behavior of local selection methods on lattice-structured evolutionary populations. These models are useful in themselves in the field of probabilistic CAs but, in conjunction with genetic operators such as mutation and crossover, they also help explain the observed behavior of evolving populations of solutions to a given problem. This has been complemented with an empirical analysis of CEAs on a test suite of discrete optimization problems and, finally, with a discussion of two hard real-life problems for which CEAs have proven their usefulness with respect to other heuristic methods. Cellular evolutionary algorithms are easy to implement and to deal with. When complemented with problem knowledge and local search methods they are competitive with other search heuristic and thus definitely belong to the toolkit of the modern engineering and scientific problem solver.

References

1. A.E. Eiben, J.E. Smith. *Introduction to Evolutionary Computing*. (Springer Heidelberg, 2003)
2. M. Gorges-Schleuter. ASPARAGOS an asynchronous parallel genetic optimisation strategy, ed. by J.D. Schaffer, *Proceedings of the Third International Conference on Genetic Algorithms*, Morgan Kaufmann, San Fransisco, CA, pp. 422–427 1989
3. B. Manderick, P. Spiessens, Fine-grained parallel genetic algorithms, ed. by J.D. Schaffer, *Proceedings of the Third International Conference on Genetic Algorithms*, Morgan Kaufmann, San Fransisco, CA, pp. 428–433 1989
4. W.D. Hillis, Co-evolving parasites improve simulated evolution as an optimization procedure. Physica D **42**, 228–234 (1990)
5. R.J. Collins, D.R. Jefferson, Selection in massively parallel genetic algorithms, ed. by R.K. Belew, L.B. Booker, *Proceedings of the Fourth International Conference on Genetic Algorithms*, Morgan Kaufmann, San Francisco, CA, pp. 249–256 1991
6. Y. Davidor, A naturally occurring niche & species phenomenon: The model and first results, ed. by R.K. Belew, L.B. Booker, *Proceedings of the Fourth International Conference on Genetic Algorithms*, Morgan Kaufmann Publishers, San Mateo, CA, pp. 257–263 1991
7. M. Tomassini, The parallel genetic cellular automata: Application to global function optimization, ed. by R.F. Albrecht, C.R. Reeves, N.C. Steele, *Proceedings of the International Conference on Artificial Neural Networks and Genetic Algorithms*, Wien, New York, NY, pp. 385–391 1993
8. D. Whitley, Cellular genetic algorithms ed. by S. Forrest, *Proceedings of the Fifth International Conference on Genetic Algorithms*, Morgan Kaufmann Publishers, San Mateo, CA, p. 658 1993
9. M. Tomassini, *Spatially Structured Evolutionary Algorithms* (Springer, Heidelberg, 2005)
10. E. Alba, B. Dorronsoro, *Cellular Genetic Algorithms* (Springer, Heidelberg, 2008)
11. B. Schönfisch, A. de Roos, Synchronous and asynchronous updating in cellular automata. BioSystems **51**, 123–143 (1999)
12. M. Giacobini, M. Tomassini, A. Tettamanzi, E. Alba. Selection intensity for cellular evolutionary algorithms for regular lattices. IEEE Trans. Evol. Comput. **9**, 489–505 (2005)
13. D.E. Goldberg, K. Deb, A comparative analysis of selection schemes used in genetic algorithms, ed. by G.J.E. Rawlins, *Foundations of Genetic Algorithms 1*, Morgan Kaufmann, San Francisco, CA, pp. 69–93 1991
14. J. Sarma, K.A. De Jong, An analysis of the effect of the neighborhood size and shape on local selection algorithms, ed. by H.M. Voigt et al., *Parallel Problem Solving from Nature (PPSN IV)*, Lecture Notes in Computer Science, (Springer Heidelberg, 1996), pp. 236–244
15. J. Sarma, K.A. De Jong, An analysis of local selection algorithms in a spatially structured evolutionary algorithm ed. by T. Bäck, *Proceedings of the Seventh International Conference on Genetic Algorithms*, Morgan Kaufmann, San Francisco, CA, pp. 181–186 1997
16. E. Alba, J.M. Troya, Cellular evolutionary algorithms: Evaluating the influence of ratio ed. by M. Schoenauer et al., *Parallel Problem Solving from Nature (PPSN VI)*, Lecture Notes in Computer Science, vol. 1917 (Springer Heidelberg, 2000), pp. 29–38
17. D.H. Wolpert, W.G. Macready, No free lunch theorems for optimization. *IEEE Trans. Evol. Comput.*, **1**(1), 67–82 (1997)
18. B. Dorronsoro, E. Alba, M. Giacobini, M. Tomassini, The influence of grid shape and asynchronicity on cellular evolutionary algorithms. In *2004 Congress on Evolutionary Computation (CEC 2004)*, IEEE Press, Piscataway, NJ, pages 2152–2158, 2004
19. E. Alba, B. Dorronsoro, Computing nine new best-so-far solutions for capacitated VRP with a cellular GA. *Inform. Process. Lett.* **98**, 225–230 (2006)
20. E. Alba, B. Dorronsoro, F. Luna, A. J. Nebro, P. Bouvry, L. Hogie, A cellular multi-objective genetic algorithm for optimal broadcasting strategy in metropolitan MANETs. *Comput. Commun.* **30**, 685–697 (2007)

Chapter 9
Artificial Evolution of Arbitrary Self-Replicating Structures in Cellular Spaces

Zhijian Pan and James A. Reggia

Self-replicating systems are systems that are capable of producing copies of themselves. The terms *replication* and *reproduction* are often considered synonymous, but in actuality there is a distinction [36]. Replication is a developmental process, involving no genetic operators and resulting in an exact duplicate of the parent organism. Reproduction, on the other hand, is an evolutionary process, involving genetic operators such as crossover and mutation, thereby contributing to the variation that is an important aspect of evolution.

The mathematician John von Neumann is credited with being the first to conduct a formal investigation of artificial self-replicating machines [19, 20]. He believed that self-replicating biological organisms could be viewed as very sophisticated machines. He argued that the important thing about a replicating organism was not the matter from which it is made, but rather the information and the complexity of the interactions between parts of the organism. In particular he asked whether we can use purely mathematical-logical considerations to discover the specific features of biological automata that make them self-replicating. Much subsequent work on artificial self-replicating machines has continued in this spirit, being motivated by the desire to understand the fundamental information processing principles and algorithms involved in self-replication, independent of how they might be physically realized. It has also been argued that a better understanding of these principles could be useful in atomic-scale manufacturing (nanotechnology), in creating robust electronic systems, in facilitating future planetary exploration, and in gaining a better understanding of the origins of life.

In the following, we give an overview of past work and recent developments involving artificial self-replication implemented as cellular automata (CA). Sections 9.1, 9.2, and 9.3 of this chapter summarize some key results and historical trends, such as the drive to produce progressively simpler replicators while at the same time giving them an added ability to carry out functions as they reproduce. The remaining

Z. Pan (✉)
IBM Pervasive Computing Lab, 1997 Annapolis Exchange Pkwy, Annapolis, MD 21401, USA
e-mail: edzpan@yahoo.com

A.G. Hoekstra et al. (eds.), *Simulating Complex Systems by Cellular Automata*, Understanding Complex Systems, DOI 10.1007/978-3-642-12203-3_9,

Sects. 9.4 and 9.5 of the chapter describe recent progress using evolutionary computation methods to automatically discover novel self-replicating structures in cellular spaces.

9.1 Self-Replicating Systems in Cellular Automata

While a variety of approaches have been taken in the past to studying self-replicating systems, including mechanical [7] and biochemical [29] systems, a central and enduring approach has focused on embedding abstract self-replicating structures in cellular spaces. In CA, a structure can simply be viewed as a configuration of contiguous active cells. Note that such a structure can also enclose empty (quiescent) cells, as long as all of the active cells in the structure remain contiguous. The number of active cells in the structure is called its size. An active cell in a structure is also called a component in the structure. A structure is called an isolated structure if no active cells which are not in it are adjacent, i.e., no active cells are in the immediate neighborhood of any active cell in the structure.

An isolated structure at time $t = 0$ is called a seed. A structure at time $t \geq 1$ is called a seed replica, or just *replica*, if the structure inherits all properties of the seed, that is, (1) it has the same configuration as the seed; (2) all of its active cells are contiguous with one another; and (3) it is isolated from other active cells. Note that, since a CA space is considered to be isotropic, meaning that the absolute directions of east, south, west, and north are indistinguishable, the replica can be displaced and perhaps rotated relative to the original.

Defining a self-replicating system embedded in a CA consists of specifying all of the following: (1) a seed and its environment; (2) a rule table; and (3) a time $t \geq 1$, such that after the rule table has been applied to the seed and recursively to subsequent configurations, n replicas are constructed in the infinite cellular space, for some positive integer n. Note that a seed itself is not a self-replicating system. A seed is only a structure, which is a static part of the self-replicating system. It is the rule table that makes the seed replicate in the given CA space.

Below, we suggest that past work on self-replicating systems in CA is best viewed as having involved two main approaches: universal constructors, and much simpler non-universal structures such as replicating loops. We then present the argument that recent work using evolutionary methods to automatically discover new types of replicating structures is likely to be a useful direction for future research, and we provide some examples supporting this hypothesis.

9.2 Universal Constructors in CA Spaces

To embed a hypothetical self-replicating machine in a CA space, von Neumann [20] envisioned that the following characteristics should be present in a self-replicating system model: (1) constructional universality, that is the ability to construct any kind of configuration in the CA space from a given description (self-replication is then only a particular case of universal construction); and (2) computational universality, that is the ability to operate as a universal Turing machine, and thus to execute any computational task.

To implement this idea, von Neumann developed his theoretical model in a CA space with tens of thousands of components in 29-state cells and using a 5-cell neighborhood [20]. His model consists of a configuration whose numerous components (non-quiescent cells) can be grouped into two functional units: a constructing unit, which constructs the new automaton, and a tape unit, which stores and reads the information needed to construct the automaton. The tape unit consists of a "tape" and a tape control, where the tape is a linear array of cells that contains a specification of the automaton to be constructed. The construction of the automaton is carried out by sending signals (in the form of propagating cell states) between the tape unit and the construction unit. The construction unit consists of a construction arm and construction control. The construction arm is an array of cells through which cell states to be constructed can be sent from the construction control to the designated places in the construction area.

Von Neumann's universal constructor model employs a complex transition rule set, with the total number of cells composing the universal constructor estimated to range from 50,000 to 200,000 [36]. In the late 1960s Codd demonstrated that if the component or cell states meet certain symmetry requirements, von Neumann's model could be reduced to a simpler 4000 component structure embedded in an 8-state, 5-neighbor 2-D cellular space [6]. Vitanyi described a sexually reproducing cellular automata model and showed that the recombination of the parents' characteristics in the offspring conforms to recombination in nature. Similarities and differences with biological systems have been discussed [38, 39].

A number of researchers also have subsequently considered the implementation of a universal constructor. Signorini discussed the implementation of the 29-state transition rule and three organs (pulser, decoder, and periodic pulser) on a SIMD (single-instruction multiple-data) computer [35]. Pesavento provided a close simulation of von Neumann's model, but self-replication is not demonstrated since the tape required to describe the universal constructor is too large to simulate [27]. Beuchat and Haenni implemented a hardware module of a 25-cell pulser using field-programmable gate arrays (FPGAs) [3]. Buckley and Mukherjee described the constructibility of signal-crossing solutions in von Neumann's 29-state cellular automata [23].

9.3 Self-Replicating Loops

In 1984, Langton observed that biological self-replicating systems are not capable of universal construction, and concluded that universal construction may be a sufficient condition for self-replication, but is not a necessity. He successfully took a loop structure from Codd's self-replicating model involving only 86 cells, in a

2-dimensional, 8-state, von Neumann neighborhood CA space, and showed that it could be modified to replicate [11, 12]. The resulting self-replicating structure is essentially a square loop, with internal and external sheaths, where the data encoding the instructions to construct a duplicated loop circulate counterclockwise. A duplicated loop is formed after 151 time steps.

Langton's self-replicating loop is strikingly simple, and can be easily simulated on computers. Further, Byl eliminated the internal sheath of Langton's loop and discovered a smaller loop, which composed of only 12 cells embedded in a six state cellular space [4]. Reggia et al. removed the external sheath, and constructed a family of yet smaller self-replicating loops, with the smallest comprising only 5 cells, embedded in a 6 state cellular space [30]. An unsheathed loop that is capable of replicating, given an appropriate set of rules, is shown in Fig. 9.1. The self-replication process is illustrated in Fig. 9.2. Loops without sheaths can be made very small, replicating in less than a dozen steps, as illustrated in Fig. 9.3.

The self-replicating models described up to this point all involved a manual design process and a trend toward producing smaller and simpler structures: from von Neumann's model which has the power of universal computation and universal construction to the simplest self-replicating loops which can do nothing but self replicate [36]. However, it was realized that a system capable of self-replication but not much else would not be very useful. In 1995, Tempesti asked whether it is possible to add additional computation capabilities to the simple self-replicating loops, and hence attain more complex replicating machines that are nevertheless completely realizable. He devised a self-replicating loop which resembles Langton's, but with the added capability of attaching an executable program that writes out the letters LSL, acronym of the Logic Systems Laboratory, while it replicates [37]. Perrier et al. further extended this approach, demonstrating the capability of constructing a self-replicating loop that could implement any program, written in a simple yet universal language [26]. This self-replicating machine includes three parts: loop, program, and data, all of which are collectively self replicating, followed by the execution of the program on the given data. In the models of Tempesti and Perrier et al., the program embedded in each loop is copied from parent to child unchanged,

```
O+-OL-OL
-      -
+      O
O      O
-      O
+      O
O      O
-+O-+O-+OOOO
```

Fig. 9.1 A self-replicating loop in a 2D cellular automata space. Read clockwise starting at the lower right, there are a series of signals (+−) embedded in the loop structure, each indicating that the arm on the bottom right should grow out one step. These are followed by two signals indicating a left turn (L− L−). These signals circulate counterclockwise around the loop, advancing one cell per time step. As they do, copies of the signals pass out the arm at the lower right, causing it to extend and turn so that a second "child" loop is constructed

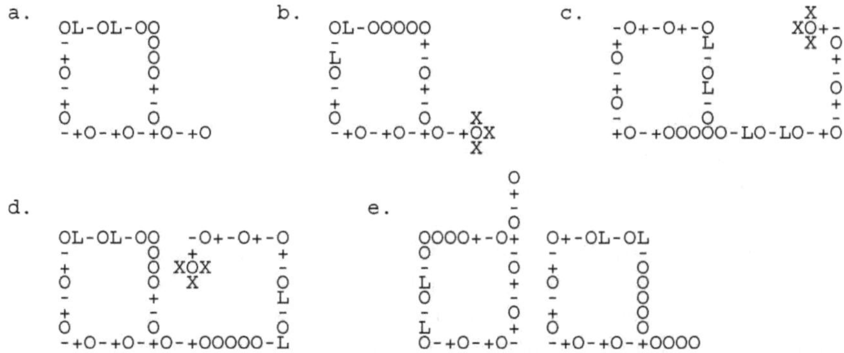

Fig. 9.2 Successive states of a self-replicating loop that started at time $t = 0$ as illustrated in the proceeding figure. The instruction sequence repeatedly circulates counterclockwise around the loop with a copy periodically passing onto the construction arm. At $t = 3$ (**a**) the sequence of instructions has circulated three positions counterclockwise with a copy also entering the construction arm. At $t = 6$ (**b**) the arrival of the first + state at the end of the construction arm produces a growth cap of X's. This growth cap, which is carried forward as the arm subsequently extends to produce the replica, is what makes a sheath unnecessary by enabling directional growth and right-left discrimination. Successive arrival at the growth tip of +'s extends the emerging structure and arrival of paired L's causes left turns, resulting in eventual formation of a new loop. Intermediate states are shown at $t = 80$ (**c**) and $t = 115$ (**d**). By $t = 150$ (**e**) a duplicate of the initial loop has formed and separated (on the right); the original loop (on the left, construction arm having moved to the top) is beginning another cycle of self-directed replication

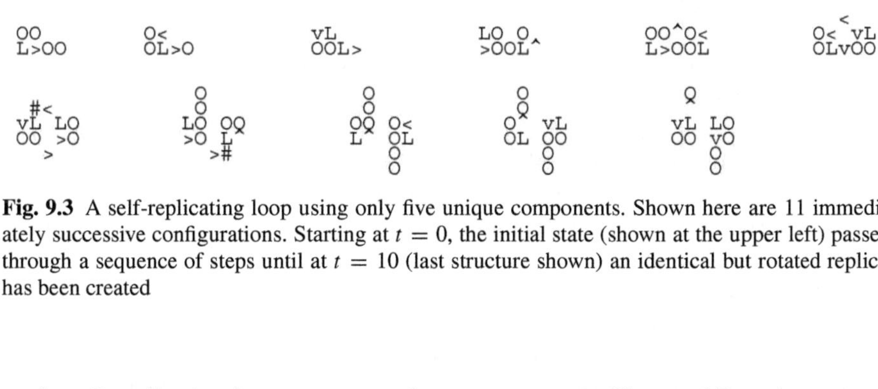

Fig. 9.3 A self-replicating loop using only five unique components. Shown here are 11 immediately successive configurations. Starting at $t = 0$, the initial state (shown at the upper left) passes through a sequence of steps until at $t = 10$ (last structure shown) an identical but rotated replica has been created

so that all replicating loops carry out the same program. Chou and Reggia reported a different approach in which each replica receives a distinct partial solution that is modified during replication [5]. Replicas with failed solutions are not allowed to continue replicating while the replicas with promising solutions will further replicate and explore finer solutions of an NP-complete problem known as SAT. These works demonstrated that simple, manually designed self-replicating loops are capable of providing some limited "secondary" function beyond simple self replication. However, such secondary constructional or computational capability is all implemented as a pre-written executable program, which is attached to the loop itself, increasing the complexity of the seed structure itself.

The discovered structures outlined above in this section all share the same property of being based on a self-replicating loop, a simple square shape that enables their replication [34]. The structures differ in size more than complexity. Morita and Imai showed that the replication of simple non-loop structures could be realized within a "reversible" cellular space [16, 17]. A reversible cellular automaton is a special, backward-deterministic type of CA in which every grid configuration of states has at most one predecessor. As a result, they created self-replicating structures like worms as well as loops. Sayama further created self-replicating worms that are capable of increasing structure complexity in terms of the length and branching frequency of the worm [34]. Chou and Reggia took a new direction and demonstrated that self-replicating loops can come about spontaneously and emerge from an initial random configuration of non-replicating components. Replication occurs in a milieu of free-floating components, replicas grow or change their sizes over time, and the transition function is based on a functional division of data fields [30]. Salzberg et al. further studied the evolutionary dynamics and diversity in a model called Evo-loops [32], a modified version of structurally dissolvable self-replicating loops [33]. Nehaniv implemented the Evo-loop model, and studied its evolution and self-replication in asynchronous cellular automata where each cell can be updated randomly and asynchronously [18].

9.4 Evolution of CA Rules

Given the local concurrent computations in CA, it is difficult to program their transition functions when the desired computation requires global communication and global integration of information across great distances in the cellular space. Thus it is extremely difficult, in general, to design local state-transition rules that, when they operate in each cell of the cellular space, produce a desired global behavior. This difficulty has contributed to limiting the number of self-replicating structures designed and studied to date.

Evolutionary computation algorithms, both genetic algorithms and genetic programming (GP), have been used to automatically evolve cellular automata rules for non-self-replication problems in cellular automata spaces [2, 8–10]. For example, various human-written algorithms have appeared for the difficult majority classification task in one-dimensional two-state cellular automata, prior to the introduction of genetic programming to evolve a rule for this task [1]. It was demonstrated that the CA rules for majority classification evolved by genetic programming achieved an accuracy exceeding all known human written rules, and that GP produced rules that are qualitatively different from all previous rules in that they employ a larger and more intricate repertoire of domains and particles to represent and communicate information across the cellular space. On the other hand, Richards et al. outline a method for extracting two-dimensional cellular automaton rules directly from experimental data [31]. This method employs genetic algorithms, to search efficiently through a space of probabilistic CA rules for rules that best reproduce the observed behavior of the data.

Inspired by the successful use of evolutionary computation methods to discover novel rule sets for other types of CA problems, Lohn et al. used a genetic algorithm to evolve rules that would support self-replication [13, 14]. This study showed that, given small but arbitrary initial configurations of non-quiescent cells ("seed structures") in a two-dimensional CA space, it is possible to automatically discover a set of rules that make the given structure replicate, and in ways quite different from self-replicating structures manually designed by investigators in the past. A rule table was generated using a genetic algorithm. When it is used to guide each cell in the cellular automata to transit its state for certain number of time steps, multiple instances of the seed structures have formed.

The evolving of rules for self-replicating CA structures using a genetic algorithm adopted a linear encoding of the rules [14]. The essential idea is that the rule table took the form of a linear listing of the entire rule set. Each rule was conceptually encoded as a string CTRBL $\rightarrow$ C$'$, where each letter specifies respectively the current states of the Center, Top, Right, Bottom, and Left cells, and the next state C$'$ of the center cell. In general, a very large rule table is needed for this, although since all possible rules are represented in a canonical order, for each rule CTRBL $\rightarrow$ C$'$ only the single state C$'$ needs to be recorded explicitly.

The effectiveness of this approach to discovering state-change rules for self-replication proved to be quite limited. To accommodate the use of a genetic algorithm, the rules governing state changes were linearly encoded, forming a large chromosome that led to very large computational costs during the evolutionary process. This created the problem that when the size of the seed structure was moderately increased, the computation cost became prohibitive for the rule table to be effectively evolved, and the yield (fraction of evolutionary runs that successfully discover self-replication) decreased dramatically. As a result, it only proved possible to evolve rule sets for self-replicating structures having no more than 4 components, even with the use of a 40-processor high performance computer, leading to some pessimism about the viability of evolutionary discovery of novel self-replicating structures.

9.5 Evolution of Self-Replicating Structure Using Genetic Programming

To resolve the computational barrier incurred in previous work, we recently explored the use of a more efficient and compact encoding of a CA's configuration and rules as tree structures. We examined whether such tree structures, combined with genetic programming (GP) methods, could dramatically improve the computational efficiency of automatic discovery of new self-replicating CA structures [25, 21, 24, 22, 23].

9.5.1 S-tree Encoding and General Structure Representation

A tree encoding provides an effective and efficient mechanism for representing arbitrary structures in a CA space that can be used by the GP model. While the approach is quite general (arbitrary neighborhoods and space dimensions), for concreteness it is developed for two-dimensional CA's and uses the 8-cell Moore neighborhood to define structure isolation. We refer to the tree used to represent a seed structure as its *structure tree* or *S-tree*.

Recall that an arbitrary structure can be viewed as a configuration of active cells in a CA space, with the conditions that the active cells inside the configuration are contiguous and isolated from all active cells outside of the configuration. Such a structure can be modeled as a connected, undirected graph, as follows.

The problem of structure encoding is converted to searching for a minimum spanning tree (MST) in order to most efficiently traverse the graph and encode its vertices (components). Figure 9.4a shows a simple seed structure in a 2-D CA space, composed of 4 oriented components. This structure is initially converted into a graph simply by adding an edge between each component and its 8 Moore neighbors, as shown in Fig. 9.4b. The quiescent cells, shown empty in Fig. 9.4a, are visualized with the symbol $*$ in Fig. 9.4b. From this example we can see such a graph has the following properties: (1) it connects every component in the structure; (2) it also includes every quiescent cell immediately adjacent to the structure (which isolates the structure from its environment); and (3) no other cells are included in the graph. We call such a graph the *Moore graph*.

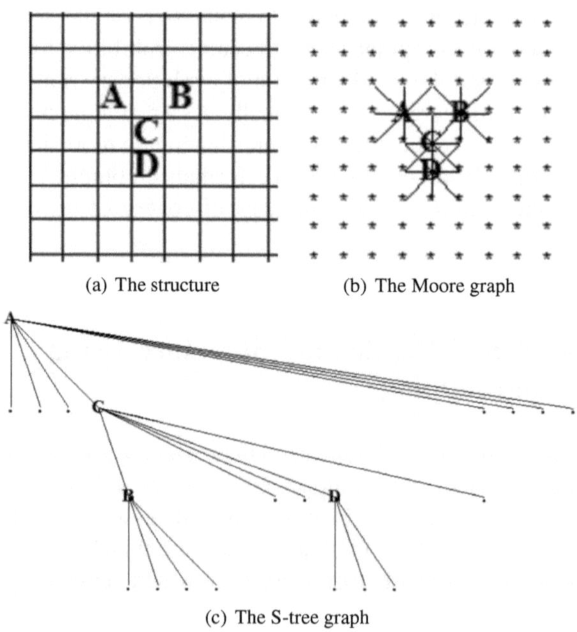

Fig. 9.4 An example structure, its Moore graph, and its S-tree graph

The Moore graph for an arbitrary structure can be converted into a minimal spanning tree which is the *S-tree*. The essential idea is as follows. After assigning a distance of 1 to every edge on the Moore graph, pick an arbitrary component of the structure as the root, and perform a breadth-first-search of the graph. The resultant S-tree for the structure shown in Fig. 9.4a is depicted in Fig. 9.4c. Starting from the root (A, in this example), explore all vertices of distance 1 (immediate Moore neighbors of the root itself); mark every vertex visited; then explore all vertices of distance 2; and so on, until all vertices are marked. The S-tree is therefore essentially a sub-graph of the initial Moore graph. It has the following desirable properties as a structural encoding mechanism: (1) it is acyclic and unambiguous, since each node has a unique path to the root; (2) it is efficient, since each node appears on the tree precisely once, and involves the shortest path from the root; (3) it is universal, since it works for arbitrary Moore graphs and arbitrary CA spaces; (4) quiescent cells can only be leaf nodes; (5) active cells may have a maximum of 8 child nodes, which can be another active cell or a quiescent cell (note the root always has 8 child nodes); and (6) its size (defined as the total number of nodes in the tree) has an upper limit which can be calculated from the size of the encoded structure. With a specific component selected as the root, is the S-tree unique for a given structure? The MST algorithm only guarantees the vertices of distance d to the root will be explored earlier than those of distance $d+1$. However, each Moore neighbor of a visited component lies the same distance from the root (such as B and D in Fig. 9.4b), which may potentially be explored by the MST algorithm in any order and therefore generate different trees. This problem may be resolved by regulating the way each active cell explores its Moore neighbors, without loss of generality. For instance, let the exploration be always in a clock-wise order starting at a specific position (for instance, the left). As a result, it is guaranteed that a specific structure always yields the same S-tree (see [24] for further details).

The S-tree representation provides an unambiguous, efficient, and universal mechanism for encoding the structural information of an arbitrary artificial machine in CA space. This makes it possible to represent arbitrary CA structures with a uniform data structure, and more importantly, enable an evolutionary model or rule learning system to be built and function without having knowledge of any details of the involved structures a priori [24]. When it is desired, an S-tree can be used to fully reconstruct the structure it represents by recursively reconstructing each Moore neighbor from the root component as guided by the S-tree. In the CA space, a structure may re-appear in a translated, rotated, and/or permuted condition. The S-tree encoding can be used to identify each of these conditions, as detailed in subsequent sections.

9.5.2 R-tree Encoding and Rule Set Representation

Just as the seed structure can be represented by an S-tree, the rules that govern state transitions of individual cells can be represented as a *rule tree* or *R-tree*. This section introduces R-tree encoding, which is much more efficient and largely resolves the limitations of an exhaustive (all rule) linear encoding previously used with genetic algorithms. In contrast to the S-tree, the R-tree formulation considered here is based on the 5-neighborhood (or von Neumann neighborhood). In other words, just as with the case of evolving self-replication with a genetic algorithm, the rules evolved are of the form CTRBL → C′.

9.5.2.1 R-tree Encoding

An *R-tree* is essentially a rooted and ordered tree that encodes every rule needed to direct the state transition of a given structure, and only those rules. The root is a dummy node. Each node at level 1 represents the state of a cell at time t (i.e., C in CTRBL → C′). Each node at level 2, 3, 4, and 5 respectively, represents the state of each von Neumann neighbor of the cell (without specifying which is top, left, bottom, and right). Each node at level 6 (the leaf nodes) represents the state of the cells at time $t{+}1$ (i.e., state C'). Therefore, the R-tree may also be viewed as similar to a decision tree, where each cell can find a unique path to a leaf by selecting each sub-branch based on the states of itself and its von Neumann neighbors. An example R-tree encoding 16 rules is shown in Fig. 9.5. Each path from the root to a leaf node corresponds to one rule. For example, the leftmost path indicates that a quiescent cell surrounded by all quiescent cells stays quiescent. In the R-tree, the actual symbol

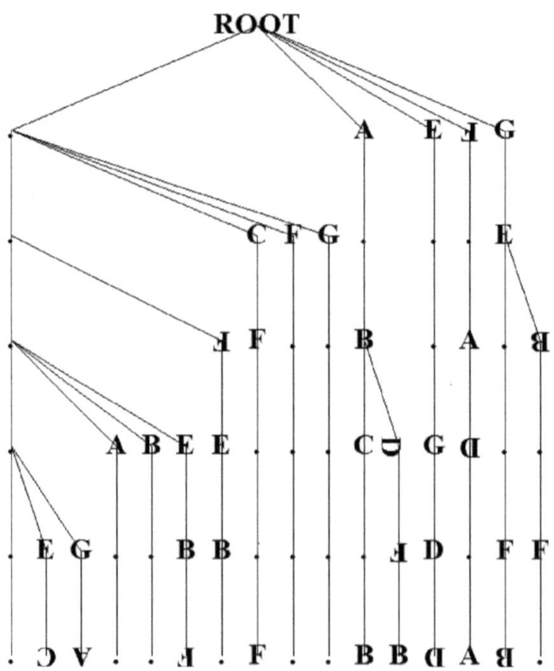

Fig. 9.5 An example R-tree representing 16 rules. Each path from the root node to a leaf represents a rule of the form CTRBLC'

in correct orientation is displayed for each state (e.g., 0; A, ᗉ, ∀, ᗆ and B, ᙠ, ꓭ, ᗷ indicate oriented components A and B in their four possible relative orientations). Each rule in the R-tree is actually applicable in four situations depending on the orientation of the state of the center cell C at level 1 in the R-tree.

The R-tree has the following properties: (1) it is a height balanced and parsimonious tree, since each branch has precisely a depth of 6; (2) taking N_s = number of possible cell states, the root and each node at level 1, 2, 3, and 4 may have a maximum of N_s child nodes, which are distinct and sorted by the state index; (3) each node at level 5 has precisely one child, which is a leaf; (4) it handles arbitrarily rotated cells with a single branch and therefore guarantees that there always exists at most one path that applies to any cell at any time, even after rotating and or permuting its orientation. Due to the R-tree properties described above, the worst search cost for a single state transition is reduced to $5\ln(N_s)$ (5 nodes on each path to leaf, each has maximum N_s child nodes, ordered for quicksort search).

9.5.2.2 R-tree Genetic Operators

R-trees also allow efficient genetic operations that manipulate sub-trees. As with regular genetic programming, the R-tree crossover operator, for instance, swaps sub-trees between the parents to form two new R-trees. However, the challenge is to ensure that the crossover operator results in new trees that remain valid R-trees. If one simply picks an arbitrary edge E_1 from R-tree$_1$ and edge E_2 from R-tree$_2$, randomly, and then swap the sub-trees under E_1 and E_2, the resulting trees may no longer be height balanced.

This problem can be resolved by restricting R-tree crossover to be a version of homologous one-point crossover, an alternative to the "standard" crossover operator in GP [28]. The essential idea is as follows. After selecting the parent R-trees, traverse both trees (in a breadth first order) jointly in parallel. Compare the states of each visited node in the two different trees. If the states match, mark the edge above that node as a potential crossover point. As soon as a mismatch is seen, stop the traversal. Next, pick an edge from the ones marked as potential crossover points, with uniform probability, and swap the sub-trees under that edge between *both* parent R-trees. An example is shown in Fig. 9.6.

R-tree crossover as defined above has clear advantages over linear representation crossover. First, R-tree crossover is potentially equivalent to a large set of linear crossovers. Second, linear crossover randomly selects the crossover point and hence is not context preserving. R-tree crossover selects a crossover point only in the common upper part of the trees. This means that until a common upper structure emerges, R-tree crossover is effectively searching a much smaller space and therefore the algorithm quickly converges toward a common (and good) upper part of the tree, which cannot be modified again without the mutation operator. Search incrementally concentrates on a slightly lower part of the tree, until level after level the entire set of trees converges.

The R-tree mutation operator simply picks an edge from the entire tree with uniform probability, and then eliminates the sub-tree below the edge. An example

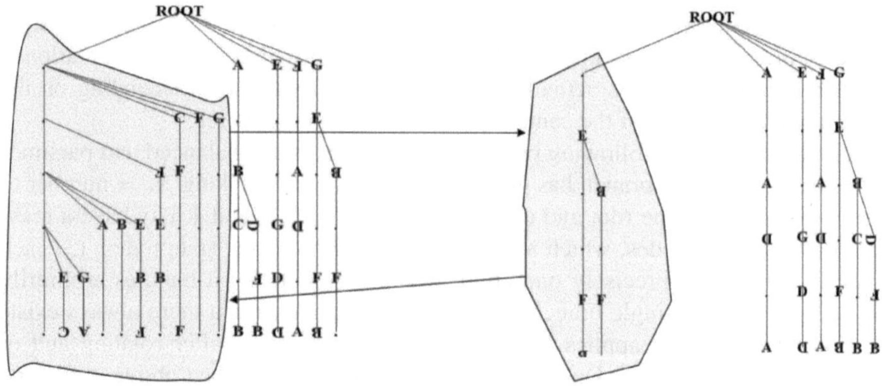

Fig. 9.6 One-point homologous crossover between parent R-trees. A crossover point is selected at the *same* location in both parent trees, ensuring that the child trees are valid R-trees. The children R-trees are formed by swapping the *shaded* sub-trees

is shown in Fig. 9.7. The R-tree *pruning operator* is an explicit mutation operator that is applied when the R-tree is used to run the CA to assess the R-trees fitness. The CA monitors the R-tree and marks inactive edges (through which no rules has been activated by any CA cell), and then the entire sub-trees below the inactive

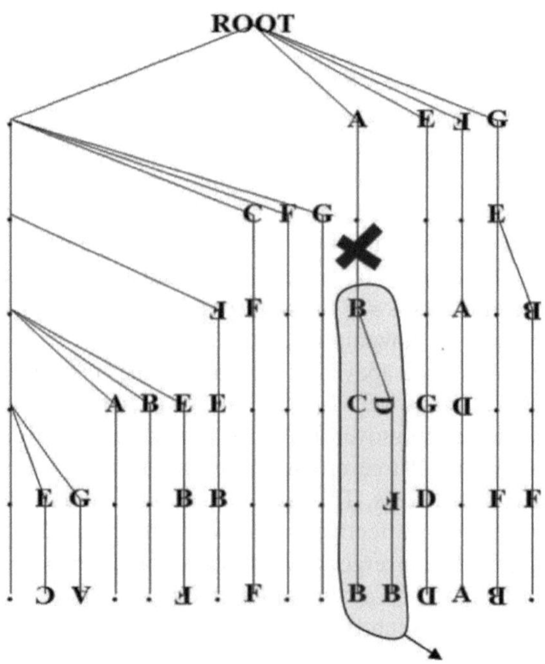

Fig. 9.7 The R-tree point mutation simply deletes a sub-tree (here, the one indicated by *shading*), allowing the CA simulation to fill it in with a new randomly-generated subtree when needed

edges will be eliminated. This helps to always keep each R-tree as parsimonious as possible. New paths in the R-tree are generated as needed, as explained in the next section. The R-tree encoding and genetic operators used allow CA rules to be constructed and evolved under a non-standard schema theorem similar to one proposed for genetic programming [28], even though R-trees do not represent conventional sequential programs.

9.5.3 Genetic Programming with S-tree and R-tree Encoding

The introduction of S-tree/R-tree encodings, with their capability for representing arbitrary structures and rules with a universal, uniform data structure, makes it possible to build a genetic programming system that can program a CA to support self-replication. To do that, the seed structure is first encoded with an S-tree. Then, an R-tree population is initialized randomly and starts to evolve as guided by a fitness function. The fitness function evaluates how well structures produced at intermediate evaluation time steps by each R-tree match the S-tree. The R-tree reproduction focuses on high fitness individuals, thus exploiting the available fitness information. This process repeats, from generation to generation, until an R-tree forms which produces a desired number of isolated structures that perfectly match the S-tree encoding, i.e., that are copies of the seed structure.

9.5.3.1 R-tree Initialization and CA Simulation with R-tree

To evaluate the fitness of each R-tree in the evolving population, first one needs to simulate the R-tree in the CA space in order to measure how well it produces self-replication. In the beginning, a population of R-trees is initialized, with each having only one default branch. Therefore, at GP generation $g = 0$, every R-tree in the population is identical, each only containing one trivial rule (00000→0). Before a simulation starts (t=0), every cell in the entire CA space is quiescent, except those cells containing the active components of a single seed structure. At each subsequent time step, $t \in T^s$, each cell c attempts to transit its state c_t to the next time step c_{t+1} by identifying and firing a specific rule in the R-tree based on the states of its von Neumann neighbors. If such a rule is not found from the current R-tree, a new rule is inserted into the R-tree with its target state (the leaf node) randomly generated. This operation is referred as *R-tree expansion*. On the other hand, at the end of current simulation, those branches in the R-tree which represent a rule or rules that were never fired by any cells at any time step are explicitly removed. This operation is referred as *R-tree pruning*. In a parallel computation platform, each R-tree could be simulated and evaluated concurrently.

The range of the time steps during which a simulation is performed is referred as the Simulation Time Steps (T^s), such as T^s = (1,2,...,12). The collection of time steps during which the configurations are considered for fitness evaluation is referred as the Evaluation Time Steps (T^v), which can be equal, or a subset of T^s. It is undesirable for T^s to be either too small or too large. If it is too small, it may be insufficient for capturing the self-replication phenomenon. If it is too big, it may lead to a significant decrease in evolution efficiency due to over-sized R-trees with initial random rules. The question that follows is how to determine an appropriate T^s. Our strategy is to let the simulation start with a small number of time steps, and thus the evolution and searching for an optimal R-tree runs fast initially. Only when evolution has made sufficient progress is it allowed to adaptively add more simulation time steps and progressively improve R-trees. This keeps the R-trees parsimonious, avoiding the GP bloating problem, and maintains effective evolutionary searching for optimal rules in a paced fashion.

9.5.3.2 S-tree Probing and R-tree Fitness Evaluation

The purpose of an R-tree's simulation is to evaluate its fitness in terms of producing duplicated seed structures. However, since every R-tree in the initial population is randomly generated, it is extremely unlikely any of them will directly lead to self-replication. When evolution begins, randomly generated R-trees produce a set of configurations during the given simulation time steps. These configurations very likely lack any clear patterns (see a sample configuration in Fig. 9.8). An essential part of any evolutionary algorithm is to reproduce more promising candidates and discard less promising candidates, as indicated by the fitness measures. It is critical to discover a universal and consistent mechanism that works with arbitrary structures and produces precise fitness measures which reflect the subtle differences leading to future self-replication. Otherwise, R-tree evolution may act as random searching, which is very unlikely to ever produce any R-tree that supports self-replication.

The introduction of the S-tree as a universal encoding mechanism of arbitrary structures gives us an unprecedented ability to perform precise fitness assignment

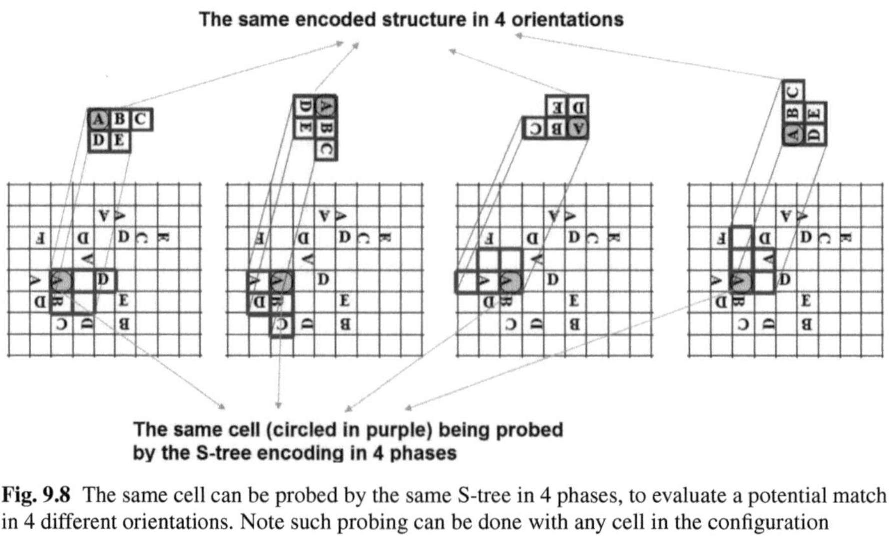

Fig. 9.8 The same cell can be probed by the same S-tree in 4 phases, to evaluate a potential match in 4 different orientations. Note such probing can be done with any cell in the configuration

for full or partial matching structures at any GP stage. This is because S-tree encoding can be exploited to retrieve complete structural information. By retrieving such information from the S-tree, and comparing it to the configuration produced by R-tree simulation, we can precisely tell how well they are matched. Figure 9.8 illustrates conceptually how this can be done. From a given configuration produced at any time step simulating a candidate R-tree, pick any active cell as the root cell (such as the cell circled in purple in Fig. 9.8). Conceptually, we use the S-tree to recover the entire encoded structure by recursively recovering every Moore neighborhood of every component from the root (see the structure shown on the top in Fig. 9.8). The state of every component in the structure is compared to the state of the corresponding cell in the current configuration, and the total number of components that matches is counted. If we then divide this result by the total number of components in the structure, we get a precise scalar measure in the range of [0, 1], indicating a complete mis-match, partial match, or a perfect match. In a real implementation, we do not actually need to recover the encoded structure as conceptually illustrated above. We only need, starting from the root cell, to recursively traverse the needed number of neighboring cells as guided by the S-tree, and compare the state in a traversed neighbor cell to the state in a corresponding node in the S-tree. Since the S-tree is a minimum spanning tree, it allows traversing every component precisely once after traveling the shortest distance. This process is hereafter referred as *S-tree probing*. Thus, S-tree probing is a process that can be used to test every cell in a given configuration, and measure how much a structure can be matched if we align the structure with that cell.

Each cell in the current configuration can be probed in 4 different ways with the same S-tree, and thus help to detect possible structures located from that cell in 4 orientations. Figure 9.8 shows an example each of these 4 probes from the same cell. Further, an S-tree contains not only the active components from a structure, but also the immediate quiescent cells surrounding the active cells. Therefore, an actual probing also traverses those surrounding cells, and can determine how many of these surrounding cells also match the states of corresponding nodes in the S-tree (which are all quiescent). This measures whether the currently probed structure is completely non-isolated, partially isolated, or fully isolated from its surrounding.

At any time step during a simulation with a candidate R-tree, we probe every possible root cell and every possible orientation with the given S-tree, return the best matching result. Let r represent a simulated (evaluated) R-tree, s an S-tree for a given structure, $\lambda(s)$ the number of nodes of s, p an infinite cellular space, $\bar{c} \in p$ a root cell being probed, $h \in (1, 2, 3, 4)$ the phase of the current probe, t a time step applying r on s in p. Define function $\kappa(r, s, p, t, \bar{c}, h)$ to be, after applying r on s in p for t time steps, and then probing s from $\bar{c}$ in phase h, the number of traversed cells which match the state of the corresponding node (active or quiescent) as guided by s. We can define a probing function as follows:

$$f_{\kappa}(r, s, p, t) = \max_{c \in p} \left(\max_{h \in (1,2,3,4)} \frac{\kappa(r, s, p, t, c, h)}{\lambda(s)} \right). \tag{9.1}$$

However, our goal is not just to produce one instance of the seed structure, but to allow self-replication to carry on sustainably. Thus, ultimately we will want to reward those R-trees which are more likely to generate a maximum number of replicas. We can use the probing function above repeatedly to identify multiple best matches, being careful to mark all of the active cells traversed by an accepted probe as "UNAVAILABLE", so that these active cells will not be counted again by subsequent probes.

The question that immediately follows is, how to determine how many probes we shall accept at each time step? One might ask, why don't we accept as many probes as possible? The answer is that, if we accept too many probes in a given time, it may have the effect of promoting the trend of forming many partially matching structures, but few would have enough room and potential to grow into full replicas, and ultimately degrade the performance of the evolution. This problem is referred as *over-probing*. To address this over-probing problem, our strategy is that, when evolution starts, each R-tree is automatically allowed to accept two probes at each time step. After some R-trees become more and more successful at generating two perfect probes, with likely higher number of simulation time steps, we can allow these R-trees to incrementally increase the number of acceptable probes (and so become more aggressive in working on additional replicas). Denote the number of acceptable probes at evaluation time $t \in \mathcal{T}^v$ for R-tree r as $\pi^t(\mathrm{r})$. We adopt a strategy in which an evolving R-tree starts with a basic goal of programming itself to find a minimum CA space just to produce two isolated seed structures. Not until this is achieved can it accept more probes. On the other hand, once this goal is achieved, it gradually raises its goal by adaptively adjusting the number of acceptable probes (π^t) at a controlled and adaptive pace.

9.5.3.3 Overall Fitness Function

Based on the above, an R-tree r at evaluation time t is allowed to accept $\pi^t(r)$ probes. Each accepted probe identifies a best probe from the cells not yet marked as "UNAVAILABLE" by previously accepted probes. Hence, we can write an overall fitness function for R-tree r:

$$f(r) = \sum_{t \in \mathcal{T}^v} \sum_{n=1}^{n=\pi^t(r)} \max_{c \in p,\ c \neg\in \bigcup_{m=1}^{m=n-1} \breve{p}_m} \max_{h \in (1,2,3,4)} \frac{\kappa(r,s,p,t,c,h)}{\lambda(s)}. \tag{9.2}$$

Equation (9.2) indicates the overall fitness measure for a candidate R-tree at a given GP generation as its accumulated result of every accepted probe at every evaluation time step. Every accepted probe finds a best probe among tested probes at every location and every orientation. The number of evaluated time steps at a given generation, $\mathcal{T}^v$, is common to all R-trees in the population, but at any specific evaluation time step, the allowable number of accepted probes varies from R-tree to R-tree. An R-tree can gain higher fitness by either earning a higher number of accepted probes, or from better individual accepted probes, or both.

9.5.4 The Replicator Factory Model and Experimental Results

The schematic view of the resulting S-tree/R-tree based GP model toward self-replication is illustrated in Fig. 9.9. First, an S-tree s is derived from the pre-specified seed structure. Then, an R-tree population of size M is initialized. Each R-tree is simulated in the given cellular space within the current T^s, while each R-tree

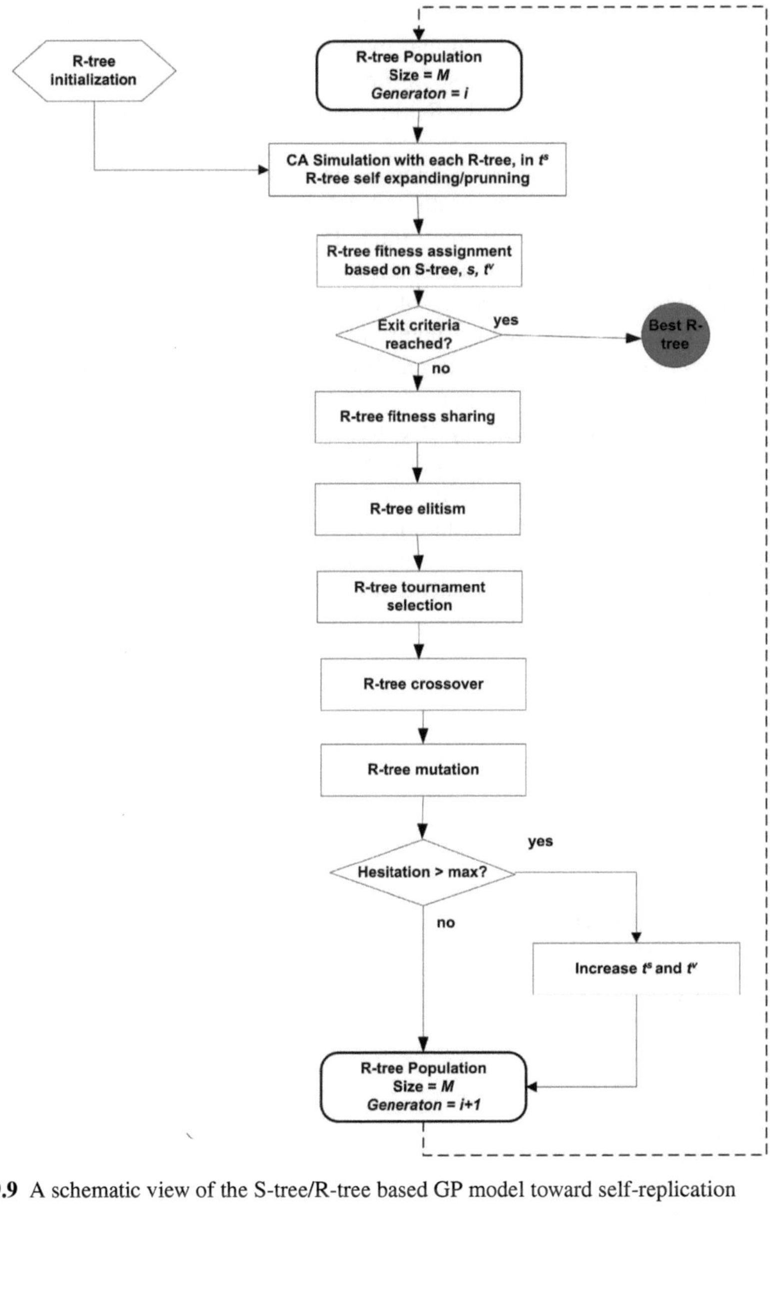

Fig. 9.9 A schematic view of the S-tree/R-tree based GP model toward self-replication

may potentially expand or prune itself as needed. Next, based on the simulation results, fitness is measured for each R-tree. If the desired fitness level is reached, the algorithm has produced the best R-tree and stops. Otherwise, the fitness values are adjusted due to fitness sharing. R-tree elitism is performed so that it is ensured that the elite R-tree in the new population will be at least as good as before. The entire population is fully ordered based on the final fitness value of each R-tree. Tournament selection is performed and *M/2* pairs of parents are selected. Each pair may perform an R-tree crossover before entering the mating pool, and each R-tree in the mating pool may be further mutated. If current hesitation (number of generations passed since last increase in fitness) has exceeded a specified $\xi_{\max}$, T^s is incrementally increased. Then, the R-tree population enters a new GP generation, and the same process repeats.

The model described above has been tested in a number of experiments, and two examples are presented here. Typically, model parameters like the following are chosen: Population Size = 100, R-tree Mutate Probability = 0.45, R-tree Crossover Probability = 0.85, R-tree GP Tournament Size = 2, and Max Hesitation = 200. Success was achieved with structures of arbitrary shape and varying numbers of components. The largest CA seed structure for which it was previously possible to evolve rules has 4 components [14]. Figure 9.10 ($t = 0$) shows one of the seed structures, consisting of 7 oriented components (29 states), for which our approach using GP finds a rule set that allowed the structure to self-replicate. With the resultant R-tree based on the von Neumann neighborhood, shown in Fig. 9.11, at time $t = 1$ (Fig. 9.10), the structure starts splitting (the original seed structure translates to the left while a rotated replica is being born to the right). At time $t = 2$ (Fig. 9.10), the splitting completes and the original and replica structures become isolated. Thus, the seed structure has replicated after only 2 time steps, a remarkably fast replication time that has not been reported before.

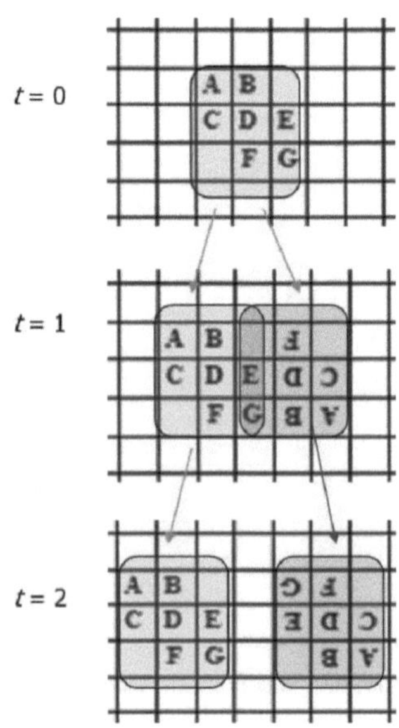

Fig. 9.10 Example replicator using the von Neumann neighborhood. The seed, a 7-oriented-component structure, evolved to self-replicate in only 2 time steps. In subsequent time steps, each replica attempts to repeat the same action, and when enough space is reached, more replicas can be isolated from each other

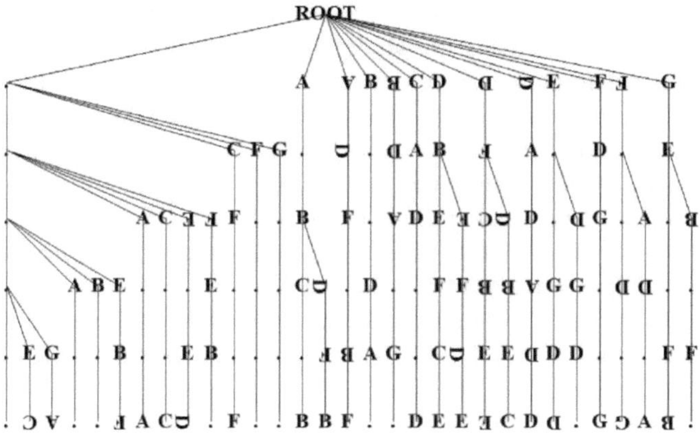

Fig. 9.11 Example R-tree evolved for the replicator in Fig. 9.10

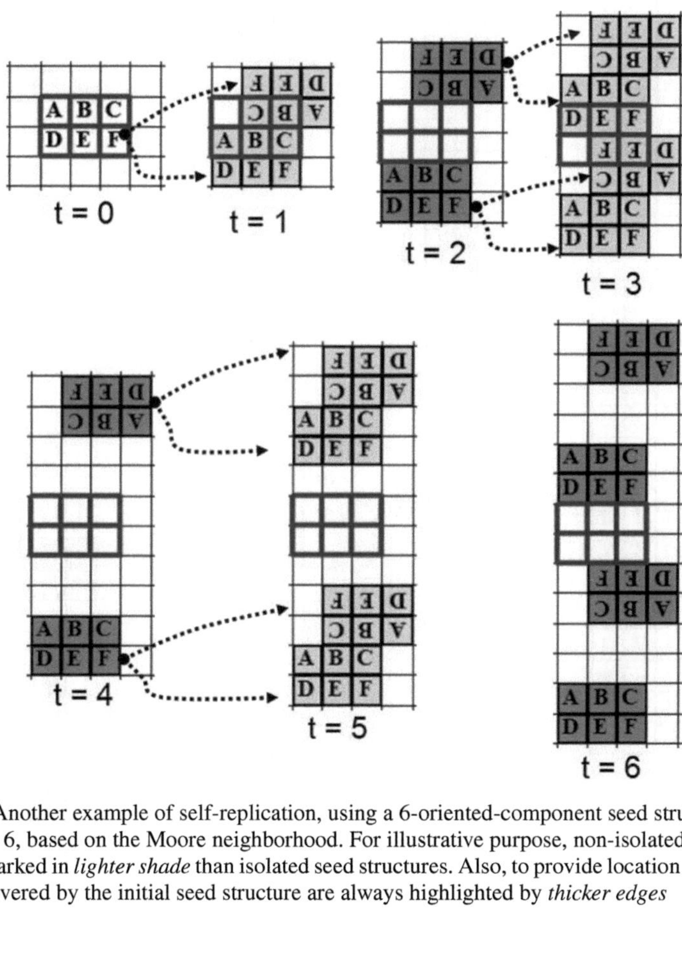

Fig. 9.12 Another example of self-replication, using a 6-oriented-component seed structure, from $t = 0$ to $t = 6$, based on the Moore neighborhood. For illustrative purpose, non-isolated seed structures are marked in *lighter shade* than isolated seed structures. Also, to provide location correlation, the cells covered by the initial seed structure are always highlighted by *thicker edges*

Another example, a structure of 6 oriented components (25 states), shown in Fig. 9.12 ($t = 0$), evolves using the Moore neighborhood. To make it easy to visualize the produced structures at each evaluation time step, color codes are used in these figures. A non-isolated seed structure is marked in yellow and isolated seed structure in blue. Also, to provide location correlation, the cells covered by the initial seed structure are always highlighted by red edges at any time step. As illustrated in Fig. 9.12, at $t = 1$, the seed expands to the right, top, and bottom, at light speed (1 cell/each time step), but not to the left. Two contiguous seed replicas are formed. At $t = 2$, the replicas move apart and get isolated. At $t = 3$, each of these isolated replicas repeats the same self-replication, but the middle ones collide, so that only two replicas are found at $t = 4$, which are now further apart than at $t = 2$, making

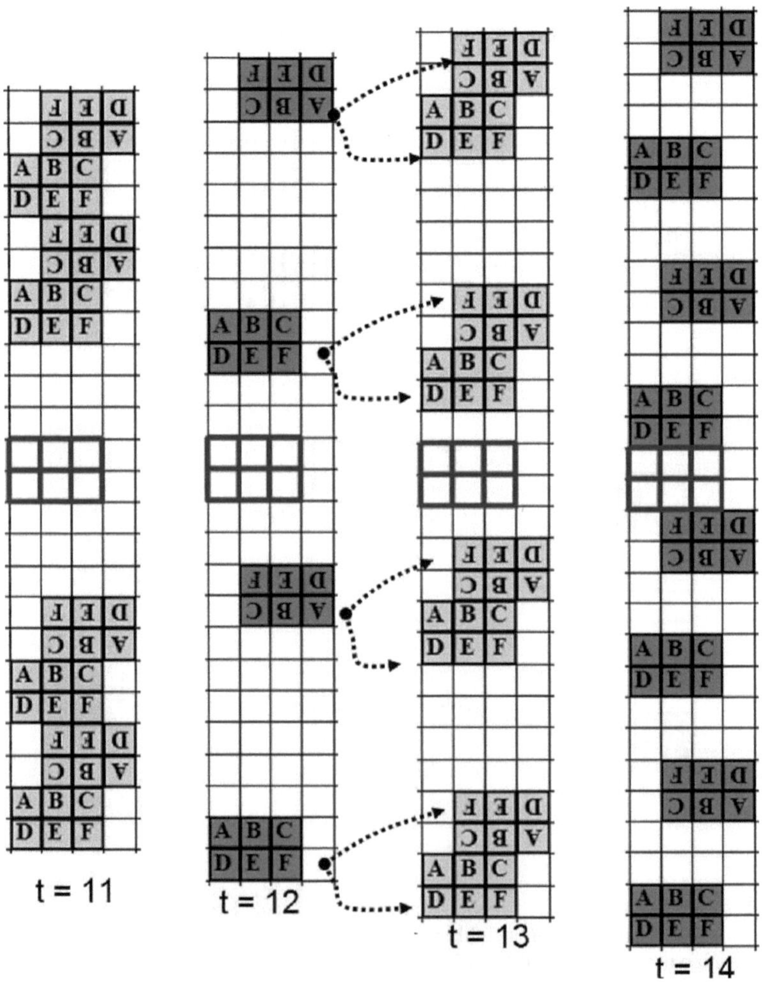

Fig. 9.13 Continuation of Fig. 9.12 from $t = 11$ to $t = 14$

more rooms to form four isolated replicas at $t = 6$. In subsequent time steps, this replication process continues indefinitely, so that, for example, by $t = 14$ there are eight distinct replicators present (see Fig. 9.13).

Figure 9.14 shows that it is possible for the same seed structure to self-replicate in different ways. A different R-tree is evolved for the same seed structure in a separate GP run. Here, very surprisingly, the evolutionary algorithm found a way to replicate the seed in only one (1) time step. The seed structure is rotated before translation, so that the space originally occupied by the seed itself can be more efficiently used, yielding enough space to form a pair of replicas within only one time step.

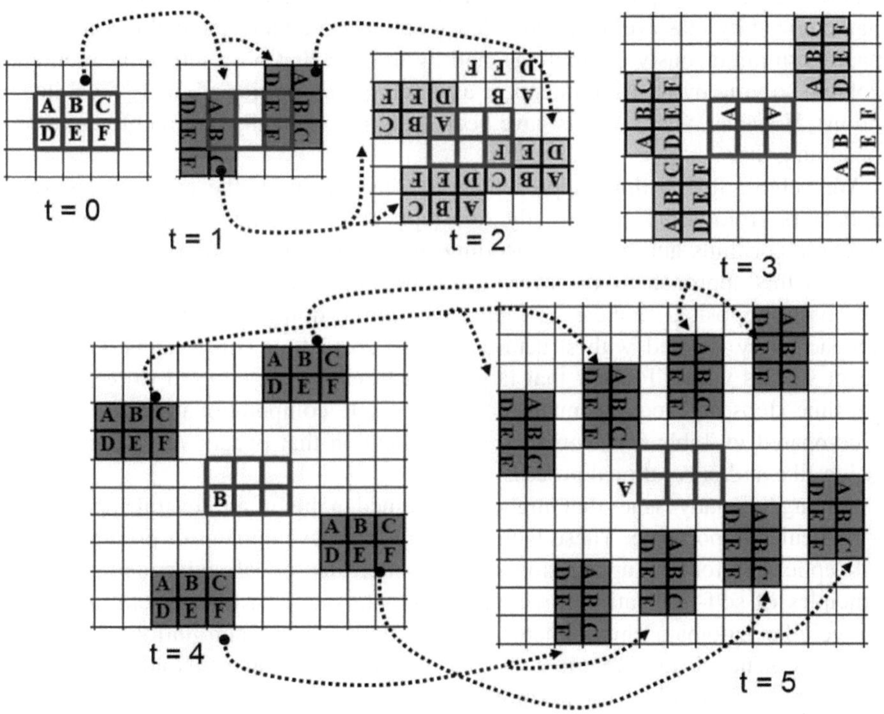

Fig. 9.14 The same seed structure as in Fig. 9.12 can evolve to replicate in a different way, as illustrated here

9.6 Discussion

Cellular automata models of self-replication have been studied for almost 50 years. In this chapter we have presented the view that past work on this topic has involved at least two different approaches. The earliest work examined large, complex universal computer-constructors that are marginally realizable. This work established the feasibility of artificial self-replication, examined many important theoretical issues, and gradually created progressively simpler self-replicating universal systems. A second and more recent approach has focused on the design of self-replicating loops and related small structures. Self-replicating loops are so small and simple that they have been readily realizable, and significant progress has been made in extending them to perform additional tasks as they replicate. This can be achieved either by attaching a set of instructions (signals) to those directing replication, or by encoding a tentative problem solution that systematically evolves into a final solution. Implementations have shown that programmed replicators are clearly capable of solving non-trivial problems. These programmed self-replicating structures are intriguing in part because they provide a novel approach to computation. This approach is characterized by massive parallelism (each cell in the underlying cellular automata space is simultaneously computing), and by the fact that both self-replication and problem-solving by replicators appear as emergent properties of solely local interactions. However, past systems have been designed manually, a difficult and time-consuming task.

We believe that a third approach merits investigation: the evolution of self-replicators from arbitrary, initially non-replicating systems. Initial work using genetic algorithms and a linear encoding of transition rules showed the potential value of this approach by discovering a new class of replicators that moved and essentially deposited replicas of themselves while doing so [14]. However, this approach only worked with structures having up to four components. Our most recent studies with GP show that larger structures can be evolved to replicate in less time steps and require much more reasonable computational times [24, 25], as compared in Table 9.1. Some exhibit replication that occurs very quickly in a fission-like and/or rotational process.

Among the many issues that might be examined in the future, several appear to be of particular importance. These include the further development of programmable self-replicators for real applications, and a better theoretical understanding of the principles of self-replication in cellular automata spaces. More general and flexible cellular automata environments, such as those having non-uniform transition functions or novel interpretations of transition functions, merit exploration. It has already proved possible, for example, to create simple self-replicating structures in which a cell can change the state of neighboring cells directly [13]. Also, from the perspective of realizing physically self-replicating devices, exchange of infor-

Table 9.1 Comparison between the initial and current approaches

	Previous approach	New approach
Structural encoding	None	S-tree
Rule encoding	Linear table	R-tree
Evolutionary algorithm	GA	GP
Fitness evaluation	Structure-specific, heuristic	Universal, precise probing
Reachable structure size	only 4 components	56+ components
No. of rules needed	1,419,857 rules	128 rules
Computation time	One week (supercomputer)	40 min (laptop)

mation between the modeling work described here and ongoing work to develop self-replicating molecules/nanotechnology is important. Closely related to this issue is ongoing investigation of the feasibility of electronic hardware directly supporting self-replication [7, 15]. If these developments occur and progress is made, we foresee a productive future for the development of a technology of self-replicating systems.

Acknowledgment This work was supported by NSF award ITS-0325089.

References

1. D. Andre, F. Bennett, J. Koza, Discovery by genetic programming of a cellular automata rule. *Proceedings First Annual Conference on Genetic Programming*, MIT Press, Cambridge 1996, pp. 3–11
2. W. Banzhaf, P. Nording, R.E. Keller, F.D. Francone, *Genetic Programming: An Introduction: On the Automatic Evolution of Computer Programs and Its Applications* (Morgan Kaufmann, San Fransisco, CA, 1997)
3. J. Beuchat, J. Haenni von Neumann's 29-state cellular automaton: A hardware implementation. IEEE Trans. Educ. **43**(3), 300–308 (August 2000)
4. J. Byl, Self-reproduction in small cellular automata. Physica D **34**, 295–299 (1989)
5. H. Chou, J. Reggia, Problem solving during artificial selection of self-replicating loops. Physica D **115**, 293–312 (1998)
6. E. Codd, *Cellular Automata* (Academic Press, New York, NY 1968)
7. R. Freitas, R. Merkle, *Kinematic Self-Replicating Machines* (Landes, Austin, TX 2004)
8. H. Haken, *Advanced Synergetics: Instability Hierarchies of Self-Organizing Systems and Devices* (Springer, Heidelberg, 1983)
9. J. Koza, *Genetic Programming: On the Programming of Computers by Means of Natural Selection* (MIT Press, Cambridge, MA 1992)
10. W. Langdon, R. Poli, *Foundations of Genetic Programming* (Springer, Heidelberg 2001)
11. C. Langton, Self-reproduction in cellular automata. Physica D **10**, 135–144 (1984)
12. C. Langton, Studying artificial life with cellular automata. Physica D **22**, 120–149 (1986)
13. J. Lohn, J. Reggia, Discovery of self-replicating structures using a genetic algorithm. *Proceedings IEEE International Conference on Evolutionary Computation*, Perth, 678–683 (1995)
14. J. Lohn, J. Reggia, Automated discovery of self-replicating structures in cellular automata. IEEE Trans. Evol. Comp., **1**, 165–178 (1997)
15. D. Mange, M. Goeke, D. Madon, et al., Embryonics. *Towards Evolvable Hardware* (Springer, Heidelberg 1996) pp. 197–200
16. K. Morita, K. Imai, Self-reproduction in a reversible cellular space. *Theor. Comp. Sci.* **168**, 337–366 (1996)
17. K. Morita, K. Imai, Simple self-reproducing cellular automata with shape-encoding mechanism, ed. by Langton C, Shimohara K, *Proceedings of the Fifth International Workshop on Synthesis and Simul. Living System*, MIT Press, Cambridge, MA, pp. 450–457, 1997
18. C. L. Nehaniv, Evolution in asynchronous cellular automata, ed. by R.K. Standish, M.A. Bedau, H.A. Abbass, *Proceedings of the Eighth International Conference on Artificial Life*, MIT Press, Cambridge, MA, 65–78, 2002
19. J. von Neumann, General and logical theory of automata, ed. by A. Tanb, *John von Neumann–Collected Works*, 5 (Macmillan, New York, NY 1961), pp. 288–328
20. J. von Neumann, in *Theory of Self-Reproducing Automata*, ed. and completed by A.W. Burks (University of Illinois Press, Champaign, IL, 1966)

21. Z. Pan, Artificial evolution of arbitrary self-replicating cellular automata, Department of Computer Science Tech. Report, http://hdl.handle.net/1903/7404, University of Maryland at College Park, August 2007
22. Z. Pan, J. Reggia, in *Evolutionary Discovery of Arbitrary Self-Replicating Structures*, ed. by V. Sundaram et al., Lecture Notes in Computer Science, Vol. 3515, (2005) pp. 404–411
23. Z. Pan, J. Reggia, *Evolutionary Discovery of Arbitrary Self-Replicating Structures*. Proceedings of the 5th International Conference in Computational Science, Atlanta, GA, USA 404–411, 2005
24. Z. Pan, J. Reggia, Artificial evolution of arbitrary self-replicating structures. J. Cell Automata, **1**(2), 105–123 (2006)
25. Z. Pan, J. Reggia, Properties of self-replicating cellular automata systems discovered using genetic programming. Adv Compl Syst., **10** (supp01), 61–84 (August 2007)
26. J. Perrier, M. Sipper, J. Zahnd, Toward a viable self-reproducing universal computer. Physica D **97**, 335–352 (1996)
27. U. Pesavento, An implementation of von Neumann's self-reproducing machine. Artif. Life **2**, 337–354 (1995)
28. R. Poli, W. Langdon, Schema theory for genetic programming with one-point crossover and point mutation. Evol. Comput. **6**, 231–252 (1998)
29. J. Rebek, Synthetic self-replicating molecules. Sci. Am. **271**(1), 48–55 (July 1994)
30. J. Reggia, S. Armentrout, H. Chou, Y. Peng, Simple systems that exhibit self-directed replication. *Science* **259**, 1282–1288 (1993)
31. F. Richards, T. Meyer, N. Packard, Extracting cellular automaton rules directly from experimental data. Physica D **45**, 189–202 (1990)
32. C. Salzberg, A. Antony, H. Sayama, Complex genetic evolution of self-replicating loops. *Artificial Life IX: Proceedings of the Ninth International Conference on the Simulation and Synthesis of Living Systems*, MIT Press Cambridge, MA, pp. 262–267, 2004
33. H. Sayama, Introduction of structural dissolution into Langton's self-reproducing loop, C. Adami, R.K. Belew, H. Kitano, C.E. Taylor, *Artificial Life VI: Proceedings of the Sixth International Conference on Artificial Life*, Los Angeles, CA, 1998, pp. 114–122
34. H. Sayama, Self-replicating worms that increase structural complexity through gene transmission, M. Bedau, J. McCaskill, N. Packard, S. Rasmussen, *Proceedings of Seventh International Conference on Artificial Life*, MIT Press, Cambridge, MA, pp. 21–30, 2000
35. J. Signorini, How a SIMD machine can implement a complex cellular automaton? A case study: von Neumann's 29-state cellular automaton. *Proceedings of the 1989 ACM/IEEE Conference on Supercomputing*, 1989 pp. 175–186
36. M. Sipper, Fifty years of research on Self-Reproduction: An overview. Artif. Life **4**, 237–257 (1998)
37. G. Tempesti, A new self-reproducing cellular automaton capable of construction and computation, ed. by F. Morn, A. Moreno, J. Merelo, P. Chacn, *Third European Conference on Artificial Life*, Lecture Notes in Computer Science, vol. 929, (Springer, Heidelberg 1995), pp. 555–563
38. P. Vitanyi, Sexually reproducing cellular automata. Math. Biosci. **18**, 23–54 (1973)
39. P. Vitanyi, Genetics of reproducing automata. In *Proceedings 1974 Conference on Biologically Motivated Automata Theory*, IEEE, New York, 1974, pp. 166–171

Part II
Applications of Cellular Automata

Chapter 10
Game Theoretical Interactions of Moving Agents

Wenjian Yu and Dirk Helbing

10.1 Introduction

Macroscopic outcomes in a social system resulting from interactions between individuals can be quite different from anyone's intent. For instance, empirical investigations [1] have shown that most colored people prefer multi-racial neighborhoods, and many white people find a certain fraction of other races in their neighborhood acceptable. So one could think that integrated neighborhoods should be widely observed, but empirically this is not true. One rather finds segregated neighborhoods, i.e. separate urban quarters, which also applies to people with different social and economic backgrounds.

This problem is scientifically addressed by mainly two streams of segregation theory [2]: the urban ecological "social distance" tradition in sociology [3, 4] and the "individual preferences" tradition in economics [5, 6]. The main idea of "social distance" theory is that the differences in culture and interests between social groups are reflected by a separation of their residential areas.

Yet, the role of social distance and individual preferences is questioned by studies of the American urban housing market, which suggest that racial discrimination and prejudices are the primary factors of residential segregation and concentration of poverty [7]. There are three stages in a housing market transaction: first, information about available housing units, second, terms and conditions of sales and financing assistance, and third, the access to units other than the advertised unit [8]. In each stage, the housing agent may behave in a discriminatory way, e.g. withhold information from customers and discourage them. Therefore, the access of minority customers to housing is severely constrained, while the theory of preference-based dynamics assumes that people can relocate freely according to their own preferences, which fails to reflect the real relocation dynamics.

W. Yu (✉)
ETH Zurich, CLU C4, Clausiusstr. 50, 8092 Zurich, Switzerland
e-mail: yuwen@ethz.ch

A.G. Hoekstra et al. (eds.), *Simulating Complex Systems by Cellular Automata*, Understanding Complex Systems, DOI 10.1007/978-3-642-12203-3_10,

Recent studies [2, 9] point out that the effect of discrimination in racial residential segregation was important in the past, but nowadays, the nature and magnitude of housing discrimination has changed. Minority households who seek to move into integrated or predominantly white areas usually will be able to do so. But the integration of ethnic groups is not widely observed and quite unstable. This is partly because, when people move to a neighborhood, where they constitute an ethnic minority, the previous inhabitants may choose to leave, while some people may be reluctant to enter integrated neighborhoods in which minorities are increasing in number, e.g. because of a decrease in the housing prices. Such migration dynamics, based on seeking preferred neighborhoods, does not necessarily presuppose the discrimination of other people, as we will shortly see.

Assuming that individuals have just a slight preference for neighborhoods, in which the same ethnic background prevails, Schelling has reproduced residential segregation by a simple model [5, 6]. Imagine that two groups of people are distributed over a one-dimensional lattice, and assume that everyone defines his/her relevant neighborhood by the four nearest neighbors on either side of him/her. Moreover, assume that each individual prefers at least four of his/her eight nearest neighbors to belong to same ethnic group as he/she does. Considering himself/herself, this implies a small majority of five out nine people in the considered neighborhood. If the condition is not met, he or she is assumed to move to the nearest site/cell that satisfies his or her desire. Hence, the model does not assume optimizing location choice, just satisficing behavior. Nevertheless, it produces segregation patterns.

10.1.1 Migration, Game Theory, and Cooperation

In game theoretical terms, this movement to a more favorable neighborhood could be reflected by a higher payoff to persons who moved, called "migrants" in the following. Migratory behavior aiming at higher payoffs is called "success-driven motion" [10–13]. As we will show in later sections, success-driven motion can reproduce residential segregation and some other observed phenomena of population dynamics as well, like population succession (i.e. cycles of changing habitation) [14]. We will also study how a change of the spatial population structure can affect the level and evolution of cooperation in a population.

It could be thought that natural selection, implying competition between individuals, would result in selfish rather than cooperative behavior [15]. Nevertheless, cooperation is widely observed also in competitive settings, from bacteria, over animals to humans [16, 17]. Game theory [18–21] has been regarded as a powerful framework to investigate this problem, as it can be used to quantify the interactions between individuals.

The mathematical description of tendencially selfish behavior is often based on the prisoner's dilemma game, in which the "reward" R represents the payoff for mutual cooperation, while the payoff for defection (cheating) on both sides is reflected by the "punishment" P. Unilateral cooperation will incur the so-called "sucker's payoff" S, while the defector gains T, the "temptation". Given the inequalities $T > R > P > S$ and $2R > T + S$, defection is tempting and cooperation is risky, so that defection is expected to be dominating, although mutual cooperation improves the average payoff.

In a well-mixed population, where an individual interacts with all the others, a defector's payoff is always higher than the average payoff, which leads to a prosperity of defectors. In an evolutionary setting as it is described by the replicator dynamics [22], only those strategies that generate above-average payoffs have the chance to spread. Therefore, from an evolutionary perspective, there will be no cooperators in the end, which contradicts the observed cooperation in reality.

A variety of mechanisms has been proposed to explain the considerable level of cooperation observed under certain circumstances. These include kin selection, direct reciprocity, indirect reciprocity, group selection and network reciprocity [15]. In particular, it is interesting that the spatial population structure can significantly change the level of cooperation in a population [23, 24]. While there could be a co-evolution of social structure and cooperation due to migratory behavior, this subject has not been well studied in the past.

10.1.2 Co-evolution of Social Structure and Cooperation

In order to model such intentional movement of individuals, spatial effects must be explicitly considered. Spatial games based on lattices [23] would allow one to study such effects. In conventional spatial games, individuals are uniformly distributed in the simulation area, and change or repeat their strategies, following a certain updating rule. For instance, an individual is assumed to unconditionally adopt the most successful strategy within the neighborhood. On the one hand, this creates typical spatio-temporal pattern. On the other hand, spatial structures play an important role in the maintenance of cooperation. In the iterative prisoner's dilemma, for example, clusters of cooperators are beneficial to cooperators [12, 13], while the evolving spatial structure in the snowdrift game [24], may inhibit cooperation. Conventional spatial games, however, neglect the possibility of individuals to move or migrate, although mobility is a well-known fact of daily social interactions. We will see that the movement of individuals, e.g. population succession, can change the spatial structure of a population significantly, and influence the evolution of cooperation dramatically.

The above mentioned social processes that are related to the migratory behavior of people, can be well integrated into the framework of spatial games. Focusing on success-driven motion, we will in the following, study the combination of migration, strategic interactions and learning (specifically imitation). Numerical simulations show that success-driven motion is a mechanism that can promote cooperation via self-organized social structures. Moreover, very surprisingly, a certain degree of fluctuations ("noise") can even enhance the level of cooperation.

10.2 Spatial Games with Mobility

10.2.1 Classification

Spatial games with mobility ("mobility games") could be classified as follows:

1. Mobility may occur in physical (geographic) space with one, two or three dimensions, or in abstract space (e.g. opinion space). It may take place in a continuous space or in a discrete space such as a grid. Rather than a regular grid, one may use an irregular grid [21], or a (fixed or dynamically changing) network structure (e.g. friendship network).
2. One may distinguish between games with continuous and with discrete motion. The first ones may be considered as particular cases of differential games [25] and shall be called "motion games". The second ones will be named "migratory games" and can be implemented, for example, in terms of cellular automata (particularly, if a parallel update is performed).
3. The mobility game may be deterministic or "noisy", i.e. influenced by fluctuations. Such "noise" may be introduced by stochastic update rules, e.g. a random sequential update or the Fermi rule [24].
4. Multiple occupation of a certain location may be possible, allowing for the agglomeration of individuals, or it may be prohibited. In the latter case, spatial exclusion requires spatial capacity constraints and/or the respect or protection of some "private territory".
5. In mobility games with spatial exclusion, empty sites are needed. Therefore, the density of free locations ("sites") is a relevant model parameter.
6. The frequency, rate, or speed of mobility may be relevant as well and should be compared with other time scales, such as the frequency of strategy changes, or the average lifetime of a spatial cluster of individuals. Depending on the specification of the game, after a transient time one may find everything from chaotic patterns [23] upto frozen patterns [26, 27]. This may be expressed by a viscosity parameter [28].
7. Mobility may be random or directed (e.g. "success-driven"). Directed mobility may depend on the expected payoff in a certain time point (iteration), or it may depend on the cumulative payoff (i.e. the accumulation of wealth by individuals).
8. If age, birth and death are considered, one speaks of "demographic games" [29].
9. One can also study different update rules such as a random sequential update, a sublattice-parallel update, or a ordered-sequential update [30]. In this contribution, we have applied a random sequential update rule, which appears to be more realistic than, for example, a parallel update [31]. Moreover, a parallel update would create conflicts regarding the occupation of empty sites in the migration step.

In the following, we will primarily focus on games with success-driven migration in two-dimensional geographic space. Furthermore, we will assume simple strategies (such as all-cooperate or all-defect), no memory and no forecasting abilities.

10.2.2 Individual Decision Making and Migration

In "migration games", each individual I is located at the position $\mathbf{X_I}$ of a discrete grid and applies a certain strategy $i = i(I, \mathbf{X}, t)$ when interacting with individuals J at locations $\mathbf{X_J}$ in the neighborhood $\mathcal{N} = \mathcal{N}(\mathbf{X_I})$, who apply strategies $j = j(J, \mathbf{X}', t)$ (see Fig. 10.1). The interactions at time t are quantified by the payoffs P_{ij}. The overall payoff for individual I resulting from interactions with all individuals J in the neighborhood $\mathcal{N}(\mathbf{X_I})$ at time t is

$$P_I(t) = P_I(\mathbf{X_I}, t) = \sum_{J:\mathbf{X_J} \in \mathcal{N}(\mathbf{X_I})} P_{ij}(J). \tag{10.1}$$

We assume that all individuals prefer places that can provide higher payoff. However, their movements are restricted by their mobilities and the number of free locations, as one can only move to empty places. In our model, the mobility is reflected by the migration range M, which could be assumed constant (see Fig. 10.2) or as a function of cumulative payoffs

$$C_I(t) = \sum_{t'=0}^{t} [P_I(t') - c_I(t')]. \tag{10.2}$$

Here, the cost of each movement $c_I(t)$ may be specified proportionally to the distance $d_I(t) \leq M$ moved:

$$c_I(t) = \beta \, d_I(t). \tag{10.3}$$

β is a constant.

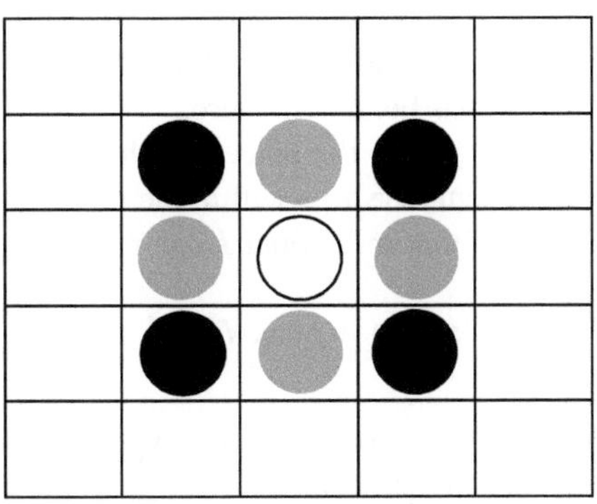

Fig. 10.1 The focal individual I is represented by the *empty circle*. It can, for example, interact with the $k = 4$ nearest individuals (*gray*) or with $k = 8$ neighbors, which includes four next-nearest neighbors (*black*)

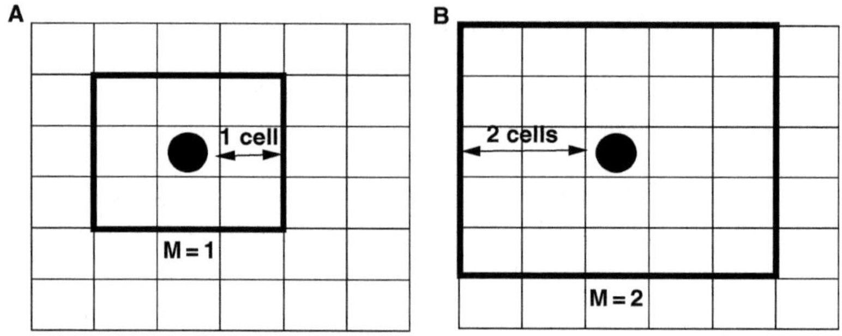

Fig. 10.2 **(a)** When $M = 1$, the focal individual (*black circle*) can migrate within the *black square* (Moore neighborhood of range 1). **(b)** Here, the migration neighborhood is chosen as Moore neighborhood of range $M = 2$

When the migration range M is restricted by the cumulative payoff, one may set

$$M(t) = \lfloor \alpha \, C_I(t) \rfloor, \tag{10.4}$$

where $\lfloor \; \rfloor$ rounds down to the next integer $\leq \alpha C_I(t)$ and α is a constant, which is set to $\frac{1}{\beta}$ in the following.

The expected payoff for individual I applying strategy i at a free location $\mathbf{X}'$ within the migration range can be calculated as

$$P_I(\mathbf{X}', t) = \sum_{J':\mathbf{X_{J'}} \in \mathcal{N}(\mathbf{X}')} P_{ij'}(J'), \tag{10.5}$$

where $j' = j'(J', \mathbf{X}', t)$ are the strategies of the individuals J' located within the neighborhood $\mathcal{N}' = \mathcal{N}(\mathbf{X}')$ centered at $\mathbf{X}'$. Here, we implicitly assume that individual I can figure out the strategies j' of the neighbors J' in a considered neighborhood $\mathcal{N}'$ by test interactions (just imagine, they visit a new neighborhood and talk to people, before they more there). When such test interactions are "cheap" as compared to the payoffs resulting after relocating, the costs of this "neighborhood testing" may be neglected, as has been done here.

We assume that individuals prefer the place $\mathbf{X}'$ which promises the highest payoff. Consequently, if $\mathcal{A}$ represents the area within the migration range M around $\mathbf{X_I}$, the maximum expected payoff that individual I can reach by relocating is

$$P_I^{\mathrm{e}}(t+1) = \max_{\mathbf{X}' \in \mathcal{A}} P_I(\mathbf{X}', t) \,. \tag{10.6}$$

In our model, individuals decide to move, if the expected payoff $P_I^{\mathrm{e}}(t+1)$ at the new location would be higher than the current one, i.e. if the short-term cost-benefit condition

$$P_I^{\mathrm{e}}(t+1) - P_I(t) > c_I(t) \tag{10.7}$$

is fulfilled. If the cost of relocating is small compared to the cumulative payoff that an individual can earn over the average time period an individual stays at the same place, the cost of movement can be neglected. In cases where two or more places promise the same maximum payoff, we assume that an individual prefers the closest one. If both, the payoff and distance are the same, individuals in our simulation choose randomly among equivalent locations.

Note that all the other individuals seek better places as well. Therefore, neighborhoods may change quickly, and the resulting payoff may fall below the expectations.

10.2.3 Learning

Learning allows individuals to adapt their behaviors in response to other people's behaviors in an interactive decision making setting. Basic learning mechanisms are, for example, unconditional imitation, best reply, and reinforcement learning [32].

Unconditional imitation means that people copy the behaviors of others. For instance, we can assume that, when i reaches a new place, it imitates the most successful strategy within the neighborhood (if it is more successful than the own strategy), or the most frequently used strategy in past interactions. It should be underlined, that it is not easy to identify the future strategy of another individual from its displayed past behavior. However, in the simplified context of a prisoner's dilemma, where individuals are assumed to play either all-defect or all-cooperate, one's behavior reveals the strategy directly. It may nevertheless change before the next iteration due to learning of the neighbors.

Reinforcement learning is a kind of backward-looking learning behavior, i.e. people tend to take the actions that yielded high payoffs in the past. Assuming that an individual has a set of strategies, among which he or she chooses with a certain probability, this probability is increased, if the gained payoff exceeds a certain threshold value ("aspiration level") [33, 34].

Best reply is a sophisticated learning model as well. According to it, people choose their strategies based on the expectation of what the others will do in a way that maximizes their expected payoff. The ability to forecast depends on one's belief about the behaviors of the others. For instance, one can try to determine the distribution of the neighbors' previous actions, and select the own strategy, which replies to it in the best way.

For simplicity, players in our model only interact with the $k = 4$ nearest neighbors and adapt their strategies by unconditional imitation. Assuming that the updating rules of all the individuals are the same, we can specify a simple migration game as follows:

1. An individual moves to a new position that is located inside the migration range M and maximizes the expected payoff.
2. The individual imitates the most successful strategy within the interaction neighborhood $\mathcal{N}$, if it is more successful than the own strategy.

Note that changing the sequence leads to a different dynamics, and the implementation of migration and learning can be varied in many ways. Therefore, spatial games with mobility promise to be a rich research field. In the following, we will restrict ourselves to some of the simplest specifications.

The above mentioned framework of spatial games with migration specifies a kind of cellular automata (CA) model, in which space and time are discrete. Furthermore, the set of actions performed in CA models usually depends on the last time step only, which allows for high-performance computing of large-scale scenarios. CA models have been successfully applied to describe a variety of social and physical systems [29, 35, 36]. The outcomes of social and physical interactions can be very well integrated into them, which can create many interesting dynamics, like the "game of life" [37].

10.3 Simulation Results and Discussion

In the following, we perform a random sequential update of the individual (i.e. their migration and learning steps), which is to be distinguished from the parallel update assumed in [23, 26]. For a discussion of some related advantages, see [31]. Our numerical simulations are performed on 49×49 grids with 40% empty sites, i.e. a density of 0.6 and periodic boundary conditions. 49×49 grids were chosen for better visibility, while the statistical evaluations (see Figs. 10.6 and 10.7) were done with 99×99 grids for comparability with Nowak's and May's spatial games [23]. In the prisoner's dilemma, the color code is chosen as follows: gray = cooperator, black = defector. In other games, gray = player of group 1, black = player of group 2. White always corresponds to an empty site.

10.3.1 Spontaneous Pattern Formation and Population Structure

In the migration game, individual preferences are reflected by the payoff matrix, which quantifies the possible outcomes of social interactions. Figure 10.3 shows a variety of patterns formed when the payoff matrix is modified. The outcomes from interactions between individuals are the driving forces changing the population structure in space. In the following, we will call all the individuals, who apply the same strategy, a "group" g. Group 1 is represented in gray, group 2 in black. The size N_g of group g is constant in true only in the migration-only case without learning. Otherwise, strategy changes imply changes of individuals between groups and changes in group size. We will distinguish the three social relations between groups: (i) both groups (or, more exactly speaking, their individuals) like each other ($P_{12} = P_{21} = 1$), (ii) group 1 is neutral with respect to group 2, but group 2 dislikes group 1 ($P_{12} = 0$, $P_{21} = -1$), and (iii) intra-group interactions are more favored than inter-group interactions ($P_{11} = P_{22} = 1$, $P_{12} = P_{21} = 0.5$). Intra-group affiliation is always preferred or at least neutral ($P_{11} = P_{22} = 1$ or $P_{11} = P_{22} = 0$). As people are allowed to move, when they like each other, a strong

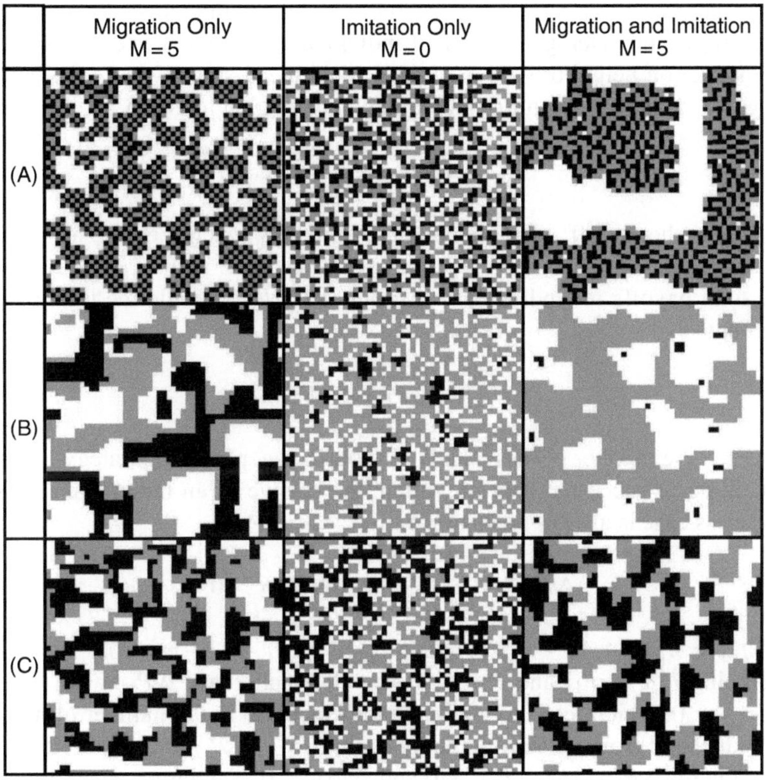

Fig. 10.3 Simulation results for 49×49 grids with a density of 0.6 for the migration-only case (*left*), the imitation-only case of conventional spatial games (*middle*) and the combination of imitation with migration (*right*). The color code is chosen as follows: *gray* = player of group 1, *black* = player of group 2, *white* = empty site. (**a**) $P_{11} = P_{22} = 0$, $P_{12} = P_{21} = 1$. Inter-group interaction is encouraged, which causes the integration of two populations. The combination of migration and imitation results in the co-existence of both groups and the formation of clusters. (**b**) $P_{11} = P_{22} = P_{12} = 1$, $P_{21} = -1$. When affinity is unilateral, one population keeps approaching the other one, which tends to evade. The combination of migration and imitation leads to the spreading of the chasing group. Co-existence is not observed. (**c**) $P_{11} = P_{22} = 1$, $P_{21} = P_{21} = 0.5$. When interactions between groups are less profitable than in the same group, residential segregation occurs as well, but in contrast to (**b**), it stabilizes

integration of populations can be observed. However, if the affinity is unilateral, then one population tends to evade the invasion of the other. Assuming the same mobility in both populations, in the current setting of our simulations the chasing population gains the majority in the end, if unconditional imitation is considered.

Residential segregation emerges through the interactions of individuals, not only when they dislike each other, but also if individuals within the same group like each other more than individual from other groups. When inter-group interactions result in smaller payoffs, people attempt to maximize their payoff by agglomerating with the same kind of people, which eventually results in the segregation of different groups as a side effect.

Some theoretical analysis can be useful to understand the micro-macro link. Let n_{g1} and n_{g2} be the average number of individuals of group 1 and 2, respectively, in the interaction neighborhood of an individual using strategy i, i.e. belonging to group $g = i$. Then, the payoff of an individual belonging to group $g \in \{1, 2\}$ is

$$P_g = n_{g1} P_{g1} + n_{g2} P_{g2}, \quad (10.8)$$

where P_{gi} is the payoff of an individual of group g when meeting an individual using strategy i.

The total payoffs for individuals of group g is $N_g P_g$, while the total payoff of both groups is

$$T' = N_1 P_1 + N_2 P_2, \quad (10.9)$$

where N_g is the number of individuals of group g, i.e. pursuing strategy $g = i$.

According to success-driven migration, the change of an individual's payoff in the noiseless case is always positive, i.e.

$$\Delta P_g = \Delta n_{g1} P_{g1} + \Delta n_{g2} P_{g2} > 0 \quad (10.10)$$

with $\Delta n_{g1} = n_{g1}(t+1) - n_{g1}(t)$ and $\Delta n_{g2} = n_{g2}(t+1) - n_{g2}(t)$.

Assuming $P_{12} \gg P_{11}$ and $P_{12} > 0$, we will usually have $\Delta P_1 \approx \Delta n_{12} P_{12} > 0$ and $\Delta n_{12} > 0$, which leads to a monotonous increase in the number of neighbors of individuals of group 1. Giving the inequality $P_{11} \gg P_{12}$ and $P_{11} > 0$, we analogously obtain $\Delta P_1 \approx \Delta n_{11} P_{11} > 0$ and $\Delta n_{11} > 0$. Therefore, when intra-group interactions of group 1 are much stronger than inter-group interactions, and positive, clusters of group 1 will expand, as each migration step will increase the average number of neighbors within group 1. However, when $P_{11} < 0$, the formation of clusters of individuals belonging to group 1 by intra-group interactions is unlikely. When $P_{12} < 0$, an individual of group 1 tends to evade members of group 2, which can be regarded as a repulsive effect attempting to keep a certain distance between individuals of different groups [38].

The total change of payoff is given by

$$\Delta T' = N_1(\Delta n_{11} P_{11} + \Delta n_{12} P_{12}) + N_2(\Delta n_{21} P_{21} + \Delta n_{22} P_{22}) > 0, \quad (10.11)$$

as long as success-driven migration is applied by all the individuals, and changes ΔN_1 and ΔN_2 in the group sizes are negligible.

Let us now consider a simple example, where $P_{11} = P_{22} = 0$. Then, we have

$$\Delta T' = N_1 \Delta n_{12} P_{12} + N_2 \Delta n_{21} P_{21}, \quad (10.12)$$

which reflects the combined effects of interactions between members of group 1 and 2. Furthermore, if $\Delta T' > 0$, we expect that the number of neighborships between individuals of both groups will monotonously increase. The corresponding macroscopic phenomenon is the spatial integration (mixture) of both groups. However, if $(P_{12} + P_{21}) < 0$, the reduction of inter-group links is likely and will result in residential segregation (see Fig. 10.3).

Here we have examined how individual preferences can change the spatial population structure in a very simple social system considering migratory behavior. The revealed social process can, to some extent, reflect the dynamics of population succession [14] in urban areas. This corresponds to case b in Fig. 10.3. Consider the payoff matrix in Fig. 10.3b, and imagine that an individual of group 1 happens to be located in the neighborhood of group 2. For the previous residents, this may not change a lot. However, it attracts other members of group 1, who are not yet living in a neighborhood of group 1 individuals. This can trigger collective migration of other group 1 members. Since interactions between members of group 1 and 2 bring positive payoffs only for members of group 1, group 2 will finally leave for new places.

One may notice that, here, we do not differentiate the intention to migrate and the actual migratory behavior. In daily life, however, the movement of people is restricted by much more factors such as wealth and time, so people do not migrate often, even if they are motivated to move. But in our simulations as well, migration activity after a few iterations is small: Starting with high migration rate, due to the artificial choice of a random initial distribution, the migration rate quickly drops to a low level [12]. Of course, other factors determining migration can be easily added to the above proposed framework of migration games.

10.3.2 Promotion of Cooperation in the Prisoner's Dilemma

The prisoner's dilemma is an important paradigm for studying the emergence of cooperation among selfish individuals. Nevertheless, studies of the prisoner's dilemma on lattices have not fully explored the effect of mobility. It would be logical, of course, for people to evade areas with a high level of defection, and to settle down in areas dominated by cooperators. But would cooperative areas be able to survive invasion attempts by defectors or even to spread in space? To answer these questions, we will now focus on the effects of success-driven migration on the spatial population structure and the level of cooperation.

Figure 10.4 compares the migration-only case with $M = 5$ (left) with the imitation only case corresponding to $M = 0$ (center) and the combined imitation-and-migration case with $M = 5$ (right). In the imitation-only case, the proportion of cooperators is greatly reduced. However, the combination of migration and imitation strikingly promotes the level of cooperation. Our explanation is that, when individuals have mobility, cooperative clusters are more likely to be promoted in the presence of invasion attempts of defectors. We can see that, in the migration-only case, cooperators manage to aggregate and to form clusters. Although defectors attempt to enter cooperative clusters, they finally end up at the boundaries of cooperative clusters, as cooperators split to evade defectors and re-aggregate in new

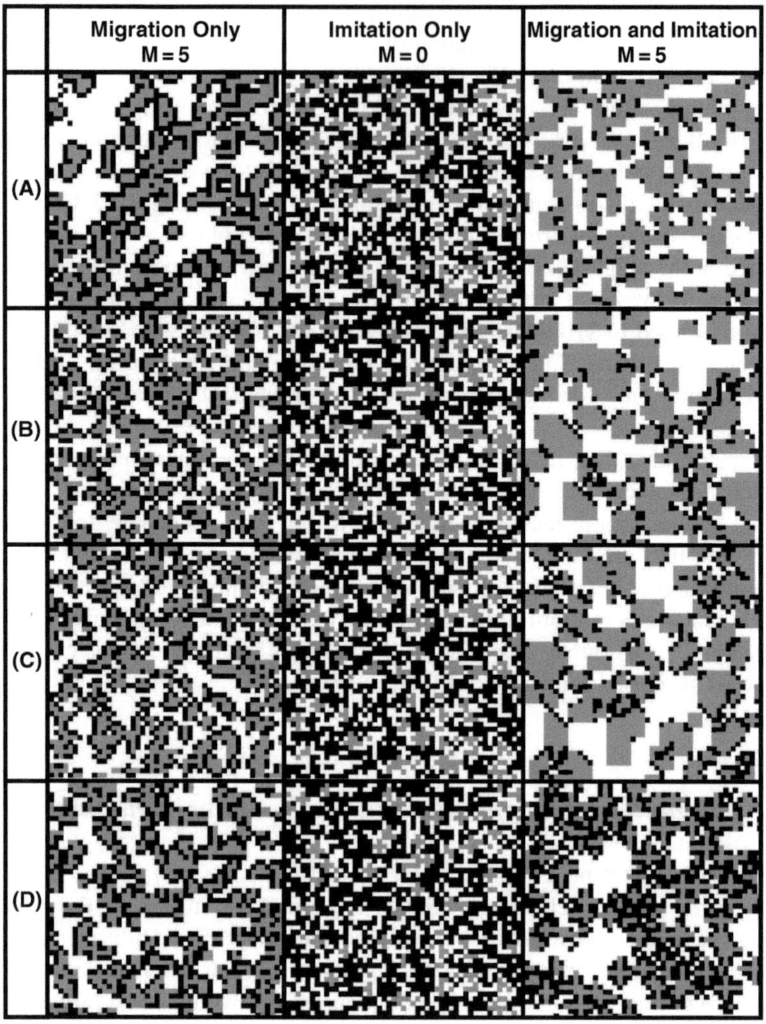

Fig. 10.4 Simulation results for 49×49 grids with a density of 0.6 for the migration-only case (*left*), the imitation-only case of conventional spatial games (*middle*) and the combination of imitation with migration (*right*). The color code is chosen as follows: *gray* = player of group 1, *black* = player of group 2, *white* = empty site. (**a**) $P_{11} = R = 1$, $P_{12} = S = -0.2$, $P_{21} = T = 1.4$, $P_{22} = P = 0$. The payoff matrix corresponds to a prisoner's dilemma. (**b**) $P_{11} = R = 1$, $P_{12} = S = 0$, $P_{21} = T = 1.4$, $P_{22} = P = 0$. The sucker's payoff is set to zero to be compatible with the payoff matrix studied by Nowak and May. (**c**) $P_{11} = R = 1$, $P_{12} = S = 0$, $P_{21} = T = 1.4$, $P_{22} = P = 0$. The migration and imitation step are inverted here, i.e. an individual first imitates, then migrates. (**d**) $P_{11} = 0.59$, $P_{12} = 0.18$, $P_{21} = 1$, $P_{22} = 0$. In the snow-drift game, similar structures are found in the migration-only and imitation-only case. Giving the possibility to move, frequent switches of strategies are observed. See main text for details

places, where defectors are excluded. In the prisoner's dilemma, it is guaranteed that $2R > T + S$, which means that the "attractive force" between cooperators is mutual and strong, while the interaction between a cooperator and a defector leads to a unilateral attractive force ($T > 0$). When $S < 0$, a cooperator replies to defectors even in a repulsive way. Therefore, defectors are less successful in joining or entering cooperative clusters than cooperators are.

Configurational analysis [12] (see Fig. 10.5), indicates that, when $P = S = 0$, and an individual only interacts with $k = 4$ nearest neighbors, a cooperative cluster can turn a defector into cooperator by unconditional imitation for $T < 1.5R$, when the defector is surrounded with one or two cooperators. When a defector is surrounded with three cooperators, cooperators can resist the invasion of a defector, even if the temptation value T is as high as $\frac{4}{3}R$. A defector can invade one of the nearby cooperators, if its neighborhood is fully occupied by cooperators. Therefore, the formation of compact clusters is important to support the spreading of cooperation. If $k > 4$, the spreading of cooperation occurs even for higher value of T.

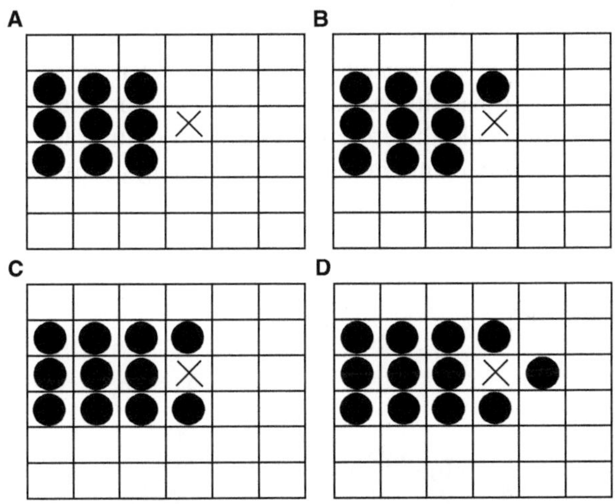

Fig. 10.5 Analysis of the invasion of a cluster of cooperators (*black circles*) by a single defector (*cross*). For the simplicity of analysis, we assume $S = P = 0$ here, but one can easily generalize the analysis to the situations with $P > S > 0$. (**a**) Scenario 1: The payoff for the defector is T, while its neighboring cooperator obtains a payoff of $3R$. Since $3R > T$, the defector will become a cooperator. (**b**) Scenario 2: The payoff for the defector is $2T$, while the maximal payoff among neighboring cooperators is $3R$. In order to make the defector become a cooperator, we need $3R > 2T$, i.e. $T < 1.5R$. (**c**) Scenario 3: The defector obtains $3T$, while the maximal payoff of neighboring cooperators is $3R$. In order to turn the defector into a cooperator, the inequality $3R > 3T$ must be satisfied, i.e. $T < R$. This condition can never be met in the prisoner's dilemma. In order that cooperators do not copy the defector, $4R > 3T$ must be satisfied, i.e. $T < \frac{4}{3}R$. (**d**) Scenario 4: The payoff for the defector is $4T$, the maximal payoff of cooperators in the whole community is $4R$. Because of $T > R$, the defector can invade the cooperators nearby. Once a cooperator becomes a defector however, the payoff for the defectors will be reduced from $4T$ to $3T$, which may stop the further invasion of defectors

In spatial games without mobility, the occurrence of a compact cooperative clusters mainly depends on the initial distribution. Giving mobilities, the migratory behavior can significantly accelerate the formation of compact clusters, and promote the level of cooperation.

Quantitative studies of how migration can promote the level of cooperation have to compare the fraction of cooperators in situations with mobility and without. Figure 10.6 shows the amplification factor, defined as

$$\delta(T, M) = \frac{f_M^T}{f_0^T}. \tag{10.13}$$

f_M^T is the fraction of cooperators, given the mobility range M and a temptation value of T. We can see that the level of cooperation in the prisoner's dilemma is promoted in a large parameter area, if the punishment P and the sucker's payoff are roughly comparable in size.

However, we have not yet studied the robustness of the migration mechanism so far. One may imagine that, once a defector would manage to invade a cooperative cluster, it may turn neighboring cooperators into defectors as well, thereby eliminating cooperation eventually. While in a noiseless environment, a defector cannot enter the center of a compact cooperative cluster, "noise" could make it happen with a certain probability. As defectors in a cooperative cluster can spread (see Fig. 10.5d), noise could therefore be thought to destroy the enhancement of cooperation by success-driven migration. Very surprisingly, this is not the case!

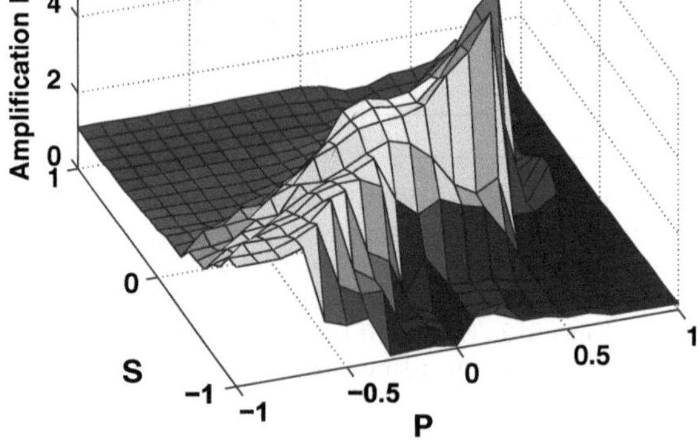

Fig. 10.6 Amplification factor of the level of cooperation by migration ($M = 5$) as a function of the sucker's payoff $P_{12} = S$ and the punishment $P_{22} = P$ in the prisoner's dilemma ($S < P$) and the snow drift game ($S > P$). The simulation was performed for 99×99 grids with a density of 0.6. P and S were varied between -1 and 1, the payoffs $R = 1$ and $T = 1.3$ were left fix

In order to verify that success-driven migration robustly promotes cooperation, we have implemented 3 kinds of noises. In each time step, a certain proportion y of individuals was selected to perform the following operations after the respective migration and imitation steps:

Noise 1: The selected players' locations were exchanged with a randomly chosen neighboring site ("neighborhood flipping").
Noise 2: The selected players' strategies were flipped, i.e. cooperation was replaced by defection and vice versa ("strategy flipping").
Noise 3: The selected players were removed from the grid, and an equal number of players was created at randomly chosen free sites, in order to mimic birth and death processes. The newly born players had a 50% chance to be cooperators and a 50% chance to be defectors.

Figure 10.7 shows the time evolution of the number of cooperators with noise strength $y = 2\%$ and $y = 10\%$ respectively. Without mobility ($M = 0$), noise reduces the number of cooperators greatly. For a mobility range $M = 5$, however, we surprisingly find that the level of cooperation can be still maintained at a high level. With 2% noise, noises 1 and 3 can even increase the number of cooperators compared to the no-noise case!

Therefore, in contrast to what one may expect, noise does not just simply destroy spatial structures and the level of cooperation. It may also overcome metastable configurations of the system, but naturally, it depends on the kind of noise, what noise levels are beneficial [10]. In order to illustrate how noise can promote cooperation to

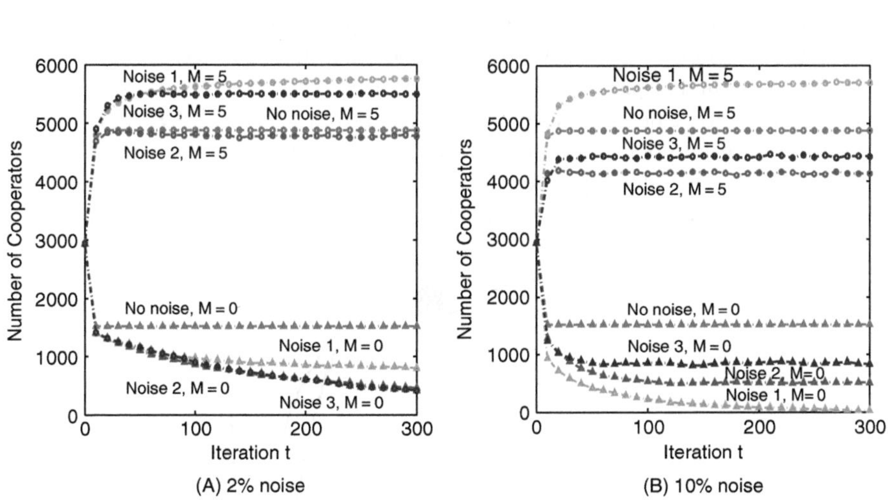

Fig. 10.7 Time evolution of the number of cooperators. Here, the simulation is performed on 99×99 grids with a density of 0.6 and payoffs $P_{11} = R = 1$, $P_{12} = S = 0$, $P_{21} = T = 1.3$, and $P_{22} = P = 0$

a level higher than the level in a noiseless system, one may define a kind of potential energy function of the system by the negative total payoff:

$$E = -T'. \tag{10.14}$$

In a noiseless system, each individual's migration step will increase T, and reduce the potential energy of the system, [see Eq. (10.11)]. Then an individual will adopt the most successful strategy within its neighborhood. The imitation step and noise can flip the strategies and increase the energy of the system. Just as for the energy functions in spin glass models [39], recurrent neural networks [40] and Boltzmann machines [41], there are many local minima of E in the migration game, one of which is reached within a few iterations. Afterwards, the noiseless system behaves stationary, but it is most likely stuck in a meta-stable state, which depends on the initial condition. Moderate noise can perturb these meta-stable states and thereby support the evolution of the system towards better configurations (see Fig. 10.8). Only if the noise level is too strong, the system behaves pretty much in a random way. In summary, randomness does not necessarily destroy a high level of cooperation. Moderate noise strengths can even support it (at least for some kinds of noise).

Our results for the conventional prisoner's dilemma with $P > S$ confirm the robustness of migration as a mechanism to promote cooperation (see Fig. 10.9). One can easily see that, without mobility, the proportion of cooperators is close to zero, while there is still a certain small number of cooperators in the environment without noise. However, when individuals can move, cooperation is significantly increased due to the spontaneous formation of clusters. Imagine that a defector is located in the center of a cooperators' cluster. In the beginning, defection can invade cooperation due to the higher payoff (see Fig. 10.5d). But once a neighboring cooperator becomes a defector, a defector's payoff is reduced from $4T$ to $3T$, if $P = 0$. If more cooperators turn into defectors, the payoff will be further reduced. Therefore, the exploitation of cooperators by defectors is self-inhibitory, if individuals copy better performing neighbors. Furthermore, a splitting of a cooperative cluster can occur,

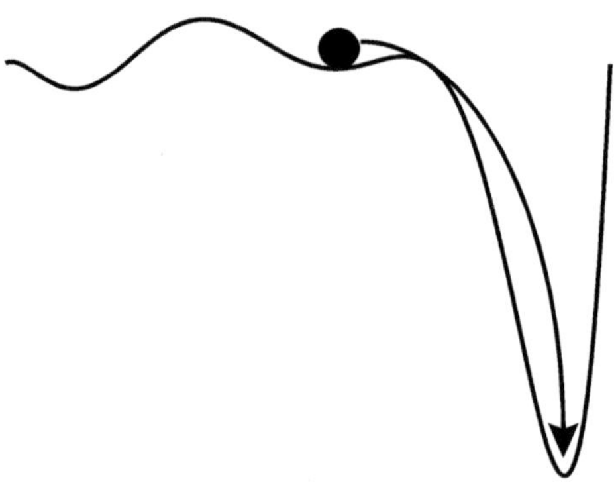

Fig. 10.8 Noise can make the system leave a sub-optimal state and reach the globally optimal state

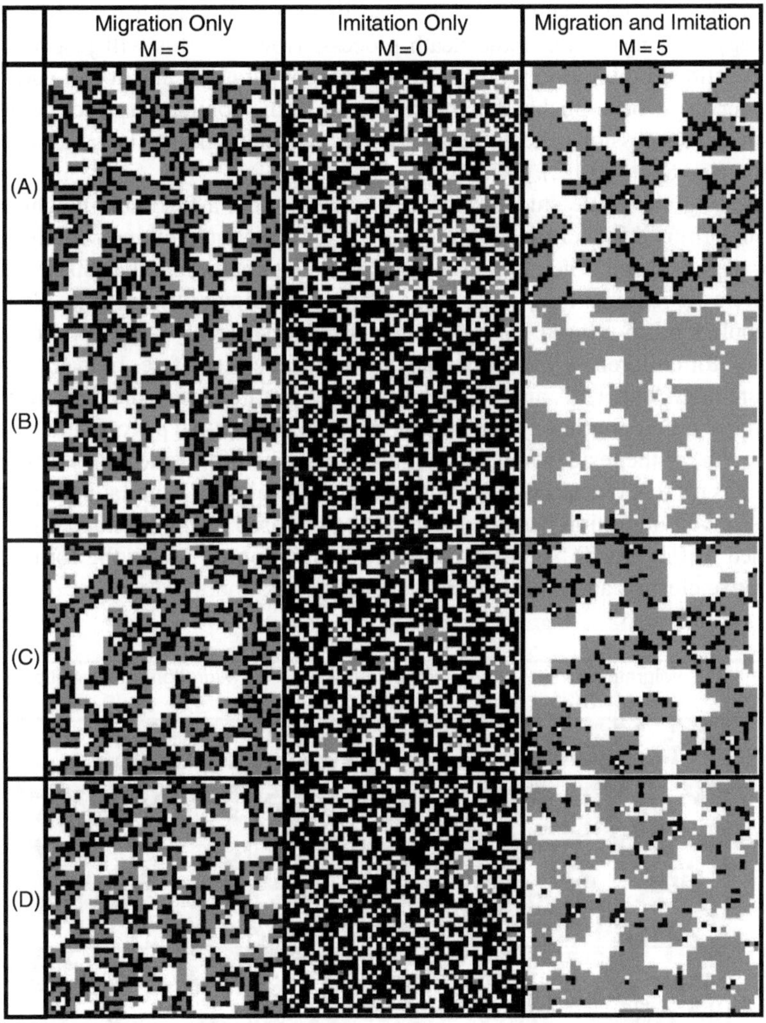

Fig. 10.9 Effect of different kinds of noises on the outcome of the spatial prisoner's dilemma with migration but no imitation (*left*), with imitation, but no migration (*center*), and with both, migration and imitation (*right*). The simulation is performed on 49×49 grids with a density of 0.6. The color code is chosen as follows: *black* = defector, *gray* = cooperator, *white* = empty site. While the resulting level of cooperation is very small in the conventional imitation-only case (*center*), the additional consideration of migration results in large levels of cooperation even in the presence of different kinds of noise. (**a**) No noise. (**b**) Noise 1 (neighborhood flipping). (**c**) Noise 2 (strategy flipping). (**d**) Noise 3 (birth and death). The payoffs were $T = 1.3$, $R = 1$, $P = 0.1$, and $S = 0$ in all cases. See main text for details

since cooperators try to move to more favorable neighborhood as soon as defectors are approaching them. The migration of those cooperators can encourage other cooperators, whose payoff depends on the mutual interactions, to move as well. That is, defectors may trigger a collective movement of cooperators. Of course, the newly formed cooperative clusters are also likely to be invaded by neighborhood or strategy flipping, or birth and death process. However, this will just repeat the above mentioned migration process. Therefore, cooperators can survive and spread even in a noisy world. Such a dynamical change of the population structure through invasion and succession reflects various features of the migratory behavior observed in reality (see Sect. 10.1.2).

In the migration rule studied above, we assume that a favorable neighborhood can be determined by fictitious play, i.e. some low-cost interactions with the people in that neighborhood ("neighborhood testing"). One may think that this is difficult in reality as it requires to reveal people's strategies. However, one may also argue that people tend to migrate to high-quality residential areas, which provide better education for children, a low crime rate, and other social welfare. In fact, neighborhoods are often "labeled", and it may be assumed that this label (the appearance and character of a neighborhood) depends on the total cumulative payoffs (the accumulated wealth) of the residents in the neighborhood. Therefore, one could assume that individuals try to move to the neighborhood with the highest cumulative payoff. The success-driven migration based on such a wealth-based "neighborhood tagging" is examined in Fig. 10.10. Again, we find that migration promotes the formation of cooperative clusters and an enhanced level of cooperation.

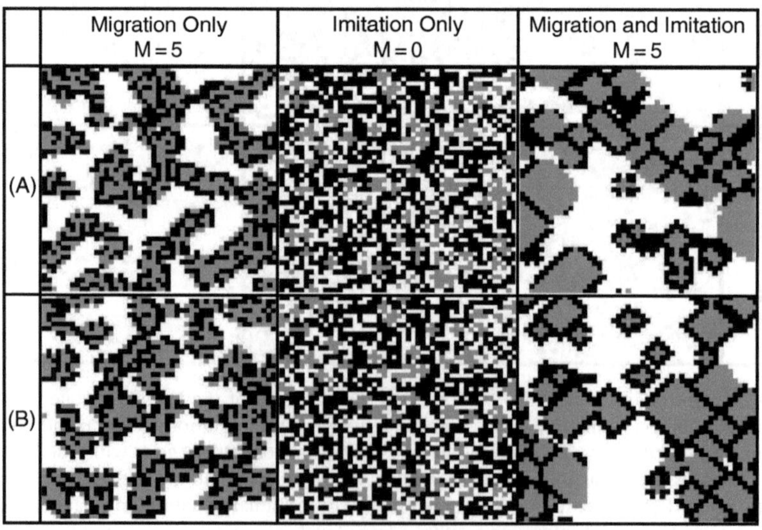

Fig. 10.10 Migratory prisoner's dilemma with wealth-based neighborhood-tagging rather than neighborhood testing as before. The simulation is performed on 49×49 grids with a density of 0.6. The color code is chosen as follows: *black* = defector, *gray* = cooperator, *white* = empty site. (**a**) $P_{11} = R = 1$, $P_{12} = S = 0$, $P_{21} = T = 1.4$, $P_{22} = P = 0$. (**b**) $P_{11} = R = 1$, $P_{12} = S = 0$, $P_{21} = T = 1.4$, $P_{22} = P = 0$, as in (**a**), but the update rule is inverted, i.e. an individual, first imitates, then migrates. In both cases, one can see that, without mobility, the proportion of cooperators becomes very low. However, when success-driven migration is possible, cooperation can spread, even if the sequence of the migration step and the unconditional imitation step is inverted

10.4 Conclusions

We have introduced the concept of migration games by considering success-driven motion. Migration games can easily reproduce macroscopic stylized facts of various social phenomena based on individual actions and interactions. Typical examples are population succession and residential segregation. These aggregate outcomes emerge from the interactions between individuals in a non-trivial way, and a theoretical analysis allows one to qualitatively understand the relation between the microscopic interactions and the emerging macroscopic phenomena. Nevertheless, further studies are required to fully elaborate the micro-macro link in a quantitative way.

For the prisoner's dilemma, we have shown that self-organized cooperative structures can promote the level of cooperation. Moreover, we have verified that the enhancement of cooperation by success-driven motion is robust to different kinds of noise. Surprisingly, we even find that moderate noise levels can promote the cooperation level further. The underlying mechanism is that, in the migration game, success-driven motion will monotonously increase the total payoff in the noiseless system, which however can lead the system into a locally optimal state. The effect of noise can drive the system out of local optima towards the globally optimal state.

The observation of pattern formation in migratory games is robust with respect to changes in the order of update steps, the initial level of cooperation, and moderate levels of randomness. It is also found for different specifications of the migration rules, but diffusive kinds of migration may destroy spatial patterns. Furthermore, our findings are robust with respect to reasonable variations in the payoffs and many other changes in the model (such as migration costs, etc.). For further details see Refs. [12, 13, 42–45]. The framework of migratory games can be easily extended. For example, one may integrate other kinds of interactions to study further social processes.

The spatial structure assumed in our simulation is very simple, and the neighborhood depends only on an individual's position. Real cities are more complex and show a co-evolutionary dynamics. For example, the city structure can reflect the distribution of social status groups. Early models like the Burgess concentric zone model [46] divides the city into specific areas separated by status rings. The city center is located in the middle of the circle, around which newer, higher-quality housing stocks tend to emerge at the perimeter of the city. Therefore, the growth of the city center will expand adjacent residential zones outwards.

Status segregation is quite obvious in such an idealized model of city structure. Poor people or new immigrants may only afford low quality housing. Middle class people live in less compacted neighborhoods. Rich people tend to accumulate in particular quarters.

Burgess's model is based on the bid rent curve. Recognizing that some poor people prefer to live near the main transportation arteries and commercial establishments, Hoyt [47] modified the concentric zone model to take this into account. In Hoyt's model, cities tend to grow in wedge-shaped patterns or sectors. Major transportation routes are emanating from the central business district (CBD). The residential areas for lower income people are located adjacent to the industrial quarters, while upper class neighborhoods are far away from industrial pollution and noise.

It would be natural to extend migration games in order to study the co-evolutionary dynamics of population structure and urban growth. On the long run, we hope this will contribute to a better understanding and planning of the population dynamics in a city.

Further research work can also study conflicts related with migratory behavior, as has been revealed by the empirical research [48, 49].

References

1. W.A. Clark, M. Fossett, Understanding the social context of the Schelling segregation model. Proc. Natl. Acad. Sci. **105**, 4109–4114 (2008)
2. M. Fossett, Ethnic preferences, social distance dynamics, and residential segregation: theoretical explorations using simulation analysis. J. Math. Socio. **30**, 185–274 (2006)
3. O.D. Duncan, B. Duncan, Residential distribution and occupational stratefication. Am. J. Sociol. **60**, 493–503 (1955)
4. S.F. Reardon, G. Firebaugh, Response: Segregation and social distance – a generalized approach to segregation measurement. Sociol. Methodol. **32**, 85–101 (2002)
5. T.C. Schelling, Dynamic models of segregation. J. Math. Socio. **1**, 143–186 (1971)
6. T.C. Schelling, *Micromotives and Macrobehavior*. (Norton, New York, 1978)
7. D.S. Massey, American apartheid: segregation and the making of the underclass. Am. J. Sociol. **96**, 329–357 (1990)
8. J. Yinger, *Closed Doors, Opportunities Lost: The Continuing Costs of Housing Discrimination*. (Russell Sage Found, New York 1995).
9. M. Macy, A.V.D. Rijt, Ethnic preferences and residential segregation: Theoretical explorations beyond Detroit. J. Math. Socio. **30**, 275–288 (2006)
10. D. Helbing, T. Platkowski, Drift- or fluctuation-induced ordering and self-organization in driven many-particle systems. Europhys. Lett. **60**, 227–233 (2002)
11. D. Helbing, T. Vicsek, Optimal self-organization. N. J. Phys. **1**, 13.1–13.17 (1999)
12. D. Helbing, W. Yu, Migration as a mechanism to promote cooperation. Adv. Compl. Syst. **11**, 641–652 (2008)
13. D. Helbing, W. Yu, The outbreak of cooperation among success-driven individuals under noisy conditions. Proc. Natl. Acad. Sci. **106**, 3680–3685 (2009), and Supplementary Information to this paper.
14. P.F. Cressey, Population succession in Chicago: 1898–1930. Am. J. Sociol. **44**, 59–69 (1938)
15. M.A. Nowak, Five rules for the evolution of cooperation. Science **314**, 1560 (2006)
16. A. Axelrod, *The Evolution of Cooperation*. (Basic Books, New York 1984)
17. A.S. Griffin, S.A. West, A. Buckling, Cooperation and competition in pathogenic bacteria. Nature **430**, 1024–1027 (2004)
18. J. Von Neumann, O. Morgenstern, *The Theory of Games and Economic Behavior*. (Princeton University Press, Princeton 1944)
19. D. Fudenberg, J. Tirole, *Game Theory*. (MIT Press, Cambridge 1991)
20. A. Diekmann, Volunteer's dilemma. J. Confl. Resolut. **29**, 605–610 (1985)
21. A. Flache, R. Hegselmann, Do irregular grids make a difference? Relaxing the spatial regularity assumption in cellular models of social dynamics. J. Artif. Soc. Soc. Simul. **4**(6) (2001)
22. P. Schuster, K. Sigmund, Replicator dynamics. J. Theor. Biol. **100**, 533–538 (1983)
23. M.A. Nowak, R.M. May, Evolutionary games and spatial chaos. Nature **359**, 826–829 (1992)

24. C. Hauert, M. Doebell, Spatial structure often inhibits the evolution of cooperation in the snowdrift game. Nature **428**, 643–646 (2004)
25. S. Hoogendoorn, P.H.L. Bovy, Simulation of pedestrian flows by optimal control and differential games. Opim. Control Appl. Mech. **24**, 153–172 (2003)
26. M.H. Vainstein, J.J. Arenzon, Disordered environments in spatial games. Phys. Rev. E **64**, 051905 (2001)
27. M.H. Vainstein, A.T.C. Silva, J.J. Arenzon, Does mobility decrease cooperation? J. Theor. Biol. **244**, 722–728 (2006)
28. E.A. Sicardi, H. Fort, M.H. Vainstein, J.J. Arenzon, Random mobility and spatial structure often enhance cooperation. J. Theor. Biol. **256**, 240–246 (2009)
29. J.M. Epstein, *Generative Social Science*. (Princeton University Press, Princeton 2006)
30. N. Rajewsky, L. Santen, A. Schadschneider, M. Schreckenberg, The asymmetric exclusion process: Comparision of update procedures. J. Statist. Phys. **92**, 151 (1998)
31. B.A. Huberman, N.S. Glance, Evolutionary games and computer simulations. Proc. Natl. Acad. Sci. **90**, 7716–7718 (1993)
32. H.P. Young, *Individual Strategy and Social Structure*. (Princeton University Press, Princeton, 1998)
33. M.W. Macy, Learning to cooperate: stochastic and tacit collusion in social exchange. Am. J. Sociol. **97**, 808–843 (1991)
34. M.W. Macy, A. Flache, Learning dynamics in social dilemmas. Proc. Natl. Acad. Sci. **99**, 7229–7236 (2002)
35. S. Wolfram, *Theory and Applications of Cellular Automata*. (World Scientific Publication, Singapore 1986)
36. B. Chopard, M. Droz, *Cellular Automata Modeling of Physical Systems*. (Cambridge University Press, Cambridge, 1998)
37. E.R. Berlekamp, J.H. Conway, R.K. Guy, *Winning Ways for Your Mathematical Plays*. Academic Press, New York, (1982)
38. D. Helbing, I. Farkas, T. Vicsek, Simulating dynamical features of escape panic. Nature **407**, 487–490 (2000)
39. S. Kirkpatrick, D. Sherrington, Infinite-ranged models of spin-glasses. Phys. Rev. B **17**, 4384–4403 (1978)
40. J.J. Hopfield, Neural networks and physical systems with emergent collective computational abilities. Proc. Natl. Acad. Sci. **79**, 2554–2558 (1982)
41. D.H. Ackley, G.E. Hinton, T.J. Sejnowski, A learning algorithm for Boltzmann machines. Cogn. Sci. **9**, 147–169 (1985)
42. D. Helbing, Pattern formation, social forces, and diffusion instability in games with success-driven motion. Eur. Phys. J. B **67**, 345–356 (2009)
43. D. Helbing, A mathematical model for the behavior of individuals in a social field. J. Math. Sociol. **19**, 189–219 (1994)
44. S. Meloni, et al., Effects of mobility in a population of Prisoner's Dilemma players. arXiv::0905.3189
45. C. Roca, *Cooperation in Evolutionary Game Theory: Effects of Time and Structure*. Ph.D. thesis, (Universidad Carlos III de Madrid, Department of Mathematics) (2009)
46. E.W. Burgess, Residential segregation in American cities. Ann Am Acad Polit Soc Sci **140**, 105–115 (1928)
47. H. Hoyt, *The Structure and Growth of Residential Neighborhoods in American Cities Washington*. Federal Housing Administration (Washion DC, 1939)
48. M. Lim, R. Metzler, Y.B. Yam, Global pattern formation and ethnic/cultural violence. Science **317**, 1540–1544 (2007)
49. N.B. Weidmann, D. Kuse, WarViews: Visualizing and animating geographic data on civil war. International Studies Perspectives **10**, 36–48 (2009)

Chapter 11
Lattice Boltzmann Simulations of Wetting and Drop Dynamics

Halim Kusumaatmaja and Julia M. Yeomans

11.1 Introduction

Recently there has been a huge effort in the scientific community to miniaturise fluidic operations to micron and nanoscales [1]. This has changed the way scientists think about fluids, and it potentially has far-reaching technological implications, analogous to the miniaturization of electronics. The goal is to engineer "lab on a chip" devices, where numerous biological and chemical experiments can be performed rapidly, and in parallel, while consuming little reagent.

An important aspect of the physics of fluids at micron and nanoscales is the increasing relevance of surface effects. Surface slip will dominate flow in nanochannels, and the movement of small drops across a substrate will be strongly affected by the interactions between the fluid and the surface. This has been exploited in the functional adaptation of many biological systems, for example lotus leaves [2], desert beetles [3] and butterfly wings [4]. Moreover, the wetting and spreading of fluids over surfaces is key to numerous technological processes, for example in oil recovery, painting, and inkjet printing.

Small liquid drops are spherical when they are in air, to minimise the surface energy. When placed on a solid the degree to which a drop spreads depends on the balance of interfacial energies between the solid, liquid, and gas phases. In equilibrium the liquid–gas interface maintains a spherical cap profile, and the liquid drop joins the solid at a contact angle θ_e, where

$$\cos\theta_{\mathrm{e}} = \frac{\sigma_{\mathrm{GS}} - \sigma_{\mathrm{LS}}}{\sigma_{\mathrm{LG}}} \tag{11.1}$$

and σ_{GS}, σ_{LS}, and σ_{LG} are the gas–solid, liquid–solid and liquid–gas surface tensions. Equation (11.1) is Young's equation and the equilibrium contact angle is

H. Kusumaatmaja (✉)
Max Planck Institute of Colloids and Interfaces, Science Park Golm, 14424 Potsdam, Germany
e-mail: kusumaatmaja@gmail.com

A.G. Hoekstra et al. (eds.), *Simulating Complex Systems by Cellular Automata*, Understanding Complex Systems, DOI 10.1007/978-3-642-12203-3_11,

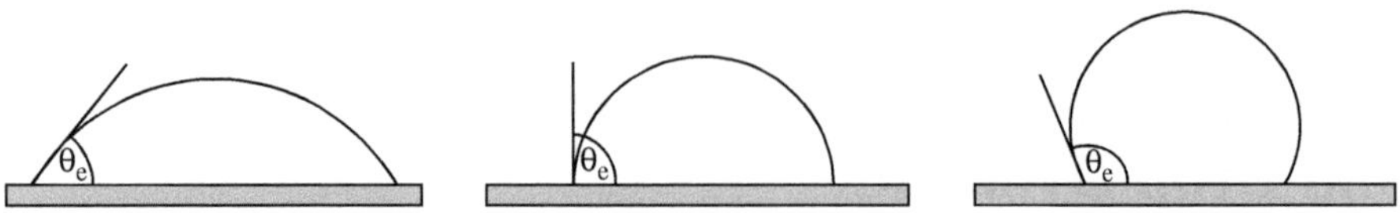

Fig. 11.1 Profile of a liquid drop on (**a**) a hydrophilic, (**b**) a neutrally wetting and (**c**) a hydrophobic surface

often called the Young angle. A solid surface is termed hydrophilic[1] when $\theta_e < 90°$, neutrally wetting when $\theta_e = 90°$ and hydrophobic when $\theta_e > 90°$. This is illustrated in Fig. 11.1. For reviews of wetting and spreading see [5–8].

Wetting phenomena are often further complicated by the fact that the solid surfaces are never perfectly homogeneous. For micron and nanometer drops, the typical length scale of surface heterogeneities can be comparable to the size of the drop itself. Random disorder on a surface is notoriously difficult to describe theoretically or numerically. However, as a result of recent and rapid developments in microfabrication techniques, it is now possible to manufacture surfaces with well controlled patterning on micron, and even nanometer, length scales. The patterning can be either chemical, with the contact angle varying from place to place, or topographical, where the relief of the surface changes. Patterning surfaces leads to a rich range of drop thermodynamics and hydrodynamics which, because the surfaces are well characterised, can now be investigated experimentally.

Analytical solutions describing the behaviour of drops on surfaces are possible in some special cases, but in general they are not tractable when the surface heterogeneities are taken into account. Therefore there is a need for powerful numerical techniques that are able to both solve the hydrodynamic equations of motion of the fluids, and to take into account the effect of surface patterning, with relative ease. To this end, in this chapter, we introduce a mesoscale numerical algorithm, the lattice Boltzmann method, and show how it may be used to investigate the physics of wetting and spreading.

Writing down an algorithm which solves the Navier-Stokes equations is rather easy. This is because these equations are based on local conservation of mass and momentum and, as long as the conservation laws are represented correctly, (and space is discretised in a sufficiently symmetric way) the hydrodynamic equations will be recovered in the continuum limit. This was pointed out by Frisch et al. [9] who wrote down the first mesoscale algorithm for the Navier-Stokes equation. This was a lattice-gas cellular automaton: particles move on a lattice and collide at the nodes according to rules which impose mass and momentum conservation. As long as the lattice has sufficient symmetry it is possible to choose collision rules that reproduce the Navier-Stokes equation in the continuum limit.

[1] Strictly, the terms hydrophilic and hydrophobic are appropriate only when the liquid is water. Nonetheless, they are often used more generally.

Cellular automata models of hydrodynamics have proved important in many contexts but can be difficult to use because of large fluctuations in macroscopic quantities such as the density and the velocity. The need to average over the fluctuations can cancel out the advantage of an algorithm which is easy to parallelise. However many other mesoscale methods have evolved from the ideas of cellular automata, and one of these is the lattice Boltzmann approach [10–13]. The discrete variables used in cellular automata are replaced by a set of distribution functions that represent the average population of fluid particles. This removes the difficulty of fluctuations: lattice Boltzmann can be viewed as a mean-field version of the cellular automata models.

Lattice Boltzmann algorithms can be extended to multiphase and complex fluids [14–16]. The algorithm solves the Navier-Stokes equations and, as we shall describe below, the thermodynamic properties of a given fluid and its interactions with a surface can be modelled rather easily by introducing a free energy functional that is minimised in equilibrium. Moreover, lattice Boltzmann algorithms are well able to handle flow in complex geometries and hence represent an efficient numerical way of treating surfaces with topographic patterning.

The chapter comprises two parts. First, in Sects. 11.2, 11.3, and 11.4 we explain the physical model and the lattice Boltzmann algorithm used to solve it. We introduce a free energy functional for a binary fluid [17, 18] and explain how it handles the essential ingredients needed to describe wetting phenomena; phase separation, surface tension and contact angles. The generalisation of the Navier-Stokes equations appropriate for the two-phase system are summarised. We then describe lattice Boltzmann algorithms that will solve the hydrodynamic equations, commenting particularly on the thermodynamic and hydrodynamic boundary conditions needed to model wetting. Next, in Sects. 11.5, 11.6, and 11.7, we illustrate the efficacy of the algorithm by describing several applications to the physics of drops on smooth and patterned surfaces. These include capillary filling, viscous fingering, controlling drop motion using chemical patterning, slip in patterned microchannels and superhydrophobic surfaces.

11.2 The Binary Model

11.2.1 Thermodynamics of the Fluid

To model drops of fluid on a surface we need first to describe their equilibrium properties, such as binary fluid coexistence, surface tension and contact angle. As we are working on micron-length scales we can use a continuum, Landau free energy [17], which is minimised in equilibrium,

$$\Psi = \int_V \left(\psi_b + \frac{\kappa}{2}(\partial_\alpha \phi)^2\right) dV + \int_S \psi_s \, dS, \tag{11.2}$$

where the bulk free energy density ψ_b is taken to have the form

$$\psi_b = \frac{c^2}{3} n \ln n + A\left(-\tfrac{1}{2}\phi^2 + \tfrac{1}{4}\phi^4\right). \qquad (11.3)$$

n is the fluid density which is chosen to be 1 everywhere, ϕ is the order parameter and $c = \Delta x/\Delta t$, where Δx and Δt represent the discretisation in space and time respectively. This choice of ψ_b gives binary phase separation into two phases with $\phi_\alpha = 1$ and $\phi_\beta = -1$, where α and β label the two coexisting bulk phases.

The second and third terms in Eq. (11.2) are needed to account for the fluid–fluid and fluid–solid surface tensions. Let us first consider the fluid–fluid surface tension and for simplicity, restrict ourselves to one dimension. Minimising the volume terms in the free energy functional with respect to ϕ leads to the condition for equilibrium

$$\mu \equiv -A\phi + A\phi^3 - \kappa \frac{d^2}{dx^2}\phi = 0 \qquad (11.4)$$

where μ is the chemical potential. Equation (11.4) allows an interface solution of the form

$$\phi = \tanh\left(\frac{x}{\sqrt{2}\xi}\right) \qquad (11.5)$$

where $\xi = \sqrt{\kappa/A}$ is defined as the interface width. Since this must typically be chosen of order a few lattice spacings in a simulation, models of this type are often called diffuse interface models [19, 20].

Using Noether's theorem, we find that

$$\psi_b - \frac{\kappa}{2}\left(\frac{d\phi}{dx}\right)^2 = \text{constant} = \psi_b|_{\phi=\pm 1}\,. \qquad (11.6)$$

We can therefore define the excess bulk free energy density as

$$W = \psi_b - \psi_b|_{\phi=\pm 1} = \tfrac{A}{2}\left(\phi^2 - 1\right)^2 = \frac{\kappa}{2}\left(\frac{d\phi}{dx}\right)^2. \qquad (11.7)$$

The surface tension, γ, of the liquid–liquid interface can be calculated by integrating the sum of the excess bulk free energy density and the second (κ) term in Eq. (11.2)

$$\gamma = \int_{\phi_\beta}^{\phi_\alpha}\left(W + \frac{\kappa}{2}\left(\frac{d\phi}{dx}\right)^2\right)dx = \int_{\phi_\beta}^{\phi_\alpha}\kappa\left(\frac{d\phi}{dx}\right)^2 dx = \sqrt{8\kappa A/9}. \qquad (11.8)$$

The second integral in Eq. (11.2) is over the system's solid surface and is used to describe the interactions between the fluid and the surface. Following Cahn [21], the surface energy density is taken to be $\psi_s = -h\phi_s$, where ϕ_s is the value of the order parameter at the surface. Minimisation of the free energy shows that the gradient in ϕ at the solid boundary is

$$\kappa \frac{d\phi}{dx} = -\frac{d\psi_s}{d\phi_s} = -h \,. \tag{11.9}$$

Equation (11.9) can be used together with Eq. (11.7) to determine ϕ_s.

The fluid–solid surface tensions can be calculated in a similar way to the fluid–fluid surface tension, except that now we also have to take into account the contributions from the surface energy term.

$$\gamma_{s\alpha} = -h\phi_{s\alpha} + \int_{\phi_{s\alpha}}^{\phi_\alpha} \kappa \left(\frac{d\phi}{dx}\right)^2 dx = \frac{\gamma}{2} - \frac{\gamma}{2}(1+\Omega)^{3/2} \,, \tag{11.10}$$

$$\gamma_{s\beta} = -h\phi_{s\beta} + \int_{\phi_{s\beta}}^{\phi_\beta} \kappa \left(\frac{d\phi}{dx}\right)^2 dx = \frac{\gamma}{2} - \frac{\gamma}{2}(1-\Omega)^{3/2} \,, \tag{11.11}$$

where $\Omega = \sqrt{\frac{2}{\kappa A}}h$. The notations $\phi_{s\alpha}$ and $\phi_{s\beta}$ stand for the values of the order parameter at the surface for phases α and β respectively.

The contact angle follows from substituting the values of the surface tensions into Young's law, Eq. (11.1), to give (with θ_e defined as the contact angle of the α-phase)

$$\cos\theta_e = \frac{\gamma_{s\beta} - \gamma_{s\alpha}}{\gamma} = \frac{(1+\Omega)^{3/2} - (1-\Omega)^{3/2}}{2} \,. \tag{11.12}$$

Equation (11.12) can be inverted to give a relation between the phenomenological parameter h and the equilibrium contact angle θ_e [17]

$$h = \sqrt{2\kappa A}\ \mathrm{sign}\left(\frac{\pi}{2} - \theta_e\right) \sqrt{\cos\left(\frac{\alpha}{3}\right)\left\{1 - \cos\left(\frac{\alpha}{3}\right)\right\}} \,, \tag{11.13}$$

where $\alpha = \cos^{-1}\left(\sin^2\theta_e\right)$ and the function sign returns the sign of its argument.

Lattice Boltzmann simulation results for the equilibrium contact angle of a liquid drop on a smooth solid surface are shown in Fig. 11.2. The exact result, given by Eq. (11.13), is also shown for comparison. Deviation is only noticeable at small contact angles. This discrepancy is mainly because the finite width of the interface, which is neglected when assuming that the drop is a spherical cap, becomes comparable to the height of the drop.

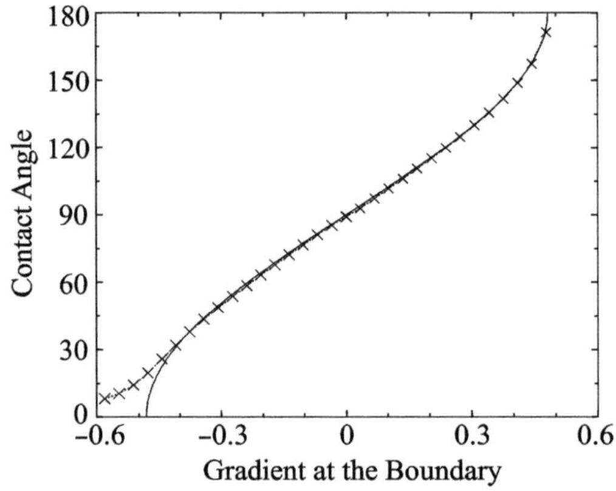

Fig. 11.2 The equilibrium contact angle as a function of the gradient in ϕ at the boundary. Crosses are lattice Boltzmann simulation results, while the *solid curve* is the theoretical expression, Eq. (11.13). We have used A = 0.04 and κ = 0.04 Reprinted figure with permission from Pooley et al. [18]. Copyright (2008) by the American Physical Society

11.3 Hydrodynamics of the Fluid

The hydrodynamic equations of motion for the binary fluid are the continuity equation (11.14), the Navier-Stokes equation (11.15) and the convection-diffusion equation (11.16)

$$\partial_t n + \partial_\alpha (n v_\alpha) = 0\,, \tag{11.14}$$

$$\partial_t (n v_\alpha) + \partial_\beta (n v_\alpha v_\beta) = -\partial_\beta P_{\alpha\beta} + \partial_\beta [n\nu(\partial_\beta v_\alpha + \partial_\alpha v_\beta) + \left(n\lambda\delta_{\alpha\beta}\partial_\gamma v_\gamma\right)] + n a_\alpha\,, \tag{11.15}$$

$$\partial_t \phi + \partial_\alpha \left(\phi v_\alpha\right) = M\nabla^2\mu \tag{11.16}$$

where $\mathbf{v}$, $\mathbf{P}$, ν, $\mathbf{a}$ and M are the local velocity, pressure tensor, shear kinematic viscosity, acceleration provided by the body force and mobility respectively. The bulk kinematic viscosity is $\lambda + \frac{d}{2}\nu$, where d is the dimension of the system.

The equilibrium properties of the fluid appear in the equations of motion through the chemical potential defined in Eq. (11.4) while the pressure can be derived from the free energy

$$\partial_\beta P_{\alpha\beta} = n\partial_\alpha \left(\frac{\delta\psi_b}{\delta n}\right) + \phi\partial_\alpha \left(\frac{\delta\psi_b}{\delta\phi}\right). \tag{11.17}$$

Using the definition of ψ_b in Eq. (11.3), it follows that [17]

$$P_{\alpha\beta} = \left(p_b - \frac{\kappa}{2}(\partial_\gamma\phi)^2 - \kappa\phi\partial_{\gamma\gamma}\phi\right)\delta_{\alpha\beta} + \kappa(\partial_\alpha\phi)(\partial_\beta\phi)\,, \tag{11.18}$$

$$p_b = \frac{c^2}{3}n + A\left(-\frac{1}{2}\phi^2 + \frac{3}{4}\phi^4\right). \tag{11.19}$$

p_b is the bulk pressure term which is related to the speed of sound in the model via $c_s^2 = \frac{dp_b}{dn} = \frac{c^2}{3}$. Equilibrium corresponds to $\partial_\beta P_{\alpha\beta} = 0$.

It is also important to note that the finite interface width allows slip to be generated close to the contact line by diffusive transport across the interface [17–20]. Slip is needed to remove the stress singularity at the moving contact line (see e.g. [22–24]). In this model it is controlled by the mobility parameter M.

11.4 The Lattice Boltzmann Algorithm

We now define a lattice Boltmann algorithm which solves Eqs. (11.14), (11.15), and (11.16). The basic idea behind lattice Boltzmann algorithms is to associate distribution functions, discrete in time and space, to a set of velocity directions $\mathbf{e}_i$. For example, for a three-dimensional, 19-velocity model, the lattice velocities are chosen to be

$$\begin{pmatrix} e_{x0\text{-}6} \\ e_{y0\text{-}6} \\ e_{z0\text{-}6} \end{pmatrix} = \begin{bmatrix} 0 & c & -c & 0 & 0 & 0 & 0 \\ 0 & 0 & 0 & c & -c & 0 & 0 \\ 0 & 0 & 0 & 0 & 0 & c & -c \end{bmatrix}, \tag{11.20}$$

$$\begin{pmatrix} e_{x7\text{-}18} \\ e_{y7\text{-}18} \\ e_{z7\text{-}18} \end{pmatrix} = \begin{bmatrix} c & -c & c & -c & 0 & 0 & 0 & 0 & c & -c & c & -c \\ c & c & -c & -c & c & -c & c & -c & 0 & 0 & 0 & 0 \\ 0 & 0 & 0 & 0 & c & c & -c & -c & c & c & -c & -c \end{bmatrix}.$$

c, the lattice velocity, is defined by $c = \Delta x/\Delta t$. The directions of the velocity vectors are shown in Fig. 11.3.

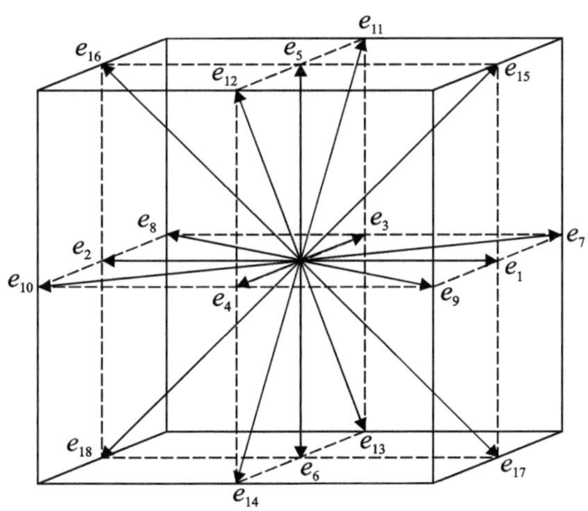

Fig. 11.3 The directions of the lattice velocity vectors in the 19-velocity lattice Boltzmann model

We need to define two distribution functions, $f_i(\mathbf{r},t)$ and $g_i(\mathbf{r},t)$, to describe a binary fluid. The physical variables are related to the distribution functions by

$$n = \sum_i f_i, \quad n u_\alpha = \sum_i f_i e_{i\alpha}, \quad \phi = \sum_i g_i, \tag{11.21}$$

where $\mathbf{u}$ is defined as $\mathbf{u} = \mathbf{v} - \mathbf{a}\Delta t/2$, and $\mathbf{a}$ is the acceleration associated with any applied body force. (This distinction between $\mathbf{u}$ and $\mathbf{v}$ is required so that the lattice Boltzmann equation recovers the continuity (11.14) and Navier-Stokes (11.15) equations in the continuum limit in a system with an applied force. In practice, the value of $\mathbf{v}$ is typically two to three order of magnitudes larger than $\mathbf{a}\Delta t$ in most simulations. Hence the distinction between $\mathbf{u}$ and $\mathbf{v}$ can usually be neglected.)

The time evolution equations for the particle distribution functions, using the standard BGK approximation [25], can be broken down into two steps

$$\begin{aligned}
\text{Collision step:}\quad & f_i'(\mathbf{r},t) = f_i(\mathbf{r},t) - \tfrac{1}{\tau}\left[f_i(\mathbf{r},t) - f_i^{\mathrm{eq}}(\mathbf{r},t)\right] + F_i(\mathbf{r},t)\,,\\
& g_i'(\mathbf{r},t) = g_i(\mathbf{r},t) - \tfrac{1}{\tau_\phi}\left[g_i(\mathbf{r},t) - g_i^{\mathrm{eq}}(\mathbf{r},t)\right],\\
\text{Propagation step:}\quad & f_i(\mathbf{r}+\mathbf{e}_i\Delta t, t+\Delta t) = f_i'(\mathbf{r},t)\,,\\
& g_i(\mathbf{r}+\mathbf{e}_i\Delta t, t+\Delta t) = g_i'(\mathbf{r},t)
\end{aligned} \tag{11.22}$$

where f_i^{eq} and g_i^{eq} are local equilibrium distribution functions, defined as a power series in the velocity, τ and τ_ϕ are the relaxation times and F_i is a term that corresponds to an external body force. It can be shown, using a Taylor expansion, that Eqs. (11.22) reproduce Eqs. (11.14), (11.15) and (11.16) in the continuum limit if the correct thermodynamic and hydrodynamic information is input to the simulation by a suitable choice of local equilibrium functions and forcing terms. Details of the derivation can be found in e.g. [10–14]. The constraints that need to be satisfied are

$$\sum_i f_i^{\mathrm{eq}} = n\,, \quad \sum_i f_i^{\mathrm{eq}} e_{i\alpha} = n v_\alpha\,, \tag{11.23}$$

$$\sum_i f_i^{\mathrm{eq}} e_{i\alpha} e_{i\beta} = P_{\alpha\beta} + n v_\alpha v_\beta\,, \tag{11.24}$$

$$\sum_i f_i^{\mathrm{eq}} e_{i\alpha} e_{i\beta} e_{i\gamma} = \frac{nc^2}{3}\left[v_\alpha \delta_{\beta\gamma} + v_\beta \delta_{\gamma\alpha} + v_\gamma \delta_{\alpha\beta}\right], \tag{11.25}$$

$$\sum_i g_i^{\mathrm{eq}} = \phi\,, \quad \sum_i g_i^{\mathrm{eq}} e_{i\alpha} = \phi v_\alpha\,, \tag{11.26}$$

$$\sum_i g_i^{\mathrm{eq}} e_{i\alpha} e_{i\beta} = \Gamma \mu \delta_{\alpha\beta} + \phi v_\alpha v_\beta\,. \tag{11.27}$$

$$\sum_i F_i = 0\,, \quad \sum_i F_i e_{i\alpha} = \Delta t \left(1 - \frac{1}{2\tau}\right) n a_\alpha\,, \tag{11.28}$$

$$\sum_i F_i e_{i\alpha} e_{i\beta} = \Delta t \left(1 - \frac{1}{2\tau}\right) (n v_\alpha a_\beta + n v_\beta a_\alpha). \tag{11.29}$$

Note that Eqs. (11.23) and the first equation in (11.26) correspond to conservation of mass, momentum and concentration.

A possible choice for f_i^{eq}, g_i^{eq} and F_i that satisfies the constraints (11.23), (11.24), (11.25), (11.26), (11.27), (11.28) and (11.29) is a power series expansion in the velocity [26, 27]

$$\begin{aligned}
f_i^{\mathrm{eq}} = {} & \frac{w_i}{c^2}\Big(p_b - \kappa\phi\nabla^2\phi + e_{i\alpha} n v_\alpha + \tfrac{3}{2c^2}\left[e_{i\alpha}e_{i\beta} - \tfrac{c^2}{3}\delta_{\alpha\beta}\right] \\
& \times \Big(n v_\alpha v_\beta + \lambda\left[v_\alpha \partial_\beta n + v_\beta \partial_\alpha n + \delta_{\alpha\beta} v_\gamma \partial_\gamma n\right]\Big)\Big) \\
& + \frac{\kappa}{c^2}\Big(w_i^{xx}\partial_x\phi\partial_x\phi + w_i^{yy}\partial_y\phi\partial_y\phi + w_i^{zz}\partial_z\phi\partial_z\phi\Big) \\
& + \frac{\kappa}{c^2}\Big(w_i^{xy}\partial_x\phi\partial_y\phi + w_i^{yz}\partial_y\phi\partial_z\phi + w_i^{zx}\partial_z\phi\partial_x\phi\Big),
\end{aligned} \tag{11.30}$$

$$g_i^{\mathrm{eq}} = \frac{w_i}{c^2}\Big(\Gamma\mu + e_{i\alpha}\phi v_\alpha + \tfrac{3}{2c^2}\left[e_{i\alpha}e_{i\beta} - \tfrac{c^2}{3}\delta_{\alpha\beta}\right]\phi v_\alpha v_\beta\Big),$$

$$F_i = \Delta t \frac{w_i}{c^2}\left(1 - \frac{1}{2\tau}\right)\left[e_{i\alpha} n a_\alpha + \frac{3}{2c^2}\left(e_{i\alpha}e_{i\beta} - \frac{c^2}{3}\delta_{\alpha\beta}\right)(n v_\alpha a_\beta + n v_\beta a_\alpha)\right].$$

where a choice for the w_i aimed at minimising spurious velocities[2] is [26]

$$\begin{aligned}
w_{1\text{-}6} &= \tfrac{1}{6}, \quad w_{7\text{-}18} = \tfrac{1}{12}, \\
w_{1,2}^{xx} &= w_{3,4}^{yy} = w_{5,6}^{zz} = \tfrac{5}{12}, \\
w_{3\text{-}6}^{xx} &= w_{1,2,5,6}^{yy} = w_{1\text{-}4}^{zz} = -\tfrac{1}{3}, \\
w_{7\text{-}10}^{xx} &= w_{15\text{-}18}^{xx} = w_{7\text{-}14}^{yy} = w_{11\text{-}18}^{zz} = -\tfrac{1}{24}, \\
w_{11\text{-}14}^{xx} &= w_{15\text{-}18}^{yy} = w_{7\text{-}10}^{zz} = \tfrac{1}{12}, \\
w_{1\text{-}6}^{xy} &= w_{1\text{-}6}^{yz} = w_{1\text{-}6}^{zx} = 0, \\
w_{7,10}^{xy} &= w_{11,14}^{yz} = w_{15,18}^{zx} = \tfrac{1}{4}, \\
w_{8,9}^{xy} &= w_{12,13}^{yz} = w_{16,17}^{zx} = -\tfrac{1}{4}, \\
w_{11\text{-}18}^{xy} &= w_{7\text{-}10}^{yz} = w_{15\text{-}18}^{yz} = w_{7\text{-}14}^{zx} = 0
\end{aligned}$$

[2] These are small velocities which remain in equilibrium. They are a consequence of discretisation errors, see Sect. 4.1.

The relaxation parameters τ and τ_ϕ in the lattice Boltzmann algorithm are related to the parameters in the hydrodynamic equations ν, λ and M through

$$\nu = (c^2 \Delta t(\tau - 1/2))/3\,, \tag{11.31}$$

$$\lambda = \nu(1 - 3c_s^2/c^2)\,, \tag{11.32}$$

$$M = \Delta t \Gamma \left(\tau_\phi - \tfrac{1}{2}\right), \tag{11.33}$$

where Γ is a tunable parameter that appears in the equilibrium distribution. Since ν, λ and M are positive quantities, the values of the relaxation times τ and τ_ϕ have to be larger than $1/2$.

In a typical binary lattice Boltzmann simulation, there are four important parameters controlling the physics: the length scale of the system L, the viscosity η, the surface tension γ, and the body force $n\mathbf{a}$. To match these to physical values we can choose only three quantities; a length scale L_o, a time scale T_o, and a mass scale M_o which are further constrained by the stability of the simulations. Therefore the simulation parameters cannot be arbitrarily matched to an experiment. In practice, a useful approach can be to determine L_o, T_o, and M_o by matching L, η and γ between simulations and experiments, and then to use these scales to determine the appropriate value of $n\mathbf{a}$. (A simulation parameter with dimensions $[L]^{n1}[T]^{n2}[M]^{n3}$ is multiplied by $L_o^{n1}\,T_o^{n2}\,M_o^{n3}$ to give the physical value.)

11.4.1 The Multiple Relaxation Time Algorithm

Figure 11.2 shows that there is excellent agreement between the theoretical value of the contact angle for a given surface field and that calculated numerically. However, these results were obtained using a relaxation time $\tau = 1$. For values of τ significantly different to unity the agreement is less good [18]. This discrepancy is caused by strong spurious velocities near the contact point which continuously push the system out of equilibrium and result in the deformation of the interface. The spurious velocities, which are a result of discretisation errors, are common to all lattice-based solutions of the Navier-Stokes equations, but are particularly pronounced near interfaces and surfaces. Taking $\tau = 1$ damps out many of the spurious contributions [18].

Since, in wetting problems, the two fluids generally have different viscosities (for example, the viscosities of water and air differ by a factor of 1000), restriction to $\tau = 1$ imposes a serious limitation. However the problem can be remedied by using a multiple relaxation time lattice Boltzmann algorithm [18].

The idea behind the multiple relaxation time lattice Boltzmann method [28–30] is that different relaxation parameters are used for different linear combinations of the distribution functions. The relaxation term $\frac{1}{\tau}\left[f_i - f_i^{\mathrm{eq}}\right]$ on the right hand side of the lattice Boltzmann equation for f_i (11.22) is replaced by

$$\mathbf{M}^{-1}\mathbf{S}\mathbf{M}\left[\mathbf{f}-\mathbf{f}^{\mathrm{eq}}\right], \tag{11.34}$$

where the particle distributions f_i and f_i^{eq} are written as column vectors and $\mathbf{M}$ is the matrix [28]

$$\mathbf{M}=\begin{pmatrix}
1 & 1 & 1 & 1 & 1 & 1 & 1 & 1 & 1 & 1 & 1 & 1 & 1 & 1 & 1 & 1 & 1 & 1 & 1\\
-30 & -11 & -11 & -11 & -11 & -11 & -11 & 8 & 8 & 8 & 8 & 8 & 8 & 8 & 8 & 8 & 8 & 8 & 8\\
12 & -4 & -4 & -4 & -4 & -4 & -4 & 1 & 1 & 1 & 1 & 1 & 1 & 1 & 1 & 1 & 1 & 1 & 1\\
0 & 1 & -1 & 0 & 0 & 0 & 0 & 1 & -1 & 1 & -1 & 0 & 0 & 0 & 0 & 1 & -1 & 1 & -1\\
0 & -4 & 4 & 0 & 0 & 0 & 0 & 1 & -1 & 1 & -1 & 0 & 0 & 0 & 0 & 1 & -1 & 1 & -1\\
0 & 0 & 0 & 1 & -1 & 0 & 0 & 1 & 1 & -1 & -1 & 1 & -1 & 1 & -1 & 0 & 0 & 0 & 0\\
0 & 0 & 0 & -4 & 4 & 0 & 0 & 1 & 1 & -1 & -1 & 1 & -1 & 1 & -1 & 0 & 0 & 0 & 0\\
0 & 0 & 0 & 0 & 0 & 1 & -1 & 0 & 0 & 0 & 0 & 1 & 1 & -1 & -1 & 1 & 1 & -1 & -1\\
0 & 0 & 0 & 0 & 0 & -4 & 4 & 0 & 0 & 0 & 0 & 1 & 1 & -1 & -1 & 1 & 1 & -1 & -1\\
0 & 2 & 2 & -1 & -1 & -1 & -1 & 1 & 1 & 1 & 1 & -2 & -2 & -2 & -2 & 1 & 1 & 1 & 1\\
0 & -4 & -4 & 2 & 2 & 2 & 2 & 1 & 1 & 1 & 1 & -2 & -2 & -2 & -2 & 1 & 1 & 1 & 1\\
0 & 0 & 0 & 1 & 1 & -1 & -1 & 1 & 1 & 1 & 1 & 0 & 0 & 0 & 0 & -1 & -1 & -1 & -1\\
0 & 0 & 0 & -2 & -2 & 2 & 2 & 1 & 1 & 1 & 1 & 0 & 0 & 0 & 0 & -1 & -1 & -1 & -1\\
0 & 0 & 0 & 0 & 0 & 0 & 0 & 1 & -1 & -1 & 1 & 0 & 0 & 0 & 0 & 0 & 0 & 0 & 0\\
0 & 0 & 0 & 0 & 0 & 0 & 0 & 0 & 0 & 0 & 0 & 1 & -1 & -1 & 1 & 0 & 0 & 0 & 0\\
0 & 0 & 0 & 0 & 0 & 0 & 0 & 0 & 0 & 0 & 0 & 0 & 0 & 0 & 0 & 1 & -1 & -1 & 1\\
0 & 0 & 0 & 0 & 0 & 0 & 0 & 1 & -1 & 1 & -1 & 0 & 0 & 0 & 0 & -1 & 1 & -1 & 1\\
0 & 0 & 0 & 0 & 0 & 0 & 0 & -1 & -1 & 1 & 1 & 1 & -1 & 1 & -1 & 0 & 0 & 0 & 0\\
0 & 0 & 0 & 0 & 0 & 0 & 0 & 0 & 0 & 0 & 0 & -1 & -1 & 1 & 1 & 1 & 1 & -1 & -1
\end{pmatrix}.$$

Each of the rows in $\mathbf{M}$ is mutually orthogonal so the inverse follows easily as

$$\mathbf{M}_{ij}^{-1}=\frac{1}{\sum_k \mathbf{M}_{jk}^2}\mathbf{M}_{ji}\,. \tag{11.35}$$

The matrix $\mathbf{M}$ performs a change of basis. The new basis is designed to contain more physically relevant variables. For example the first row corresponds to the density n. Similarly, the fourth, sixth and eighth lines calculate the momentum densities nu_x, nu_y and nu_z respectively. These are the conserved moments. The 10th, 12th, 14th, 15th, 16th lines correspond to the components of the symmetric, traceless, viscous stress tensor $3\sigma_{xx}$, $\sigma_{yy}-\sigma_{zz}$, σ_{xy}, σ_{yz} and σ_{xz}. The other terms do not contain any real physical meaning and they are often called the ghost modes.

The matrix $\mathbf{S}$ in Eq. (11.34) is diagonal and contains the information about how fast each variable relaxes at each time step. A useful choice is [28]

$$\mathbf{S}=\mathrm{diag}\left(0,1,1,0,1,0,1,0,1,\omega,1,\omega,1,\omega,\omega,\omega,1,1,1\right), \tag{11.36}$$

where $\omega = 1/\tau$ now determines the fluid viscosities ν and λ. Note that some of the elements of $\mathbf{S}$ are zero. This choice is arbitrary as these modes correspond to the conserved moments, for which $M_{ji}\left[f_i - f_i^{\text{eq}}\right] = 0\,(j = 0, 3, 5, 7)$. Using unity for the remaining, ghost, modes minimises the spurious velocities. This is an acceptable choice because these modes do not correspond to physical variables.

For a system with variable viscosity it would seem necessary to recalculate the collision matrix $\mathbf{C} = \mathbf{M}^{-1}\mathbf{SM}$ at each lattice node and at each time-step. This is very demanding computationally. One practical approach to overcome this difficulty is to create a lookup table for various values of the viscosity.

11.4.2 Boundary Conditions

In a typical lattice Boltzmann simulation of a wetting problem there are two important boundary conditions: the wetting boundary condition, given by Eq. (11.9), and the no-slip boundary condition on the fluid velocity. While these boundary conditions are simple conceptually, their implementation can be tricky for complex geometries.

One way to implement the no-slip condition is a linear interpolation bounce back rule proposed by Bouzidi et al. [31]. A schematic diagram illustrating this approach, for the one dimensional case and the $f_i(\mathbf{r}, t)$ distribution function, is shown in Fig. 11.4a. In one dimension, there are two distribution functions, f_1[k] and f_2[k], for a given lattice node k. When the node k is located to the right of a wall, as shown in Fig. 11.4a, the function f_1^*[k] is undetermined after the propagation step. (To clarify notation we use * to denotes distribution functions after propagation.) To determine f_1^*[k], Bouzidi et al. consider two cases. If the distance of the wall from the fluid node, d_{wall}, is less than half of a lattice spacing, f_1^*[k] is chosen to be a weighted average of f_2[k] and f_2[k+1]. If, however, d_{wall} is more than half of a lattice spacing f_1^*[k] is interpolated from f_2[k] and f_1[k]:

$$\begin{aligned} d_{\text{wall}} < 0.5 \;:\; f_1^*[\text{k}] &= f_2[\text{k}] \times 2d_{\text{wall}} + f_2[\text{k}+1] \times (1 - 2d_{\text{wall}})\,, \\ d_{\text{wall}} > 0.5 \;:\; f_1^*[\text{k}] &= f_2[\text{k}]/(2d_{\text{wall}}) + f_1[\text{k}] \times (1 - 1/(2d_{\text{wall}}))\,. \end{aligned} \tag{11.37}$$

The bounce back rules for the other lattice directions in higher dimensions, and for the $g_i(\mathbf{r}, t)$ distribution function, are applied in exactly the same way (with d_{wall} normalised to the lattice spacing in the relevant direction). This ensures that there is no momentum flux parallel to the wall and that the no-slip boundary condition is satisfied at the position of the wall. It was shown by Ginzburg and d'Humières [32] that this no-slip boundary is accurate to the second order.

When the velocity of the wall is non-zero, for example when one wants to investigate a shear flow, the bounce back rule should be modified by adding the following terms [31]:

$$\begin{aligned} d_{\text{wall}} < 0.5 \;:\; \Delta f &= -2\,n\,w_i\,(\mathbf{e}_i \cdot \mathbf{v}_{\text{wall}}) \\ \Delta g &= -2\,\phi\,w_i\,(\mathbf{e}_i \cdot \mathbf{v}_{\text{wall}}) \end{aligned}$$

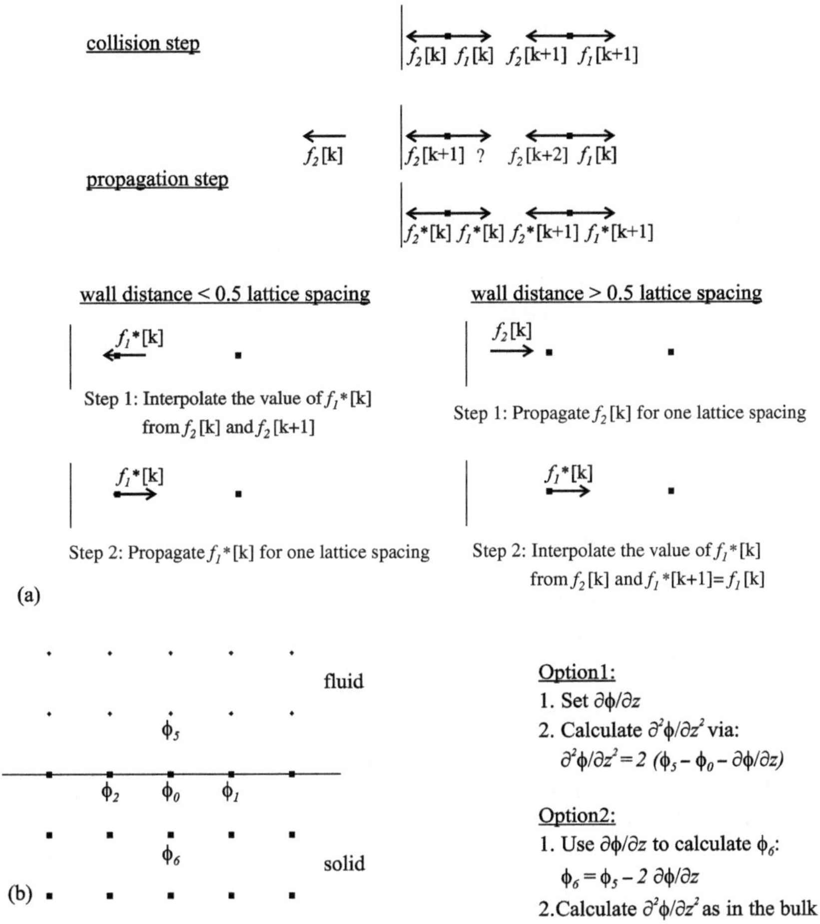

Fig. 11.4 Schematic diagram of the no-slip and wetting boundary conditions. (**a**) Link bounce back rule. (**b**) Implementations of the wetting boundary condition

$$d_{\text{wall}} > 0.5 \;: \Delta f = -n\, w_i / d_{\text{wall}}\, (\mathbf{e}_i \cdot \mathbf{v}_{\text{wall}})$$
$$\Delta g = -\phi\, w_i / d_{\text{wall}}\, (\mathbf{e}_i \cdot \mathbf{v}_{\text{wall}})$$

where w_i and $\mathbf{e}_i$ are the weight coefficient and the lattice velocity direction of the distribution functions before being bounced off the wall (e.g. in Eq. (11.37), this would correspond to w_2 and $\mathbf{e}_2$).

We next describe two ways to implement the wetting boundary condition. Equation (11.9) sets the value of the first derivative $\partial\phi/\partial z|_0$ at the surface, but an estimate of the second derivative is also required to calculate the equilibrium distribution function (11.30). Our explanations refer to the labelling of lattice nodes in Fig. 11.4b, for an interface perpendicular to the z-axis.

In a first method, $\partial\phi/\partial z$ is set to take the value given by Eq. (11.9) at the ϕ_0 lattice node and $\partial^2\phi/\partial z^2$ is calculated by Taylor expanding ϕ_5 with respect to ϕ_0 and neglecting third and higher derivatives in ϕ,

$$\partial^2\phi/\partial z^2|_0 = 2 \times (\phi_5 - \phi_0 - \partial\phi/\partial z|_0) \,. \tag{11.38}$$

The main advantage of this implementation is it is not necessary to simulate any solid nodes.

An alternative implentation of the wetting boundary condition is to assign appropriate density values to the solid nodes neighbouring the boundary, so that Eq. (11.9) is satisfied. In the schematic diagram shown in Fig. 11.4c, this corresponds to assigning[3]

$$\phi_6 = \phi_5 - 2\,\partial\phi/\partial z|_0 \,. \tag{11.39}$$

The main advantage of this approach is that $\nabla^2\phi$ can be calculated in exactly the same way at the surface as in the bulk. Furthermore, since all the nearest and next nearest neighbour nodes of any surface site have appropriate density values, better accuracy can be achieved by choosing the best stencil to calculate derivatives [26].

For more complex geometries, for example surfaces which do not follow a lattice axis or corners, the wetting boundary conditions can be implemented in a similar way. This typically gives a set of linear equations that must be solved simultaneously.

Finally we summarise an algorithm that we have found to work well for simulating the dynamics of the contact line in fluids where the two components have different viscosities [18]:

Step 1: Calculate the density, concentration and velocity using the moments described in Eqs. (11.21).

Step 2: Set the velocity of the boundary nodes to zero, or more generally to the velocity of the wall. This reduces spurious velocities introduced by the bounce-back boundary conditions.

Step 3: Implement the wetting boundary condition by setting the first and second derivatives of the order parameter.

Step 4: Calculate the equilibrium distribution function and use the multiple relaxation time lattice Boltzmann method to perform the collision step.

Step 5: Perform the streaming step with the bounce back rule at the boundaries (Eq. (11.37)).

---

[3] If the wall is located at the mid-link between ϕ_5 and ϕ_0, appropriate wetting boundary conditions can be implemented by setting $\phi_0 = \phi_5 - \partial\phi/\partial z|_0$.

11.4.3 Other Lattice Boltzmann Algorithms

The lattice Boltzmann implementations we have described in this section are not unique, and many authors have proposed alternative approaches to solve the equations of motion of multiphase fluids. For example:

1. Phase ordering can be imposed by using an effective interaction, rather than a free energy [15, 33].
2. The thermodynamics leading to phase ordering can be included in the lattice Boltzmann scheme as a forcing term, rather than as a correction to the pressure in the second moment of the equilibrium distribution function [33, 34].
3. Different sets of velocity vectors can be defined [35, 36].
4. The forms of f^{eq}, g^{eq}, and F that satisfy the hydrodynamic equations of motion in the continuum limit are not unique [27, 35, 36]. It is useful to exploit this to minimise spurious currents [26].
5. There are many ways of implementing the hydrodynamic boundary conditions [32, 37–39].

11.5 Smooth Walls

In the next three sections, we shall describe a number of examples where lattice Boltzmann simulations have proved successful in providing insights to wetting phenomena. We start with two problems where the solid boundaries are assumed to be flat and homogeneous. Firstly, we discuss the capillary penetration of a wetting fluid [40], and secondly, we look at the classical problem of fingering instabilities in narrow channels [41]. These are both relevant in many industrial and biological systems, and they play an increasingly important role in many microfluidic devices. We then explain, in Sects. 11.6 and 11.7 how chemical and topographical heterogeneities on a surface may lead to complex drop morphologies that depend sensitively on the details of the surface patterning, as well as the path by which the system is prepared[4].

11.5.1 Capillary Filling

When a liquid is brought into contact with a small capillary tube, it will penetrate the capillary provided that this lowers its surface energy i.e. when the capillary is hydrophilic with respect to the liquid. The classical analysis of the dynamics of

[4] Some of the results in Sects. 11.5, 11.6, and 11.7 were obtained using a lattice Boltzmann algorithm for a one-component, liquid–gas system rather than a two-component fluid. Details of this algorithm are given in [39, 42]. In the one-component model contact line slip occurs because of evaporation and condensation, which is rapid because of the unphysically wide interface. This can lead to unphysical dynamics [43–45].

capillary filling is due to Lucas [46] and Washburn [47]. Consider a capillary of height h with an infinite reservoir of liquid of dynamic viscosity $\eta = n\nu$ at one end. Assuming that the penetrating liquid adopts a parabolic profile, it will fill the capillary with a mean velocity

$$\bar{v} = -\frac{h^2}{12\eta}\frac{dp}{dx} \tag{11.40}$$

where $\frac{dp}{dx}$ is the pressure gradient that sets up the flow. The driving force for the filling is provided by the decrease in free energy as the fluid wets the walls or, equivalently, by the Laplace pressure across the curved liquid–gas interface. Hence

$$\frac{dp}{dx} = -\frac{\gamma}{Rl} \tag{11.41}$$

where $R = h/2\cos\theta^a$ is the radius of curvature of the interface and l is the length of liquid in the tube. Eliminating $\frac{dp}{dx}$ from Eqs. (11.40) and (11.41) and identifying $\bar{v} = dl/dt$ gives the Lucas-Washburn law

$$l = \left(\sigma_{\mathrm{LG}} h \cos\theta^a / 3\eta\right)^{1/2} (t + t_0)^{1/2} \tag{11.42}$$

where t_0 is an integration constant.

In Eq. (11.42) it is appropriate to use, not the static, but the advancing contact angle θ^a, as this controls the curvature of the interface and hence the Laplace pressure. The Lucas-Washburn law assumes that there is no resistance to motion from any fluid already in the capillary. Therefore it applies only if the dynamic viscosity of the invading phase η_A is large compared to that of the displaced fluid η_B. If the dissipation in the displaced fluid is taken into account the modified Lucas-Washburn law becomes

$$\eta_A \frac{l^2}{2} + \eta_B \left(Ll - \frac{l^2}{2}\right) = \frac{\sigma_{\mathrm{LG}} h \cos\theta^a}{6}(t + t_0) \tag{11.43}$$

where L is the total length of the capillary.

Numerical results showing capillary filling of a smooth channel are presented in Fig. 11.5. The plot is for a channel of length $L = 640$, infinite width and height $h = 50$. Reservoirs (480×200) of components A and B are attached at each end of the capillary. The two reservoirs are connected to ensure that they have the same pressure. The parameters of the model are chosen so that $\theta_{\mathrm{e}} = 60°$, $\gamma = 0.0188$, $\eta_{\mathrm{A}} = 0.83$, $\eta_{\mathrm{B}} = 0.03$ and $M = 0.05$. The solid line in Fig. 11.5a is a fit to the Lucas-Washburn law using the measured value of the advancing contact angle and correcting for the small viscosity of the displaced B-component. The fit is excellent, except very close to the beginning of the simulation, where deviations due to inertial effects and a non-Poiseuille flow profile are expected.

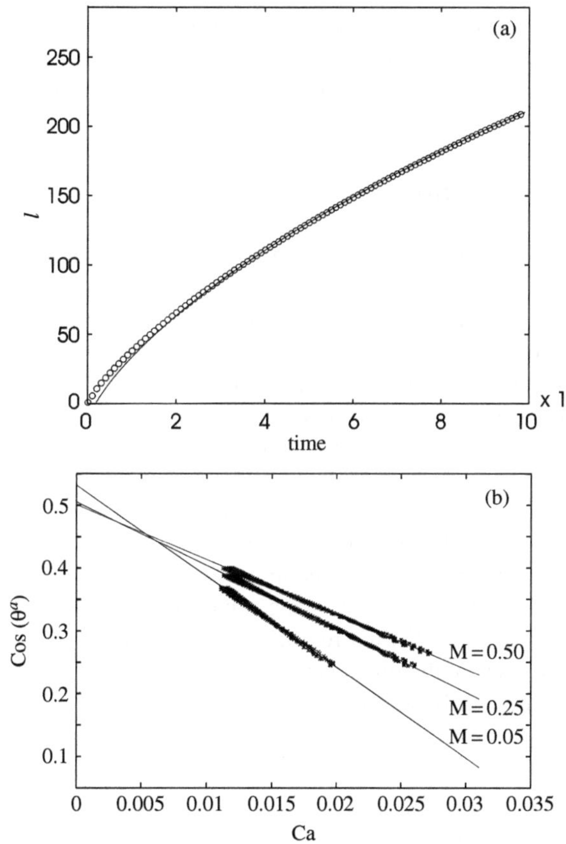

Fig. 11.5 (**a**) Distance a fluid–fluid interface moves along a capillary as a function of time. *Circles* are the lattice Boltzmann simulation results. The *solid line* is a fit to the Lucas-Washburn law using the measured advancing contact angle and correcting for the small viscosity of the displaced component. (**b**) The advancing contact angle of the liquid–liquid interface as a function of the capillary number. The *crosses* are simulation results and the *solid lines* are linear fits of $\cos\theta^a$ to the capillary number [48]. Reprinted figures with permission from Kusumaatmaja, [40] and Pooley, [18]. Copyright (2008) by the American Physical Society

To lowest order in the capillary number, Ca= $v^I \nu/\gamma_{LG}$ where v^I is the interface velocity, the advancing contact angle is related to the equilibrium angle and the capillary number by [48]

$$\cos\theta^a = \cos\theta^{\mathrm{eq}} - \mathrm{Ca}\log(KL/l_{\mathrm{s}}) \qquad (11.44)$$

where K is a constant, L is the length scale of the system and l_{s} is the effective slip length at the three phase contact line. Figure 11.5b shows the expected linear dependence, and that the advancing contact angle tends to the correct value as $\mathrm{Ca} \to 0$; We obtain $\theta^a|_{Ca\to 0} = 58°$, $60°$ and $60°$ for $M = 0.05$, 0.1 and 0.5 respectively.

Note that the slope of the graph, and hence the slip length, depend on the mobility M. This occurs because in diffuse interface models of binary fluids the contact line singularity is relieved by inter-diffusion of the two fluid components [17].

11.5.2 Viscous Fingering

We have just considered the rate at which a viscous fluid displaces a liquid of low viscosity when the driving force is the hydrophilic nature of the channel walls. In such a case, the fluid–fluid interface is stable and has the form of a meniscus. The situation is, however, more complicated when a less viscous fluid is driven to displace a more viscous one, as the interface can now be unstable. If the fluids are moving in the narrow gap between two parallel plates this instability gives rise to the well-known Saffman-Taylor [49] fingers. A typical experiment showing the development of a finger is shown in Fig. 11.6 [50].

Usually the Saffman-Taylor instability is treated as a two dimensional problem, taking an average over the distance between the bounding plates. However the third dimension can affect the way in which the finger forms. Ledesma-Aguilar et al. [41] studied the three dimensional motion, using binary lattice Boltzmann simulations, and found that there are two distinct regimes. If the contact line is able to keep up with the leading interface of the finger (which will happen, at higher Peclet numbers, if the diffusion is sufficiently strong [51]), the fluid-fluid interface retains the form of a meniscus in the direction, z say, between the plates and it is possible to treat

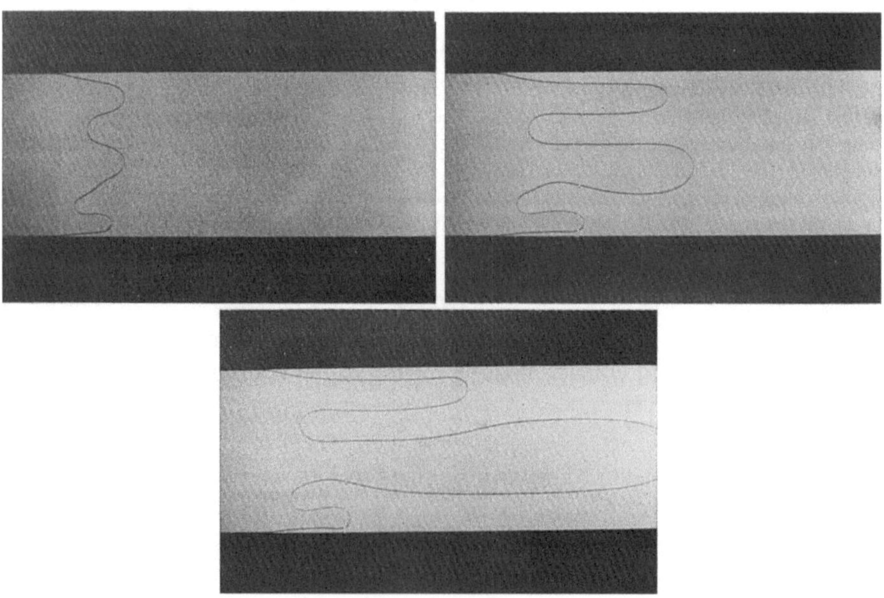

Fig. 11.6 Time evolution of a Saffman-Taylor finger. Reprinted with permission from Tabeling et al. [50, pp. 67–82]. Copyright 1987 Cambridge University Press

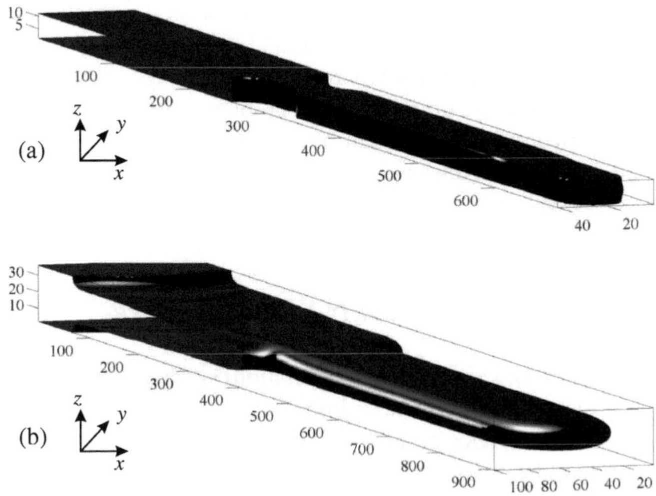

Fig. 11.7 Saffman-Taylor instabilities in the (**a**) meniscus and (**b**) surface film regimes. (**a**) When diffusion is sufficiently strong (high Peclet number), the fluid–fluid interface retains the form of a meniscus in the direction between the plates and the problem is essentially two-dimensional. (b) At low Peclet number, the contact line falls behind and a layer of the displaced phase is formed close to the plates. This means that the advancing fluid forms a finger-like structure in both the x-y and the x-z planes (We thank Ioannis Zacharoudiou for these figures.)

the problem two-dimensionally. At low Peclet numbers the contact line falls behind and a layer of the displaced phase is formed close to the plates. This means that the advancing fluid forms a finger-like structure in both the x-y and the x-z planes. Simulation results showing the shape of the interface in the meniscus and surface film regime are shown in Fig. 11.7.

11.6 Chemical Patterning

We now describe examples where a lattice Boltzmann approach has been used to model drops spreading on chemically patterned surfaces. This is particularly exciting at present because it is becoming increasingly feasible to fabricate surfaces with heterogeneities in a controlled and reproducible manner, allowing surface patterning to be used as a part of a designer toolbox to control the shapes and dynamics of small liquid drops [8, 52, 53]. Variation in the surface wettability can be implemented easily in the lattice Boltzmann simulations by applying different values of the phenomenological parameter h in Eq. (11.13) at different surface lattice sites. However, it is important to note that the typical length scale of the variation in h has to be larger than the interface width of the model.

We first look at a drop spreading on a chemically patterned surface. For a homogeneous surface, the final state is a spherical cap with a contact angle equals to the Young angle. This is not the case for heterogeneous surfaces. Depending on the initial conditions of the system, the drop can take several metastable states, with shapes that may vary considerably from spherical [8, 52–54].

We then consider two examples where simulations suggest how chemical patterning might be applied to solve industrial problems. In the first [55], we show how a (relatively) hydrophobic grid can be used to alleviate mottle [56] in ink-jet printing. In the second example, we demonstrate that chemical patterning can be used to control drop size and polydispersity [57].

11.6.1 Spreading on a Chemically Striped Surface

Figure 11.8 compares experiments and simulations of drops on a chemically patterned substrate. The surface is lined with relatively hydrophilic and hydrophobic stripes with contact angles 5° and 64° and widths 26 and 47 μm respectively. Figure 11.8a shows the final state of drops jetted onto the surface. The drops' volumes were chosen so that their final diameters were comparable to the stripe width. It is apparent from the figure that the drops can take two final configurations, "diamond"-like and "butterfly"-like.

Figure 11.8b shows simulations of the same system, with parameters chosen so that length scales, surface tension, contact angles, fluid viscosity and liquid density correspond to those of the experiment. Again the diamond and butterfly configurations are observed at long times. The simulations allowed us to follow the dynamics of the liquid drops' motion in detail. In particular, we found that the final drop shape is selected by the initial impact position and velocity. If the drop can touch two neighbouring hydrophilic stripes as it spreads, it will reach the butterfly configuration; if not it will retract back to the diamond pattern, spanning a single stripe.

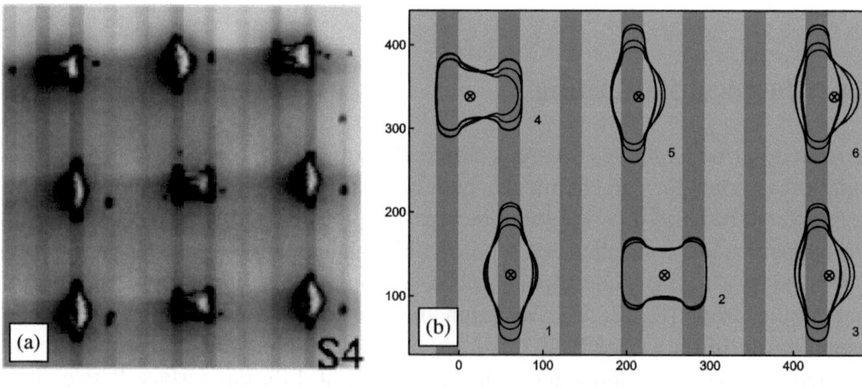

Fig. 11.8 Drops spreading on a chemically striped surface. (**a**) Scanning electron micrographs of ink-jetted drops. (**b**) Numerical simulations of drops hitting the surface at various impact points, indicated by *encircled crosses*. For each drop the faint lines represent the extent of the base of the drop as it evolves and the *bold line* depicts its final shape. Relatively hydrophilic and hydrophobic stripes appear dark and pale, respectively. Reprinted with permission from Léopoldès et al. [54, pp. 9818–9822]. Copyright 2003 American Chemical Society

This can be seen in Fig. 11.8b, where the faint lines show the time evolution of the base of the drop and the solid lines its final shape. Both states are free energy minima but one of the two is a metastable minimum: which one is sensitive to the exact choice of the physical parameters.

11.6.2 Using Chemical Patterning to Control Drop Positioning

An inkjet printed image is produced by jetting an array of micron-scale liquid drops onto a surface. To achieve a solid colour the aim is that the drops, which are jetted at a distance apart comparable to their diameter, should coalesce to form a uniform covering of ink. However, in practice, irregular coalescence due to surface imperfections and randomness in the positions of the jetted drops can dominate. This leads to the formation of large, irregular drops with areas of bare substrate between them as shown in the upper part of Fig. 11.9b. Such configurations lead to poor image quality, called mottle [56].

Figure 11.9a shows that irregular coalescence can be overcome by using a grid of (relatively) hydrophobic chemical stripes. Here the drop has an initial radius of 15 μm and the substrate has contact angle 5°. The hydrophobic grid has stripes of width 6 μm, separated by 66 μm, and contact angle 65°. The simulation shows that the drop is confined even when its initial point of impact is close to the corner of a square.

Results from an experiment demonstrating a similar effect are shown in Fig. 11.9b. The ink drops have a radius $R = 30\,\mu\text{m}$ and they are jetted in a $50\,\mu\text{m} \times 50\,\mu\text{m}$ array. In the upper part of the figure there is no hydrophobic grid and a mottled final configuration is observed. The lower part of Fig. 11.9b carries hydrophobic stripes of 5 μm width forming squares of side 40 μm. The drops now form a more regular pattern determined by the grid.

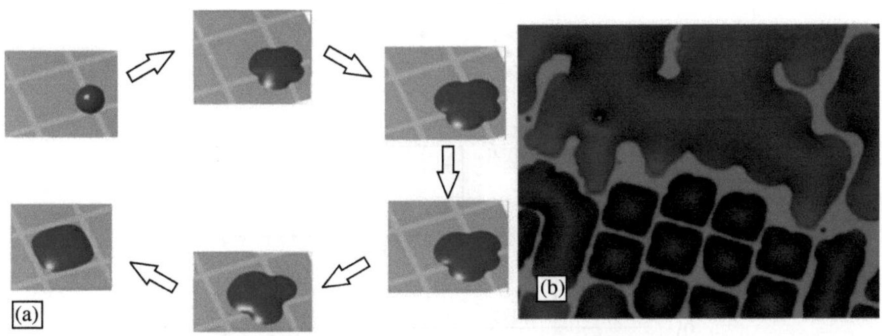

Fig. 11.9 Control of drop position using chemical patterning. (**a**) Time evolution of a drop jetted onto a substrate patterned by a grid. Relatively hydrophobic and hydrophilic areas are *light grey* stripes (65°) and *dark grey* areas (5°) respectively. (**b**) Inkjet drops jetted onto a substrate and cured: (*top*) homogeneous surface and (*bottom*) surface patterned by a relatively hydrophobic grid. Reprinted with permission from Dupuis et al. [55]. Copyright 2005 American Institute of Physics

11.6.3 Using Chemical Patterning to Sort Drop by Size

It is often desirable in microfluidic devices to be able to manipulate and control the motion of liquid drops (see [1] and the references therein). Here we demonstrate a particular example where chemical patterning may be used to sort drops according to their size. The schematic diagram of the system is shown in Fig. 11.10. The surface is patterned with a rectangular grid of hydrophilic (relative to the background) stripes, and a drop is input to the device at A and subject to a body force at an angle $< 45°$ to the x-axis.

The path taken by the drop through the device depends on the drop contact angles with the substrate and on the strength of the body force. It also, of particular relevance to us here, depends on the width of the stripes relative to the drop radius. Figure 11.11a–c show simulations of the paths of drops of initial radius $R = 25, 26$ and 29 moving through such a device. In cases where the drops are confined in the δ_1 stripe, they will move in the x-direction from A to the cross-junction B, where their paths may diverge. In order for a drop to move in the y-direction, the capillary force in this direction must be large enough to overcome the sum of the capillary force and the excess external body force in the x-direction (recall $a_x > a_y$). This is where the asymmetry of the drop shape comes into play. As the volume of the drop is increased, a larger fraction of it overhangs the stripes and hence a larger fraction will interact with the hydrophilic stripe along the y-direction at the junction. This increases the capillary force along y and means that larger drops (e.g. $R > 26$) will move in the y-direction to point C, whereas smaller drops (e.g. $R = 25$) will continue to move along x. By choosing the stripes along the y direction to be of equal widths, but those along x to increase in width with increasing y, it is possible to move the larger drops further along y. As one can see from Fig. 11.11, the drops

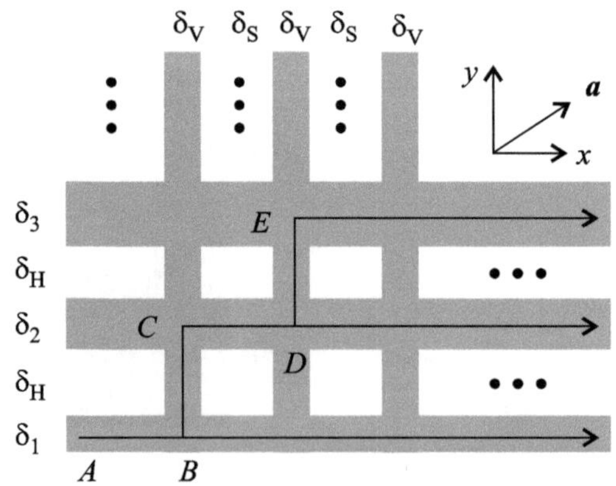

Fig. 11.10 Schematic diagram of a drop sorter. The *grey* stripes on the surface are hydrophilic with respect to the background. δ labels the widths of the stripes and ***a*** the imposed acceleration. The *arrows* show possible paths of a drop through the device. Reprinted with permission from Kusumaatmaja and Yeomans [57, pp. 956–959]. Copyright 2007 American Chemical Society

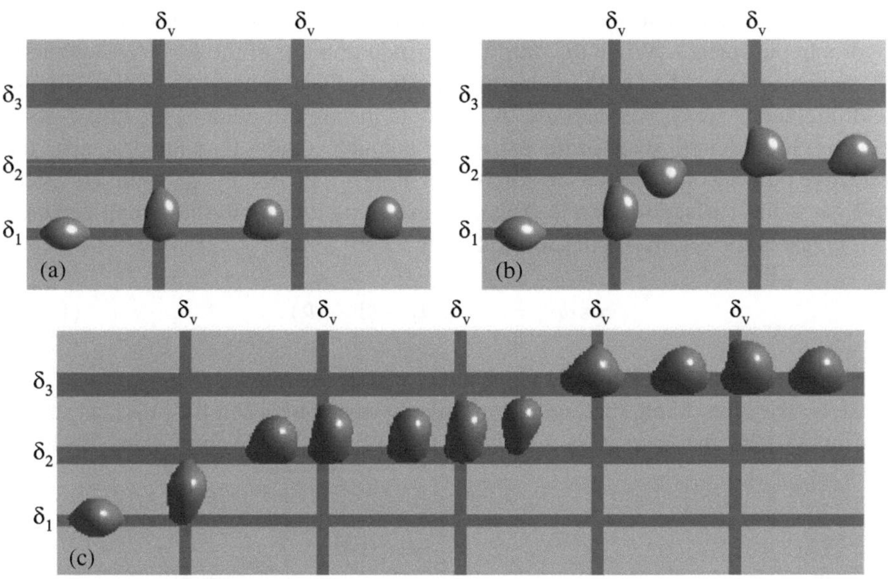

Fig. 11.11 Paths taken by drops of radius (**a**) $R = 25$, (**b**) $R = 26$, and (**c**) $R = 29$ through the drop sorter. $\delta_1 = 20$, $\delta_2 = 30$, $\delta_3 = 40$, and $\delta_V = 20$. Reprinted with permission from Kusumaatmaja and Yeomans [57, pp. 956–959]. Copyright 2007 American Chemical Society

of initial radius $R = 26$ and $R = 29$ are finally confined in the second and third stripe respectively.

These simulations suggest that by increasing the number of stripes and carefully controlling their widths it may be possible to sort polydisperse drops into collections of monodisperse drops. Two other parameters, the wettability contrast and the external body force, could also be adjusted to fine-tune the device.

11.7 Topographical Patterning: Superhydrophobic Surfaces

Superhydrophobic surfaces are a prime example of how heterogeneities can alter the wettability of a surface. On a smooth hydrophobic surface, the highest contact angle that can be achieved is of order 120–130° [6, 58] attainable for, for example, a water drop spreading on fluorinated solids. When the hydrophobic surface is made rough, however, higher contact angles are possible. Several natural materials exhibit this, so-called, superhydrophobicity. Examples include the leaves of the lotus plant [2], butterfly wings [4], water strider legs [59] and duck feathers [60]. Many research groups have now fabricated superhydrophobic surfaces by patterning hydrophobic surfaces with regular posts [58, 61, 62] or with nano-hairs [63]. Indeed superhydrophobicity is a surprisingly robust phenomenon, which requires neither careful patterning nor intrinsic hydrophobicity of the surface material [64, 65].

It is possible to distinguish two ways in which a drop can behave on a superhydrophobic surface. When the drop is suspended on top of the surface roughness, as shown in Fig. 11.12b, the substrate is effectively a composite of liquid–solid and liquid–gas areas. We shall use Φ to denote the area fraction of the liquid–solid contact. If the length scale of the patterning is much smaller than the drop size, the effective liquid–solid surface tension is then $\Phi\, \gamma_{LS} + (1 - \Phi)\, \gamma_{LG}$, while the effective gas–solid surface tension is $\Phi\, \gamma_{GS}$. Substituting these into the Young equation (11.1), gives the Cassie-Baxter formula [60]

$$\cos\theta_{\mathrm{CB}} = \Phi \cos\theta_{\mathrm{e}} - (1 - \Phi)\,. \tag{11.45}$$

This configuration is called the suspended or Cassie-Baxter state.

If, on the other hand, the liquid drop fills the space between the posts, as shown in Fig. 11.12b, the drop is said to lie in the collapsed or Wenzel state. Both the

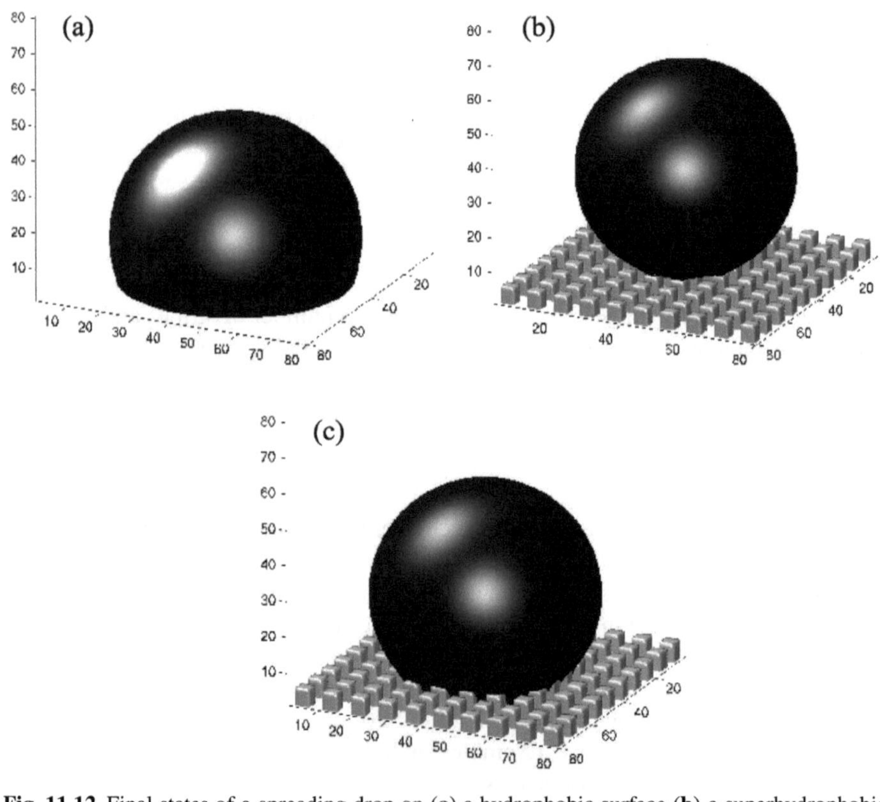

Fig. 11.12 Final states of a spreading drop on (**a**) a hydrophobic surface (**b**) a superhydrophobic surface with the drop suspended (**c**) a superhydrophobic surface with the drop collapsed. Reprinted with permission from Dupuis and Yeomans [39, pp. 2624–2629]. Copyright 2005 American Chemical Society

liquid–solid and gas–solid contact areas are increased by a roughness factor r and the macroscopic contact angle is therefore given by the Wenzel equation [66]

$$\cos\theta_W = r\cos\theta_e\,. \tag{11.46}$$

Figure 11.12 shows simulation results for the final state of a drop of radius $R = 30$ which has spread on a smooth (Fig. 11.12a) and a superhydrophobic surface (Fig. 11.12b and c). A contact angle $\theta_e = 110°$ is set on every surface site. The resultant macroscopic contact angles in the simulations are 110°, 156° and 130° for the flat surface, suspended drop and collapsed drop respectively. The values for the suspended and collapsed drop are compatible with the ones obtained from the Cassie-Baxter and Wenzel formulae, but they are not exactly the same. There are two reasons for this. Firstly, the drop only covers a small number of posts in the simulations. Secondly, the surface inhomogeneities result in the existence of multiple local free energy minima, not just that prescribed by the Cassie-Baxter or Wenzel formulae. This can cause pinning of the contact line and lead to values of contact angles which depend not only on the thermodynamic variables describing the state of the drop, but also on the path by which that state was achieved. This phenomenon, contact angle hysteresis, is well known [5, 67–70], but has suprising consequence for drops on superhydrophobic substrates. We now describe these in more detail.

11.7.1 Contact Line Pinning and Contact Angle Hysteresis

Both chemical and topographical surface patterning may pin the contact line. This can results in variation in the value of the contact angle around a drop. It can also lead to hysteresis, a dependence of the drop shape on its dynamical history. A useful approach to quantify contact angle hysteresis is to slowly increase the volume of a drop until it starts to spread. The contact angle at this moment is termed the advancing angle. Similarly, if the drop volume is slowly reduced, it will start to retreat across the surface at the receding contact angle. The difference between the advancing and receding angles is termed the contact angle hysteresis. However, it should be cautioned that this is not a unique definition; the advancing and receding angles will depend on the direction, relative to the surface patterning, in which they are measured. Moreover the difference in contact angles between the advancing and receding edge of a moving drop will not necessarily be the same as the value measured quasistatically.

This concept of pinning, and of the resulting advancing and receding contact angles, is illustrated in Fig. 11.13 for a drop crossing a ridge. For the contact line to advance, it has to wet the sides of the grooves (Fig. 11.13a) which, according to the Gibb's criterion [71], occurs when the contact angle is locally equal to the Young angle. Therefore the advancing angle (measured with respect to the surface) is $\theta^a = \theta_e + 90°$ for rectangular ridges and, more generally $\theta^a = \theta_e + \alpha$ [70–72] for a surface of maximum inclination α. Similarly, for the contact line to recede,

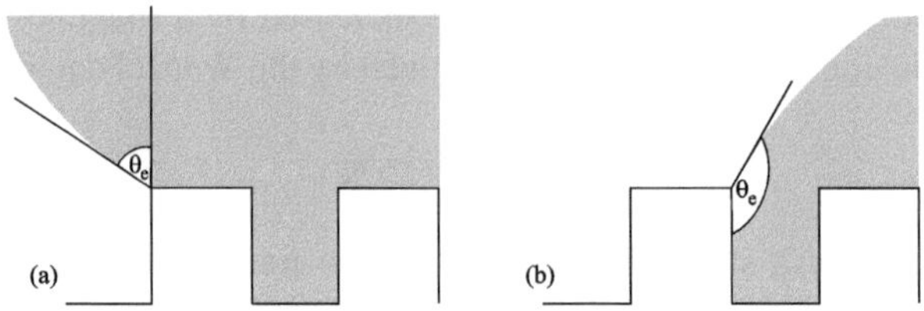

Fig. 11.13 Graphical illustration of the pinning of an (**a**) advancing and (**b**) receding contact line on a surface patterned with *square* ridges

the drop has to dewet the sides of the posts (Fig. 11.13b). This is possible when $\theta^r = \theta_e - 90°$ for rectangular ridges and $\theta^r = \theta_{\rm e} - \alpha$ [70–72] in general.

Applying these criteria in the context of a two-dimensional drop on a superhydrophobic surface patterned with square posts gives surprising results. For the suspended state $\theta^a = 180°$, the upper limit for the value of the contact angle, and $\theta^r = \theta_e$. For the collapsed drop $\theta^a = 180°$ and $\theta^r = \theta_{\rm e} - 90°$. In three dimensions we obtain the same qualitative behaviour, though there may be a decrease in the value for the advancing angle and an increase in that for the receding angle because of curvature contributions to the free energy [73].

We now consider Boltzmann simulations [73], showing that they are able to capture contact line pinning and hysteresis. Figure 11.14a and b show the simulation results for a cylindrical (two-dimensional) suspended drop on a superhydrophobic surface comprising regularly spaced posts. In this set of simulations, we used post width = 7, post separation = 13, and an equilibrium contact angle $\theta_{\rm e} = 120°$. Even after the drop volume was increased quasistatically by a factor ~ 4, and the drop contact angle had reached 162°, no interface depinning transition was observed. After this point, it was no longer possible to continue running the simulations, as the drop filled the simulation box. As the drop volume was slowly decreased, however, the contact line depinned and jumped back across the posts at $\theta^r = 120°$ as predicted analytically.

We now discuss hysteresis for a cylindrical collapsed drop, where the gaps between the posts are filled with liquid. When the drop volume is increased, the drop behaves in the same way as for the suspended state and no contact line motion between posts is observed during the simulation. This is because locally, in the vicinity of the contact line, the drop has no information as to whether it is in the collapsed or suspended state. Typical behavior as the drop volume is decreased is shown in Fig. 11.14c. As for the suspended drop, the contact line is pinned at the outer edge of a post until $\theta = \theta_{\rm e}$. It then retreats smoothly across the post. However, unlike the suspended case, the contact line is pinned again, at the inner edge of the posts. At this point, the drop is found to recede at 32°, consistent with the expected analytical result $\theta_{\rm e} - 90° = 30°$.

Even in this simple two-dimensional model, the contact angle hysteresis is much larger for the collapsed state than for the suspended state. This result has an important consequence that, although the static contact angle is increased in both the Wenzel and the Cassie-Baxter states, their dynamical behaviors are very different.

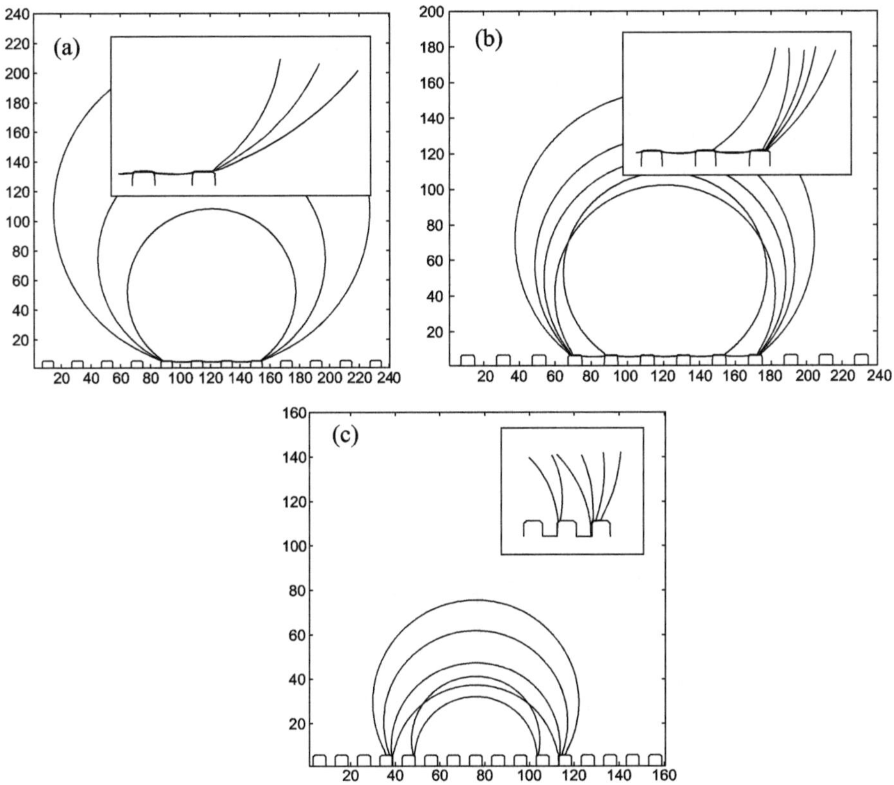

Fig. 11.14 Drop shape as a function of time from lattice Boltzmann simulations of a cylindrical drop (**a**-**b**) suspended and (**c**) collapsed on a topographically patterned surface. (**a**) The advancing contact line remains pinned during the simulation. (**b**) The receding contact line is pinned until $\theta^r \sim 120°$. (**c**) In the collapsed state, the receding contact line is pinned strongly at the inner edge of the posts. The position of the contact lines can be seen more clearly in the *insets*. Reprinted with permission from Kusumaatmaja and Yeomans [73, pp. 6019–6032]. Copyright 2007 American Chemical Society

A liquid drop in the suspended state is very mobile, while that in the collapsed state is very immobile [6, 58].

11.7.2 The Slip Length of Superhydrophobic Surfaces

Another aspect where the dynamics of fluids moving across superhydrophobic surfaces differs between the suspended and the collapsed states is in the value of the slip length. Consider a single phase moving across a solid surface: the slip length, which is defined as the ratio of slip velocity to shear rate at the wall, is a measure of the drag of the surface on the fluid. Slip lengths are typically of order a few nanometers and therefore can be taken as zero in a macroscopic channel (the no-slip boundary condition). However the degree of slip becomes increasingly important as channels are miniaturised. Recall that the average velocity of a liquid flowing through a channel $v \propto h^2 \nabla P$, where h is the height of the channel and ∇P is the pressure gradient that sets up the flow. As channel sizes are reduced an increasingly large pressure gradient is needed for a given throughput velocity. This can be alleviated by increasing the slip length at the channel walls.

For a smooth solid surface, the slip length increases as the wettability of the surface decreases [74]. However, its magnitude remains of order nanometers and therefore is of no real significance except for tiny channels. In this subsection, we

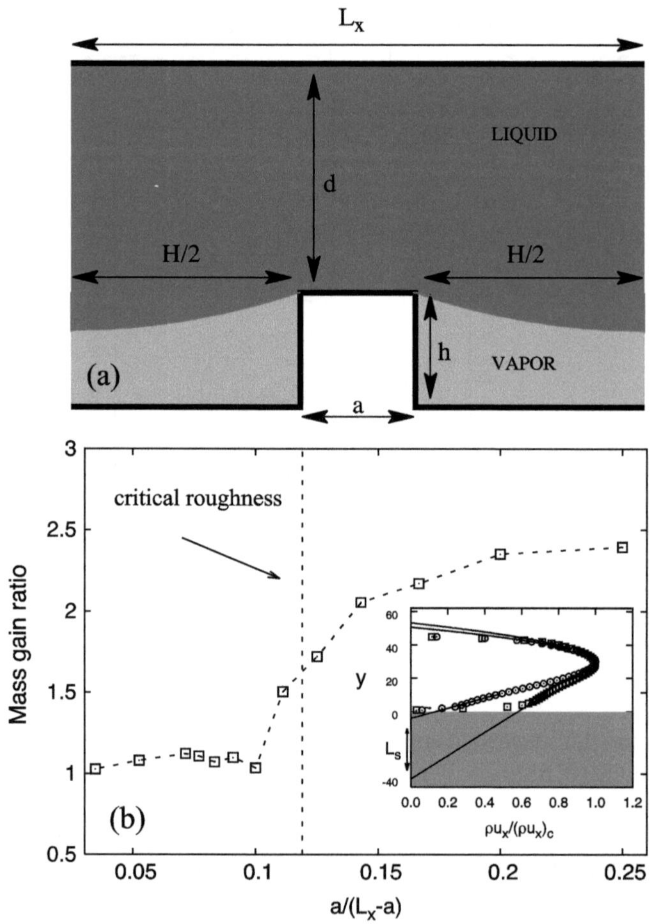

Fig. 11.15 (**a**) The geometry used to simulate flow over a superhydrophobic surface. The simulation parameters were: $h = 14$, $L_y = 45$, $L_x = 90$, and $\theta_e = 160°$. (**b**) Mass flow rate (normalised to the collapsed state) as a function of the effective roughness $a/(L_x - a)$. *inset*: Momentum profile for a suspended and a collapsed state. Both momentum profiles are shown for $x/L_x = 0.1$ and normalised to their center channel values. The *straight lines* correspond to extrapolations of the profiles to beyond the boundaries. Adapted figures with permission from Sbragaglia, et al. [77]. Copyright (2006) by the American Physical Society

will present results that show that a slip length of the order of microns might be induced by trapping the flowing fluid in the suspended state [75, 76]. The crucial idea is that the substrate acts as a composite of liquid–gas (perfect slip) and liquid–solid (no slip) areas and hence, the larger the liquid–gas section, the larger the slip length.

Results from simulations by Sbragaglia et al. [77] are shown in Fig. 11.15. They found that there is a critical roughness above which the mass flow rate through a microchannel increases significantly. This is because the fluid is in the suspended or collapsed state, above or below the critical roughness. The inset in Fig. 11.15 depicts the typical velocity profiles in the two states.

Further research [78, 79], however, has found that the shape of the liquid–gas interface plays an important role in determining the value of the slip length. The curvature of this interface leads to extra viscous dissipation which negates any advantage it might provide in the first place. Designing surface geometries where the slip length can be increased remains a major challenge.

11.7.3 The Transition from the Suspended to the Collapsed State on Superhydrophobic Surfaces

Given that the suspended and collapsed states have different dynamical behaviours, it is important to understand how and when the collapsed and suspended states are metastable or stable and to describe mechanisms for transitions between them.

For a given drop volume, the drop free energy increases with contact angle. This implies that the Cassie-Baxter state has the lowest energy when $\theta_{\text{CB}} < \theta_{\text{W}}$ and, similarly, that the Wenzel state is the ground state for $\theta_{\text{W}} < \theta_{\text{CB}}$. However, in many cases, both states are local minima of the free energy and there is a finite energy barrier opposing the transition between them. The origin of the energy barrier is pinning of the contact line, similar to that discussed in Sect. 11.7.1. For a transition from the suspended to the collapsed state to occur, the contact angle formed by the liquid drop on the sides of the posts has to become equal to the advancing contact angle.

There are several ways in which the collapse transition can be induced. Firstly, one can apply an external pressure or force [39]. Alternatively, the work required to overcome the energy barrier may be provided by a finite impact velocity of the drop [80, 81].

The collapse transition can also be initiated by reducing the volume of the drop by, for example, evaporation. This increases the Laplace pressure inside the drop (recall that $\Delta P \propto 1/R$ where R is the drop radius) and hence the curvature of the interface beneath it. For short posts the interface then touches the surface beneath the posts and the transition can take place. For longer posts collapse occurs when the interface curvature becomes sufficiently large that the interface reaches the equilibrium contact angle on the post sides, and hence depins [82, 83].

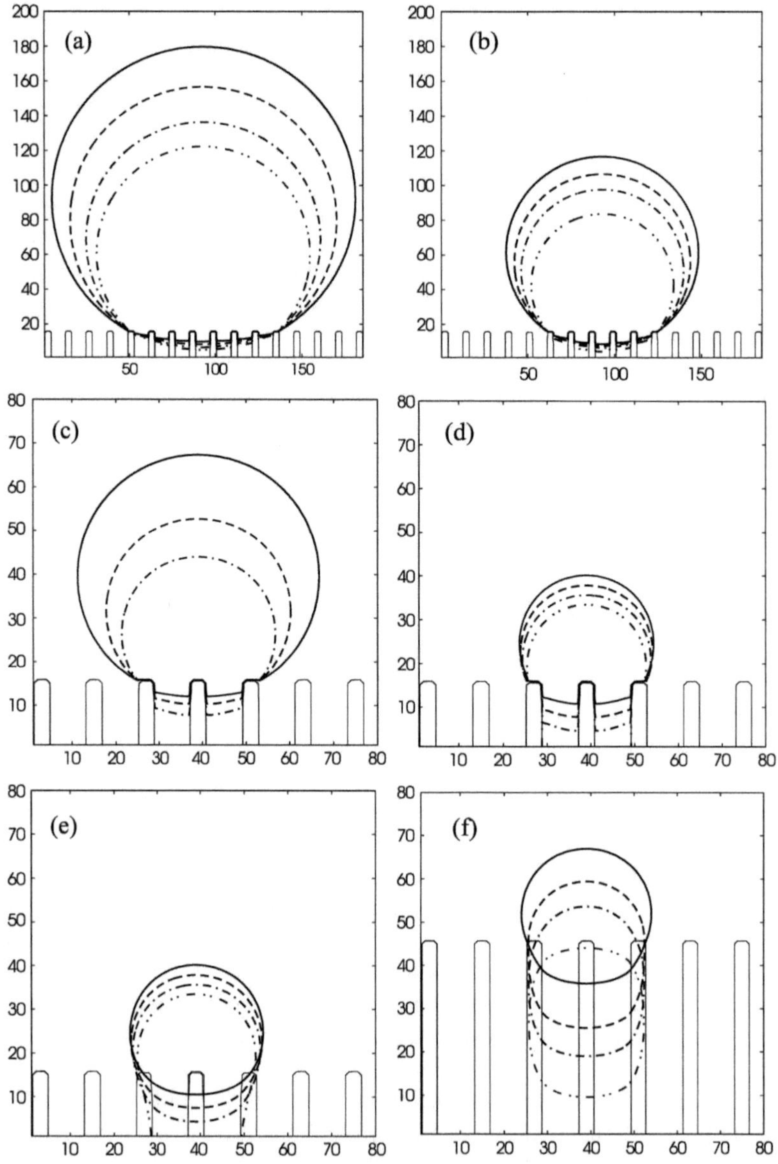

Fig. 11.16 Evolution of a cylindrical drop on a square array of posts of width $a = 3$, spacing $b = 9$ and height $l = 15$. (**a–c**) Evolution before collapse showing depinning of the receding contact line (note the scale change between (**b**) and (**c**)). (**d–f**) Motion of the collapsing drop: (**d**) cross sections in the plane bisecting the posts. (**e**) Same times as (**d**), but in the plane bisecting the gap between the posts. (**f**) Cross sections in the plane bisecting the gap, but with $l = 45$ to enable the collapse to be followed to later times. Adapted figures with permission from Kusumaatmaja et al. [82]

Figure 11.16 shows simulations indicating how the collapse transition proceeds for long posts as the liquid evaporates slowly. As the drop volume decreases it penetrates further into the gaps between the posts. However, movement down the posts is preempted by movement across the surface. The drop depins to lie on less posts, and the penetration is reduced. This continues until the drop lies on only three posts, when it eventually collapses.

It is useful to note that, to obtain the results in Fig. 11.16, we simulated a cylindrical drop on a square array of posts rather than a full, three dimensional, spherical drop. This allowed us to exploit the translational symmetry to reduce the system size to the repeat distance of the lattice in the third dimension, while preserving the important physics, in particular a two dimensional curvature of the interface between the posts.

11.8 Discussion

In this chapter we have concentrated on the use of lattice Boltzmann algorithms to study wetting and spreading. There are many other applications and areas for future research. We give some examples, inevitably selective, of interesting problems:

1. Different choices for the free energy can allow for new physics. A fruitful extensions is to include curvature terms which give lamellar phases [84] and vesicles[5] [85, 86].
2. Algorithmic advances, in particular those aimed at greater stability and the reduction of spurious velocities, will improve the ease of implementation of lattice Boltzmann codes. For example, hybrid algorithms, where the Navier-Stokes equation for the velocity field is solved using a lattice Boltzmann approach, but the convection-diffusion equation is treated using conventional finite difference techniques, are being developed. There has been work to develop the use of non-uniform grids [87, 88]. Entropic lattice Boltzmann models, which are unconditionally stable, are also possible [89, 90].
3. The lattice Boltzmann evolution equation can be viewed as the discretisation of a simplified Boltzmann equation and there is discussion as to whether it includes physics beyond that of the Navier Stokes equations [91–93]. Recent work has been successful in matching lattice Boltzmann and molecular dynamics simulations of simple fluids [94].
4. Including thermal fluctuations in a multiphase lattice Boltzmann method is still a major challenge [95, 96]. A simple approach is to include momentum-conserving random noise in the stress tensor. However, it was recently pointed out [95] that this method breaks down on small length scales.

[5] This work is in the context of phase field models, but the same free energy could be used within a binary lattice Boltzmann simulation.

5. The wetting and bounce back boundary conditions can be extended to cases where the solid surfaces themselves are mobile [31, 37]. The algorithm can then be used to study the dynamics of colloids in single- and multi-phase fluids [37, 97, 98].
6. A recent algorithm, coupling a lattice Boltzmann solvent to a molecular dynamics simulation of polymers is proving an exciting new tool for polymer hydrodynamics [99, 100]. Lattice Boltzmann has also been coupled to elastic filaments and membranes [101, 102]
7. Because lattice Boltzmann can handle tortuous boundaries it is particularly suited to simulating flow in porous materials [103–105] and to solving realistic models of blood flow [106, 107].
8. Lattice Boltzmann algorithms can be used to solve the equations of motion of more complex fluids, such as liquid crystals [108] and biologically active materials [109]. They provide a natural way of incorporating viscoelasticity.

The hydrodynamic equations of motion, together with an equilibrium corresponding to the minimum of a free energy, provide a realistic and elegant model of the wetting and spreading properties of multiphase fluids. Lattice Boltzmann algorithms are an effective tool to solve the continuum equations, helping us to understand wetting problems too complicated to be tractable analytically, and to motivate and interpret experiments.

References

1. T.M. Squires, S.R. Quake, Rev. Mod. Phys. **77**, 977 (2005)
2. W. Barthlott, C. Neinhuis, Planta **202**, 1 (1997)
3. A.R. Parker, C.R. Lawrence, Nature **414**, 33 (2001)
4. Y.M. Zheng, X.F. Gao, L. Jiang, Soft Matter **3**, 178 (2007)
5. P.G. de Gennes, Rev. Mod. Phys. **57**, 827 (1985)
6. D. Quéré, Annu. Rev. Fluid Mech. **38**, 71 (2008)
7. S. Herminghaus, M. Brinkmann, R. Seemann, Annu. Rev. Fluid Mech. **38**, 101 (2008)
8. R. Lipowsky, M. Brinkmann, R. Dimova, T. Franke, J. Kierfeld, X. Zhang, J. Phys.: Condens. Matter **17**, S537 (2005)
9. U. Frisch, B. Hasslacher, P. Pomeau, Phys. Rev. Lett. **56**, 1505 (1986)
10. S. Succi, *The Lattice Boltzmann Equation; for Fluid Dynamics and Beyond* (Oxford University Press, Oxford, 2001)
11. S. Chen, G.D. Doolen, Annu. Rev. Fluid Mech. **30**, 329 (1998)
12. R. Benzi, S. Succi, M. Vergassola, Phys. Rep. **222**, 145 (1992)
13. J.M. Yeomans, Physica A **369**, 159 (2006)
14. M.R. Swift, E. Orlandini, W.R. Osborn, J.M. Yeomans, Phys. Rev. E **54**, 5041 (1996)
15. X. Shan, H. Chen, Phys. Rev. E **49**, 2941 (1994)
16. A.K. Gunstensen, D.H. Rothman, S. Zaleski, G. Zanetti, Phys. Rev. A **43**, 4320 (1991)
17. A.J. Briant, J.M. Yeomans, Phys. Rev. E **69**, 031603 (2004)
18. C.M. Pooley, H. Kusumaatmaja, J.M. Yeomans, Phys. Rev. E **78**, 056709 (2008)
19. D. Jacqmin, J. Fluid Mech. **402**, 57 (2000)
20. P. Seppecher, Int. J. Eng. Sci. **34**, 977 (1996)
21. J. Cahn, J. Chem. Phys. **66**, 3667 (1977)
22. R. Cox, J. Fluid Mech. **168**, 169 (1986)

23. C. Huh, L. Scriven, J. Colloid Interf. Sci. **35**, 85 (1971)
24. T. Qian, X. Wang, P. Sheng, J. Fluid Mech. **564**, 333 (2006)
25. P. Bhatnagar, E. Gross, M. Krook, Phys. Rev. **94**, 511 (1954)
26. C. Pooley, K. Furtado, Phys. Rev. E **77**, 046702 (2008)
27. Z. Guo, C. Zheng, B. Shi, Phys. Rev. E **65**, 046308 (2002)
28. D. d'Humieres, I. Ginzburg, M. Krafczyk, P. Lallemand, L. Luo, Philos. Trans. R. Soc. A **360**, 437 (2002)
29. K. Premnath, J. Abraham, J. Comput. Phys. **224**, 539 (2007)
30. R. Du, B. Shi, X. Chen, Phys. Lett. A **359**, 564 (2006)
31. M. Bouzidi, M. Firdaouss, P. Lallemand, Phys. Fluid **13**, 3452 (2001)
32. I. Ginzburg, D. d'Humieres, Phys. Rev. E **68**, 066614 (2003)
33. R. Benzi, L. Biferale, M. Sbragaglia, S. Succi, F. Toschi, Phys. Rev. E **74**, 021509 (2006)
34. A. Wagner, Q. Li, Physica A **362**, 105 (2006)
35. D. Wolf-Gladrow, Lecture Notes in Mathematics, vol. 1725, chapter 5 (Springer-Verlag, Berlin, 2000)
36. S.S. Chikatamarla, I.V. Karlin, Comp. Phys. Comm. **179**, 140 (2008)
37. A. Ladd, R. Verberg, J. Stat. Phys. **104**, 1191 (2001)
38. J. Latt, B. Chopard, Phys. Rev. E **77**, 056703 (2008)
39. A. Dupuis, J.M. Yeomans, Langmuir **21**, 2624 (2005)
40. H. Kusumaatmaja, C.M. Pooley, J.M. Yeomans, Phys. Rev. E. **77**, 067301 (2008)
41. R. Ledesma-Aguilar, I. Pagonabarraga, A. Hernández-Machado, Phys. Fluid **19**, 102113 (2007)
42. A.J. Briant, A.J. Wagner, J.M. Yeomans, Phys. Rev. E **69**, 031602 (2004)
43. F. Diotallevi, L. Biferale, S. Chibbaro, G. Pontrelli, F. Toschi, S. Succi, Eur. Phys. J. Special Topics **171**, 237 (2009)
44. H. Kusumaatmaja, D.Phil. Thesis, University of Oxford, (2008)
45. H. Kusumaatmaja, A. Dupuis, J.M. Yeomans Europhys. Lett. **73**, 740 (2006)
46. R. Lucas, Kolloid-Z **23**, 15 (1918)
47. E. Washburn, Phys. Rev. **17**, 273 (1921)
48. M. Latva-Kokko, D.H. Rothman, Phys. Rev. Lett. **98**, 254503 (2007)
49. P. Saffman, G. Taylor, Proc. R. Soc. London Ser. A **245**, 312 (1958)
50. P. Tabeling, G. Zocchi, A. Libchaber, J. Fluid Mech. **177**, 67 (1987)
51. R. Ledesma-Aguilar, A. Hernández-Machado, I. Pagonabarraga, Phys. Fluid 19, 102112 (2007)
52. H. Gau, S. Hermingaus, P. Lenz, R. Lipowsky, Science **283**, 46 (1999)
53. A.A. Darhuber, S.M. Troian, S.M. Miller, S. Wagner, J. Appl. Phys. **87**, 7768 (2000)
54. J. Léopoldès, A. Dupuis, D.G. Bucknall, J.M. Yeomans, Langmuir **19**, 9818 (2003)
55. A. Dupuis, J. Léopoldès, J.M. Yeomans, Appl. Phys. Lett. **87**, 024103 (2005)
56. N.P. Sandreuter, Tappi J. **77**, 173 (1994)
57. H. Kusumaatmaja, J.M. Yeomans Langmuir **23**, 956 (2007)
58. D. Quéré, Rep. Prog. Phys. **68**, 2495 (2005)
59. X. Gao, L. Jiang, Nature **432**, 36 (2004)
60. A.B.D. Cassie, S. Baxter, Trans. Faraday Soc. **40**, 546 (1944)
61. D. Öner, T.J. McCarthy, Langmuir **16**, 7777 (2000)
62. J. Bico, C. Marzolin, D. Quéré, Europhys. Lett. **47**, 220 (1999)
63. U. Mock, R. Förster, W. Menz, R. Jürgen, J. Phys.: Condens. Matter **17**, S639 (2005)
64. A. Tuteja, W. Choi, M. Ma, J. Mabry, S. Mazzella, G. Rutledge, G. McKinley, R. Cohen, Science **318**, 1618 (2007)
65. L. Cao, T. Price, M. Weiss, D. Gao, Langmuir **24**, 1640 (2008)
66. R.N. Wenzel, Ind. Eng. Chem. **28**, 988 (1936)
67. J.F. Joanny, P.G. de Gennes, J. Chem. Phys. **81**, 552 (1984)
68. R.E. Johnson, R.H. Dettre, Adv. Chem. Ser. **43**, 112 (1964)
69. C. Huh, S.G. Mason, J. Coll. Int. Sci. **60**, 11 (1977)

70. J.F. Oliver, C. Huh, S.G. Mason, J. Coll. Int. Sci. **59**, 568 (1977)
71. J.W. Gibbs, Scientific Papers 1906. Dover reprint, Dover, NewYork (1961)
72. R. Shuttleworth, G.L.J. Bailey, Discuss. Faraday Soc. **3**, 16. (1948)
73. H. Kusumaatmaja, J.M. Yeomans, Langmuir **23**, 6019 (2007)
74. J.-L. Barrat, L. Bocquet, Faraday Discuss **112**, 119 (1999)
75. C. Cottin-Bizonne, E. Charlaix, L. Bocquet, J.-L. Barrat, Nature Mat. **2**, 237 (2003)
76. J. Ou, J.B. Perot, J.P. Rothstein, Phys. Fluids **17**, 103606 (2005)
77. M. Sbragaglia, R. Benzi, L. Biferale, S. Succi, F. Toschi, Phys. Rev. Lett. **97**, 204503 (2006)
78. J. Hyväluoma, J. Harting, Phys. Rev. Lett. **100**, 246001 (2008)
79. A. Steinberger, C. Cottin-Bizonne, P. Kleimann, E. Charlaix, Nature Mater. **6**, 665 (2007)
80. J. Hyväluoma, J. Timonen, Europhys. Lett. **83**, 64002 (2008)
81. D. Bartolo, F. Bouamrirene, É. Verneuil, A. Buguin, P. Silberzan, S. Moulinet, Europhys. Lett. **74**, 299 (2006)
82. H. Kusumaatmaja, M.L. Blow, A. Dupuis, J.M. Yeomans, Europhys. Lett. **81**, 36003 (2008)
83. M. Reyssat, J.M. Yeomans, D. Quéré, Europhys. Lett. **81**, 26006 (2008)
84. G. Gonnella, E. Orlandini, J.M. Yeomans, Phys. Rev. E **58**, 480 (1998)
85. D. Jamet, C. Misbah, Phys. Rev. E **76**, 051907 (2007)
86. D. Jamet, C. Misbah, Phys. Rev. E **78**, 031902 (2008)
87. X.Y. He, L.S. Luo, M. Dembo, J. Comp. Phys. **129**, 357 (1996)
88. O. Filippova, D. Hanel, J. Comp. Phys. **147**, 219 (1998)
89. B.M. Boghosian, J. Yepez, P.V. Coveney, A. Wagner, Proc. R. Soc. Lond. A **457**, 717 (2001)
90. S. Chikatamarla, S. Ansumali, I.V. Karlin, Phys. Rev. Lett. **97**, 010201 (2006)
91. X.B. Nie, G.D. Doolen, S.Y. Chen, J. Stat. Phys. **107**, 279 (2002)
92. F. Toschi, S. Succi, Europhys. Lett. **69**, 549 (2005)
93. Y.H. Zhang, R.S. Qin, D.R. Emerson, Phys. Rev. E **71**, 047702 (2005)
94. J. Horbach, S. Succi, Phys. Rev. Lett. **96**, 224503 (2006)
95. R. Adhikari, K. Stratford, M.E. Cates, A.J. Wagner, Europhys. Lett. **71**, 473 (2005)
96. B. Dünweg, U.D. Schiller, A.J.C. Ladd, Phys. Rev. E **76**, 036704 (2007)
97. K. Stratford, R. Adhikari, I. Pagonabarraga, J.C. Desplat, M.E. Cates, Science **2198**, 30 (2005)
98. M.E. Cates, J.C. Desplat, P. Stansell, A.J. Wagner, K. Stratford, R. Adhikari, I. Pagonabarraga, Phil. Trans. R. Soc. A **363**, 1917 (2005)
99. O.B. Usta, A.J.C. Ladd, J.E. Butler, J. Chem. Phys. **122**, 094902 (2005)
100. P. Ahlrichs, B. Dunweg, J. Chem. Phys. **111**, 8225 (1999)
101. A. Alexeev, R. Verberg, A.C. Balazs, Macromolecules **38**, 10244 (2005)
102. G.A. Buxton, R. Verberg, D. Jasnow, A.C. Balazs, Phys. Rev. E **71**, 056707 (2005)
103. B. Ferréol, D.H. Rothman, Transport in Porous Media **20**, 3 (1995)
104. R.J. Hill, D.L. Koch, A.J.C. Ladd, J. Fluid Mech. **448**, 243 (2001)
105. A. Koponen, D. Kandhai, E. Hellén, M. Alava, A. Hoekstra, M. Kataja, K. Niskanen, P. Sloot, J. Timonen, Phys. Rev. Lett. **80**, 716 (1998)
106. B. Chopard, R. Ouared, Int. J. Mod. Phys. C **18**, 712 (2007)
107. A.M. Artoli, A.G. Hoekstra, P.M.A. Sloot, J. Biomech. **39**, 873 (2006)
108. C. Denniston, D. Marenduzzo, E. Orlandini, J.M. Yeomans, Philos. Trans. R. Soc. A **362**, 1745 (2004)
109. D. Marenduzzo, E. Orlandini, J.M. Yeomans, Phys. Rev. Lett. **98**, 118102 (2007)

Chapter 12
CA Modeling of Ant-Traffic on Trails

Debashish Chowdhury, Katsuhiro Nishinari, and Andreas Schadschneider

12.1 Introduction

There has been significant progress in modelling complex systems by using cellular automata (CA) [1, 2]; such complex systems include, for example vehicular traffic [3] and biological systems [4, 5]. In most cases, particle-hopping CA models have been used to study the spatio-temporal organization in systems of interacting particles driven far from equilibrium [2, 3]. In traffic systems, vehicles are represented by particles while their mutual influence is captured by the inter-particle interactions. Generically, these inter-particle interactions tend to hinder their motions which leads a monotonic decrease of the average speed as function of the particle density [6, 7].

Physicists, applied mathematicians, statisticians and traffic engineers have developed a variety of models which can reproduce the empirically observed properties of vehicular traffic rather accurately [3, 8, 9]. Here we describe an extension of a particularly successful approach based on CA to a seemingly different problem, namely the traffic-like collective movements of ants on trails [10, 11] (see Fig. 12.1). Following this approach, we develop a model which predicts a counter-intuitive result [6]. More specifically, the model predicts a non-monotonic variation of the average speed of ants with their density on the trail. Some of the predictions of this model of ant-traffic have been tested in recent empirical investigation [12]. Most of this chapter is a review of our earlier papers published elsewhere. However, we present this critical overview from our current perspective in the light of the developments over the last few years.

The similarity between ant traffic and vehicular traffic on highways has also been noted by biologists. Burd et al. [13] were the first to measure the average speed of ants on a trail as a function of density. In traffic engineering this corresponds to the so-called *fundamental diagram* which is – as indicated by the name – the most important characteristics of traffic dynamics.

D. Chowdhury (✉)
Department of Physics, Indian Institute of Technology, Kanpur 208016, India
e-mail: debch@iitk.ac.in

A.G. Hoekstra et al. (eds.), *Simulating Complex Systems by Cellular Automata*, Understanding Complex Systems, DOI 10.1007/978-3-642-12203-3_12,

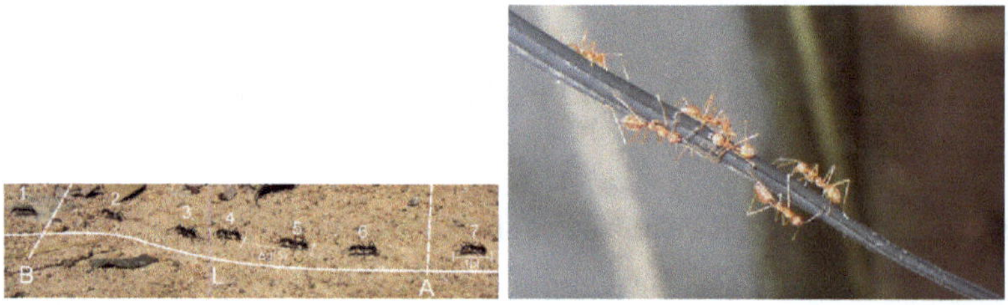

Fig. 12.1 Examples for ant trails

This chapter is organized as follows: First, in Sect. 12.2, we define the basic model for uni-directional traffic on an existing single-lane ant trail. In Sect. 12.4, we extend this model to capture bi-directional traffic on a two-lane ant-trail. We discuss a model of bi-directional traffic on a single-lane ant-trail in Sect. 12.5. Then, in Sect. 12.6 we report results from a recent empirical study and compare with the predictions of the model. Finally, in Sect. 12.7 we summarize the main conclusions.

12.2 A Model of Unidirectional Traffic on a Single-Lane Ant Trail

The models which will be described in the following assume the existence of a fully developed ant trail on which a steady flow takes place. The formation of the trail itself is a problem of self organization which has been studied quite extensively in the past (see e.g. [14] for an overview).

Let us first define the ant trail model (ATM) which is a simple model for a unidirectional collective movement of ants on a pre-existing single-lane ant trail. The ants communicate with each other by dropping a chemical (generically called *pheromone*) on the substrate as they move forward [15, 16]. The pheromone sticks to the substrate long enough for the other following ants to pick up the chemical signal and follow the trail. This mechanism is captured in the ATM by extending the *asymmetric simple exclusion process (ASEP)* [17–20], the simplest and most studied model for driven diffusive systems.

ASEP can be interpreted as a cellular automaton model. It describes the directed motion of particles on a discrete one-dimensional lattice of sites each of which represents the center of a cell. Each cell can be occupied by at most one particle at a time. Identifying the particles in the ASEP with ants we need to incorporate pheromone-mediated interactions among the ants. This leads to an ASEP-like model where the hopping probability of a particle depends on the state of the target cell (see Fig. 12.2).

The cells are labelled by the index i ($i = 1, 2, \ldots, L$) where L is the length of the lattice. One associates two binary variables S_i and σ_i with each site i where S_i takes the value 0 or 1 depending on whether the cell is empty or occupied by an ant. Similarly, $\sigma_i = 1$ if the cell i contains pheromone; otherwise, $\sigma_i = 0$. The

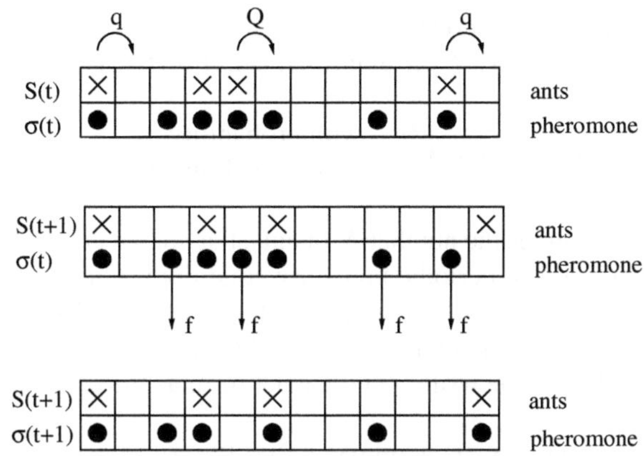

Fig. 12.2 Illustration of the update procedure in the ATM. *Top*: Configuration at time t, i.e. *before* stage *I* of the update. The non-vanishing hopping probabilities of the ants are shown explicitly. *Middle*: Configuration *after* one possible realisation of *stage I*. Also indicated are the pheromones that may evaporate in stage *II* of the update scheme. *Bottom*: Configuration *after* one possible realization of *stage II*. Two pheromones have evaporated and one pheromone has been created due to the motion of an ant

instantaneous state (i.e., the configuration) of the system at any time is specified completely by the set $(\{S\}, \{\sigma\})$.

Since a unidirectional motion is assumed, ants do not move backward. Their forward-hopping probability is higher if there is pheromone ahead of it. The state of the system is updated at each time step in *two stages*. In stage I ants are allowed to move. Here the subset $\{S(t+1)\}$ at the time step $t+1$ is obtained using the full information $(\{S(t)\}, \{\sigma(t)\})$ at time t. Stage II corresponds to the evaporation of pheromone. Here only the subset $\{\sigma(t)\}$ is updated so that at the end of stage II the new configuration $(\{S(t+1)\}, \{\sigma(t+1)\})$ at time $t+1$ is obtained. In each stage the dynamical rules are applied *in parallel* to all ants and pheromones, respectively.

Stage I: Motion of ants

An ant in cell i that has an empty cell in front of it, i.e., $S_i(t) = 1$ and $S_{i+1}(t) = 0$, hops forward with

$$\text{probability} = \begin{cases} Q & \text{if } \sigma_{i+1}(t) = 1, \\ q & \text{if } \sigma_{i+1}(t) = 0, \end{cases} \tag{12.1}$$

where, to be consistent with real ant-trails, we assume $q < Q$.

Stage II: Evaporation of pheromones

At each cell i occupied by an ant after stage I a pheromone will be created:

$$\sigma_i(t+1) = 1 \quad \text{if} \quad S_i(t+1) = 1. \tag{12.2}$$

In addition, any "free" pheromone at a site i, which is not occupied by an ant, will evaporate with the probability f per unit time, i.e., if $S_i(t+1) = 0, \sigma_i(t) = 1$, then

$$\sigma_i(t+1) = \begin{cases} 0 & \text{with probability } f, \\ 1 & \text{with probability } 1 - f. \end{cases} \tag{12.3}$$

The dynamics conserves the number N of ants, but not the number of pheromones. We first discuss the case of periodic boundary conditions which simplifies the theoretical analysis. An extension to the case of the open boundary conditions [21], which is more relevant for the application to real trails, is briefly discussed later.

Formally, the rules can be written in compact form as the coupled equations

$$\begin{aligned} S_j(t+1) &= S_j(t) + \min(\eta_{j-1}(t), S_{j-1}(t), 1 - S_j(t)) \\ &\quad - \min(\eta_j(t), S_j(t), 1 - S_{j+1}(t)), \end{aligned} \tag{12.4}$$

$$\sigma_j(t+1) = \max(S_j(t+1), \min(\sigma_j(t), \xi_j(t))), \tag{12.5}$$

where ξ and η are stochastic variables defined by

$$\xi_j(t) = \begin{cases} 0 & \text{with probability } f \\ 1 & \text{with probability } 1 - f \end{cases}, \tag{12.6}$$

$$\eta_j(t) = \begin{cases} 1 & \text{with probability } p = q + (Q - q)\sigma_{j+1}(t) \\ 0 & \text{with probability } 1 - p \end{cases}. \tag{12.7}$$

In the limits $f = 0$ and $f = 1$ the model reduces to the ASEP. For $f = 0$ a pheromone, once created, will never evaporate. So in the stationary state all cells are occupied by pheromone and the ants will always move with rate Q. For $f = 1$, on the other hand, each pheromone will evaporate immediately. Therefore, in the stationary state, the ants will always move with rate q.

This is reflected in Eqs. (12.4) and (12.5) which reduce to the ASEP if we choose p as a constant, i.e., p does not depend on σ. If we further consider the deterministic limit $p = 1$, then this model reduces to the Burgers CA [22], which is also known as an exactly solvable CA. It should also be mentioned that the ATM is closely related to the bus-route models [23, 24].

12.2.1 Computer Simulation Results

The ASEP [17–20] with parallel updating has been used often as an extremely simple model of vehicular traffic on single-lane highways. The most important quantity of interest in the context of flow properties of the traffic models is the *fundamental diagram*, i.e., the flow-versus-density relation, where flow J is the product of the density ρ and the average speed v. It is especially relevant if one wants to compare the properties of ant traffic with those of vehicular traffic. The flow (or current) J and the average speed v of vehicles are related by the *hydrodynamic relation*

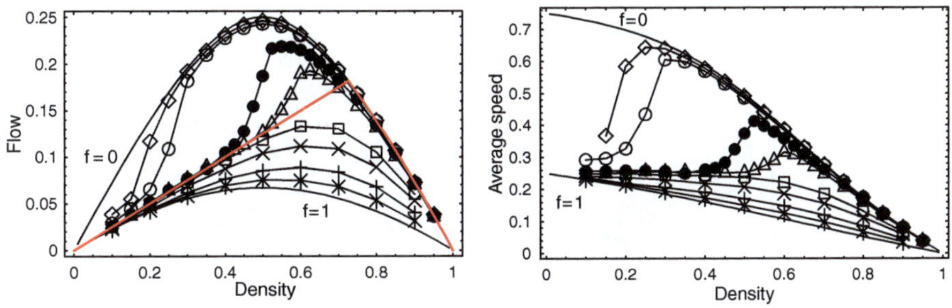

Fig. 12.3 The average flow (*left*) and speed (*right*) of the ants, extracted from computer simulation data, are plotted against their densities for the parameters $Q = 0.75, q = 0.25, L = 500$ and evaporation probabilities $f = 0.0005(\diamond)$, $0.001(\circ)$, $0.005(\bullet)$, $0.01(\triangle)$, $0.05(\square)$, $0.10(\times)$, $0.25(+)$, $0.50(*)$. The *lines* connecting these data points merely serve as the guide to the eye. The cases $f = 0$ and $f = 1$ are also displayed, which are identical to the ASEP corresponding to the effective hopping probabilities Q and q, respectively. The analytical curve corresponds to $f \to 0$ in the thermodynamic limit, which is discussed in Sect. 12.2.2 is also depicted (the thick *red curve* without ornaments)

$J = \rho v$ and, therefore, either of the functions $J(\rho)$ and $v(\rho)$ can be used to express the effects of interactions of the ants on the flow.

Fundamental diagrams of the ATM obtained by computer simulations are shown in Fig. 12.3. In the ASEP the flow remains invariant under the interchange of ρ and $1 - \rho0$; this particle-hole symmetry leads to a fundamental diagram that is symmetrical about $\rho = \frac{1}{2}$. In the ATM, a particle-hole symmetry is only observed in the special cases $f = 0$ and $f = 1$ where it reduces to the ASEP.

However, the most surprising observation is that over a range of small values of f, the fundamental diagram exhibits an anomalous behaviour. Unlike common vehicular traffic, v is not a monotonically decreasing function of the density ρ (Fig. 12.3). Instead, a relatively sharp crossover can be observed where the speed *increases* with the density. In the usual form $J(\rho)$ of the fundamental diagram this transition leads to the existence of an inflection point (Fig. 12.3).

By a detailed analysis of the spatio-temporal organizations in the steady-state, we were able to distinguish three different regimes of density. At low densities a loosely assembled cluster is formed (Fig. 12.4) that propagates with the probability q. These clusters are rather different from jam clusters encountered in highway traffic which have a density close to the maximal density and move in a direction opposite to the direction of motion of the particles (cars). In contrast, a *loose cluster* has a typical density $\rho_{\ell c}$ which is larger than the average density ρ, but smaller than the maximal density $\rho = 1$, i.e. $\rho < \rho_{\ell c} < 1$. It moves in the same direction as the ants. The leading ant in the cluster which typically hops with probability q will determine the velocity of the cluster.

In the intermediate density regime, the leading ant occasionally hops with probability Q instead of q, because sometimes it feels the pheromone dropped by the last ant in the cluster. This arises from the fact that, because of the periodic boundary conditions, the gap size between the last and leading ant becomes shorter as the

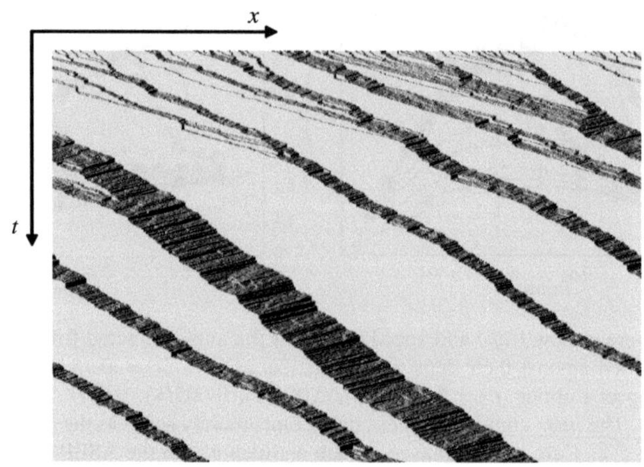

Fig. 12.4 Spatial-temporal behaviour of loose clusters in the low density case ($\rho = 0.16$). Parameters are $Q = 0.75$, $q = 0.25$, $f = 0.005$). Loose clusters emerge from the random initial configuration and will eventually merge into one big loose cluster after sufficiently long time

cluster becomes larger, so that the leading ant is likely to find the pheromone in front of it. This increase of the average speed in the intermediate-density region (Fig. 12.3) leads to the anomalous fundamental diagram.

Finally, at high densities, the mutual hindrance against the movements of the ants dominates the flow behaviour. This leads to a homogeneous state similar to that of the ASEP. In this regime loose clusters no longer exist and ants are uniformly distributed after a long time. Thus homogeneous mean-field theories can account for the qualitative features of the fundamental diagram only in this regime.

12.2.2 Analytical Results

So far, an exact solution for the stationary state of the ATM has not been achieved. However, it is possible to describe the dynamics rather well using approximate analytical theories.

First, mean-field type approaches have been suggested [6, 7]. However, homogeneous mean-field theories fail in the intermediate density regime. Here the loose cluster dominates the dynamics which can not be described properly by the mean-field theories which assume a uniform distribution of the ants.

12.2.2.1 ZRP and ATM

A better theoretical treatment of ATM can be formulated by realizing [7, 21] that the ATM is closely related to the *zero-range process* (*ZRP*), which is one of the exactly solvable models of interacting Markov processes [25–27]. The ZRP consists of the moving particles of the exclusion process, but in contrast, these particles do

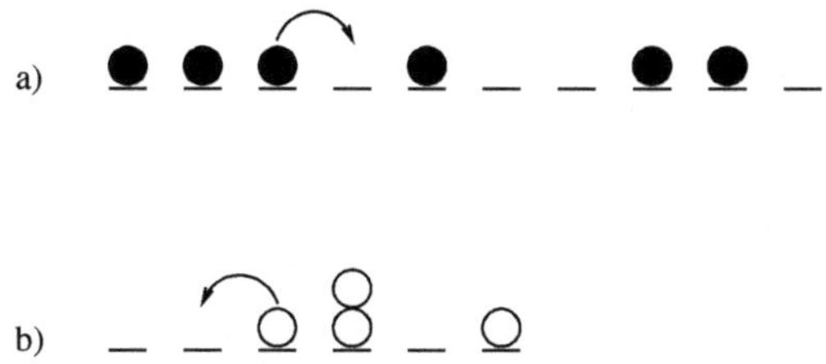

Fig. 12.5 Illustration of the mapping between (**a**) the ASEP and (**b**) a ZRP. Particles of the ASEP become lattice sites in the ZRP. The number of particles at a site in the ZRP corresponds to the headway in ASEP picture. Particles in the ZRP representation move in the opposite direction than in the ASEP representation

not obey an exclusion principle. Therefore each lattice site can be occupied by an arbitrary number of particles. A distinct characteristic of the ZRP is the special form of the transition probabilities. The hopping probability of a particle to its nearest neighbour site depends only on the number of the starting site, not that at the target site.

The ASEP can be interpreted as a special ZRP where the particles in the ASEP are identified with the sites in the ZRP [26, 27]. The number of particles present at a site j is then given by the number of empty cells in front of particle j in the ASEP, also known as *headway* in traffic engineering. This mapping is shown in Fig. 12.5.

In ATM representation, the hopping probability u can be expressed as

$$u = q(1 - g) + Qg, \tag{12.8}$$

where g is the probability that there is a surviving pheromone on the first site of a gap. Assume that the gap size is x and the average velocity of ants is v. Since $g(t+1) = (1-f)g(t)$ holds at each time step, we obtain $g(x) = (1-f)^{x/v}$ after iterating it by x/v times, which is the time interval of between the passage of successive ants through any arbitrary site. Note that in this argument we have implicitly used a mean field approximation: that the ants move with the mean velocity v maintaining equal spacing x. Thus, in the ATM the hopping probability u is related to gaps x by [6]

$$u(x) = q + (Q - q)(1 - f)^{x/v}. \tag{12.9}$$

Using the formal mapping between the ZRP and ASEP we conclude that the steady state of the ATM can be well described by that of the ZRP with parallel dynamics. This allows one to translate the known exact results of the stationary state of the ZRP to the ATM case.

The average velocity v of ants is calculated by

$$v = \sum_{x=1}^{L-N} u(x)p(x) \tag{12.10}$$

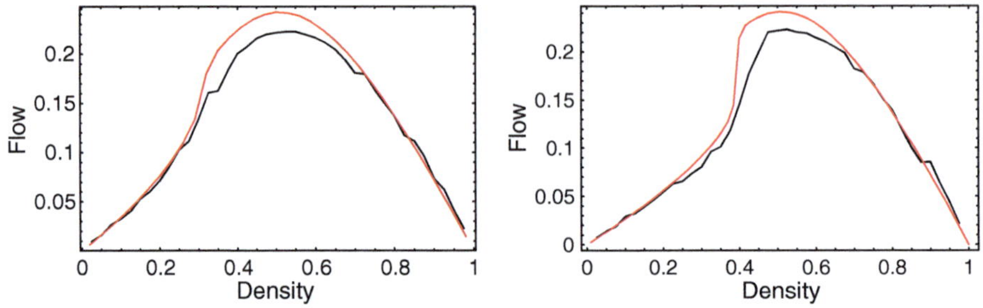

Fig. 12.6 The fundamental diagram of the ATM for system sizes $L = 100$ (*left*) and $L = 200$ (*right*). The dynamical parameters are $Q = 0.75$, $q = 0.25$, $f = 0.005$. The smooth *red curve* has been obtained from the ZRP description while the zigzaged *black one* is the numerical data

since the number of particles in the ZRP picture is $L - N$. $p(x)$ is the probability of finding a gap of size x, which is given by

$$p(x) = h(x)\frac{Z(L-x-1, N-1)}{Z(L,N)}, \tag{12.11}$$

where $Z(L, N)$ is usually called *partition function* since it appears as normalization factor in the probability distribution of headway configurations [21].

Since the ATM is formulated with parallel update, the form of $h(x)$, as calculated in (12.11), is given by [28]

$$h(x) = \begin{cases} 1 - u(1) & \text{for} \quad x = 0 \\ \dfrac{1-u(1)}{1-u(x)} \displaystyle\prod_{y=1}^{x} \frac{1-u(y)}{u(y)} & \text{for} \quad x > 0 \end{cases} . \tag{12.12}$$

The partition function Z is obtained by the recurrence relation

$$Z(L, N) = \sum_{x=0}^{L-N} Z(L-x-1, N-1)h(x), \tag{12.13}$$

with $Z(x, 1) = h(x-1)$ and $Z(x, x) = h(0)$, which is easily obtained by (12.11) with the normalization $\sum p(x) = 1$.

The fundamental diagram of the ATM can be derived by using (12.10) with changing ρ from 0 to 1. The velocity v in (12.9) can be set to $v = q$, which is known to be a good approximation for v [23]. Strictly speaking, v should be determined self-consistently by (12.9) and (12.10). Figure 12.6 shows results for $L = 100$ and $L = 200$. The good agreement with the simulation data confirms that the ZRP provides an accurate description of the steady state of the ATM.

12.2.2.2 Thermodynamic Limit of ATM

Next we discuss the thermodynamic limit of the ATM, that is, the case $L \to \infty$ with $\rho = N/L$ fixed. From Fig. 12.3 we see that the curve shows sharp increase near the density region $0.4 < \rho < 0.5$, and the tendency is expected to be even stronger with the increase of L. This indicates the possibility of a phase transition in the thermodynamic limit. The mapping to the ZRP is exploited to explore this possibility in an analytic way by determining the behaviour of $Z(L, N)$ in the thermodynamic limit. First $Z(L, N)$ is represented as an integral which can be evaluated by the saddle point method [28]. A central role is played by the generating function $G(s)$ of h defined by

$$G(s) = \sum_{x=0}^{\infty} h(x) s^x . \tag{12.14}$$

It can be shown [21] that G converges in the range

$$0 < z < z_c = \frac{q}{1-q} , \tag{12.15}$$

where z is defined by

$$\frac{1}{\rho} - 1 = z \frac{\partial \ln G(z)}{\partial z} . \tag{12.16}$$

In the case $f > 0$, $G(z_c)$ diverges which implies that there is no phase transition. This is because from (12.16), we have $\rho = 1$ when $z = 0$, and $\rho = 0$ at $z = z_c$ if $G(z_c)$ diverges. Thus, in the entire density region $0 \le \rho \le 1$ there is no singularity in G and, hence, no phase transition in the ATM.

The situation drastically changes in the limit $f \to 0$ since then $G(z_c)$ becomes finite. Thus there is a phase transition in the case $f = 0$ at a critical density

$$\rho_c = \frac{Q-q}{Q-q^2} \tag{12.17}$$

obtained from (12.16). The corresponding average velocity at $z = z_c$ is found as

$$v_c = q . \tag{12.18}$$

It should be noted [7] that (12.17) is also obtained by the intersection point of the line $J = v_c \rho$ and the ASEP curve [3]

$$J = \frac{1}{2}\left(1 - \sqrt{1 - 4Q\rho(1-\rho)}\right) \tag{12.19}$$

in the flow-density diagram. Note that the limits $L \to \infty$ and $f \to 0$ do not commute [23]. If one takes $f \to 0$ before $L \to \infty$, then the flow is given by the ASEP result (12.19). This order is relevant for the case of numerical simulations. On the other hand, if $f \to 0$ is taken after $L \to \infty$, one obtains the thick curve in Fig. 12.3 and the anomalous variation of the average velocity with density disappears.

12.2.2.3 Open Boundary Conditions

So far we have considered the ATM with only periodic boundary conditions. However, for ant trails the open boundary conditions are more realistic.

Suppose α and β denote the probabilities of incoming and outgoing particles at the open boundaries per unit time. The phase diagram of the ASEP in the $\alpha-\beta$-plane is well understood [19]. Here we summarize the effects of varying the pheromone evaporation probability f on this phase diagram.

Just as in the case of the ASEP, for all f one finds three different phases, namely, the *high-density phase*, the *low-density phase* and the *maximal current phase* (see Fig. 12.7). In the low-density phase the current is limited by the low input probability α and is therefore independent of β. In contrast, in the high-density phase the particle removal at the end of the chain is the current-limiting factor and thus the current is independent of α. Finally, in the maximal current phase the current is limited by the bulk transport capacity. Here J is independent of both α and β.

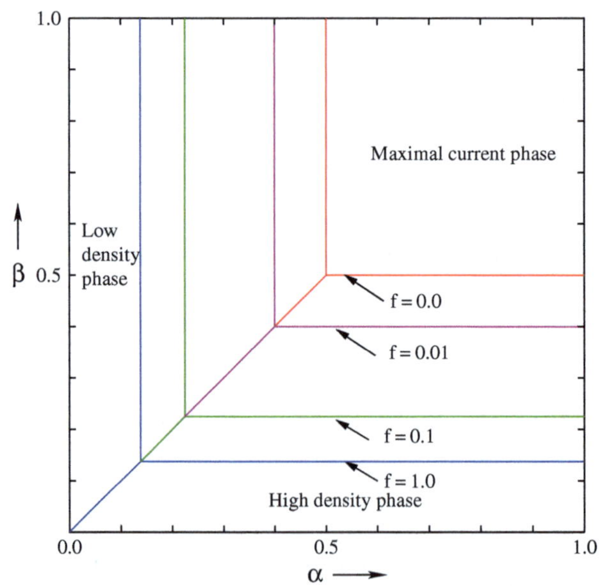

Fig. 12.7 The phase diagram of the ATM with open boundary conditions in the $\alpha - \beta$-plane for several values of the pheromone evaporation probability f $(0 \leq f \leq 1)$. The values of the hopping parameters are $Q = 0.75$, $q = 0.25$

The location $\alpha_c(f)$ and $\beta_c(f)$ of the phase transition lines in the ATM depends on the evaporation rate f. A more quantitative understanding of the phase diagram [21] can be obtained by extending the domain wall theory [29, 30] developed for driven diffusive systems with open boundaries to the ATM case. Many properties of the open system can be extracted from those of the corresponding system under periodic boundary conditions.

12.3 The Multi Robots Implementation

An alternative to the cellular automaton computer simulation is the implementation of the ant trail model as a robotic system [31]. Here social insects are used as a source of inspiration. Aggregation patterns for example are the basis for collective action or collaboration.

12.3.1 Experimental Setup

Here we will focus on the unidirectional case where all robots move on a circular track. The ant-pheromone interaction is incorporated by a virtual pheromone system (V-DEAR). A virtual pheromone field is projected onto the circular track and perceived and modified by the robots (see Fig. 12.8). The concentration of pheromones $p(x,t) \in [0, P_0]$ is indicated by the color and brightness of the projected field. Depending on the concentration, two velocities can be assumed by the robots:

$$v(x) = \begin{cases} V_Q & \text{for } p(x,t) \geq p_{\text{th}} \\ V_q & \text{for } p(x,t) < p_{\text{th}} \end{cases} \tag{12.20}$$

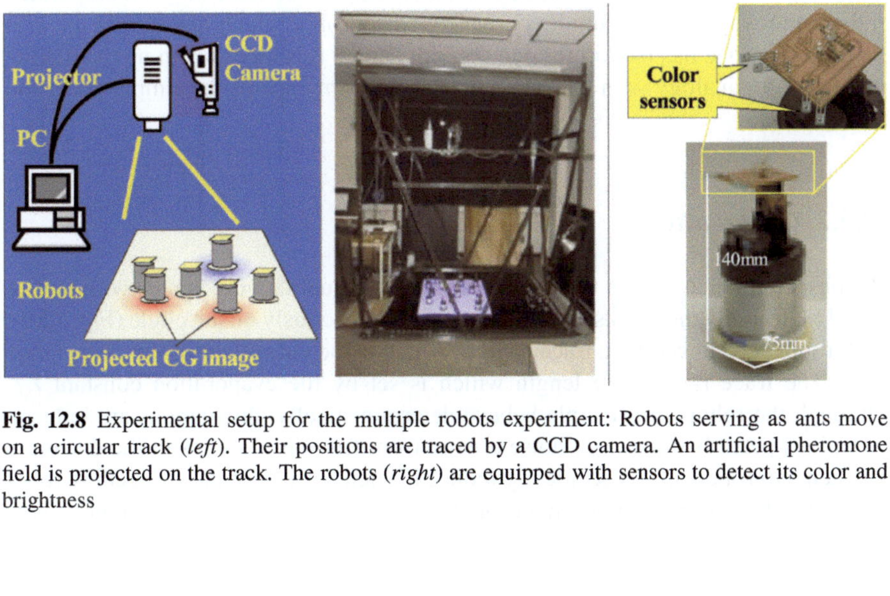

Fig. 12.8 Experimental setup for the multiple robots experiment: Robots serving as ants move on a circular track (*left*). Their positions are traced by a CCD camera. An artificial pheromone field is projected on the track. The robots (*right*) are equipped with sensors to detect its color and brightness

Table 12.1 The parameters for the experimental setup are shown here. For a better comparison with cellular automaton models lengths are additionally given in units of the body-length (*bl*) of one single robot. At $\rho = 1$ the circular track is occupied by 22 robots. The dimensionless pheromone concentration ranges from 0 to 250

V_Q (cm/s)	V_q (cm/s)	L (robot)	p_{th}	P_0
7 (0.94 bl/sec)	1.4 (0.19 bl/sec)	22 (22 bl)	127	250

Unlike in the ATM, the gradient of the pheromone concentration does not lead to a corresponding gradient of velocity depending on the distance to the preceding ant. Therefore only the difference to the threshold value p_{th} is of importance.

The dynamics of the virtual pheromone field is very much the same as in the ATM. The positions of the robots are detected by a CCD camera and the pheromone field is modified accordingly. A robot occupying site x leads to the maximal pheromone concentration P_0 at that site. An unoccupied site is described by a decaying pheromone concentration:

$$p(x,t) = \begin{cases} P_0 & \text{if } x \text{ is occupied} \\ P_0 \exp(-\lambda_f t) & \text{else} \end{cases} . \tag{12.21}$$

For $\lambda_f = 0$ the pheromone concentration always exceeds $p_{\text{th}} < P_0$ once a site has been occupied by a robot leading to V_Q. Unlike in the undirectional model $\lambda_f = 1$ does not lead to an instantaneous evaporation. For the corresponding case $\lambda_f \longrightarrow \infty$ or at least $\lambda_f \gg 1$ is needed.

For the experiment two different intrinsic velocities V_q and V_Q were used (see Table 12.1). The threshold value for the pheromone concentration p_{th} was chosen such that the pheromone trace is of finite length. For our system this means that the pheromone trace following each robot reaches a concentration $p < p_{\text{th}}$ before it assumes $p = P_0$ due to the presence of a succeeding robot. Blocking is incorporated in the same way as in the unidirectional ATM. A robot which has caught up with the preceding one has to stop for one second. Due to this mechanism no overtaking is possible. In analogy with the ATM, this mechanism ensures the simple exclusion principle.

12.3.2 Observations

The observed behaviour in the experimental setup qualitatively agrees with that of the ATM. Initially the robots are distributed homogeneously on the circle (Fig. 12.9). Each robot is followed by a trace of light which is the equivalent of the pheromone trace. The trace is of finite length which is set by the evaporation constant λ_f. Although the robots in principle behave deterministically, fluctuations are induced by various perturbations, e.g., noise arising from friction. As a result the distances between two robots might accidently reduce in such a way that the succeeding robot is affected by the pheromone trace of the preceding one. Generally some robots

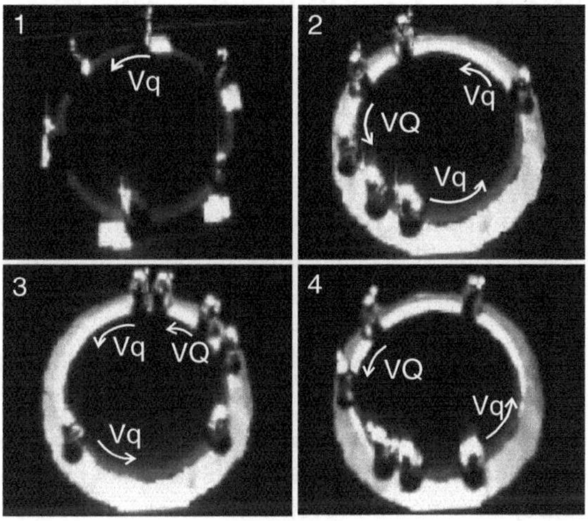

Fig. 12.9 Platoon formation in the robot experiment: (1) In the initial state the robots are placed homogeneously on the circular track. (2) Each robot is followed by a pheromone trace of finite length. Robots perceiving this trace move with V_Q and with $V_q < V_Q$ otherwise. As a result a platoon is formed as robots tend to collect behind a slow robot moving with V_q. (3) Robots within a platoon catch up with V_Q. (4) At later times only one platoon moving with V_q is left

will move with V_Q whereas others move with V_q. In the final stage a stable platoon moving with a velocity V_q is formed (Fig. 12.9).

Flow is measured directly by the number of robots passing a fixed point on the track in a certain time interval. From this the average velocity is calculated using the hydrodynamic relation. For $f = 0$ the behaviour of the system is roughly identical to that of the ASEP (Fig. 12.10). At low densities, flow increases almost linearly. For densities larger than $\rho = \frac{1}{2}$ flow finally decreases. Overall this shows that effects arising from mutual blocking are also present in the robot experiment. Unlike in the ATM, $\lambda_f = 1$ does not correspond to an instantaneous evaporation of the pheromones. Therefore the ASEP-case is not recovered in this limit and a nearly constant average velocity is observed (Fig. 12.10). Furthermore, similar to the ATM, non-monotonic behaviour of the velocity is observed for small values of λ_f.

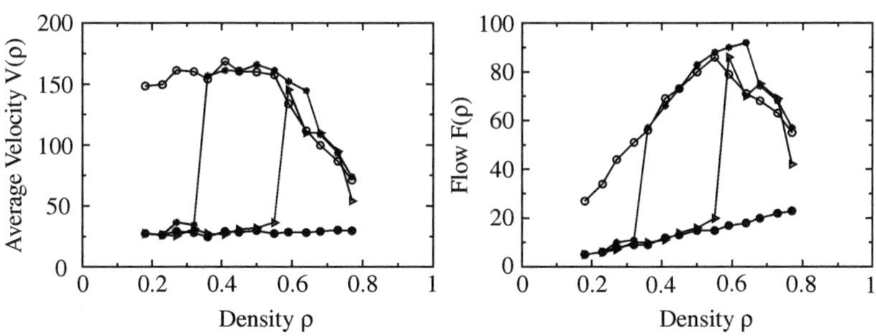

Fig. 12.10 Fundamental diagrams for the robot implementation of the ATM: $V_Q = 7\,\text{cm/s}$, $V_q = 1.4\,\text{cm/s}$ and $\lambda_f = 0(\circ), 0.03(*), 0.05(\triangleright), 1(\bullet)$. For $\lambda_f = 0.03$ and $\lambda_f = 0.05$ the same non-monotonic behaviour as for the computer simulation of the ATM is observed in the density-dependence of the average velocity (left)

12.4 A model of Bidirectional Traffic on a Two-Lane Ant Trail

Real ant trails are often not unidirectional. Instead the trail is shared by ants moving in opposite directions. Therefore an extension of the ATM as described in Sect. 12.2.1 to bidirectional ant traffic is required. We will discuss two different approaches. In this section a two-lane model is introduced with separate lanes for the opposite directions. In Sect. 12.5 a simpler single-lane model is discussed where ants moving in opposite direction share the same lane.

12.4.1 Extensions of the Uni-Directional Model

In this model of bidirectional ant traffic [32] the trail consists of *two* lanes of cells (see Fig. 12.11). These two lanes need not be physically separate rigid lanes in real space; these are, however, convenient for describing the movements of ants in two opposite directions. In the initial configuration, under periodic boundary conditions, a randomly selected subset of the ants move in the clockwise direction in one lane while the others move counterclockwise in the other lane. However, ants are allowed neither to take U-turn [33] nor to change lane. Thus, the ratio of the populations of clockwise-moving and anti-clockwise moving ants remains unchanged as the system evolves with time. All results discussed in the following correspond to the *symmetric* case where equal number of ants move in the two directions. Therefore, the average flux of outbound and nestbound ants are identical.

The rules governing the dropping and evaporation of pheromone in the model of bidirectional ant traffic are identical to those in the model of uni-directional traffic. The *common* pheromone trail is created and reinforced by both the outbound and nestbound ants. The probabilities of forward movement of the ants in the model of bidirectional ant traffic are also natural extensions of the similar situations in the unidirectional traffic. When an ant (in either of the two lanes) *does not* face any other ant approaching it from the opposite direction the likelihood of its forward movement onto the ant-free cell immediately in front of it is Q or q, respectively, depending on whether or not it finds pheromone ahead. Finally, if an ant finds another oncoming ant just in front of it, as shown in Fig. 12.11, it moves forward onto the next cell with probability K.

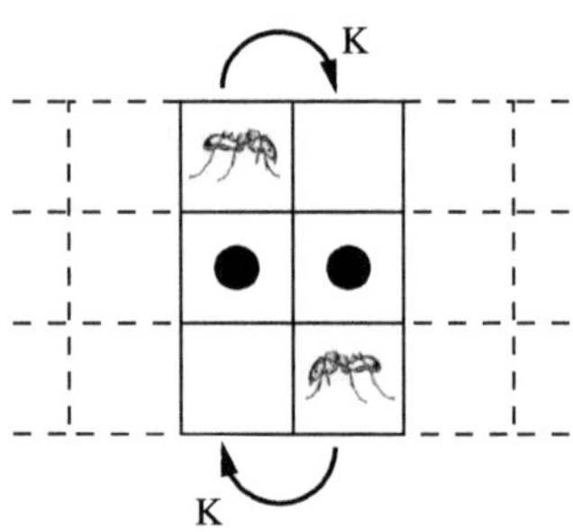

Fig. 12.11 A typical head-on encounter of two oppositely moving ants in the model of *bidirectional* ant traffic

Since ants do not segregate in perfectly well defined lanes, head-on encounters of oppositely moving individuals occur quite often although the frequency of such encounters and the lane discipline varies from one species of ants to another. In reality, two ants approaching each other feel the hindrance, turn by a small angle to avoid head-on collision [34] and, eventually, pass each other. At first sight, it may appear that the ants in the model follow perfect lane discipline. However, that is not true. In the model, the violation of lane discipline and head-on encounters of oppositely moving ants is captured, effectively, in an indirect manner by assuming $K < Q$. But, a left-moving (right-moving) ant *cannot* overtake another left-moving (right-moving) ant immediately in front of it in the same lane. Even in the special limit $K = Q$ the traffic dynamics on two lanes would remain coupled because the pheromone dropped by the outbound ants also influence the nestbound ants and vice versa.

Since for realistic model one has $q < Q$ and $K < Q$, this leaves two interesting parameter regions, namely $q < K < Q$ and $K < q < Q$.

12.4.1.1 Results and Physical Interpretations

The variations of flux with density of ants, for a set of biologically relevant values of the parameters, are shown in Figs. 12.12a and 12.13a; the corresponding average speeds are plotted against density in Figs. 12.12b and 12.13b, respectively. In the Fig. 12.12, the non-monotonic variation of the average speed with density gives rise to the unusual shape of the flux-versus-density diagram over a range of values of f. This feature of the model of bidirectional traffic is similar to that of the unidirectional ant traffic.

An additional feature of the density-dependence of the flux in the bidirectional ant traffic model is the occurrence of a plateau region. This plateau is more pronounced in Fig. 12.13a than in Fig. 12.12a. Such plateaus in the flux-versus-density diagram have been observed earlier [35, 36] in models related to vehicular traffic where randomly placed bottlenecks slow down the traffic in certain locations along

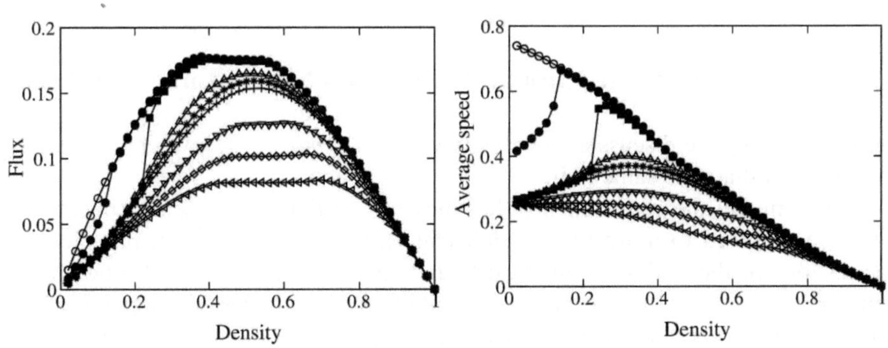

Fig. 12.12 Typical fundamental diagrams for the bidirectional 2-lane model in the case $q < K < Q$. The parameters are $Q = 0.75$, $q = 0.25$ and $K = 0.5$. The symbols ○, •, ■, △, ∗, +, ▽, ◇ and ◁ correspond, respectively, to $f = 0, 0.0005, 0.005, 0.05, 0.075, 0.10, 0.25, 0.5$ and 1

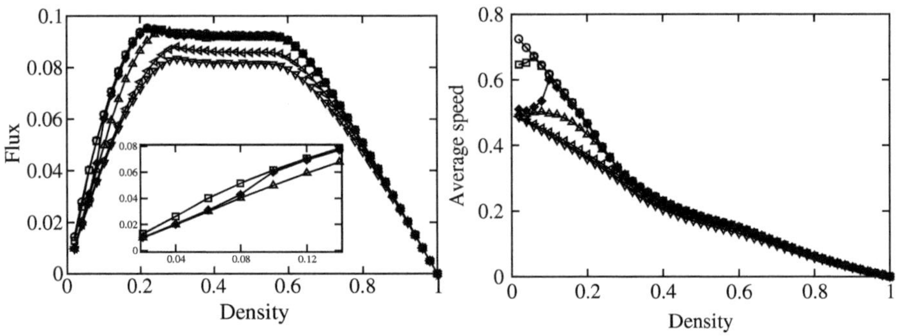

Fig. 12.13 Typical fundamental diagrams for the bidirectional 2-lane model in the case $K < q < Q$. The parameters are $Q = 0.75, q = 0.50$ and $K = 0.25$ The symbols ○, □, ♦, △, ◁ and ▽ correspond, respectively, to $f = 0, 0.0005, 0.005, 0.05, 0.5$ and 1. The *inset* in (**a**) is a magnified re-plot of the same data to emphasize the fact that the unusual trend of variation of flux with density in this case is similar to that observed in unidirectional model

the route. Note that in Fig. 12.12a the plateaus appear only in the two limits $f \to 0$ and $f \to 1$ but not for an intermediate range of values of f. In the limit $f \to 0$, most often the likelihood of the forward movement of the ants is $Q = 0.75$ whereas they are forced to move with a smaller probability $K = 0.5$ at those locations where they face another ant immediately in front approaching from the opposite direction (like the situations depicted in Fig. 12.11). Thus, such encounters of oppositely moving ants have the same effect on ant traffic as bottlenecks on vehicular traffic.

But why do the plateaus re-appear in the Fig. 12.12a also in the limit $f \to 1$? At sufficiently high densities, oppositely moving ants facing each other move with probability $K = 0.5$ rather than $q = 0.25$. In this case, locations where the ants have to move with the lower probability q will be, effectively bottlenecks and hence the re-appearance of the plateau. As f approaches unity there will be larger number of such locations and, hence, the wider will be the plateau. This is consistent with our observation in Fig. 12.12a.

12.5 A Model of Bidirectional Traffic on a Single-Lane Ant Trail

In the following we discuss a model, known as *PRL model* [37], where oppositely moving ants share the same trail.

In this model the right-moving (left-moving) particles, represented by R (L), are never allowed to move towards left (right); these two groups of particles are the analogs of the outbound and nest-bound ants in a *bidirectional* traffic on the same trail. Thus, no U-turn is allowed. In addition to the ASEP-like hopping of the particles onto the neighboring vacant sites in the respective directions of motion, the R and L particles on nearest-neighbour sites and facing each other are allowed to exchange their positions, i.e., the transition $RL \xrightarrow{K} LR$ takes place, with the probability K. This might be considered as a minimal model for the motion of ants on a hanging cable as shown in Fig. 12.1. When a outbound ant and a nest-bound ant face each other on the upper side of the cable, they slow down and, eventually, pass each other after one of them, at least temporarily, switches over to the lower side of the cable. Similar observations have been made for normal ant-trails where ants pass each other after turning by a small angle to avoid head-on collision [34, 38]. In our model, as commonly observed in most real ant trails, none of the ants is allowed to overtake another moving in the same direction.

One then introduces a third species of particles, labelled by P, corresponding to the pheromone. The P particles are deposited on the lattice by the R and L particles when the latter hop out of a site; an existing P particle at a site disappears when a R or L particle arrives at the same location. The P particles cannot hop but can evaporate, with a probability f per unit time. None of the lattice sites can accomodate more than one particle at a time.

The state of the system is updated in a random-sequential manner. Using periodic boundary conditions, the densities of the R and the L particles are conserved. In contrast, the density of the P particles is a non-conserved variable. The distinct initial states and the corresponding final states for pairs of nearest-neighbor sites are shown in Fig. 12.14 together with the respective transition probabilties.

Suppose N_+ and $N_- = N - N_+$ are the total numbers of R and L particles, respectively. For a system of length L the corresponding densities are $\rho_\pm = N_\pm/L$ with the total density $\rho = \rho_+ + \rho_- = N/L$. Of the N particles, a fraction $\phi = N_+/N = \rho_+/\rho$ are of the type R while the remaining fraction $1-\phi$ are L particles. The corresponding fluxes are denoted by $J_\pm$. In both the limits $\phi = 1$ and $\phi = 0$ this model reduces to the ATM model [6, 7].

One unusual feature of this PRL model is that the flux does *not* vanish in the dense-packing limit $\rho \to 1$. In fact, in the full-filling limit $\rho = 1$, the *exact* non-vanishing flux $J_+ = K\rho_+\rho_- = J_-$ at $\rho_+ + \rho_- = \rho = 1$ arises only from the exchange of the R and L particles, irrespective of the magnitudes of f, Q and q.

In the special case $Q = q =: q_h$ the hopping of the ants become independent of pheromone. This special case of the PRL model is identical to the AHR model [39] with $q_- = 0 = \kappa$. A simple mean-field approximation yields the estimates

$$J_\pm \simeq \rho_\pm \left[q_h(1-\rho) + Kc_\mp\right] \tag{12.22}$$

initial	final	rate
RL	RL	$1-K$
	LR	K
RP	RP	$(1-f)(1-Q)$
	R0	$f(1-Q)$
	0R	fQ
	PR	$(1-f)Q$
R0	R0	$1-q$
	0R	fq
	PR	$(1-f)q$

initial	final	rate
PR	PR	$1-f$
	0R	f
P0	P0	$1-f$
	00	f
PP	PP	$(1-f)^2$
	P0	$f(1-f)$
	0P	$f(1-f)$
	00	f^2

Fig. 12.14 Nontrivial transitions and their transition rates. Transitions from initial states PL, $0L$ and $0P$ are not listed. They can be obtained from those for LP, $L0$ and $P0$, respectively, by replacing $R \leftrightarrow L$ and, then, taking the mirror image

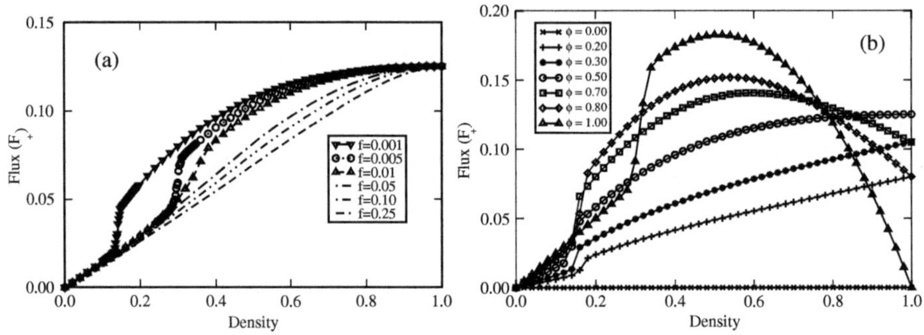

Fig. 12.15 The fundamental diagrams in the steady-state of the PRL model for several different values of (left) f (for $\phi = 0.5$) and (right) ϕ (for $f = 0.001$). The other common parameters are $Q = 0.75$, $q = 0.25$, $K = 0.5$ and $L = 1000$

irrespective of f, for the fluxes $J_{\pm}$ at any arbitrary ρ. These results agree reasonably well with the exact values of the flux [40] for all $q_h \geq 1/2$ but deviate more from the exact values for $q_h < 1/2$, indicating the presence of stronger correlations at smaller values of q_h.

For the generic case $q \neq Q$, the flux in the PRL model depends on the evaporation rate f of the P particles. Figure 12.15 shows fundamental diagrams for wide ranges of values of f (in Fig. 12.15a) and ϕ (in Fig. 12.15b). The data in Fig. 12.15 are consistent with the physically expected value of $J_{\pm}(\rho = 1) = K\rho_{+}\rho_{-}$, because in the dense packing limit only the exchange of the oppositely moving particles contributes to the flux. Moreover, the sharp rise of the flux over a narrow range of ρ observed in both Fig. 12.15a and b arise from the nonmonotonic variation of the average s peed with density observed in the unidirectional ATM [6, 7].

As we have seen earlier in Sect. 12.2.1, in the special limits $\phi = 0$ and $\phi = 1$, over a certain regime of density (especially at small f), the particles form loose (i.e., non-compact) clusters [7]. If the system evolves from a random initial condition at $t = 0$, then during coarsening of the cluster, its size $R(t)$ at time t is given by $R(t) \sim t^{1/2}$ [23, 24] see Fig. 12.16. Therefore, in the absence of encounter with oppositely moving particles, $\tau_{\pm}$, the coarsening time for the right-moving and left-moving particles would grow with system size as $\tau_{+} \sim \phi^2 L^2$ and $\tau_{-} \sim (1-\phi)^2 L^2$.

In the PRL model with periodic boundary conditions, the oppositely moving loose clusters "collide" against each other periodically where the time gap τ_g between the successive collisions increases linearly with the system size ($\tau_g \sim L$). During a collision each loose cluster "shreds" the oppositely moving cluster; both clusters shred the other equally if $\phi = 1/2$ (Fig. 12.17a). However, for all $\phi \neq 1/2$, the minority cluster suffers more severe shredding than that suffered by the majority cluster (Fig. 12.17b) because each member of a cluster contributes in the shredding of the oppositely moving cluster. In small systems the "shredded" clusters get opportunity for significant re-coarsening before getting shredded again in the next encounter with the oppositely moving particles. But, in sufficiently large systems, shredded appearance of the clusters persists as demonstrated by the space-time plots for two different system sizes in Fig. 12.18.

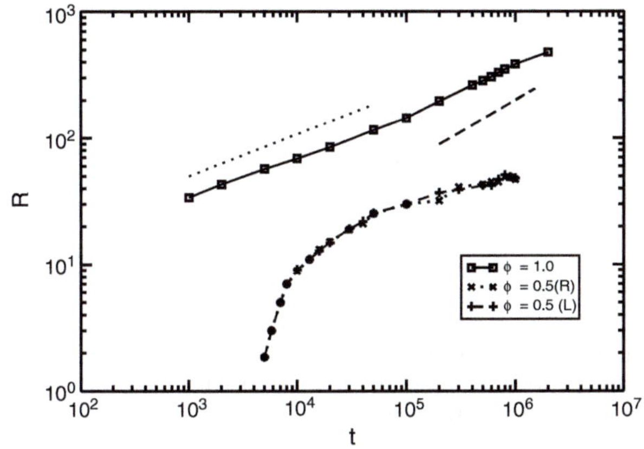

Fig. 12.16 Average size of the cluster R plotted against time t for $\phi = 1.0$, and $\phi = 0.5$, both for the same total density $\rho = 0.2$; the other common parameters being $Q = 0.75$, $q = 0.25$, $K = 0.50$, $f = 0.005$, $L = 4000$

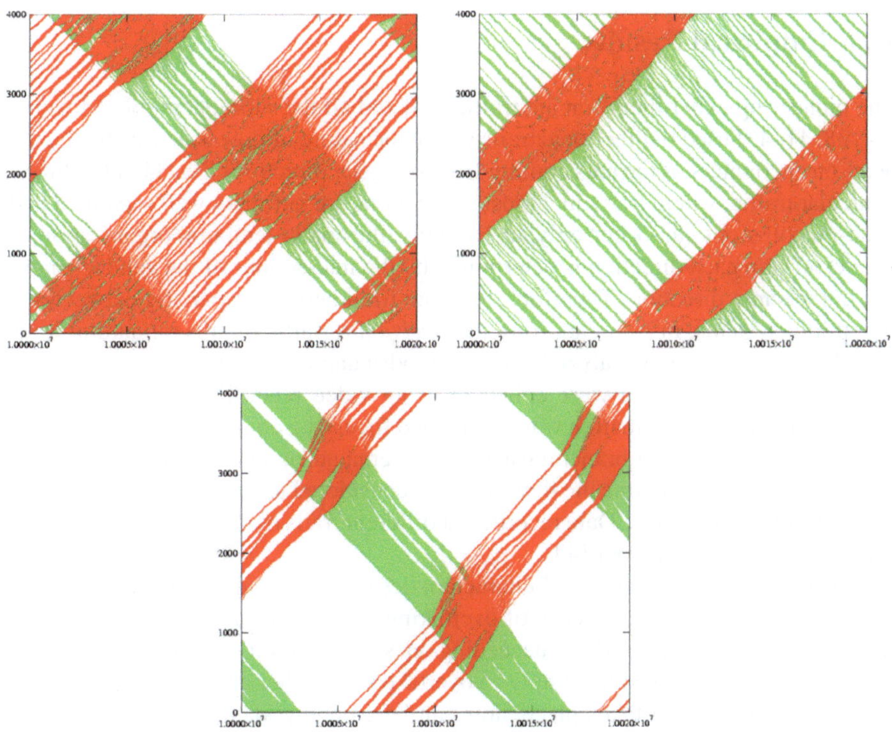

Fig. 12.17 Space-time plot of the PRL model for $Q = 0.75$, $q = 0.25$, $f = 0.005$, $L = 4000$, $\rho = 0.2$ and (a) $\phi = 0.5$, $K = 0.2$, (b) $\phi = 0.3$, $K = 0.2$, (c) $\phi = 0.3$, $K = 0.5$. The *red* and *green* dots represent the right-moving and left-moving ants, respectively

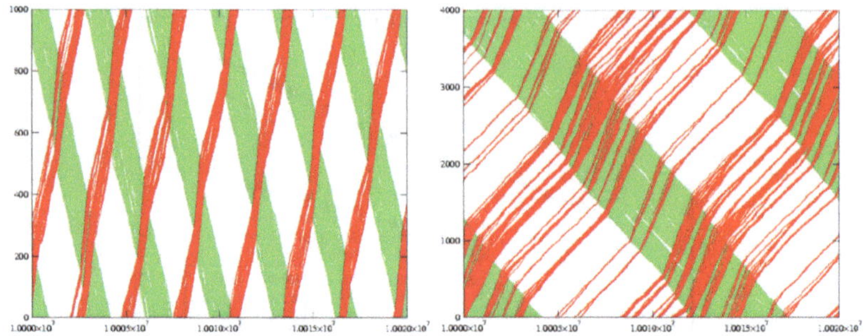

Fig. 12.18 Space-time plot of the PRL model for $Q = 0.50$, $q = 0.25$, $f = 0.005$, $\rho = 0.2$, $\phi = 0.3$, $K = 1.0$ and (**a**) $L = 1000$, (**b**) $L = 4000$. The *red* and *green dots* represent the right-moving and left-moving ants, respectively

Thus, coarsening and shredding phenomena compete against each other [37] and this competition determines the overall spatio-temporal pattern. Therefore, in the late stage of evolution, the system settles to a state where, because of alternate occurrence of shredding and coarsening, the typical size of the clusters varies periodically.

12.6 Empirical Results

The pioneering experiments on ant traffic [13] and all the subsequent related works [34, 38, 41–43] used bidirectional trails where the nature of flow is dominated by the head-on encounters of the ants coming from opposite directions [38, 34, 43]. But, in vehicular traffic, where flows in opposite directions are normally well separated and head-on collisions can occur only accidentally, the spatio-temporal organization of the vehicles in each direction is determined by the interactions of the vehicles moving in the same direction. Therefore, to allow for a better comparison between the two traffic systems, in [12] data from *unidirectional* traffic of ants on natural trails have been collected and analyzed using methods adapted from traffic engineering.

The experimental data were collected using video recordings of natural trails [44] where the natural situation was maintained focussing on sections of trails which had neither crossings nor branching and which remained unaltered for several hours. During the observation time the flow could be considered to be stationary and undisturbed by external factors. Data recorded at different trails of the same type basically revealed the same behaviour [44].

One of the distinct behavioral characteristics of individual ants observed in the course of recording is the absence of overtaking. Although some ants (temporarily) left the trail and were passed by succeeding ones, it was never observed any incident where an ant would speed up simply to overtake some other ant in front.

Since no overtaking takes place, ants can be uniquely identified by the ordered sequence in which they enter the observed section of the trail. Suppose, the nth ant enters the section at A at time $t_+(n)$ and leaves the section at B at time $t_-(n)$ (see Fig. 12.1). An efficient tool for analyzing such data is the *cumulative plot* (Fig. 12.19) [45]; it shows the numbers $n_+(t)$ and $n_-(t)$ of ants which have passed the point A and B, respectively, up to time t. The two resulting curves, which are sometimes called *arrival function* and *departure function* can be obtained by inverting $t_+(n)$ and $t_-(n)$, respectively.

Using this strategy, the data could be analyzed very efficiently. The travel time $\Delta T(n)$ of the nth ant in the section between the points A and B is given by

$$\Delta T(n) = t_-(n) - t_+(n) \tag{12.23}$$

and the time-averaged speed of the nth ant during the period $\Delta T(n)$ is

$$v(n) = \frac{L}{\Delta T(n)} \tag{12.24}$$

The time-headway of two succeeding ants can be obtained easily at the entrance and exit points A and B (Fig. 12.19, left inset). Since $v(n)$ is, by definition (Eq. (12.24)), the time-averaged velocity $v(n)$ of the n-th ant, the distance-headway between the n-th ant and the ant in front of it is given by

$$\begin{aligned} \Delta d(n) &= \Delta t_+(n)\, v(n-1)\,, \\ \Delta t_+(n) &= t_+(n) - t_+(n-1)\,. \end{aligned} \tag{12.25}$$

Entry and exit of each ant changes the instantaneous number $N(t)$ of the ants in the trail section between A and B by one unit (Fig. 12.19, right inset). Therefore

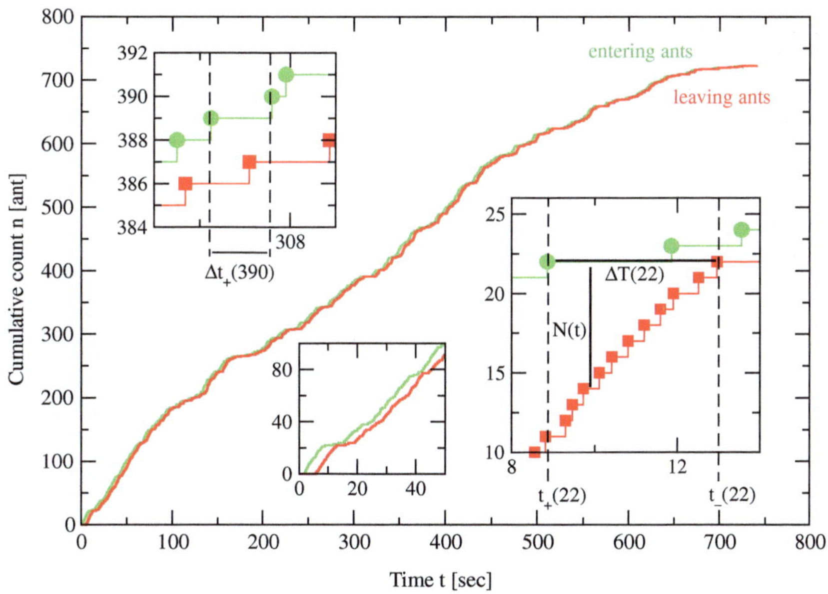

Fig. 12.19 Figure illustrating the technique employed for data extraction. The cumulative count of the ants which have entered $n_+(t)$ (•) and left $n_-(t)$ (□) the trail section between A and B. The *right inset* shows the travel time ΔT for the 22th ant. On the *left inset* the time-headway Δt_+ of the 390th ant is shown

$N(t)$ fluctuates, but stays constant in between two events of entry or exit. Sorting the counts of these events by time one obtains a chronological list $\{t_i\} = \{t_\pm(n)\}$ of the changes of the instantaneous particle number

$$N(t) = n_+(t) - n_-(t) = const. \quad \text{while } t \in [t_i, t_{i+1}[. \tag{12.26}$$

Next we estimate the effective local density experienced by the n-th ant at a given instant of time. During the time interval $\Delta T(n)$ it spends within the observed trail section, the average number of ants in the same section is given by

$$\langle N \rangle_{t(n)} = \frac{1}{\Delta T(n)} \sum_{t_i = t_+(n)}^{t_i < t_-(n)} N(t_i)(t_{i+1} - t_i) \tag{12.27}$$

During the same time interval, the (dimensionless) density $\rho(n)$ affecting the movement of the n-th ant is given by

$$\rho(n) = \frac{\langle N \rangle_{t(n)}}{N_{\max}} = \frac{\tilde{\rho}(n)}{\tilde{\rho}_{\max}} \qquad \text{with} \quad \tilde{\rho}(n) = \frac{\langle N \rangle_{t(n)}}{L}, \tag{12.28}$$

where $N_{\max} = 17 = L/(1\ \text{bl})$ and $\tilde{\rho}_{\max} = N_{\max}/L$; *bl* being the body length of an ant ($1bl \approx 18$ mm). The instantaneous particle numbers and the single-ant velocity are averaged over the same time-interval $\Delta T(n)$.

Figure 12.20 shows the fundamental diagram obtained from this analysis. The most unusual feature is that, unlike vehicular traffic, there is no significant decrease of the average velocity with increasing density. Consequently, the flux increases approximately linearly over the entire regime $\rho \in [0, 0.8]$ of observed density. The jammed branch of the fundamental diagram, which is commonly observed in vehicular traffic and which is characterized by a monotonic decrease of flow with increas-

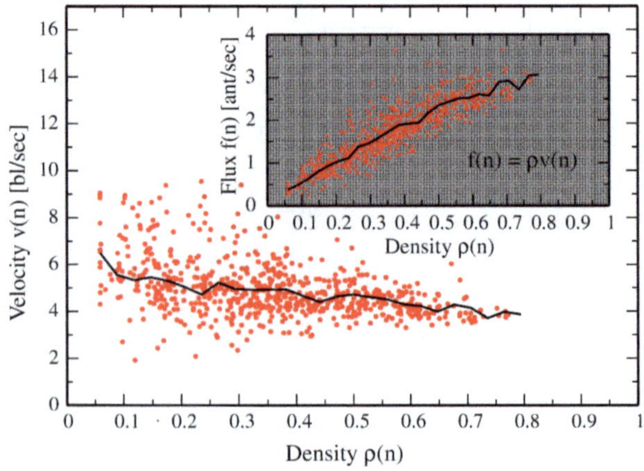

Fig. 12.20 Average velocity (*solid line*) and single-ant velocities (*dots*) for unidirectional single-lane trail section of length $L = 17$ bl. The corresponding flux-velocity relation is plotted in the inset. Mutual blocking is obviously suppressed as the average velocity is almost independent of the density. Consequently, the flux increases almost linearly with the density in the fundamental diagram (see *inset*)

ing density, is completely missing. Obviously effects of mutual blocking, which are normally expected to become dominant at high densities are strongly suppressed in ant traffic.

From the time-series of the single-ant velocities also their distributions in different density regimes can be determined. The most striking feature is that it becomes much sharper with increasing global density whereas the most probable velocity decreases only slighty [12].

Another important quantity that characterizes the spatial distribution of the ants on the trail is the distance-headway distribution [12]. Time-series show clustering of small distance-headways whereas larger headways are much more scattered. The distribution of these headways becomes much sharper with increasing density while the maximum shifts only slightly to smaller headways. At low densities, predominantly large distance-headways are found; the corresponding distribution for sufficiently long distance-headways is well described by a negative-exponential distribution which is characteristic of the so-called random-headway state [46]. In contrast, at very high densities mostly very short distance-headways are found; in this regime, the log-normal distribution appears to provide the best fit to the empirical data.

The absence of a jammed phase in the fundamental diagram is closely related to the characteristic features of the distributions of the distance-headways of the ants along the observed section of the trail. The dominant, and directly observable, feature of this spatial distribution is the platoons formed by the ants, as predicted by the models discussed before. Ants inside a platoon move with almost identical velocities maintaining small distance-headways. These intra-platoon distance-headways are responsible for the clustering of data observed in the corresponding time-series [12]. In contrast, larger distance-headways are inter-platoon separations. The full distribution of distance-headways has an average of $D = 2.59$ bl which is quite close to the value $D = 1.66$ bl found for very high densities. This indicates the existence of a density-independent distance-headway for the ants moving inside platoons.

The interpretations of the observed trends of variations of the flux, average velocity and distance-headway distribution with increasing density is consistent with the corresponding variation of the distribution of the velocities of the ants. Ants within a platoon move at a slower average velocity whereas solitary ants can move faster if they detect a strong pheromone trace created by a preceeding platoon. Moreover, since fluctuations of velocities of different platoons are larger than the intra-platoon fluctuations, the distribution becomes sharper at higher densities because the platoons merge thereby reducing their number and increasing the length of the longest one. The maximum of the velocity distribution is almost independent of the density. Its position at sufficiently large densities can be interpreted as platoon velocity, $v_p \approx 4.6$ bl/s.

It is worth pointing out that physical origin of the occurrence of the nearly constant average velocity of the vehicles in highway traffic is very different from the constant velocity of ants in ant traffic. In vehicular traffic, the average velocity of the vehicles remains practically unaffected by increasing density, provided the density is sufficiently low, because at those densities the vehicles are well separated from each other and, therefore, can move practically unhindered in the so-called free-flow state. On the other hand, in ant traffic, this constant velocity regime is a reflection of the fact that ants march together collectively forming platoons which reduce the effective density.

Thus, in spite of some superficial similarities, the characteristic features of ant traffic seem to be rather different from those of vehicular traffic. Perhaps, ant traffic is analogous to human pedestrian traffic [47, 42, 43], as was conjectured beautifully by Hölldobler and Wilson in their classic book [10].

12.7 Concluding Discussions

Several theoretical investigations have been carried out earlier, in terms of CA, to study the emergence of the trail patterns in ant colonies [4]. However, to our knowledge, our work is the first attempt to understand the traffic-like flow of ants on well formed trails using the language of CA. Here we have reviewed, in some detail, the stochastic cellular automaton model of an ant trail introduced in [6] and some of its generalizations. The model is characterized by two coupled dynamical variables, representing the ants and the pheromone. Under periodic boundary conditions, one of the variables (ants) is conserved, whereas the other variable (pheromone) is always a non-conserved variable. The dynamics of these two variables are coupled to each other. This coupling leads to surprising results, especially an anomalous fundamental diagram which arises from an unusual non-monotonic variation of the average speed of the ants with their density on the trail in the intermediate regime of the ant density. We would like to emphasize that this surprising result could not be anticipated as a trivial consequence of the dynamical prescriptions of the model. It is only over a range of pheromone-evaporation rate that the surprising increase of average velocity of ants with increasing density was observed.

Our experimental investigations on real ant trails have, indeed, exposed some unusual features of the fundamental diagram. The average velocity is practically independent of density. This, in turn, leads to the complete absence of a jammed branch in the fundamental diagram. In order to test the possibility of the non-monotonic variation of the average velocity with density, our experiment should be repeated with a circular trail which would mimic the trail under period boundary conditions. Our first set of experiments have already unveiled a totally unexpected feature of the fundamental diagram of the ant trafic. Clearly there is a need for extension of our model to account for the observed features of the fundamental diagram. At present we are actively considering several possible realistic extensions of our model. These include (a) different levels of concentration of pheromone at each site, (b) diffusional spread of pheromone, (iii) tendency of persistent movement of ants, etc.

Acknowledgements We thank our collaborators V. Guttal, A. Kunwar, K. Sugawara, T. Kazama and especially Alexander John for many discussions. Research of the group of one of the authors (DC) is supported by CSIR (India). DC also thanks Alexander von Humboldt Foundation for re-invitation, under the Humboldt Alumni program, to the university of Köln where part of this work was completed.

References

1. S. Wolfram, *Theory and Applications of Cellular Automata* (World Scientific, Singapore, 1986)
2. B. Chopard, M. Droz, *Cellular Automata Modelling of Physical Systems* (Cambridge University Press, 1998)
3. D. Chowdhury, L. Santen, A. Schadschneider, Statistical physics of vehicular traffic and some related systems. Phys. Rep. **329**, 199–329 (2000)
4. D. Chowdhury, K. Nishinari, A. Schadschneider, Self-organized patterns and traffic flow in colonies of organisms: From bacteria and social insects to vertebrates. Phase Transit. **77**, 601–624 (2004)
5. D. Chowdhury, K. Nishinari, A. Schadschneider, Physics of transport and traffic phenomena in biology: From molecular motors and cells to organisms. Phys. Life Rev. **2**, 318 (2005)
6. D. Chowdhury, V. Guttal, K. Nishinari, A. Schadschneider, A cellular-automata model of flow in ant trails: Non-monotonic variation of speed with density. J. Phys. A: Math. Gen. **35**, L573–L577 (2002)
7. K. Nishinari, D. Chowdhury, A. Schadschneider, Cluster formation and anomalous fundamental diagram in an ant trail model. Phys. Rev. E **67**, 036120 (2003)
8. D. Helbing, Traffic and related self-driven many-particle systems. Rev. Mod. Phys. **73**, 1067 (2001)
9. B. Kerner: *The Physics of Traffic* (Springer, Heidelberg, 2004)
10. B. Hölldobler, E.O. Wilson, *The Ants* (Belknap, Cambridge, 1990)
11. B. Hölldobler, E.O. Wilson: *The Superorganism: The Beauty, Elegance, and Strangeness of Insect Societies* (W.W. Norton, New York, 2008)
12. A. John, A. Schadschneider, D. Chowdhury, K. Nishinari, Trafficlike collective movement of ants on trails: Absence of jammed phase. Phys. Rev. Lett. **102**, 108001 (2009)
13. M. Burd, D. Archer, N. Aranwela, D.J. Stradling, Traffic dynamics of the leaf cutting ant. American Natur. **159**, 283 (2002)
14. F. Schweitzer, *Brownian Agents and Active Particles*, Springer Series in Synergetics (Springer, Heidelberg, 2003)
15. S. Camazine, J.L. Deneubourg, N.R. Franks, J. Sneyd, G. Theraulaz, E. Bonabeau, *Self-organization in Biological Systems* (Princeton University Press, Princeton, 2001)
16. A.S. Mikhailov, V. Calenbuhr, *From Cells to Societies* (Springer, Berlin, 2002)
17. B. Derrida, An exactly soluble non-equilibrium system: The asymmetric simple exclusion process. Phys. Rep. **301**, 65 (1998)
18. B. Derrida, M.R. Evans, in: *Nonequilibrium Statistical Mechanics in One Dimension*, ed. by V. Privman (Cambridge University Press, Cambridge, 1997)
19. G.M. Schütz, Exactly solvable models for many-body systems far from equilibrium, ed. by C. Domb, J.L. Lebowitz, *Phase Transitions and Critical Phenomena*, Vol. 19, (Academic Press, London, UK, 2000)
20. R.A. Blythe, M.R. Evans, Nonequilibrium steady states of matrix product form: a solver's guide. J. Phys. A **40**, R333 (2007)
21. A. Kunwar, A. John, K. Nishinari, A. Schadschneider, D. Chowdhury, Collective traffic-like movement of ants on a trail – dynamical phases and phase transitions. J. Phys. Soc. Jpn. **73**, 2979 (2004)
22. K. Nishinari, D. Takahashi, Analytical properties of ultradiscrete Burgers equation and rule-184 cellular automaton. J. Phys. A: Math. Gen. **31**, 5439 (1998)
23. O.J. O'Loan, M.R. Evans, M.E. Cates, Jamming transition in a homogeneous one-dimensional system: The bus route model. Phys. Rev. E **58**, 1404 (1998)
24. D. Chowdhury, R.C. Desai, Steady-states and kinetics of ordering in bus-route models: Connection with the Nagel-Schreckenberg model. Eur. Phys. J. B **15**, 375 (2000)
25. F. Spitzer, Interaction of Markov processes. Adv. Math. **5**, 246–290 (1970)
26. M.R. Evans, Phase transitions in one-dimensional nonequilibrium systems. Braz. J. Phys. **30**, 42 (2000)

27. M.R. Evans, T. Hanney, Nonequilibrium statistical mechanics of the zero-range process and related models. J. Phys. A **38**, R195 (2005)
28. M.R. Evans, Exact steady states of disordered hopping particle models with parallel and ordered sequential dynamics. J. Phys. A **30**, 5669 (1997)
29. A.B. Kolomeisky, G. Schütz, E.B. Kolomeisky, J.P. Straley, Phase diagram of one-dimensional driven lattice gases with open boundaries. J. Phys. A **31**, 6911 (1998)
30. V. Popkov, G. Schütz, Steady-state selection in driven diffusive systems with open boundaries. Europhys. Lett. **48**, 257 (1999)
31. K. Nishinari, K. Sugawara, T. Kazama, A. Schadschneider, D. Chowdhury, Modelling of self-driven particles: Foraging ants and pedestrians. Physica A **372**, 132 (2006)
32. A. John, A. Schadschneider, D. Chowdhury, K. Nishinari, Collective effects in traffic on bi-directional ant trails. J. Theor. Biol. **231**, 279 (2004)
33. R. Beckers, J.L. Deneubourg, S. Goss, Trails and U-turns in the selection of a path by the ant *Lasius niger*. J. Theor. Biol. **159**, 397 (1992)
34. I.D. Couzin, N.R. Franks, Self-organized lane formation and optimized traffic flow in army ants. Proc. Roy Soc. London B **270**, 139 (2003)
35. S.A. Janowsky, J.L. Lebowitz, Finite-size effects and shock fluctuations in the asymmetric simple-exclusion process. Phys. Rev. A **45**, 618 (1992)
36. G. Tripathy, M. Barma, Steady state and dynamics of driven diffusive systems with quenched disorder. Phys. Rev. Lett. **78**, 3039 (1997)
37. A. Kunwar, D. Chowdhury, A. Schadschneider, K. Nishinari, Competition of coarsening and shredding of clusters in a driven diffusive lattice gas. J. Stat. Mech. (2006) P06012
38. M. Burd, N. Aranwela, Head-on encounter rates and walking speed of foragers in leaf-cutting ant traffic. Insect. Sociaux **50**, 3 (2003)
39. P. F. Arndt, T. Heinzel, V. Rittenberg, Spontaneous breaking of translational invariance in one-dimensional stationary states on a ring. J. Phys. A **31**, L45 (1998); J. Stat. Phys. **97**, 1 (1999)
40. N. Rajewsky, T. Sasamoto, E.R. Speer, Spatial particle condensation for an exclusion process on a ring. Physica A **279**, 123 (2000)
41. K. Johnson, L.F. Rossi, A mathematical and experimental study of ant foraging trail dynamics. J. Theor. Biol. **241**, 360 (2006)
42. A. Dussutour, J.L. Deneubourg, V. Fourcassié, Temporal organization of bi-directional traffic in the ant Lasius niger (L.), Jrl. Exp. Biol. **208**, 2903 (2005)
43. A. John, A. Schadschneider, D. Chowdhury, K. Nishinari, Characteristics of ant-inspired traffic flow – Applying the social insect metaphor to traffic models. Swarm Intelligence **3**, 199 (2008)
44. A. John, *Physics of Traffic on Ant Trails and Related Systems*, Doctoral Thesis, (Universität zu Köln, Cologne, Germany, 2006)
45. P. Chakroborty, A. Das, *Principles of Transportation Engineering* (Prentice Hall of India, Englewood Cliffs, NJ, 2003)
46. A.D. May, *Traffic Flow Fundamentals* (Prentice Hall Englewood Cliffs, NJ, 1990)
47. C. Burstedde, K. Klauck, A. Schadschneider, J. Zittartz, Simulation of pedestrian dynamics using a 2-dimensional cellular automaton. Physica A **295**, 507 (2001)

Chapter 13
Lattice-Gas Cellular Automaton Modeling of Emergent Behavior in Interacting Cell Populations

Haralambos Hatzikirou and Andreas Deutsch

13.1 Introduction

Biological organisms are complex systems characterized by collective behavior emerging out of the interaction of a large number of components (molecules and cells). In complex systems, even if the basic and local interactions are perfectly known, it is possible that the global (collective) behavior obeys new laws that are not obviously extrapolated from the individual properties. Only an understanding of the dynamics of collective effects at the molecular, and cellular scale allows answers to biological key questions such as: what enables ensembles of molecules to organize themselves into cells? How do ensembles of cells create tissues and whole organisms? Key to solving these problems is the design and analysis of appropriate mathematical models for spatio-temporal pattern formation. Early models of spatio-temporal pattern formation focused on the dynamics of diffusible morphogen signals and have been formulated as partial differential equations (e.g. [25]). Today, it is realized that, in addition to diffusible signals, the role of cells in morphogenesis can not be neglected. Living cells possess migration strategies that go far beyond the merely random displacements of non-living molecules (diffusion). More and more evidence has been collected how populations of interacting and migrating cells can in a self-organized manner contribute to the formation of order in a developing organism. It has been realized, that both the particular type of cell interaction and migration are crucial and suitable combinations allow for a wide range of patterns. The question is: What are appropriate mathematical models for analyzing organization principles of moving and interacting discrete cells? It has turned out that cellular automata (CA), in particular lattice-gas cellular automata (LGCA) can model the interplay of cells with themselves and their heterogeneous environment [15]. These models describe interaction at a cell-based (microscopic) scale. Cell-based models (for a review see [18]) are required if one is attempting to extract the organization

H. Hatzikirou (✉)
Center for Information Services and High Performance Computing, Technische Universität Dresden, Nöthnitzerstr. 46, 01069 Dresden, Germany
e-mail: haralambos.hatzikirou@tu-dresden.de

A.G. Hoekstra et al. (eds.), *Simulating Complex Systems by Cellular Automata*, Understanding Complex Systems, DOI 10.1007/978-3-642-12203-3_13,

principles of interacting cell systems down to length scales of the order of a cell diameter in order to link the individual (microscopic) cell dynamics with a particular collective (macroscopic) phenomenon.

Cellular automata (CA) are discrete dynamical systems. They were introduced by J. von Neumann and S. Ulam in the 1950s in an attempt to model biological self-reproduction [31]. Since then, it has become clear that CA have a much broader potential as models for physical, chemical and biological self-organization. In particular, CA models have been proposed for a large number of biological applications for studying the emergence of collective macroscopic behavior emerging from the microscopic interaction of individual components, such as molecules, cells or organisms [15]. However, currently there exists a huge jungle of different rules for often the same or similar processes (e.g. for random walk or proliferation). Therefore, there is need for a specification and classification of CA rules. Such a classification approach has comprehensively been performed for one-dimensional automata [33]. Furthermore, examples of successful analysis of CA models beyond purely visual inspection of simulation outcomes are still rare.

Here, we introduce lattice-gas cellular automata (LGCA) as models for collective behavior emerging from microscopic migration and interaction processes [15, 20]. LGCA represent a class of CA whose structure facilitates mathematical analysis. Implementing movement of individuals in traditional cellular automaton models is not straightforward, as one site in a lattice can typically only contain one individual, and consequently movement of individuals cause collisions when two individuals move to the same empty site. In a lattice-gas model this problem is avoided by having separate channels for each direction of movement and imposing an exclusion principle. Furthermore, the update rule is split into two parts which are called interaction and propagation, respectively. The interaction rule of LGCA can be compared with the update rule for CA in that it assigns new states to each particle based on the states of the sites in a local neighborhood. After the interaction/collision step the state of each node is propagated to a neighboring node. This split of the update rule allows for transport of particles while keeping the rules simple. The emergent collective behavior, e.g. spatio-temporal pattern formation in a LGCA shows up in the macroscopic limit which can be derived from a theory of statistical mechanics on a lattice. In place of discrete particles, Lattice Boltzmann (LB) models deal with continuous distribution functions which interact locally and which propagate after collision to the next neighbor node. LB models can be interpreted as mean-field approximations of LGCA. LGCA and LB models have been originally introduced as models of fluid flow [20]. Meanwhile, LGCA and LB models have found numerous applications in physics, chemistry and more recently biology [13, 15, 17, 28, 32].

In particular, we present two examples for LGCA models. The first example focuses on the collective behavior of moving and proliferating cells which is characterized by the emergence of a traveling wavefront. We derive a macroscopic description and, by means of a cut-off mean-field analysis, we calculate the wavefront speed. This analysis enables us to estimate (macroscopic) cell population spreading based on established microscopic cell properties, such as cell motility and proliferation rate. The second example addresses the precise interplay of moving cells with their typically heterogeneous environment which is crucial for central biological processes as embryonic morphogenesis, wound healing, immune reactions or tumor growth. We introduce a LGCA model of cell migration in different biological environments. Then we analyze the emergent migration features of the cell population under specific environmental constraints.

13.2 Lattice-Gas Cellular Automata

We define a d-dimensional regular lattice $\mathcal{L} = L_1 \times \cdots \times L_d \subset \mathbb{Z}^d$, where $L_1, ..., L_d$ are the numbers of nodes in each lattice dimension. Here, we will refer to two-dimensional models ($d = 2$). Particles move on the discrete lattice with discrete velocities, i.e. they hop at discrete time steps $k \in \mathbb{N}$ from a given node to a neighboring one. A set of velocity channels $(\mathbf{r}, \mathbf{c}_i)$, $i = 1, \ldots, b$, is associated with each node $\mathbf{r} \in \mathcal{L} \subset \mathbb{Z}^d$ of the lattice. The parameter b is the *coordination number*, i.e. the number of velocity channels on a node which coincides with the number of nearest neighbors on a given lattice. In particular, the set of velocity channels for the square lattice as considered here, is represented by the two-dimensional channel velocity vectors $\mathbf{c}_1 = \begin{pmatrix} 1 \\ 0 \end{pmatrix}$, $\mathbf{c}_2 = \begin{pmatrix} 0 \\ 1 \end{pmatrix}$, $\mathbf{c}_3 = \begin{pmatrix} -1 \\ 0 \end{pmatrix}$, $\mathbf{c}_4 = \begin{pmatrix} 0 \\ -1 \end{pmatrix}$ (see Fig. 13.1). In addition, there is a variable number $\beta \in \mathbb{N}_0 = \mathbb{N} \cup \{0\}$ of rest channels (zero-velocity channels), $(\mathbf{r}, \mathbf{c}_i)$, $b < i \leq b + \beta$. Furthermore, an exclusion principle is imposed. This requires, that not more than one particle can be at the same node within the same channel. As a consequence, each node $\mathbf{r}$ can host up to $\tilde{b} = b + \beta$ particles, which are distributed in different channels $(\mathbf{r}, \mathbf{c}_i)$ with at most one particle per channel. Accordingly, node state $\eta(\mathbf{r})$ is given by

$$\eta(\mathbf{r}) := \big(\eta_1(\mathbf{r}), \ldots, \eta_{\tilde{b}}(\mathbf{r})\big),$$

where $\eta(\mathbf{r})$ is called *node configuration* and the quantities $\eta_i(\mathbf{r}) \in \{0, 1\}, i = 1, \ldots, \tilde{b}$ are called *occupation numbers*, which are Boolean variables that indicate the presence ($\eta_i(\mathbf{r}) = 1$) or absence ($\eta_i(\mathbf{r}) = 0$) of a particle in the respective channel $(\mathbf{r}, \mathbf{c}_i)$. Therefore, the set of elementary states $\mathcal{E}$ of a single node is given by

$$\mathcal{E} = \{0, 1\}^{\tilde{b}}.$$

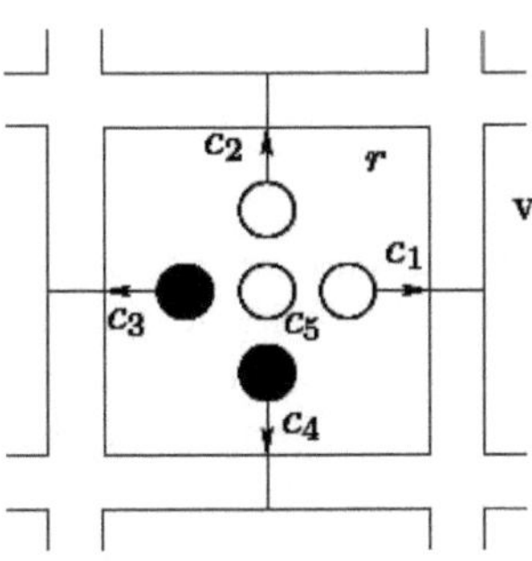

$$\begin{aligned}
\tilde{b} &= 5 \\
\text{velocity channels:} &\quad (r, c_1), (r, c_2), (r, c_3), (r, c_4) \\
\text{rest channel:} &\quad (r, c_5) \\
\eta(r) &= (\eta_1(r), \eta_2(r), \eta_3(r), \eta_4(r), \eta_5(r)) \\
&= (0, 0, 1, 1, 0) \\
\rho(\mathbf{r}) &= 2
\end{aligned}$$

Fig. 13.1 Node configuration: channels of node $\mathbf{r}$ in a two-dimensional square lattice ($b = 4$) with one rest channel ($\beta = 1$). *Filled dots* denote the presence of a particle in the respective channel

The *node density* is the total number of particles present at a node $\mathbf{r}$ and time $k \in \mathbb{N}$ denoted by

$$n(\mathbf{r}, k) := \sum_{i=1}^{\tilde{b}} \eta_i(\mathbf{r}, k).$$

For any node $\mathbf{r} \in \mathcal{L}$, the nearest lattice neighborhood $\mathcal{N}_b(\mathbf{r})$ is a finite list of neighboring nodes and is defined as

$$\mathcal{N}_b(\mathbf{r}) := \{\mathbf{r} + \mathbf{c}_i \; : \; \mathbf{c}_i \in \mathcal{N}_b \, , \; i = 1, \ldots, b\} \; .$$

Figure 13.1 gives an example of the representation of a node on a two-dimensional lattice with $b = 4$ and $\beta = 1$, i.e. $\tilde{b} = 5$.

13.2.1 Dynamics in Lattice-Gas Cellular Automata

The dynamics of a LGCA arises from the application of superpositions of local (probabilistic) *interaction* and deterministic *propagation* (transport) steps applied simultaneously to all lattice nodes and at each discrete time step. The definitions of these steps have to satisfy the exclusion principle, i.e. two or more particles are not allowed to occupy the same channel.

According to a model-specific *interaction* rule ($\mathcal{R}^{\mathrm{C}}$), particles can change channels (see Fig. 13.2) and/or are created or destroyed. The temporal evolution of a state $\eta(\mathbf{r}, k) \in \{0, 1\}^{\tilde{b}}$ in a LGCA is determined by the temporal evolution of the occupation numbers $\eta_i(\mathbf{r}, k)$ for each $i \in \{1, \ldots, \tilde{b}\}$ at node $\mathbf{r}$ and time k. Accordingly, the pre-interaction state $\eta_i(\mathbf{r}, k)$ is replaced by the post-interaction state $\eta_i^{\mathrm{C}}(\mathbf{r}, k)$ determined by

$$\eta_i^{\mathrm{C}}(\mathbf{r}, k) = \mathcal{R}_i^{\mathrm{C}}\big(\{\eta(\mathbf{r}, k) | \mathbf{r} \in \mathcal{N}_b(\mathbf{r})\}\big), \tag{13.1}$$

$$\eta^{\mathrm{C}}(\mathbf{r}, k) = \mathcal{R}^{\mathrm{C}}\big(\{\eta(\mathbf{r}, k) | \mathbf{r} \in \mathcal{N}_b(\mathbf{r})\}\big) = \Big(\mathcal{R}_i^{\mathrm{C}}\big(\{\eta(\mathbf{r}, k) | \mathbf{r} \in \mathcal{N}_b(\mathbf{r})\}\big)\Big)_{i=1}^{\tilde{b}},$$

realized with probability $\mathbb{P}\big(\eta \rightarrow \eta^{\mathrm{C}}\big)$ and $\eta^{\mathrm{C}} \in (0, 1)^{\tilde{b}}$, which is the time-independent probability for transition from the pre-interaction to the post-interaction node state.

In the deterministic *propagation* or streaming step (P), all particles are moved simultaneously to nodes in the direction of their velocity, i.e. a particle residing in channel $(\mathbf{r}, \mathbf{c}_i)$ at time k is moved to another channel $(\mathbf{r} + m\mathbf{c}_i, \mathbf{c}_i)$ during one time step (Fig. 13.3). Here, $m \in \mathbb{N}_0$ determines the *single particle speed* and $m\mathbf{c}_i$ the *translocation* of the particle. Because all particles residing at the same velocity channel move the same number m of lattice units, the exclusion principle is maintained. Particles occupying rest channels do not move since they have "zero

Fig. 13.2 Example of a possible interaction of particles at a node **r**; *filled dots* denote the presence of a particle in the respective channel. *Arrows* indicate channel directions

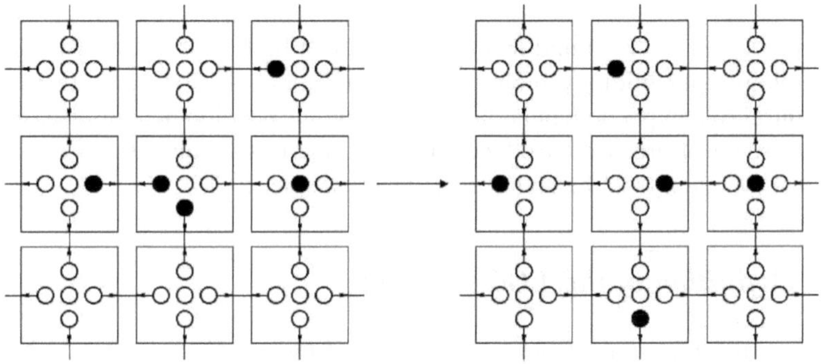

Fig. 13.3 Propagation in a two-dimensional square lattice with speed $m = 1$; lattice configurations before and after the propagation step; *filled dots* denote the presence of a particle in the respective channel

velocity". In terms of occupation numbers, the state of channel $(\mathbf{r} + m\mathbf{c}_i, \mathbf{c}_i)$ after propagation is given by

$$\eta_i(\mathbf{r} + m\mathbf{c}_i, k + \tau) = \eta_i^{\mathrm{P}}(\mathbf{r}, k), \tag{13.2}$$

where $\tau \in \mathbb{N}$ is the automaton's time-step. We note that the propagation operator is mass and momentum conserving. Hence, if only the propagation step was applied then particles would simply move along straight lines in directions corresponding to particle velocities.

Combining interactive dynamics (C), Eq. (13.1) with propagation (P), Eq. (13.2) implies that

$$\eta_i(\mathbf{r} + m\mathbf{c}_i, k + \tau) = \eta_i^{\mathrm{CP}}(\mathbf{r}, k) \,. \tag{13.3}$$

This can be rewritten as the *microdynamical difference equations*

$$\eta_i(\mathbf{r}+m\mathbf{c}_i, k+\tau) - \eta_i(\mathbf{r}, k) = \eta_i^{\mathrm{CP}}(\mathbf{r}, k) - \eta_i(\mathbf{r}, k) =: C_i(\eta_{\mathcal{N}(\mathbf{r})}(k)), \; i = 1, \ldots, \tilde{b}, \tag{13.4}$$

where we define C_i as the *change in the occupation number* due to interaction. It is given by

$$C_i(\eta_{\mathcal{N}(\mathbf{r})}(k)) = \begin{cases} 1, \text{ creation of a particle in channel } (\mathbf{r}, \mathbf{c}_i) \\ 0, \text{ no change in channel } (\mathbf{r}, \mathbf{c}_i) \\ -1, \text{ annihilation of a particle in channel } (\mathbf{r}, \mathbf{c}_i). \end{cases} \tag{13.5}$$

13.3 A LGCA Model for Growing Cell Populations

Growth processes can be found in almost any scientific field, such as physics, ecology, sociology, epidemiology, biology etc. In particular in biology, growth processes play a central role in phenomena related to embryonic development or diseases such as tumor growth. Here, we introduce a microscopic birth/death cell process which results in a traveling front behavior at the macroscopic level.

13.3.1 Definition of the LGCA Model

Automaton dynamics arise from the repetition of three rules (operators): Propagation (P), reorientation (O) and growth (R). In particular, cell motion is defined by the combination of the reorientation and the propagation operators while the growth operator controls the change of the local number of cells at a node.

The reorientation operator is responsible for the redistribution of cells within the velocity channels of a node, providing a new node velocity distribution (see Fig. 13.4). Here, we assume that individual cells perform random walks. The corresponding transition probabilities are

$$\mathbb{P}(\eta \to \eta^{\mathrm{O}})(\mathbf{r}, \cdot) = \frac{1}{Z}\delta\big(n(\mathbf{r}, \cdot), n^{\mathrm{O}}(\mathbf{r}, \cdot)\big), \tag{13.6}$$

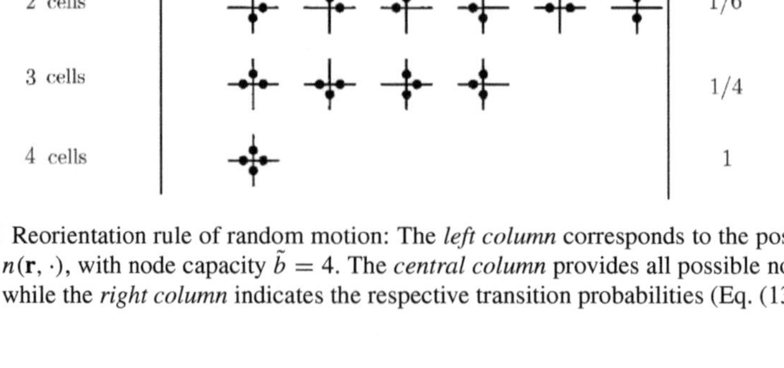

Fig. 13.4 Reorientation rule of random motion: The *left column* corresponds to the possible node densities $n(\mathbf{r}, \cdot)$, with node capacity $\tilde{b} = 4$. The *central column* provides all possible node configurations, while the *right column* indicates the respective transition probabilities (Eq. (13.6))

where the normalization factor $Z = \sum_{\eta^{O}(r,\cdot)} \delta\big(n(\mathbf{r},\cdot), n^{O}(\mathbf{r},\cdot)\big)$ corresponds to the equivalence class defined by the value of the pre-interaction node density $n(\mathbf{r},\cdot)$.

13.3.1.1 Growth (R)

We define a stochastic birth/death process for the cells as follows:

- **Birth**: We assume that the proliferation rule depends on the node capacity $\tilde{b}$, which is interpreted as a microscopic volume exclusion. For the creation of a new cell on a node, the existence of at least one cell and at least one free channel are required, i.e.:

$$\mathcal{R}_i(\mathbf{r},\cdot) = \xi_i(\mathbf{r},\cdot)(1 - \eta_i(\mathbf{r},\cdot)), \tag{13.7}$$

  where $\xi_i(\mathbf{r},\cdot)$'s are random Boolean variables, with $\sum_{i=1}^{\tilde{b}} \xi_i(\mathbf{r},\cdot) = 1$, and the corresponding probabilities are:

$$\mathbb{P}(\xi_i(\mathbf{r},\cdot) = 1) = r_M \frac{\sum_{i=1}^{\tilde{b}} \eta_i(\mathbf{r},\cdot)}{\tilde{b}}. \tag{13.8}$$

  Here, r_M is the probability of occupying a channel, if at least one cell exists on the node. The growth law, as defined above, is also known as *carrying capacity-limited* or *contact-inhibited* growth.
- **Death**: We assume that a certain nutrient availability implies a maximum node occupancy C, i.e. the node nutrient supply cannot support more than $C \leq \tilde{b}$ living cells. Thus, we define a death rate for each cell that ensures the existence of at most C cells per node:

$$r_d = \frac{\tilde{b} - C}{\tilde{b}} r_M, \tag{13.9}$$

  where the factor $\frac{\tilde{b}-C}{\tilde{b}}$ is a dimensionless quantity.

13.3.2 Microdynamical Equations

The above defined dynamics is fully specified by the following microdynamical equations:

$$\eta_i^{R}(\mathbf{r},k) = \eta_i(\mathbf{r},k) + \mathcal{R}_i(\mathbf{r},k), \tag{13.10}$$

$$\eta_i(\mathbf{r} + m\mathbf{c}_i, k + \tau) = \sum_{j=1}^{\tilde{b}} \mu_j(\mathbf{r},k)\eta_j^{R}(\mathbf{r},k). \tag{13.11}$$

Equation (13.10) refers to the application of the *growth* operator (R), which assigns a new occupation number for a given channel through a stochastic growth process. The second equation (13.11) refers to the *redistribution* of cells on the velocity channels and the *propagation* to the neighboring nodes, corresponding to the random walk as introduced in the previous chapter.

The $\mu_j(\mathbf{r}, k) \in \{0, 1\}$ are Boolean random variables which select only one of the $\tilde{b}$ terms of the rhs of Eq. (13.11). Therefore, they should satisfy the relation $\sum_{j=1}^{\tilde{b}} \mu_j(\mathbf{r}, k) = 1$. As stated above, we implement the random walk as a simple reshuffling of the cells within the node channels that leads to the probability of choosing a channel: $\langle \mu_j \rangle = 1/\tilde{b}$, for $j = 1, ..., \tilde{b}$. The terms $\mathcal{R}_i(\mathbf{r}, k) \in \{0, 1\}$, for $i = 0, \ldots \tilde{b}$ (Eq. (13.7)) represent birth/death processes, i.e. creation/annihilation of cells in channel i defined by the growth rule, which are applied to each channel independently.

13.3.3 Simulations

We have simulated our LGCA model on a two-dimensional 100×100 lattice for 150 time steps. In Fig. 13.5, we show simulations for different times, for fixed maximum

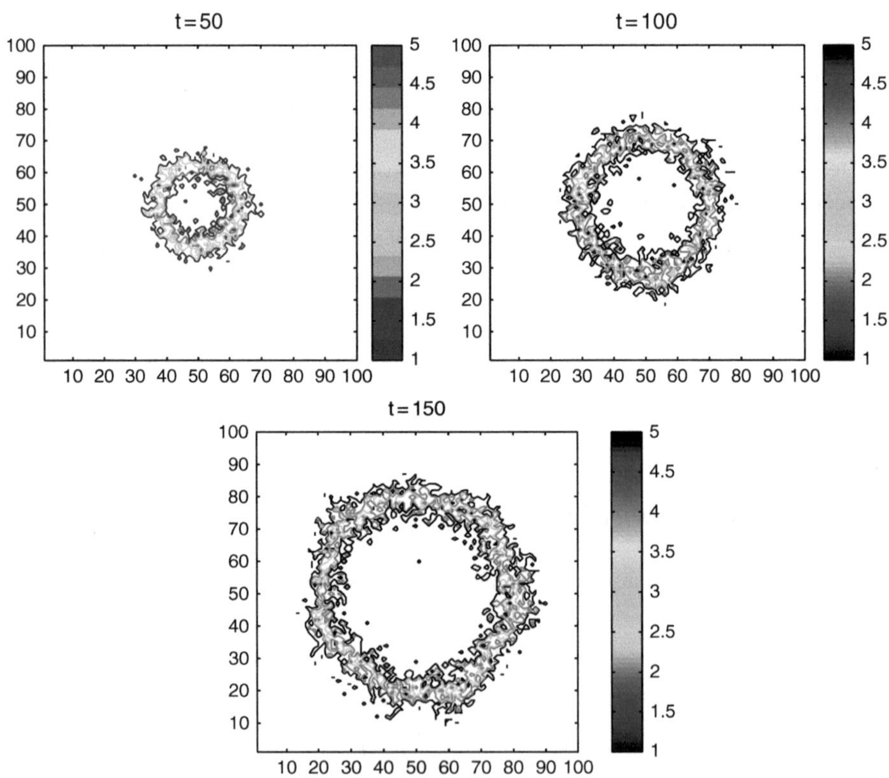

Fig. 13.5 Typical simulations of the spatio-temporal evolution of the LGCA growth process starting from an initial fully occupied cluster of nodes in the center of the lattice. The three figures show snapshots of the same simulation at different times. The different grey levels encode the node density

occupancy $C = \tilde{b}$ and for fixed proliferation rate $r_M = 0.01$. The initial condition is just a small disc. From the simulations, we conclude the following:

(01) The pattern evolving in simulations from a localized initial occupation is an isotropically growing disc.
(02) Furthermore, simulations indicate a moving front along which the occupancy of the initially empty nodes is increasing from zero particles to the maximum occupancy C.

In order to get further insight into the macroscopic behavior of the growth process, we use a different simulation setup. We consider a "tube", especially a 2000×10 lattice with periodic boundary condition on the L_2-axis, and a thin stripe of cells as initial condition (Fig. 13.6). A typical simulation time lasts for 2000 time steps. The result of our simulations is a propagating 2D traveling front along the L_1-axis, mimicking a "growing tube". This setting has the following advantages:

- One can project the system to one dimension by averaging the concentration profile along the L_2-axis, i.e. $n(r_x, k) = \frac{1}{|L_2|} \sum_{r_y \in |L_2|} n(\mathbf{r}, k)$.
- The front is well-defined as the mean position of the foremost cells.

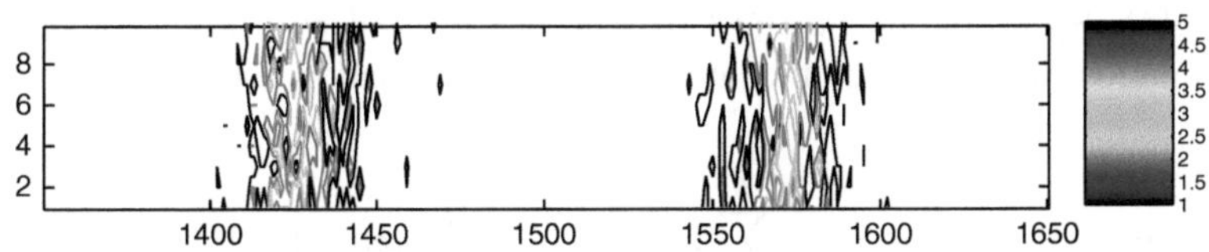

Fig. 13.6 Typical simulation on a "tubular" lattice, i.e. with periodic boundary condition along the y-axis. The different grey levels denote the node density. In the central region of the figure, the *white part* denotes nodes with maximum density

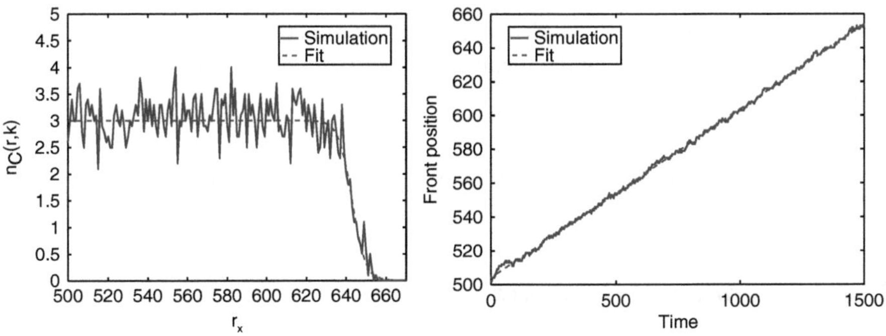

Fig. 13.7 *Left*: Snapshot of the average concentration profile along the L_1-axis, i.e. $n_x(k) = n(r_x, k) = \frac{1}{|L_2|} \sum_{r_y \in |L_2|} n(\mathbf{r}, k)$. Here, the maximum occupancy is considered as $C = 3$. *Right*: Linear growth of the front distance from its initial position, denoted as front position. The slope of the line defines the speed of the invasion

- The diffusive dynamics of the front relaxes faster than the discoidal 2D evolution.
- The front profile relaxes to an almost steady state shape, which moves almost uniformly along the L_1-axis.

The goal is to predict the front velocity. In the following section, we provide the details of the front analysis. Finally, we observe that the front evolves linearly in time, as shown in Fig. 13.7 (right).

13.4 Analysis

In this section, we analyze the behavior of our growth LGCA model. By means of a mean-field approximation, we derive a partial differential equation that describes the automaton's macroscopic behavior. Subsequently, we introduce a cut-off in the mean-field description and we calculate the speed of the invasive front.

13.4.1 Mean-Field Approximation

As seen above, our LGCA is governed by the microdynamical equations (13.10) and (13.11). By averaging Eqs. (13.10) and (13.11) and by using the mean-field approximation, we can obtain the lattice Boltzmann equation (LBE)

$$f_i(\mathbf{r}+m\mathbf{c}_i, k+\tau) - f_i(\mathbf{r}, k) = \sum_{j=1}^{\tilde{b}} \Omega_{ij} f_j(\mathbf{r}, k) + \sum_{j=1}^{\tilde{b}} (\delta_{ij} + \Omega_{ij}) \tilde{\mathcal{R}}_j(\mathbf{r}, k), \quad (13.12)$$

where the matrix $\Omega_{ij} = 1/\tilde{b} - \delta_{ij}$ is the transition matrix of the underlying shuffling process. Moreover, we assume that the mean-field reaction term is independent of the particle direction, i.e. $\tilde{\mathcal{R}}_i = F(\rho)/\tilde{b}$, where $F(\rho)$ is the mean-field cell reaction term for a single node. Using the mean-field approximation, we obtain the reaction term $\tilde{\mathcal{R}}_i$:

$$\tilde{\mathcal{R}}_i(\mathbf{r}, k) = r_M f_i(\mathbf{r}, k)\left(1 - \frac{r_D}{r_M} - f_i(\mathbf{r}, k)\right). \quad (13.13)$$

13.4.2 Macroscopic Dynamics

In order to derive a macroscopic description, we use the Chapman-Enskog methodology. Here, we assume diffusive scaling as

$$\mathbf{x} = \varepsilon \mathbf{r} \text{ and } t = \varepsilon^2 k, \quad (13.14)$$

where $(\mathbf{x}, t)$ are the continuous variables as $\varepsilon \to 0$. Using the spatio-temporal scaling relation Eq. (13.14) and replacing the first part of Eq. (13.12) by its Taylor expansion leads to:

$$f_i(\mathbf{r} + m\mathbf{c}_i, k + \tau) - f_i(\mathbf{r}, k) = \big(\varepsilon^2 \tau \partial_t + \varepsilon^4 \frac{\tau^2}{2} \partial_{tt} + \varepsilon m (\mathbf{c}_i \cdot \nabla) + \varepsilon^2 \frac{m^2}{2} (\mathbf{c}_i \cdot \nabla)^2 + \varepsilon^3 \tau m \partial_t (\mathbf{c}_i \cdot \nabla)\big) f_i(\mathbf{r}, k). \tag{13.15}$$

Furthermore, we assume an asymptotic expansion of f_i:

$$f_i = f_i^{(0)} + \varepsilon f_i^{(1)} + \varepsilon^2 f_i^{(2)} + \mathcal{O}(\varepsilon^3). \tag{13.16}$$

An important aspect is the scaling of the growth term. We argue that the birth of cells is taking place at a much slower time scale than the motion. The idea is that growth can be considered as a perturbation of cell motion. That means that the dominant process is random cell motion (as it is shown below). The growth rate is assumed to be scaled according to the macroscopic time scaling, i.e.

$$\bar{\mathcal{R}}_i \to \varepsilon^2 \tilde{\mathcal{R}}_i. \tag{13.17}$$

Equation (13.17) implies that the macroscopic rate should be scaled as $r_M = \varepsilon^2 \tilde{r}_M \ll 1$, where $\tilde{r}_M = \mathcal{O}(1)$. Therefore, our approximation is valid only for very low growth rates.

Collecting the equal $\mathcal{O}(\varepsilon)$ terms, we can formally derive a spatio-temporal mean-field macroscopic approximation (for detail see [13]):

$$\partial_t \rho = \frac{m^2}{\tilde{b}\tau} \nabla^2 \rho + \frac{1}{\tau} F(\rho), \tag{13.18}$$

where the term $F(\rho(\mathbf{r}, k)) = \sum_i^{\tilde{b}} \tilde{\mathcal{R}}_i(\mathbf{r}, k)$ is the macroscopic reaction law and using the definitions (13.7) and (13.9) we obtain:

$$F(\rho) = r_M \rho (C - \rho), \tag{13.19}$$

Accordingly, Eq. (13.19) is a kind of *Fisher-Kolmogorov* equation.

13.4.2.1 Cut-off Mean-Field Approximation

The spatio-temporal mean-field approximation (13.18) agrees qualitatively with the system's linearized macroscopic dynamics. However, it fails to provide satisfactory quantitative predictions because it neglects the correlations arising from the local fluctuating dynamics. Studies on chemical fronts have shown that these fluctuations may significantly affect the propagation velocity of the wave front [7, 30].

In order to improve the mean-field approximation (here we characterize it as "naive"), we introduce the *cut-off mean-field approach* [9, 14]. The idea is that the mean-field continuous equation (13.18) fails to describe the behavior of individual cells due to their strong fluctuations at the tip of the front [7]. Therefore, we introduce the cut-off continuous approach which describes the system up to a threshold density δ of the order of magnitude of one cell, i.e. $\delta \sim \mathcal{O}(1/\tilde{b})$. Let's assume that the full non-linear reactive dynamics can be described by a term $F(\rho)$. Then, the fully non-linear cut-off MF equation reads

$$\partial_t \rho = D\nabla^2 \rho + F(\rho)\Theta(\rho - \delta), \tag{13.20}$$

where $\Theta(\cdot)$ is a Heaviside function. Obviously, if we set $\delta = 0$ then the cut-off PDE will coincide with the naive mean-field approximation.

The cut-off macroscopic description (13.20) adds an extra fixed point, i.e. $\rho(x_i) = \{0, \delta, C\}$, $i = 0, \delta, C$ which breaks the front into three well-defined regions (see Fig. 13.8).

In order to characterize the linearized growth dynamics at the front, we modify the LBE for the cells:

$$\begin{aligned} f_i(\mathbf{r}+\mathbf{c}_i, k+1) - f_i(\mathbf{r}, k) &= \sum_{j=1}^{\tilde{b}} \left(\frac{1}{\tilde{b}} - \delta_{ij}\right) f_j(\mathbf{r}, k) \\ &\quad + \frac{1}{\tilde{b}} \sum_{j=1}^{\tilde{b}} \left[\langle \eta_j^{\mathrm{R}}(\mathbf{r}, k)\rangle - f_j(\mathbf{r}, k)\right]\Theta(\rho - \delta), \end{aligned} \tag{13.21}$$

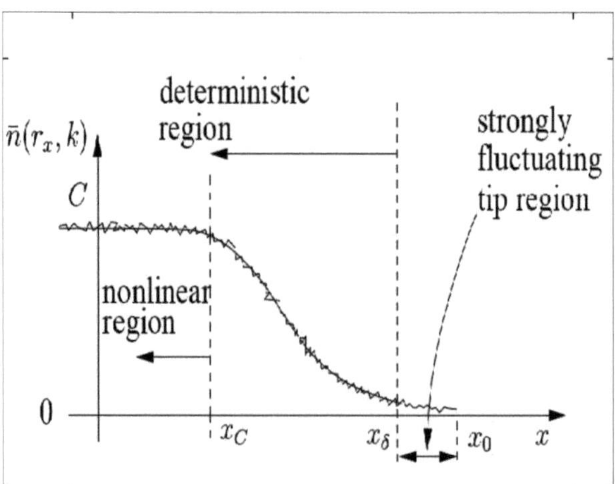

Fig. 13.8 A sketch of the wavefront as shown in Fig. 13.7 (*left*). We distinguish three regions: (i) $x \in [x_\delta, x_0]$, where $0 < \rho(x) < \delta$: this region represents a highly fluctuating zone, where the cells perform a random walk with almost no proliferation, (ii) $x \in [x_C, x_\delta]$, where $\delta < \rho(x) < C$: this region is a result of non-linear proliferation and cell diffusion and (iii) $x \in [0, x_C]$, where $\rho(x) \simeq C$: this regime represents the bulk of the front (saturated lattice) where no significant changes are observed

where the first summation of the rhs accounts for the reorientation dynamics and the second term is the reactive term of the LBE. Intuitively, the Θ function "cuts off" the reaction term for local densities lower than the threshold δ. Therefore, for $\rho < \delta$ the cells are influenced only by the random walk dynamics. Moreover from Eq. (13.21), we can easily deduce the nonlinear reaction term of Eq. (13.20):

$$F(\rho) = \sum_{j=1}^{\tilde{b}} \left[\langle \eta_j^{\mathrm{R}}(\mathbf{r}, k) \rangle - f_j(\mathbf{r}, k) \right]. \tag{13.22}$$

13.4.3 Traveling Front Analysis

In this subsection our goal is to analyze and characterize analytically the observed traveling front behavior. We assume that our system evolves in a "tube", as in Fig. 13.6. Moreover, we make the following assumptions:

(A1) the isotropic evolution of the system allows for the dimension reduction of the analysis to one dimension,
(A2) the system evolves for asymptotically long times, and
(A3) the initial front is sufficiently steep.

Under the assumptions (A1)–(A3), we can conclude that the front relaxes to a time invariant profile. Thus, assuming the translational invariance of the system along the front propagation axis L_1, we investigate the steady-state front solutions. The main observable is the average density profile along the axis L_1, i.e.

$$\rho(x,t) = \frac{1}{|L_2|} \int_0^{|L_2|} \rho(x,y,t)\, dy \, \in [0, \tilde{b}]. \tag{13.23}$$

Plugging the traveling front solution, $\rho(x,t) = U(x - vt)$, where $x \in L_1$ and v the front velocity into Eq. (13.18), we obtain:

$$DU'' + vU' + \left. \frac{d\tilde{F}}{dU} \right|_{U=0} = 0, \quad \lim_{\xi \to -\infty} u = U^{\max}, \quad \lim_{\xi \to +\infty} U = 0, U' < 0, \tag{13.24}$$

in terms of the comoving coordinate $\xi = x - vt$ and the prime denotes the derivative with respect to the variable ξ. The term $\tilde{F}$ represents the reaction terms in the naive MF approximation expressed in terms of U. The front speed for the naive MF can be calculated following the classical methodology [5, 26], i.e.

$$v_{\mathrm{n}} = 2\sqrt{Dr_m}. \tag{13.25}$$

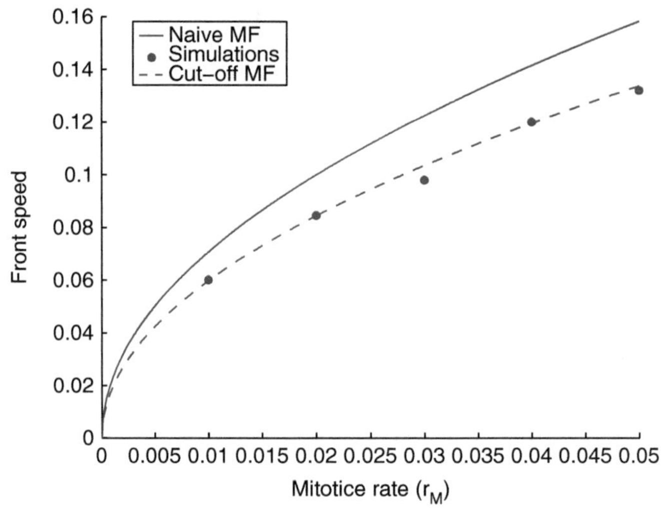

Fig. 13.9 Comparison of the calculated front speed for the naive and the cut-off MF, i.e. v_n and v_c respectively, against simulations. We observe that the cut-off MF predicts closely the front speed calculated from the simulations for $K \simeq 0.85$

The above speed estimation overestimates the actual front speed found in the simulations. In particular, it is the maximum asymptotic value that the discrete front speed can acquire [9] (see also Fig. 13.9).

The calculation of the front speed under the cut-off MF approximation is more challenging. Following the results proposed by Brunet et al. [9], we can obtain an estimate for the cut-off front speed

$$v_c = 2\sqrt{Dr_M}\left(1 - \frac{K}{\ln^2(\delta)}\right). \tag{13.26}$$

The cut-off front speed estimation includes a correction factor $1 - \frac{K}{\ln^2(\delta)}$, which allows for a better approximation of the actual front speed calculated from the LGCA simulations. The above equation provides a satisfactory description of the system up to the resolution of δ, i.e. to the order of one cell. A reasonable choice of the cut-off would be $\delta = 1/\tilde{b}$. The parameter K is fitted to match quantitatively the simulation results. Several studies have attempted to find an analytical estimate of K but till now this remains an open problem [10]. The cut-off mean-field approximation is a heuristic-phenomenological approach which mimics the leading-order effect of finite population number fluctuations by introducing a cut-off in the MF equation. In Fig. 13.9, we show a comparison of the front speed for varying proliferation rates r_M calculated by the naive MF and the cut-off MF against the front speed obtained from simulations. We observe that for an appropriate choice of K the cut-off MF predicts quantitatively the simulated front speed for all parameter values.

13.5 Modelling the Influence of the Microenvironment on Cell Migration

Active migration of tissue cells is essential for a number of biological processes such as inflammation, wound healing, embryogenesis and tumor cell metastasis [6]. Both in natural tissues and artificial environments, such as in vitro tissue cultures, cells can exhibit migratory behavior. In particular, the cellular microenvironment provides the substrate for cell migration. In the following, we provide more details about different cell migration strategies in various environments. Environmental heterogeneity contributes to the complexity of the resulting cellular behaviors. In particular, the cellular microenvironment can either enhance collective motion of cells or direct cell dispersion. Subsequently, we show how a suitable microscopical mathematical model (a LGCA) can contribute to understand the interplay of moving cells with their heterogeneous environment.

13.5.1 Cell Migration Strategies

The cellular microenvironment is a highly heterogeneous medium including the extracellular matrix (ECM) composed of fibrillar structures, collagen matrices, diffusible chemical signals as well as other mobile and immobile cells. Cells move within their environment by responding to their surrounding's stimuli. In addition, cells change their environment locally by producing or absorbing chemicals and/or by degrading the neighboring tissue. This feedback establishes a dynamic relationship between individual cells and the surrounding substrate.

One can distinguish two distinct strategies of cells responding to environmental stimuli: either the cells are following a certain direction and/or the environment imposes only an orientational preference. For example the graded spatial distribution of adhesion ligands along the ECM is thought to influence the direction of cell migration [24], a phenomenon known as haptotaxis [12]. Chemotaxis mediated by diffusible chemotactic signals provides a further example of directed cell motion in a dynamically changing environment. On the other hand, amoeboid and mesenchymal strategies imply an alignment of cells to fibrillar structures. Mesenchymal cells use additionally proteolysis to facilitate their movement and remodel the neighboring tissue (dynamic environment). Table 13.1 summarizes the different cell migration strategies.

Table 13.1 In this table, we relate the environmental effects to different cell migration strategies. One can distinguish static and dynamic environments. In addition, we identify environments that impart directional or only orientational information for migrating cells (see text for explanations)

	Static	Dynamic
Direction	Haptotaxis	Chemotaxis
Orientation	Amoeboid	Mesenchymal

13.5.2 LGCA Models of Cell Motion in a Static Environment

In this subsection, we define two LGCA models that describe cell motion in different environments. The mathematical entity that allows for the modeling of such environments is a *tensor field*, which is a collection of different tensors distributed over a spatial domain (for details see [22]). To model cell motion in a given tensor field (environment), we use a special kind of interaction rule for the LGCA dynamics, firstly introduced by Alexander et al. [1]. We consider biological cells as random walkers that are reoriented by maximizing a potential-like term. Assuming that the cell motion is affected by cell–cell and cell–environment interactions, we can define the potential as the sum of these two interactions.:

$$G(\mathbf{r},\cdot) = \sum_j G_j(\mathbf{r},\cdot) = G_{cc}(\mathbf{r},\cdot) + G_{ce}(\mathbf{r},\cdot), \tag{13.27}$$

where $G_j(\mathbf{r},\cdot)$, $j = cc, ce$ is the sub-potential that is related to cell–cell and cell–environment interactions, respectively.

Interaction rules are formulated in such a way that cells preferably reorient into directions which maximize (or minimize) the potential, that is according to the gradients of the potential $\mathbf{G}'(\mathbf{r},\cdot) = \nabla G(\mathbf{r},\cdot)$.

Consider a lattice-gas cellular automaton defined on a two-dimensional lattice with b velocity channels ($b = 4$ or $b = 6$). Let the *flux* be denoted by

$$\mathbf{J}(\eta(\mathbf{r},\cdot)) = \sum_{i=1}^{b} \mathbf{c}_i \eta_i(\mathbf{r},\cdot).$$

The probability that η^C is the outcome of an interaction at node $\mathbf{r}$ is defined by

$$\mathbb{P}(\eta \to \eta^{\mathrm{C}} | G(\mathbf{r},\cdot)) = \frac{1}{Z} \exp\left[\alpha F\big(\mathbf{G}'(\mathbf{r},\cdot), \mathbf{J}(\eta^{\mathrm{C}}(\mathbf{r},\cdot))\big)\right] \delta\big(n(\mathbf{r},\cdot), n^{\mathrm{C}}(\mathbf{r},\cdot)\big), \tag{13.28}$$

where η is the pre-interaction state at $\mathbf{r}$ and the Kronecker's δ assumes the mass conservation of this operator. The sensitivity is tuned by the positive, real parameter α. The normalization factor is given by

$$Z = Z(\eta(\mathbf{r},\cdot)) = \sum_{\eta^{\mathrm{C}} \in \mathcal{E}} \exp\left[\alpha F\big(\mathbf{G}'(\mathbf{r},\cdot), \mathbf{J}(\eta^{\mathrm{C}})\big)\right] \delta\big(n(\mathbf{r},\cdot), n^{\mathrm{C}}(\mathbf{r},\cdot)\big).$$

$F(\cdot)$ is a functional that defines the effect of the $\mathbf{G}'$ gradients on the new configuration. A common choice of $F(\cdot)$ is the inner product $< \cdot, \cdot >$, which favors (or penalizes) the configurations that tend to have the same (or inverse) direction of the gradient $\mathbf{G}'$. Accordingly, the dynamics is fully specified by the following microdynamical equation (for more details see the previous section)

$$\eta_i(\mathbf{r}+\mathbf{c}_i, k+1) = \eta_i^{\mathrm{C}}(\mathbf{r}, k).$$

In the following, we present two stochastic potential-based interaction rules that correspond to the motion of cells in a vector field (i.e. rank 1 tensor field) and a rank 2 tensor field, respectively. We exclude any other cell-cell interactions and we consider that the population consists of a fixed number of cells (mass break conservation).

13.5.3 Model I

This model describes cell motion in a static environment that carries directional information expressed by a vector field **E**. Biologically relevant examples are the motion of cells that respond to fixed integrin[1] concentrations along the ECM (haptotaxis). The spatial concentration differences of integrin proteins constitute a gradient field that creates a kind of "drift" **E** [16]. We choose a two dimensional LGCA without rest channels and the stochastic interaction rule of the automaton follows the definition of the potential-based rules (Eq. (13.27) with $\alpha = 1$):

$$\mathbb{P}(\eta \to \eta^{\mathrm{C}})(\mathbf{r}, \cdot) = \frac{1}{Z}\exp\big(\langle \mathbf{E}(\mathbf{r}), \mathbf{J}(\eta^{\mathrm{C}}(\mathbf{r}, \cdot))\rangle\big)\delta\big(n(\mathbf{r}, \cdot), n^{\mathrm{C}}(\mathbf{r}, \cdot)\big), \qquad (13.29)$$

where the vector field $\mathbf{G}'(\mathbf{r}) = \mathbf{E}(\mathbf{r})$ is independent of time, and the functional F is defined as:

$$F\big(\mathbf{G}'(\mathbf{r}), \mathbf{J}(\eta^{\mathrm{C}}(\mathbf{r}, \cdot))\big) = \langle \mathbf{E}(\mathbf{r}), \mathbf{J}(\eta^{\mathrm{C}}(\mathbf{r}, \cdot))\rangle. \qquad (13.30)$$

We simulate our LGCA for spatially homogeneous **E** for various intensities and directions. In Fig. 13.10, we observe the time evolution of a cell cluster under the influence of a given field. We see that the cells collectively move towards the gradient direction and they roughly keep the shape of the initial cluster. The simulations in Fig. 13.11 show the evolution of the system for different fields. It is evident that the "cells" follow the direction of the field and their speed responds positively to an increase of the field intensity.

13.5.4 Model II

We now focus on cell migration in environments that promote alignment (orientational changes). Examples of such motion are provided by neutrophil or leukocyte movement through the pores of the ECM, the motion of cells along fibrillar tissues

---

[1] Integrins are receptors that mediate attachment between a cell and the tissues surrounding it, which may be other cells or the extracellular matrix (ECM).

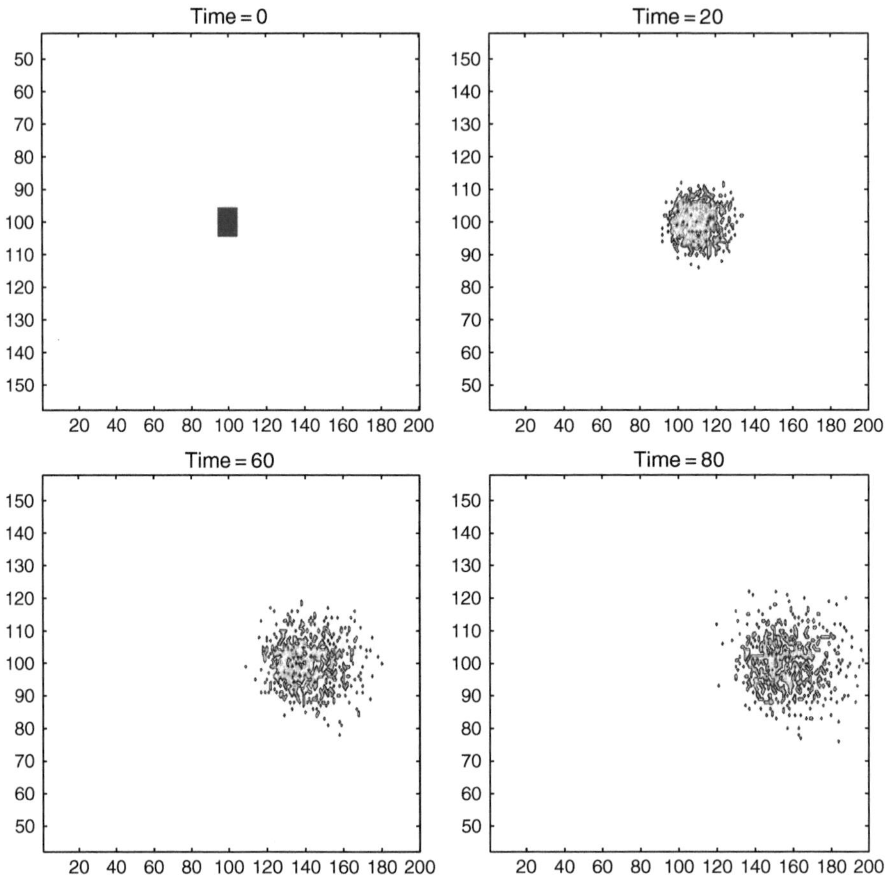

Fig. 13.10 Time evolution of a cell population under the effect of a field $\mathbf{E} = (1,0)$. One can observe that the environmental drive moves all the cells of the cluster into the direction of the vector field. Different grey levels represent different cell densities

or the motion of glioma cells along fiber tracts. Such an environment can be modeled by a second rank tensor field representing a spatial anisotropy along the tissue. In each point, a tensor (i.e. a matrix) informs the cells about the local orientation and strength of the anisotropy and proposes a principle (local) axis of movement. For instance, the brain's fibre tracts impose a spatial anisotropy and their strength of alignment affects the strength of anisotropy.

Here, we use the information of the principal eigenvector of the tensor (that encodes the environmental influence) which defines the local principle axis of cell movement. Thus, we end up again with a vector field but in this case we exploit only the orientational information of the vector. The new rule for cell movement in an "oriented environment" is:

$$\mathbb{P}(\eta \to \eta^{\mathrm{C}})(\mathbf{r}, \cdot) = \frac{1}{Z} \exp\left(\left|\langle \mathbf{E}(\mathbf{r}), \mathbf{J}(\eta^{\mathrm{C}}(\mathbf{r}, \cdot))\rangle\right|\right) \delta\left(n(\mathbf{r}, \cdot), n^{\mathrm{C}}(\mathbf{r}, \cdot)\right). \tag{13.31}$$

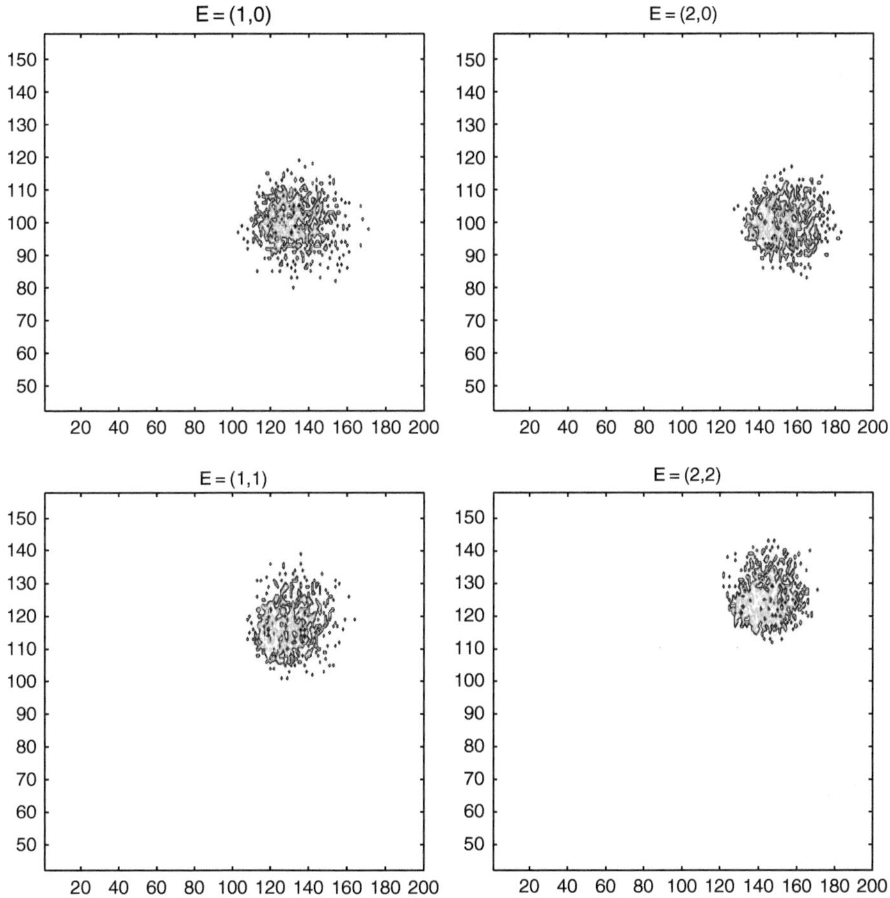

Fig. 13.11 Time evolution of the cell population under the influence of different fields (after 100 time steps). Increasing the strength of the field, we observe that the cell cluster is moving faster in the direction of the field. This behavior is characteristic of a haptotactically moving cell population. The initial condition is a small cluster of cells in the center of the lattice. Different grey levels indicate different cell densities (as in Fig. 13.10)

where the vector field $\mathbf{G}'(\mathbf{r}) = \mathbf{E}(\mathbf{r})$, is independent of time, and the functional F is defined as:

$$F\big(\mathbf{G}'(\mathbf{r}), \mathbf{J}(\eta^{\mathrm{C}}(\mathbf{r}, \cdot))\big) = \big|\langle \mathbf{E}(\mathbf{r}), \mathbf{J}(\eta^{\mathrm{C}}(\mathbf{r}, \cdot))\rangle\big|. \tag{13.32}$$

In Fig. 13.12, we show the time evolution of a simulation of model II for a given field. Figure 13.13 displays the typical resulting patterns for different choices of tensor fields. We observe that the anisotropy leads to the creation of an ellipsoidal

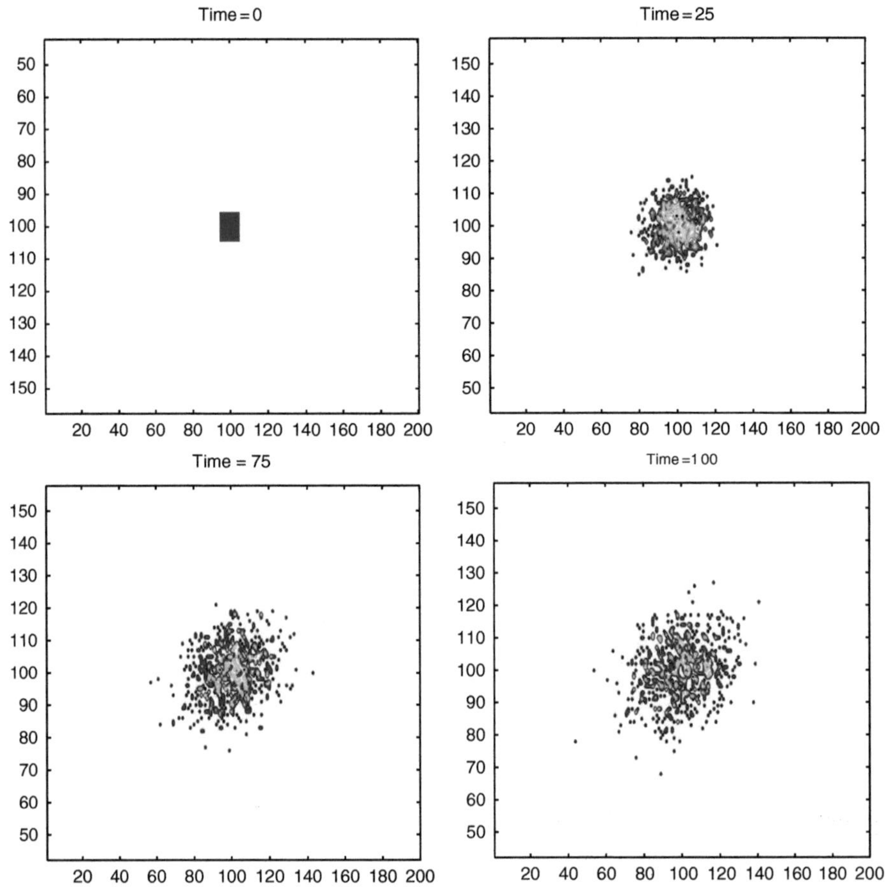

Fig. 13.12 Time evolution of a cell population under the effect of a tensor field with principal eigenvector (principal orientation axis) $\mathbf{E} = (2,2)$. We observe cell alignment along the orientation of the axis defined by E, as time evolves. Moreover, the initial rectangular shape of the cell cluster is transformed into an ellipsoidal pattern with principal axis along the field **E**. Different grey levels indicate different cell densities (as in Fig. 13.10)

pattern, where the length of the main ellipsoid's axis correlates positively with the anisotropy strength.

This rule can be used to model the migration of glioma cells within the brain. Glioma cells tend to spread faster along fiber tracts. Diffusion Tensor Imaging (DTI) is a Magnetic Resonance Imaging (MRI) based method that provides the local anisotropy information in terms of diffusion tensors. High anisotropy points belong to the brain's white matter, which consists of fiber tracks. A preprocessing of the diffusion tensor field allows the extraction of the principle eigenvectors of the diffusion tensors, that provides us with the local principle axis of motion. By considering a proliferative cell population, as in [21], and using the resulting eigenvector field we can model and simulate glioma cell invasion. In Fig. 13.14, we simulate an example

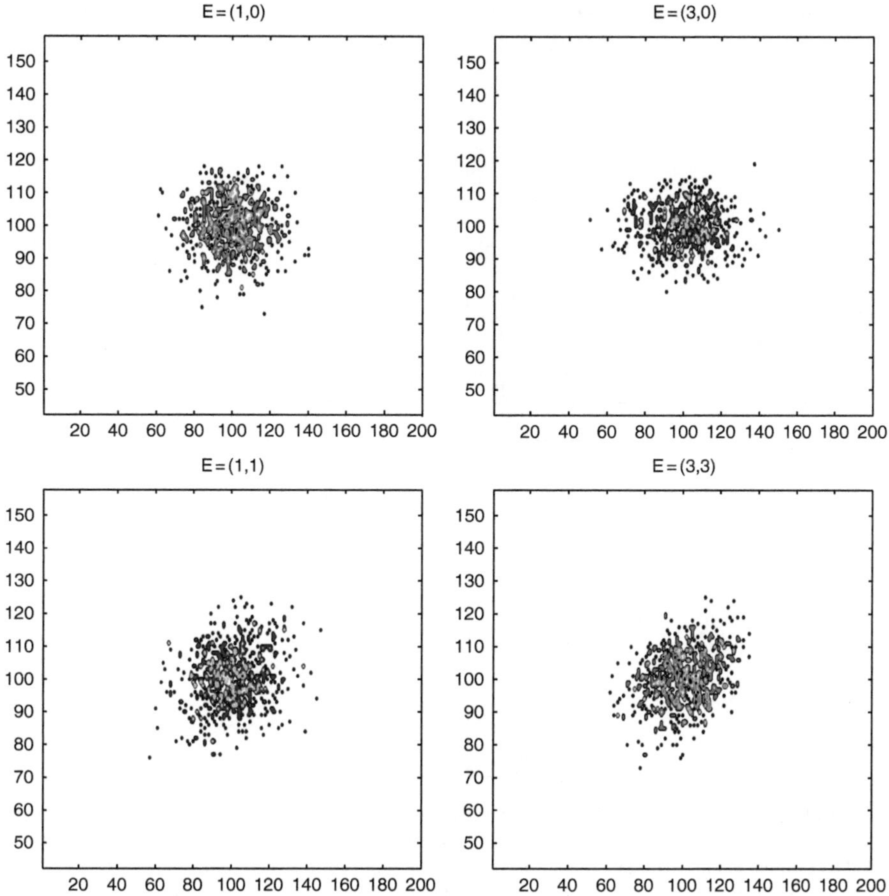

Fig. 13.13 Time evolution of the pattern for four different tensor fields (after 100 time steps). We observe the elongation of the ellipsoidal cell cluster when the field strength is increased. Above each figure the principal eigenvector of the tensor field is denoted. The initial conditions consist always of a small cluster of cells in the center of the lattice. Different grey levels indicate different cell densities (as in Fig. 13.10)

of brain tumor growth and show the effect of fiber tracts on tumor growth using the DTI information.

13.6 Analysis of the LGCA Models for Motion in Static Environments

In this section, we provide a theoretical analysis of the proposed LGCA models. Our aim is to calculate the equilibrium cell distribution and to estimate the speed of cell dispersion under different environments. Finally, we compare our theoretical results with the computer simulations.

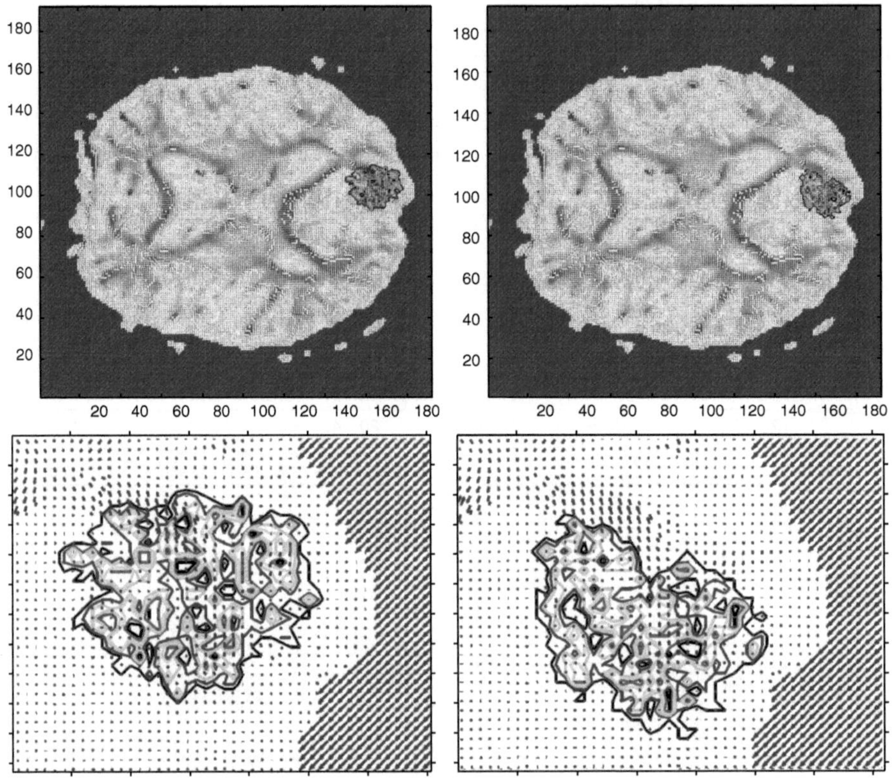

Fig. 13.14 The effect of the brain's fiber tracts on brain tumor growth: We use a LGCA model of a proliferating (glioma) cancer cell population (for definition see [21]) moving in a tensor field provided by clinical DTI (Diffusion Tensor Imaging) data, representing the brain's fiber tracts. *Top*: the *left figure* shows a simulation without any environmental bias of the cell motion (i.e. cells perform random walks). In the *top right* figure, DTI information is incorporated; the simulation exhibits the anisotropy of a brain tumor due to the effect of the fiber tracts. *Bottom*: Magnifications of the tumor region in the simulations above. Simulations indicate how environmental heterogeneities can affect cell migration and invasion

13.6.1 Model I

In this subsection, we analyze model I and we derive an estimate of the cell spreading speed in dependence of the environmental field strength. The first idea is to choose a macroscopically accessible observable that can be measured experimentally. A reasonable choice is the mean lattice flux $\langle \mathbf{J}(\eta^{\mathrm{C}}) \rangle_{\mathbf{E}}$, which characterizes the mean motion of the cells, with respect to changes of the field's strength $|\mathbf{E}|$:

$$\langle \mathbf{J}(\eta^{\mathrm{C}}) \rangle_{\mathbf{E}} = \sum_i \mathbf{c}_i f_i^{\mathrm{eq}}, \tag{13.33}$$

where f_i^{eq}, $i = 1, ..., b$ is the equilibrium density distribution of each channel which, in this case, depends on **E**. Mathematically, this is the mean flux *response* to changes of the external vector field **E**. The quantity that measures the linear response of the system to the environmental stimuli is called *susceptibility*:

$$\chi = \frac{\partial \langle \mathbf{J} \rangle_{\mathbf{E}}}{\partial \mathbf{E}}. \tag{13.34}$$

We expand the mean flux in terms of small fields as:

$$\langle \mathbf{J} \rangle_{\mathbf{E}} = \langle \mathbf{J} \rangle_{\mathbf{E}=\mathbf{0}} + \frac{\partial \langle \mathbf{J} \rangle_{\mathbf{E}}}{\partial \mathbf{E}} \mathbf{E} + O(\mathbf{E}^2). \tag{13.35}$$

For the zero-field case, the mean flux is zero since the cells are moving randomly within the medium (diffusion). Accordingly, for small fields $\mathbf{E} = \begin{pmatrix} e_1 \\ e_2 \end{pmatrix}$ the linear approximation reads

$$\langle \mathbf{J} \rangle_{\mathbf{E}} = \frac{\partial \langle \mathbf{J} \rangle_{\mathbf{E}}}{\partial \mathbf{E}} \mathbf{E}.$$

The *general linear response relation* is

$$\langle \mathbf{J}(\eta^{\mathrm{C}}) \rangle_{\mathbf{E}} = \chi_{\alpha\beta} e_\beta = \chi e_\alpha, \tag{13.36}$$

where the second rank tensor $\chi_{\alpha\beta}$ is assumed to be isotropic, i.e. $\chi_{\alpha\beta} = \chi \delta_{\alpha\beta}$. Note that we have used Einstein's notation for the sums (summation is implied for repetitive indices) and tensors.

The aim is to estimate the stationary mean flux for fields **E**. At first, we have to calculate the equilibrium distribution that depends on the external field. The external drive destroys the detailed balance (DB) conditions[2] that would lead to a Gibbs equilibrium distribution. In the case of non-zero external field, the system is out of equilibrium. The external field (environment) induces a breakdown of the spatial symmetry which leads to non-trivial equilibrium distributions depending on the details of the transition probabilities. The (Fermi) exclusion principle allows us to assume that the equilibrium distribution follows a kind of Fermi-Dirac distribution [20]:

[2] The detailed balance (DB) and the semi-detailed balance (SDB) impose the following condition for the microscopic transition probabilities: $\mathbb{P}(\eta \to \eta^{\mathrm{C}}) = \mathbb{P}(\eta^{\mathrm{C}} \to \eta)$ and $\forall \eta^{\mathrm{C}} \in \mathcal{E} : \sum_\eta \mathbb{P}(\eta \to \eta^{\mathrm{C}}) = 1$. Intuitively, the DB condition means that the system jumps to a new micro-configuration and comes back to the old one with the same probability (micro-reversibility). The relaxed SDB does not imply this symmetry. However, the SDB guarantees the existence of steady states and the sole dependence of the Gibbs steady state distribution on the invariants of the system (conserved quantities).

$$f_i^{\mathrm{eq}} = \frac{1}{1+e^{x(\mathbf{E})}}, \tag{13.37}$$

where $x(\mathbf{E})$ is a quantity that depends on the field $\mathbf{E}$ and the mass of the system (if the DB conditions were fulfilled, the argument of the exponential would depend only on the invariants of the system). Moreover, the sigmoidal form of Eq. (13.37) ensures the positivity of the probabilities $f_i^{\mathrm{eq}} \geq 0,\ \forall x(\mathbf{E}) \in \mathbb{R}$. Thus, one can write the following *ansatz*:

$$x(\mathbf{E}) = h_0 + h_1 \mathbf{c}_i \mathbf{E} + h_2 \mathbf{E}^2. \tag{13.38}$$

After some algebra (the details can be found in [22]), for small fields $\mathbf{E}$, one finds that the equilibrium distribution looks like:

$$f_i^{\mathrm{eq}} = d + d(d-1)h_1 \mathbf{c}_i \mathbf{E} + \frac{1}{2} d(d-1)(2d-1)h_1^2 \sum_\alpha c_{i\alpha}^2 e_\alpha^2 + d(d-1)h_2 \mathbf{E}^2, \tag{13.39}$$

where $d = \rho/b$ and $\rho = \sum_{i=1}^{b} f_i^{\mathrm{eq}}$ is the mean node density (which coincides with the macroscopic cell density) and the parameters h_1, h_2 have to be determined. Using the mass conservation condition, we find a relation between the two parameters:

$$h_2 = \frac{1-2d}{4} h_1^2. \tag{13.40}$$

Finally, the equilibrium distribution can be explicitly calculated for small driving fields:

$$f_i^{\mathrm{eq}} = d + d(d-1)h_1 \mathbf{c}_i \mathbf{E} + \frac{1}{2} d(d-1)(2d-1)h_1^2 Q_{\alpha\beta} e_\alpha e_\beta, \tag{13.41}$$

where $Q_{\alpha\beta} = c_{i\alpha} c_{i\beta} - \frac{1}{2}\delta_{\alpha\beta}$ is a second order tensor.

If we calculate the mean flux, using the equilibrium distribution up to first order terms of $\mathbf{E}$, we obtain from Eq. (13.33) the linear response relation:

$$\langle \mathbf{J}(\eta^C) \rangle = \sum_i c_{i\alpha} f_i^{\mathrm{eq}} = \frac{b}{2} d(d-1)h_1 \mathbf{E}. \tag{13.42}$$

Thus, the susceptibility reads:

$$\chi = \frac{1}{2} bd(d-1)h_1 = -\frac{1}{2} b g_{\mathrm{eq}} h_1, \tag{13.43}$$

where $g_{\mathrm{eq}} = f_i^{\mathrm{eq}}(1 - f_i^{\mathrm{eq}})$ is the equilibrium single particle fluctuation. In [11], the equilibrium distribution is directly calculated from the non-linear lattice Boltzmann equation corresponding to a LGCA with the same rule for small external fields. In

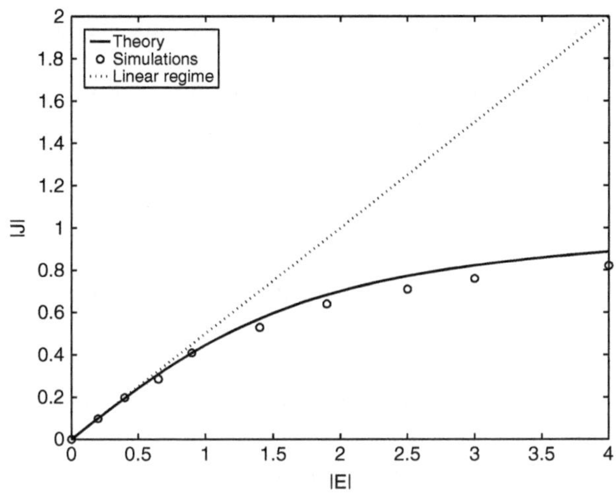

Fig. 13.15 This figure shows the variation of the normalized measure of the total lattice flux $|\mathbf{J}|$ against the field intensity $|\mathbf{E}|$, where $\mathbf{E} = (e_1, e_2)$. We compare the simulated values with the theoretical calculations (for the linear and non-linear theory). We observe that the linear theory predicts the flux strength for low field intensities. Using the full distribution, the theoretical flux is close to the simulated values also for larger field strengths

the same work, the corresponding susceptibility is determined and this result coincides with ours for $h_1 = -1$. Accordingly, we consider $h_1 = -1$ in the following.

Our method allows us to proceed beyond the linear case, since we have explicitly calculated the equilibrium distribution of our LGCA:

$$f_i^{\mathrm{eq}} = \frac{1}{1 + \exp\left(ln(\frac{1-d}{d}) - \mathbf{c}_i\mathbf{E} + \frac{1-2d}{4}\mathbf{E}^2\right)}. \tag{13.44}$$

Using the definition of the mean lattice flux Eq. (13.33), we can obtain a good theoretical estimation for larger values of the field. Figure 13.15 shows the behavior of the system's normalized flux obtained by simulations and a comparison with our theoretical findings. For small values of the field intensity $|\mathbf{E}|$ the linear approximation performs rather well and for larger values the agreement of our non-linear estimate with the simulated values is more than satisfactory. One observes that the flux response to large fields saturates. This is a biologically plausible result, since the cell speed is finite and an infinite increase of the field intensity cannot lead to infinite fluxes (the mean flux is proportional to the mean velocity). Experimental findings in systems of cell migration mediated by adhesion receptors, such as ECM integrins, support the model's behavior [27, 34].

13.6.2 Model II

In the following section, our analysis characterizes cell motion by a different measurable macroscopic variable and provides an estimate of the cell dispersion for model II. In this case, it is obvious that the average flux, defined in Eq. (13.33), is zero (due to the symmetry of the interaction rule). In order to measure the anisotropy, we introduce the flux difference between $\mathbf{v}_1$ and $\mathbf{v}_2$, where the $\mathbf{v}_i$'s are eigenvectors of the anisotropy matrix (they are linear combinations of the $\mathbf{c}_i$'s). For simplicity of the calculations, we consider $b = 4$ and X-Y anisotropy. We define:

$$|\langle \mathbf{J}_{\mathbf{v}_1} \rangle - \langle \mathbf{J}_{\mathbf{v}_2} \rangle| = |\langle \mathbf{J}_{x^+} \rangle - \langle \mathbf{J}_{y^+} \rangle| = |c_{11} f_1^{\text{eq}} - c_{22} f_2^{\text{eq}}|. \tag{13.45}$$

As before, we expand the equilibrium distribution around the field $\mathbf{E} = \mathbf{0}$ and we obtain equation

$$f_i = f_i(\mathbf{E} = \mathbf{0}) + (\nabla_{\mathbf{E}}) f_i \mathbf{E} + \frac{1}{2} \mathbf{E}^T (\nabla_{\mathbf{E}}^2) f_i \mathbf{E}. \tag{13.46}$$

With similar arguments as for the previous model I, we can assume that the equilibrium distribution follows a kind of Fermi-Dirac distribution (compare with Eq. (13.37)). This time our *ansatz* has the following form,

$$x(\mathbf{E}) = h_0 + h_1 |\mathbf{c}_i \mathbf{E}| + h_2 \mathbf{E}^2, \tag{13.47}$$

because the rule is symmetric under the inversion $\mathbf{c}_i \rightarrow -\mathbf{c}_i$. Conducting similar calculations as in the previous subsection, one can derive the following expression for the equilibrium distribution:

$$\begin{aligned} f_i^{\text{eq}} &= d + d(d-1) h_1 |\mathbf{c}_i \mathbf{E}| \\ &\quad + \frac{1}{2} d(d-1)(2d-1) h_1^2 \sum_\alpha c_{i\alpha}^2 e_\alpha^2 \\ &\quad + d(d-1)(2d-1) h_1^2 |c_{i\alpha} c_{i\beta}| e_\alpha e_\beta \\ &\quad + d(d-1) h_2 \mathbf{E}^2. \end{aligned} \tag{13.48}$$

Here, we identify a relation between h_1 and h_2 using the microscopic mass conservation law. To simplify the calculations we assume a square lattice (similar calculations can also be carried out for the hexagonal lattice case) and using $c_{11} = c_{22} = 1$, we derive the difference of fluxes along the X-Y axes (we restrict ourselves here to the linear approximation):

$$|f_1^{\text{eq}} - f_2^{\text{eq}}| = d(d-1) h_1 \left| \sum_\alpha |c_{1\alpha}| e_\alpha - \sum_\alpha |c_{2\alpha}| e_\alpha \right| = d(d-1) h_1 |e_1 - e_2|. \tag{13.49}$$

We observe that the parameter h_1 is still free and we should find a way to calculate it. Using a method similar to the work of [11] and we find that $h_1 = -1/2$. Substituting this value into the last relation and comparing with simulations (Fig. 13.16), we observe again a very good agreement between the linear approximation and the simulations.

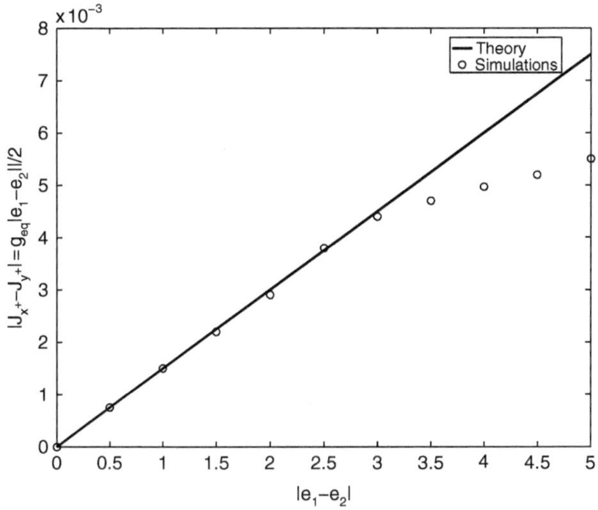

Fig. 13.16 The figure shows the variation of the X-Y flux difference against the anisotropy strength (according to Model II). We compare the simulated values with the linear theory and observe a good agreement for low anisotropy strength ($|e_1 - e_2| \leq 3$). We observe that the range of agreement, in the linear theory, is larger than in the case of model I

13.7 Discussion

In this chapter, we focus on the collective behavior emerging in interacting cell populations. The analysis of collective behavior of interacting cell systems is important for the understanding of phenomena such as morphogenesis, wound healing, tissue growth, tumor invasion etc. We are interested in finding appropriate mathematical models that allow for the description and the analysis of populations composed of discrete, interacting cells. Cellular automata, and particularly LGCA, provide a discrete modeling approach, where a micro-scale investigation is allowed through a stochastic description of the dynamics at the cellular level [15]. In this chapter, we have provided two examples of LGCA models: (i) the collective dynamics of a growing cell population and (ii) the macroscopic behavior of a cell population interacting with its microenvironment.

The first example addresses the collective behavior of a proliferating cell population. Simulations show that growing populations trigger a traveling invasion front, i.e. the growing cell population can be viewed as a wavefront that propagates into its surrounding environment. Via the cut-off mean-field analysis of the discrete LBE, we derive a reaction-diffusion equation that describes our system macroscopically. This cut-off reaction-diffusion equation enables us to calculate accurately the speed of the wavefront. We predict the front velocity to scale with the square root of the product of rates for mitosis and migration. This means that we are able to derive the expansion speed of growing cell populations by incorporating experimentally accessible parameters, as the mitotic and cell motility rates, respectively.

To study and analyze the effects of the microenvironment on cell migration, we have introduced a further LGCA model. We have identified and modeled the two main effects of static environments on cell migration:

- Model I addresses motion in an environment providing directional information. Such environments can be mediated by integrin (adhesive ECM molecules) density gradient fields or diffusible chemical signals leading to haptotactical or chemotactical movement, respectively. We have carried out simulations for different static fields, in order to understand the environmental effect on pattern formation. The main conclusion is that such an environment favors the collective motion of the cells in the direction of the gradients. Interestingly, we observe in Fig. 13.10 that the cell population approximately keeps the shape of the initial cluster and moves in the same direction. This suggests that collective motion is not necessary an alternative cell migration strategy, as described in [19]. Collective motion can be interpreted as emergent behavior in a population of amoeboidly moving cells in a directed environment. Finally, we have calculated theoretically an estimator of the cell spreading speed, i.e. the mean flux for variations of the gradient field strength. The results exhibit a positive response of the cell flux to an increasing field strength. The saturation of the response for large stimuli emphasizes the biological relevance of the model.
- Model II describes cell migration in an environment that influences the orientation of the cells (e.g. alignment). Fibrillar ECMs induce cell alignment and can be considered as an example of an environment that affects cell orientation. Simulations show that such motion produces alignment along a principal orientation (i.e. fiber) and the cells tend to disperse along it (Fig. 13.12). We have calculated the cell response to variations of the field strength, in terms of the flux difference between the principal axis of motion and its perpendicular axis. This difference gives us an estimate of the speed and the direction of cell dispersion. Finally, we observe a similar saturation plateau for large fields, as in model I. Moreover, we gave an application of the second model for the case of brain tumor growth using DTI data (Fig. 13.14).
- The microenvironment plays also a crucial role in the evolutionary dynamics (as a kind of selective pressure) of evolving cellular systems, in particular cancer [2–4].

In the above examples we have seen that LGCA provide an appropriate modeling framework for the analysis of emergent behavior since they allow for:

- The LGCA rules can mimic the microscopic processes at the cellular level (coarse-grained sub-cellular dynamics). Here we focused on the analysis of two selected microscopic interaction rules. Moreover, we showed that with the help of methods motivated by statistical mechanics, we can estimate the macroscopic behavior of the whole population (e.g. mean flux).
- Cell motion through heterogeneous media involves phenomena at various spatial and temporal scales. These cannot be captured in a purely macroscopic modeling approach. In macroscopic models of heterogeneous media diffusion is treated by

using powerful methods that homogenize the environment by the definition of an effective diffusion coefficient (the homogenization process can be perceived as an intelligent averaging of the environment in terms of diffusion coefficients). Continuous limits and effective descriptions require characteristic scales to be bounded and their validity lies far above these bounds [23]. In particular, it is found that in motion through heterogeneous media, anomalous diffusion (subdiffusion) describes the particles' movement over relevant experimental time scales, particularly if the environment is fractal [29]; existing macroscopic continuum equations can not describe such phenomena. On the other hand, discrete microscopic models, like LGCA, can capture different spatio-temporal scales and they are well-suited for simulating such phenomena.

- Moreover, the discrete structure of the LGCA facilitates the implementation of complicated environments (in the form of tensor fields) without any of the computational problems characterizing continuous models.
- LGCA are examples of parallel algorithms. This fact makes them computationally very efficient.

The mean-field (Boltzmann) equation characterizing a given LGCA model arises under the assumption that the probability of finding two cells at specific positions is given by the product of corresponding single particle distribution functions, i.e. any correlations are neglected and distributions fully factorize. It is a challenge to include two-, three-, etc. particle distribution functions which will allow a systematic study of correlation effects. This analysis could particularly improve our understanding of short and long time behavior. In particular, in the case of a traveling front expansion (see above) we have indicated the importance of such correlations at the tip of the front.

The need for discrete models, especially cellular automata, goes beyond the analysis of collective behavior in interacting cell populations. A discrete cell-oriented approach is also required if the dynamic system behavior depends on fluctuations at the individual cell level. This is, for example, the case at the front of invading tumors and crucial for the formation of metastases. Lately, experimental findings of Bru et al. [8] indicate that many tumors share the same surface dynamics. This finding motivated the analysis of the tumor interface by means of a fractal scaling analysis. Obviously, corresponding cancer models have also to be of a discrete nature and CA models are promising candidates to identify growth mechanisms that lead to a particular scaling.

Based on the variability in the local dynamics, an "interaction-module oriented" cellular automaton modeling provides an intuitive and powerful approach to capture essential aspects of complex phenomena at various scales [15]. In conclusion, there are both challenging future perspectives with regards to interesting biological applications of the lattice-gas cellular automaton idea and possible refinements of analytical tools for the investigation of lattice-gas cellular automata. The potential of cellular automata for modeling essential aspects of biological systems will be further exploited in the future.

References

1. F.J. Alexander, I.Edrei, P.L. Garrido, J.L. Lebowitz, Phase transitions in a probabilistic cellular automaton: growth kinetics and critical properties. J. Statist. Phys. **68**(3/4), 497–514, (1992)
2. A.R. Anderson, A.M. Weaver, P.T. Cummings, V. Quaranta, Tumor morphology and phenotypic evolution driven by selective pressure from the microenvironment. Cell **127**(5), 905–915, (2006)
3. D. Basanta, H. Hatzikirou, A. Deutsch, The emergence of invasiveness in tumours: A game theoretic approach. Eur. Phys. J. B **63**, 393–397, (2008)
4. D. Basanta, M. Simon, H. Hatzikirou, A. Deutsch, An evolutionary game theory perspective elucidates the role of glycolysis in tumour invasion. Cell Prolif. **41**, 980–987, (2008)
5. R.D. Benguria, M.C. Depassier, V. Mendez, Propagation of fronts of a reaction-convection-diffusion equation. Phys. Rev. E **69**, 031106, (2004)
6. D. Bray, *Cell Movements* (Garland Publishing, New York, 1992)
7. H.P. Breuer, W. Huber, F. Petruccione, Fluctuation effects on wave propagation in a reaction-diffusion process. Phys. D **73**, 259, (1994)
8. A. Bru, S. Albertos, J.L. Subiza, J. Lopez Garcia-Asenjo, I. Bru, The universal dynamics of tumor growth. Bioph. J. **85**, 2948–2961, (2003)
9. I. Brunet, B. Derrida Shift in the velocity of a front due to a cutoff. Phys. Rev. E **56**(3), 2597–2604, (1997)
10. I. Brunet, B. Derrida Effect of microscopic noise in front propagation. J. Stat. Phys. **103**(1/2), 269–282, (2001)
11. H. Bussemaker, Analysis of a pattern forming lattice gas automaton: Mean field theory and beyond. Phys. Rev. E **53**(4), 1644–1661, (1996)
12. S.B. Carter, Principles of cell motility: the direction of cell movement and cancer invasion. Nature **208**(5016), 1183–1187, (1965)
13. B. Chopard, M. Droz, *Cellular Automata Modeling of Physical Systems* (Cambridge University Press, Cambridge, 1998)
14. E. Cohen, D. Kessler, H. Levine, Fluctuation-regularized front propagation dynamics in reaction-diffusion systems. Phys. Rev. Lett. **94**, 158302, (2005)
15. A. Deutsch, S. Dormann, *Cellular Automaton Modeling of Biological Pattern Formation* (Birkhäuser, Basel 2005)
16. R.B. Dickinson, R.T. Tranquillo, A stochastic model for cell random motility and haptotaxis based on adhesion receptor fuctuations. J. Math. Biol. **31**, 563–600, (1993).
17. G.D. Doolen, *Lattice Gas Methods for Partial Differential Equations* (Addison-Wesley, New York, 1990)
18. D. Drasdo, S. Höhme, Individual-based approaches to birth and death in avascular tumors. Math. Comp. Model. **37**, 1163–1175, (2003)
19. P. Friedl, Prespecification and plasticity: shifting mechanisms of cell migration. Curr. Opin. Cell. Biol. **16**(1), 14–23, (2004)
20. U. Frisch, D. d'Humieres, B. Hasslacher, P. Lallemand, Y. Pomeau, J.P. Rivet, Lattice gas hydrodynamics in two and three dimensions. Compl. Syst. **1**, 649–707, (1987)
21. H. Hatzikirou, L. Brusch, C. Schaller, M. Simon, A. Deutsch, Prediction of traveling front behavior in a lattice-gas cellular automaton model for tumor invasion. Comput. Math. Appl. **59**, 2326–2339, (2010)
22. H. Hatzikirou, A. Deutsch, Cellular automata as microscopic models of cell migration in heterogeneous environments. Curr. Top. Dev. Biol. **81**, 401–434, (2008)
23. A. Lesne, Discrete vs continuous controversy in physics. Math. Struct. Comp. Sc. **17**(2), 185–223, (2007)
24. J.B. McCarthy, L.T. Furcht, Laminin and fibronectin promote the haptotactic migration of b16 mouse melanoma cells. J. Cell Biol. **98**(4), 1474–1480, (1984)
25. H. Meinhardt, *Models of Biological Pattern Formation* (Academic New York, 1982)
26. J. Murray, *Mathematical Biology I: An Introduction* (Springer, Heidelberg 2001)

27. S.P. Palecek, J.C. Loftus, M.H. Ginsberg, D.A. Lauffenburger, A. F. Horwitz, Integrin-ligand binding governs cell-substratum adhesiveness. Nature **388**(6638), 210, (1997)
28. D.H. Rothman, S. Zaleski, Lattice-gas models of phase separation: interfaces, phase transitions, and multiphase flow. Rev. Mod. Phys. **66**(4), 1417–1479, (1994)
29. M. Saxton, Anomalous diffusion due to obstacles: a Monte Carlo study. Biophys. J. **66**, 394–401, (1994)
30. M.V. Velikanov, R. Kapral, Fluctuation effects on quadratic autocatalysis fronts. J. Chem. Phys. **110**, 109–115, (1999)
31. J. von Neumann, *Theory of Self-Reproducing Automata* (University of Illinois Press, Urbana, IL, 1966)
32. D.A. Wolf-Gladrow, *Lattice-gas Cellular Automata and Lattice Boltzmann Models: An Introduction* (Springer, Heidelberg 2005)
33. S. Wolfram, *A New Kind of Science* (Wolfram Media, Inc., Champaign, IL 2002)
34. M.H. Zaman, P. Matsudaira, D.A. Lauffenburger, Understanding effects of matrix protease and matrix organization on directional persistence and translational speed in three-dimensional cell migration. Ann. Biomed. Eng. **35**(1), 91–100, (2006)

Chapter 14
Cellular Automata for Simultaneous Analysis and Optimal Structural Topology Design

Zafer Gürdal and Ramzi Zakhama

14.1 Introduction

The Cellular Automata (CA) paradigm has been finding more and more applications in engineering and sciences in the past decade, but nevertheless its use for engineering design has not been widely popular. The proposed chapter is a special and unique implementation of the paradigm for combined analysis and design of continuum structures made of isotropic and fiber reinforced orthotropic materials. In particular, the use of a computational approach for topology design of the structural domain together with its local field and design variables is discussed.

Topology design of load carrying engineering structures has been one of the areas that have gaining popularity in the design engineering community. Current state of the art design tools enable engineers to define the boundaries of solid domains, providing them with useful tools for preliminary design. However, such boundaries are still rather course, even for two-dimensional domains let alone three-dimensional parts, due to computational efficiency of such tools. Of course the need and the drive for future development of these tools is to achieve a high level of resolution in part details, and be able to address not only structural load carrying functionality during design but also address other features, be it response to other non-mechanical and multidisciplinary loads (wind loads and thermal, magnetic and electric fields) or manufacturability.

The lack of computational efficiency mentioned above can be attributed to two restrictive features of most currently used design tools. The first is the inherently serial nature of both the analysis and the design algorithms. Most available topology optimization algorithms do not benefit from the computational efficiency that can be achieved from massively parallel implementations. This is primarily due to the difficulties associated with parallelizing the analysis routines that need to be executed after design changes. Developers of the optimization tools typically rely on off-the-shelf finite element analyses that either cannot be parallelized due to restricts

Z. Gürdal (✉)
Faculty of Aerospace Engineering, Delft University of Technology, Delft, The Netherlands
e-mail: z.gurdal@tudelft.nl

A.G. Hoekstra et al. (eds.), *Simulating Complex Systems by Cellular Automata*, Understanding Complex Systems, DOI 10.1007/978-3-642-12203-3_14,

access to their solvers, or simply do not scale well in terms of parallelization. Moreover, the number of design variables increases as the domain is discretized into smaller and smaller parts to achieve higher design resolution. The immediate outcome of this restriction makes the design optimization extremely computationally demanding due to what is commonly referred to as the *curse of dimensionality*. The second limitations is the perceived notion that a design process is simply performing repetitive analyses and monitoring the performance of the design while maintaining constraint satisfaction during the process. Of course engineers currently have very fast analysis tools that can evaluate the performance of a design to a large degree of precision. Nevertheless, even though they are fast, it is probably not justified to perform an engineering analysis of a part to such high precisions if one is going to toss away the result as soon as the design is modified to improve it. It can be argued that an efficient engineering design requires its own computational paradigm that enables computation of topologies without the use of time consuming high precision analyses.

The growing interest in solving complex problems using CA has recently found its implementation in complex structural design problems. The paradigm appears to address most of the limitations described above based on features well described in the earlier chapters of this book. Kita and Toyoda [1] were among the first to use the cellular automata paradigm for solving topology optimization problems. They constructed CA design rules to obtain two-dimensional topologies based on an Evolutionary Structural Optimization (ESO) approach [2, 3]. In their approach the analysis of the structure however is performed using the Finite Element method.

Another pioneering work is attributed to Gürdal and Tatting [4] who used the CA paradigm to perform an integrated analysis and design. They solved the topology and sizing design of trusses that exhibit linear and geometrically nonlinear responses. The analysis rules are derived from local neighborhood equilibrium, while a simple design rule that is based on fully stressed design (a Stress Ratio (SR) method) [5] is used to size the truss members, in which the cross sectional areas of the members that connect the neighboring cells were the design variables.

The concept was later extended to preliminary implementation of the design of two-dimensional continuum structures by Tatting and Gürdal [6]. At the cell level the two-dimensional continuum is modeled by a truss layout that is equivalent to the continuum cell according to an energy criterion. The relationship between the thickness of the continuum structure and the cross sectional areas of the truss members is established by equating the strain energy of the continuum cell and that of the truss cell for given nodal displacements. The local analysis rules are again derived from the equilibrium condition of the cell, and a fully stressed material condition is selected to construct the local design rule. Numerical examples are carried out and results compared to an iterative Finite Element Analysis based design scheme that used GENESIS software to demonstrate the efficiency of the combined CA analysis and design.

Encouraged by the success of applying the CA paradigm to structural design, Abdalla and Gürdal [7] extended CA to the design of an Euler-Bernoulli column for minimum weight under buckling constraint, which is an eigenvalue problem. Global nature of eigenvalue problems, and its reduction to local analysis and design rules were the principal contribution of the work. The analysis rule is derived by minimization of the total potential energy in a cell neighborhood and the design rule is formulated as a local mini-optimization problem involving force resultants. The proposed CA algorithm is shown to converge correctly to an analytical optima for a number of classical test cases.

A more formed treatment of the problem of topology optimization of an elastic two-dimensional plate appeared first in Abdalla and Gürdal [8]. In their work, CA design rule is formulated for the first time using rigorous optimality criteria based on SIMP material [9–13] approach. The CA analysis rule was derived from the principle of minimal total potential energy. An extension of this work was made by Setoodeh et al. [14] to combine fiber angle and topology design of an anisotropic fiber reinforced laminae. Fibre angles and density measures at each cell of a domain are updated based on the optimality criteria for the minimum compliance. Topology optimization of 2-D elastic continuum structures subject to in plane loads and exhibiting geometric nonlinearities was performed by Zakhama et al. [15].

The cellular automata paradigm is also well known to be an inherently massively parallel algorithm. Slotta et al. [16] have implemented Gürdal and Tatting's [4] work using standard programming languages and parallelization libraries. The domain is decomposed into different groups of cells. Each group is assigned to a processor and the same local rules are applied for all the processors. Results demonstrate that the CA method is perfectly suited for parallel computation. Setoodeh et al. [17] proposed solving topology optimization for a continuum structure using a pipeline parallel implementation of cellular automata on distributed memory architecture. Numerical results show that the pipeline implementation converges successfully and generates optimal designs.

For the above mentioned structural analysis and design studies it has been observed that the CA convergence rate deteriorates considerably as the cell density is refined. This is due to the slow propagation of cell level field variables across the structural domain governed by elliptic partial differential equations. Additionally when a CA algorithm is implemented on a serial machine it looses its most attractive feature- parallelism [16, 17]. A methodology based on the Multigrid scheme can be used to accelerate the CA convergence process on serial machines. It has been demonstrated that the CA method takes advantage of the acceleration effect of multigrid schemes [18, 19]. The main idea in the multigrid concept is to use different discretization levels of cell grids, where the iterations of a classical iterative method on the finer grid are coupled with the iterations for the correction of the solution on the coarser grids. This concept is illustrated in depth by Wesseling [20].

Tovar et al. [21] have proposed another alternative to accelerate the CA convergence. The authors proposed a scheme based on Finite Element method to accelerate the analysis process followed by CA design rule. This type of strategy is often called

Hybrid Cellular Automata (HCA). In their paper, the CA local design rules are based on control theory, which minimizes the error between a local Strain Energy Density (SED) and the averaged SED value. More recently, Tovar et al. [22] have derived the CA local design rules based on optimality criteria interpreted as Kuhn-Tucker conditions of a multi-objective problem in addition to the control theory defined earlier in [21].

In the following basic elements of the definitions that are specific to topology design optimization of structural domains is described. In particular, sections on local rules that will ensure local equilibrium for analysis purposes and optimality for design purposes are described. Numerical implementation for the cellular automata paradigm for different type of structures are presented. Examples describing engineering applications are provided starting with two- and three-dimensional isotropic domains with local density design variables, followed by anisotropic medium in which local design variables in the form of fiber orientation angles is used in addition to the density variables.

14.2 Modeling for Structural Analysis and Design

For structural analysis and design local cell state will include physical and geometric properties of a solid domain. Principal response quantities (i.e., unknown field variables) of the analysis effort of a solid domain is typically the displacements. Local geometry of the cell are typically cross-sectional areas, cell densities, fiber-angle orientation, etc., which represent the design variables associated with the cell. In addition, local tractions applied to the cell and material properties of the solid may be needed for computations, and constitute part of the cell state, even if they may not be changing during the calculations.

A structural domain can be 1-,2- or 3-Dimensional, In the present chapter the trivial 1-D structures such as beams and columns [7, 18] are ignored, putting the emphasis on 2- and 3-D domains. Such domains can be of discrete nature, such as a truss type structure [4, 16, 23, 24], or a continuum type [6, 8, 14, 15, 17, 19, 25], such as plate and shell type structures. In the following basic description of the CA representation of these different kinds of structural domains are provided.

14.2.1 Truss Domain

Following the basic elements of the CA methodology described in Chap. 1, the cell representation of a simple 2-D discrete structural domain is a ground truss structure shown in Fig. 14.1a. In this representation, each cell is made up of eight truss members extending from the cell center at every 45° orientation. The Moore neighborhood with radius $r = 1$ is selected as shown in Fig. 14.1b. This neighborhood is composed of the eight adjacent cells which are marked by NW, N, NE, W, E, SW, S, and SE (see Fig. 14.1b) following the traditional compass representation

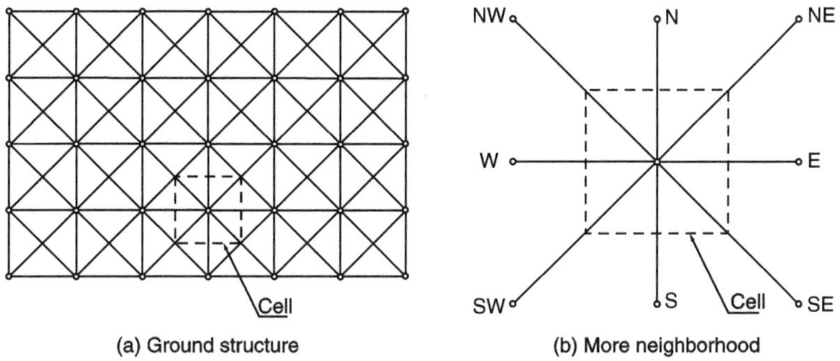

Fig. 14.1 CA ground structure and Moore neighborhood for trusses

of directions. For 2-D structural analysis the primary field variables are the nodal displacement components at truss junctions. The cell state also depends on the geometric and material properties of the truss members as well as the applied external loads at the truss nodes. Hence the cell state in the present application consists of two displacement components in mutually orthogonal directions, cross sectional areas of eight bars attached to the cell, and external loads in the two primary direction of the domain. In the present example the material type of the members is assumed to be fixed and kept outside the state of the cell that will change iteratively. Following the notation introduced in Chap. 1, the definition of the cell state at a given time iteration can be defined as

$$\Sigma(i) = \left\{(u_i, v_i), \left(f_i^x, f_i^y\right), \left(A_i^{\mathrm{NW}}, A_i^{\mathrm{N}}, A_i^{\mathrm{NE}}, A_i^{\mathrm{W}}, A_i^{\mathrm{E}}, A_i^{\mathrm{SW}}, A_i^{\mathrm{S}}, A_i^{\mathrm{SE}}\right)\right\}, \tag{14.1}$$

where u_i and v_i are the horizontal and vertical displacements, respectively. The reaction forces are denoted by f_i^x and f_i^y in the x and y directions respectively, and the member areas are represented by $A_i^{\mathrm{NW}}, A_i^{\mathrm{N}}, \ldots, A_i^{\mathrm{SE}}$.

The boundary condition mentioned in Chap. 1 is chosen as fixed for this example. To accomplish that, for cells at the boundary, cross sectional areas of the members which lie outside the structural domain are set to zero, which removes those truss members and provides a finite boundary for the truss ground structure. For the cell locations where the structure is physically restrained to prevent rigid body motion or restrained because of functional requirements, the appropriate displacement components are set to be zero and unchanging.

14.2.2 Isotropic Continuum Domain

In this section, the CA discretization of two and three dimensional structural domains is considered. The elastic continuum domain is discretized by a lattice of regular cells which are equally spaced in the x and y directions (see Fig. 14.2a), or

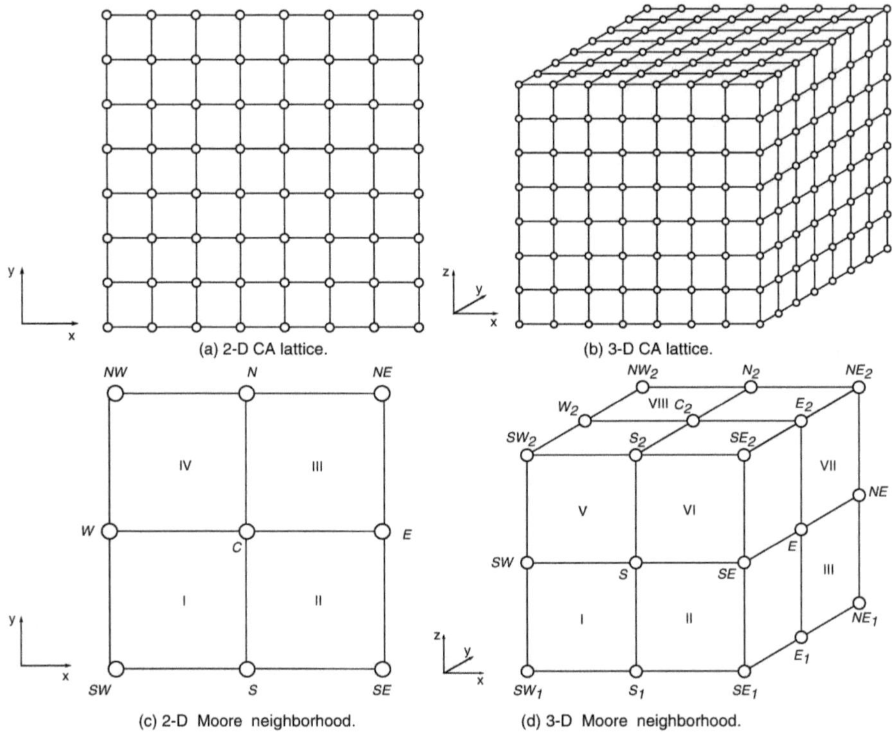

Fig. 14.2 CA lattices and Moore neighborhood for continuum structures

x, y and z for a three-dimensional structural domain (see Fig. 14.2b). Traditional Moore neighborhood is used to define the connectivity of the lattice as shown in Figs. 14.2c and 14.2d. In this case, the neighborhood includes the entire are, numbered by roman numerals, between the cell points again represented by the compass directions.

Each center cell C_i communicates with its neighbors by a local rule and its state is denoted as $\Sigma(i)^{(t)}$ where t is the iteration number. For isotropic continuum topology structures in two and three dimensions, the state of the ith cell is defined by

$$\Sigma(i) = \left\{ \left(u_i^{(1...m)}\right), \left(f_i^{(1...m)}\right), \rho_i \right\}, \tag{14.2}$$

where m corresponds to the dimensionality of the domain, with $m = 2$ or 3 for two or three dimensional domains, respectively. The components $\left(u_i^{(1...m)}\right)$ are the cell displacements in the directions $(1...m)$, and $\left(f_i^{(1...m)}\right)$ the external forces acting on the ith cell in the respective $(1...m)$ directions. Each cell of the discretized domain has its own density measure ρ_i at the node point independently of the densities of the elements numbered by roman numerals that define the neighborhood.

14.2.3 Composite Lamina Continuum Domain

A special case of a 2-D continuum is orthotropic fiber reinforced composite laminates in which the fibers provide stiffness and strength in preferred directions. Hence, determining the fiber orientation angle is an important part of the design; activity that is commonly referred to as *tailoring*. For combined topology and fiber-angle design, the basic CA elements of the isotropic continuum domain remain almost the same, however, the state of the ith cell is modified as follows:

$$\Sigma(i) = \left\{ \left(u_i^{(1...m)}\right), \left(f_i^{(1...m)}\right), (\rho_i, \theta_i), \overline{\mathbf{Q}}_i \right\}, \tag{14.3}$$

where θ_i is the fiber angle of the ith cell and $\overline{\mathbf{Q}}_i$ is the reduced transformed stiffness in which, due to symmetry, only the upper half diagonal of the matrix is stored.

14.3 Analysis Update Rule

14.3.1 Truss Structures

Local analysis rule is derived from the equilibrium condition of a cell with its neighbors. Within a cell, each truss member of the neighborhood structure ($k = 1, ..., 8$) has a Young modulus E, a length L_i^k before deformation, and a cross-sectional area A_i^k. The total potential energy associated the cell is the sum of the strain energy in each of the eight truss member of the neighborhood structure, as well as the potential energy of the external forces applied to the cell:

$$\Pi_i = \sum_{k=1}^{8} \frac{E\ A_i^k\ L_i^k\ (\varepsilon_i^k)^2}{2} - f_i^x\ u_i - f_i^y\ v_i, \tag{14.4}$$

where ε_i^k is the truss member strain which depends on the relative displacements of the neighboring cells. The strain is evaluated using the Green's strain definition for a truss members:

$$\varepsilon_i^k = \frac{(u_i - u_i^k)cos\theta^k - (v_i - v_i^k)sin\theta^k}{L_i^k}, \tag{14.5}$$

where (u_i^k, v_i^k) are the neighboring displacements, and θ^k is the orientation angle of the kth truss element member from the cell center.

Thus, the equilibrium equations are obtained by minimizing the total potential energy with respect to the cell displacements u_i and v_i:

$$\frac{\partial \Pi_i}{\partial u_i} = 0, \qquad \frac{\partial \Pi_i}{\partial v_i} = 0. \tag{14.6}$$

14.3.2 Isotropic Continuum Structures

The equilibrium of the neighborhood structure (see Figs. 14.2c,d) is again used to formulate the local analysis update rule. The total potential energy associated with a cell is the sum of the strain energy in each element of the neighborhood structure and the potential energy due to the external forces applied directly to the cell. Note that in this discretization scheme there will only be a few cells in the domain at which external forces are applied. Most cell equilibrium will only involve interaction of the local cells through the solid domain between them, which following the terminology of finite element analysis are referred to as the elements in this chapter:

$$\Pi_i = \sum_{k=1}^{N_{\text{element}}} U_i^k - \mathbf{f}_i \cdot \mathbf{u}_i, \tag{14.7}$$

where N_{element} is the number of elements surrounding a cell represented by the roman numerals in the figure, U_i^k is the strain energy for the kth element, $\mathbf{f}_i$ is the applied force vector and $\mathbf{u}_i$ is the displacement vector for all the cell's neighborhood including the cell itself.

The strain energy of an element is expressed in terms of the strain energy of the base material as follows:

$$U^k = \bar{\rho}^p \tilde{U}, \tag{14.8}$$

where

$$\tilde{U} = \frac{1}{2} \int_{\text{element}} \Gamma \cdot \overline{\mathbf{Q}} \cdot \Gamma \; dxdydz, \tag{14.9}$$

is the strain energy of the base material, Γ is the small-strain tensor, and $\overline{\mathbf{Q}}$ is the reduced in-plane stiffness matrix. The symbol p in the equation is called penalization parameter and is used for design purposes, its role will be explained later in the design rules.

The elements densities $\bar{\rho}$ are obtained by an average density interpolation [8] given by

$$\frac{1}{\bar{\rho}^p} = \frac{1}{N_{\text{cell}}} \sum_{i=1}^{N_{\text{cell}}} \frac{1}{\rho_i^p}, \tag{14.10}$$

where ρ_i's are the density measures of the cells surrounding the element, and N_{cell} is the number of cells defining the element. For the two-dimensional neighborhood structure $N_{\text{cell}} = 4$ and for the three-dimensional neighborhood structure $N_{\text{cell}} = 8$.

The density interpolation scheme in the previous equation is chosen such that any node with a density measure below a threshold value would *turn off* all four elements in which the node participates. Using this scheme checkerboard patterns are suppressed automatically during the optimization process.

Equilibrium equations are obtained by minimizing the total potential energy with respect to the cell displacements:

$$\min_{\mathbf{u}_C} \Pi_i. \tag{14.11}$$

The resulting equilibrium equations for each cell are written in a residual form:

$$\mathbf{R}_C(\mathbf{u}_C, \mathbf{u}_N) = \begin{Bmatrix} \mathbf{G}_C(\mathbf{u}_C, \mathbf{u}_N) \\ \mathbf{G}_N(\mathbf{u}_C, \mathbf{u}_N) \end{Bmatrix} + \begin{Bmatrix} \mathbf{f}_C \\ \mathbf{f}_N \end{Bmatrix} = 0, \tag{14.12}$$

where $\mathbf{u}_C$ and $\mathbf{u}_N$ are the displacement vectors of the cell and the neighborhood, respectively, $\mathbf{G}_C$ and $\mathbf{G}_N$ are the vectors of the internal forces, $\mathbf{f}_C$ and $\mathbf{f}_N$ are the vector of the applied forces relative to the cell and the vector of the internal forces relative to the neighborhood, respectively.

Differentiating the vector $\mathbf{R}_C$ with respect to the components of $\mathbf{u}_C$, the linear stiffness matrix can be written as

$$\mathbf{K} = -\frac{\partial \mathbf{R}_C}{\partial \mathbf{u}_C}(\mathbf{u}_C, \mathbf{u}_N). \tag{14.13}$$

The stiffness matrix $\mathbf{K}$ can also be expressed as the Hessian of the total potential energy:

$$\mathbf{K}_{pq} = \frac{\partial^2 \Pi_i}{\partial u_p \, \partial u_q}. \tag{14.14}$$

Thus, the cell displacements are updated as follows:

$$\mathbf{u}_C^{t+1} = \mathbf{u}_C^t + \Delta\mathbf{u}_C, \tag{14.15}$$

$$\Delta\mathbf{u}_C = (\mathbf{K}_C)^{-1} \cdot \left(\mathbf{G}_C(\mathbf{u}_N^{t+1}) + \mathbf{f}_C\right), \tag{14.16}$$

where $\mathbf{K}_C$ is a (2×2) or (3×3) cell stiffness matrix for two or three dimensional cases, respectively.

14.3.3 Composite Lamina Continuum Structures

When considering fiber-angle in the topology optimization problem the same formulation described above is used, with the only exception that now the fiber orientation is allowed to change from cell to cell. This changes the computation of the reduced in-plane stiffness, which is obtained as follows:

$$\overline{\mathbf{Q}} = \frac{1}{N_{\mathrm{cell}}} \sum_{i=1}^{N_{\mathrm{cell}}} \overline{\mathbf{Q}}_i, \tag{14.17}$$

where $\overline{\mathbf{Q}}_i$ is the in-plane transformed reduced stiffness of the four nodes of the element.

Thus, the analysis update rule is performed as described earlier using (14.15).

14.4 Design Update Rule

Structural analysis is based on a fairly well established principles and mathematical formulation that results in well know partial differential equations. Hence, the analysis rules described above are derived using the same principles. The design world on the other hand is much less restrictive, and there are variety of possibilities that one can implement design changes during an iterative scheme. The possibilities range from purely heuristic changes to, simple pattern matching, or to formal mathematical formulation. In the following, implementation of the design rules for the the three cases that we are discussing are presented.

14.4.1 Truss Structures

The design update rule in this case is derived from resizing the truss element members of the neighborhood structure based on full utilization of load carrying capability of the material. The cross sectional update formula is commonly referred to as the *fully stressed design* or *stress ratio* approach [5]. This scheme consists on computing a new cross sectional area $(A_i^k)^{(t+1)}$ which is based on the previous cross sectional area $(A_i^k)^{(t)}$ and the allowable stress σ_{all} chosen by the user as the maximum stress that the material can carry:

$$(A_i^k)^{(t+1)} = (A_i^k)^{(t)} \frac{E\ |\varepsilon_i^k|}{\sigma_{\mathrm{all}}}. \tag{14.18}$$

14.4.2 Continuum Structures

The structural topology design problem is posed according to the minimal compliance formulation. Its aim is to minimize the elastic strain energy of the structure, or equivalently maximize its total potential energy Π at equilibrium, subject to a limitation on the material volume. Thus, the design problem is written as

$$\min_{\rho} W_c(\rho, \mathbf{u}^*) \quad \text{or} \quad \max_{\rho} \Pi(\rho, \mathbf{u}^*), \tag{14.19}$$

under the constraints:

$$\mathbf{g}(\rho) \leq 0, \tag{14.20}$$

and the volume constraint:

$$\int_{\Omega} \rho \, d\Omega \leq \eta \cdot V_{\Omega}, \tag{14.21}$$

where ρ is the local density distribution of material which is chosen as the design variable, Ω is the prescribed design domain, $\mathbf{u}^*$ is the displacement vector at equilibrium, and $\mathbf{g}$ is a vector of local constraints which set bounds on the density distribution. The volume V of the structure is limited to an available fraction η of the total volume of the design material domain V_{Ω}. From the optimality conditions of the system level design problem (14.19), (14.20), and (14.21), local optimality conditions are derived which are associated with the cell level optimization problem. According to the specialization of the SIMP method, the local stiffness of the structure is expressed as a function of a fictitious local density distribution ρ. The local optimization problem takes on the form [8, 14]:

$$\min_{\rho} \frac{\Phi^*}{\rho^p} + \mu \, \rho, \tag{14.22}$$

$$\varepsilon \leq \rho \leq 1, \tag{14.23}$$

where

- $\varepsilon > 0$ is a very small number, set as a lower bound on ρ to avoid numerical instability that may result from structural discontinuities when zero density is allowed,
- $p \geq 1$ is a penalization parameter that is introduced in order to lead the design to a black or white topology, by assigning sufficiently high values to p, typically $p = 3$,
- $\Phi^* = \rho^p \, \hat{\Phi}$, is an approximately invariant local quantity, and $\hat{\Phi}$ is the complementary energy density,
- μ is the Lagrange multiplier associated with the global volume constraint (14.21). It is the only global quantity that is involved in this local problem. It serves in updating the material densities in the domain. It is updated at the global level by satisfying the total volume constraint [8, 14].

The update of each cell density of the continuum structure is obtained from the solution of this one-dimensional convex problem. The analytically solution [8, 14] of this local optimization problem is as follows:

$$\begin{cases} \hat{\rho} & \text{for} \quad \varepsilon < \hat{\rho} < 1 \\ \varepsilon & \text{for} \quad \hat{\rho} \leq \varepsilon, \\ 1 & \text{for} \quad \hat{\rho} \geq 1 \end{cases} \tag{14.24}$$

where

$$\hat{\rho} = \left(\frac{\Phi^*}{\bar{\mu}}\right)^{\frac{1}{1+p}}, \qquad \bar{\mu} = \frac{\mu}{p}. \tag{14.25}$$

The energy density Φ^* for each cell of the domain can be written as an average among the N_{element} elements of the Moore neighborhood structure:

$$\Phi^* = \frac{1}{n\,v} \sum_{i=1}^{N_{\text{element}}} \bar{\rho}_i^{2\,p}\, \tilde{U}_i, \tag{14.26}$$

where n is the number of non-shadow elements with nonzero density, v is the volume of a cell, which is $v = h^2$ or h^3 for two or three dimensional cases, respectively, and h is the distance between two immediate neighbor cells.

14.4.3 Composite Lamina Continuum Structures

By considering θ and ρ to be the design variables, we can convert the problem of combined topology and fibre-angle design to a local optimization problem through the general formulation (14.19), (14.20), and (14.21) as

$$\min_{\rho,\theta} \frac{\Phi(\theta)}{\rho^p} + \mu\,\rho, \tag{14.27}$$

$$\varepsilon \le \rho \le 1. \tag{14.28}$$

The value of $\Phi(\theta)$ is evaluated based on the current value of the cell density ρ and the strain vector Γ, and then used to update the local density through the solution of (14.27) and (14.28). Due to its special mathematical form, this local optimization problem can be easily split into two subproblems:
one for fibre-angle design,

$$\Phi^* = \min_{\theta} \Phi(\theta), \tag{14.29}$$

$$\Phi^* = \frac{1}{n\,v} \sum_{i=1}^{N_{\text{element}}} \bar{\rho}_i^{2\,p}\, \tilde{U}_i(\theta), \tag{14.30}$$

and the second one for topology,

$$\min_{\rho} \frac{\Phi^*}{\rho^p} + \mu\,\rho, \tag{14.31}$$

$$\varepsilon \le \rho \le 1. \tag{14.32}$$

It is well known that the optimal fiber-angle orientation for "shear weak" materials coincides with the principal stress direction [26, 27]. For "shear strong" materials there exists a closed form solution for which, depending on the principal strain ratio and material properties, the orientation might again coincide with principal stress direction or be different from the principal stress direction [26, 27].

In addition to the mathematical formulation of the fiber orientation angle update, it is also possible to use schemes that are less formal. For example, it is known that more than often incorporation of manufacturing requirements into mathematical formulations is not possible or will result in computationally expensive schemes that will be unaffordable. One such consideration is the continuity of the fiber orientation angle from one cell to another. In real life, fibers are continuous strands and abrupt change in orientation angle from one cell to another is not feasible. To account for such a requirement Setoodeh et al. [25] implemented a pattern matching technique, which re-updated the fiber orientation computed using the mathematical expressions with orientation angle patterns of the neighboring cells forming uniform orientation angle in the neighborhoods, with only well defined boundaries in the domain with different fiber orientation angles.

14.5 Cellular Automata Implementation Schemes

The update of the cells for trusses and continuum structures can be done simultaneously, which corresponds to the Jacobi scheme, as follows:

$$\Sigma(i)^{(t+1)} = \phi\left(\Sigma(i)^{(t)}, \Sigma(NM)^{(t)}\right), \tag{14.33}$$

or sequentially, which corresponds to Gauss-Seidel scheme:

$$\Sigma(i)^{(t+1)} = \phi\left(\Sigma(i)^{(t)}, \Sigma(M)^{(t+1)}, \Sigma(NM)^{(t)}\right), \tag{14.34}$$

where M is the set of neighboring cells whose states have been modified in the current iteration and NM is the set of remaining cells, which have not yet been modified.

The Gauss-Seidel method is used for the analysis update. For the design update, the Jacobi method is found to be the appropriate one to use to preserve the symmetry of the solution [8].

14.5.1 Truss Structures

The ground truss structure algorithm is based on the repeats of the analysis and design update rules for each cell of a domain. The algorithm starts from updating the displacement for a given structure until the norm of the force imbalance (residual) reaches a pre-specified tolerance ε_r. Then, the cross sectional areas are updated using the design update rule. The algorithm has deemed to converge when the structural design no longer changes.

14.5.2 Continuum Structures

For the topology of continuum structures, the analysis and design iterations are nested. A flowchart of the CA design algorithm is presented in Fig. 14.3. Starting from a structure with zero displacements and from densities set to volume fraction η, analysis updates are performed repeatedly until the norm of the force imbalance (residual) reaches a pre-specified tolerance ε_r. Next, the design is updated over the whole domain, then the volume constraint is checked. If the volume constraint is not satisfied, the Lagrange multipliers are updated and so is the design. The process continues until the relative difference between five successive compliance values is less than a pre-specified tolerance ε_c and the variation in cell densities is less than a tolerance ε_d.

From a computational perspective, the attractive feature of CA is its inherent parallelism. This feature appears to be particulary effective with regard to the analysis update. When it is not fully exploited, CA algorithms can be quite slow to converge. This is because communication between cells is limited only to immediate neighbors. The information from the cells where the loads are applied has to travel by neighbor-to-neighbor interaction throughout the domain. As the lattice is refined, the number of lattice updates needed to reach equilibrium significantly increases manifesting the deterioration in the rate of convergence alluded to above. Thus,

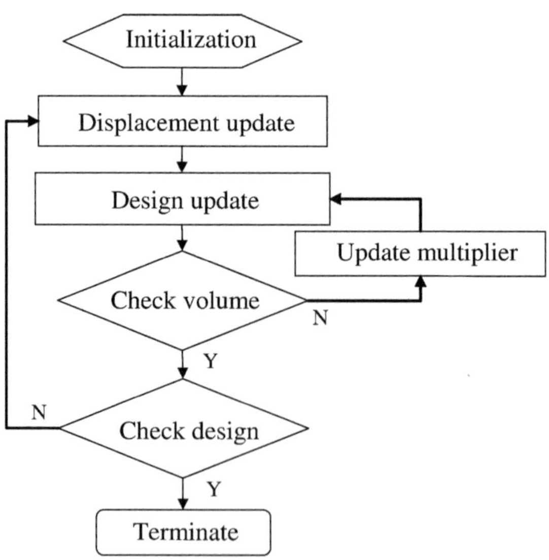

Fig. 14.3 CA continuum design algorithm

when CA is implemented on a serial machine it loses its most attractive feature as far as the analysis update is concerned. The design features of CA, though, remain effective.

An alternative methodology based on multigrid scheme [18–20] is used to accelerate the CA design algorithm on a serial machine. The multigrid acceleration scheme uses different discretization levels of grids. The CA iterations on the finest grid are coupled with the iterations of the correction solution on the coarse grids. The multigrid accelerated CA algorithm is demonstrated to be a powerful tool for solving topology optimization problems compared to other algorithms based on traditional finite element analysis [19]. The computational cost using this scheme is numerically found to be proportional to the number of cells.

14.6 Numerical Examples

In this section, some examples of topology optimization of continuum structures are considered to illustrate the robustness of the CA based combined analysis and design algorithm. As mentioned earlier the CA-based analysis is computationally expensive compared to an analysis using modern tools on a serial machine. Therefore, it is essential to implement CA in a parallel environment to exploit the true merits of a CA-based structural analysis and design. However, massively parallel computing machines that are most suited to this kind of computations are not as easily accessible as serial ones. To accelerate the convergence of CA iterations, two schemes are used in the present chapter. The first scheme is based on multigrid accelerated CA [19]. The second scheme is based on HCA, which uses a global finite element analysis instead of iterative updates of cell displacements followed by local update rules used for the design.

14.6.1 Example 1: 2-D Plate Topology Design

To demonstrate the performance and efficiency of the multigrid accelerated CA algorithm in solving the topology optimization problem, its results are compared with an existing method that is based on iterative finite element analysis solutions. Since the same CA design update rule is used in all tested algorithms, the comparison concerns design algorithms based on different analysis processes, namely the multigrid scheme and the commercial NASTRAN finite element code. The example studied is a symmetric cantilever (see Fig. 14.4) plate which is 1,000 mm long, 250 mm high, and 1 mm thick. The penalization parameter p is set to 3, the volume fraction is set to 0.5, the Poisson ratio is 0.4 and the Young modulus E is 1,000 N/mm^2. The tip load considered is $P = 100$ N acting at the center point of the free end of the cantilever.

Different discretization levels are used for the comparison; the results are generated for 11 grid levels, starting from the coarsest grid level of 9×3 cells, up to

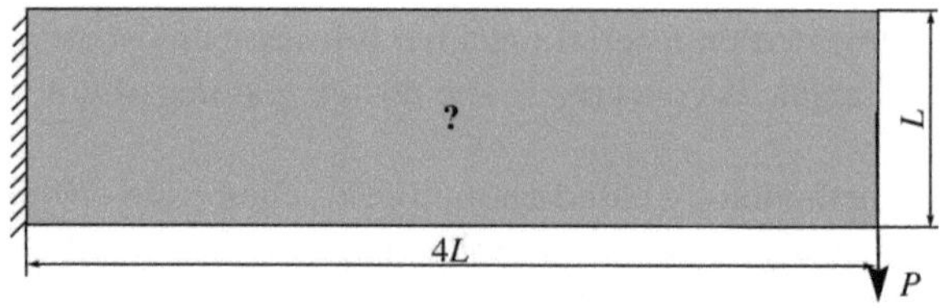

Fig. 14.4 Geometry and loading

the finest grid level of 4,097 × 1,025 cells. Convergence time for the HCA solution using the commercial NASTRAN code and for the multigrid accelerated CA algorithm are illustrated in Fig. 14.5. The vertical and horizontal axes represent the convergence time and the number of cells, respectively, on a log-log scale. First, it is observed that the commercial NASTRAN code showed a higher convergence time than the other algorithm. Moreover, the commercial NASTRAN code suffers lack of memory while running the grid level of 2,049 × 513 cells. On the contrary, the cellular automata paradigm can handle large problems because of its local nature which makes the storage of the global stiffness matrix unnecessary. The run time to convergence relative to the multigrid algorithm appears to be nearly proportional to the number of cells, which reveals a computational effort in the order of $O(N)$. As for the optimal topologies, it can be seen from Table 14.1 that those obtained by the multigrid algorithm and by the use of NASTRAN for analysis are practically the same with a slightly (0.005%–0.03%) but persistently lower compliance in the multigrid results.

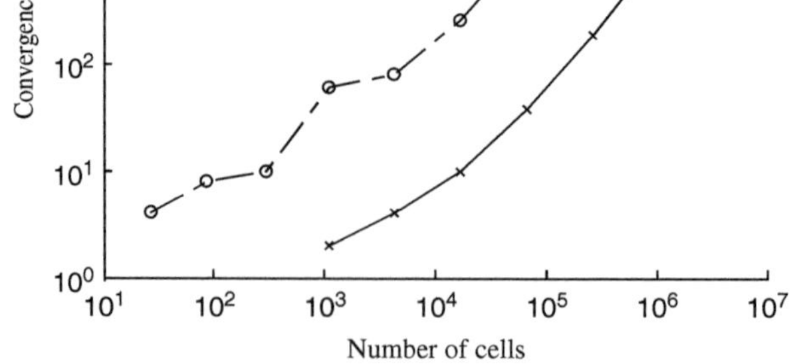

Fig. 14.5 Convergence time using NASTRAN and Multigrid accelerated CA

Table 14.1 Optimal topologies and compliances

Cell number	Optimal topology using NASTRAN	Optimal topology using Multigrid
129×33	 4,273.6	 4,258.7
257×65	 4,064.1	 4,062.7
513×129	 3,985.7	 3,984.2
1,025×257	 3,983	 3,980.9
2,049×513	 3,994	 3,992
4,097×1,025	Lack of memory	 3,998.4

14.6.2 Example 2: 2- and 3-D Compression Bridge

In this example, the objective is to find an optimal topology for a bridge which crosses a river and supports a uniformly distributed traffic loading. The design domain, the loading and the boundary conditions of the bridge problem are represented in Fig. 14.6. Requirements of waterway traffic underneath and road traffic on the bridge translate into the definition of imposed zones: empty (void) zones for the waterway and vehicle traffic through the bridge, and a dense (black) one for the deck and supports, as represented in Fig. 14.6. The design domain is discretized with 257 × 65 cells for the two-dimensional case and with 257 × 65 × 33 for the three-dimensional case including the empty zone. The penalization parameter p is set to 3, the volume fraction is set to 0.1 and the Poisson ratio to 0.3.

The final topology for the two-dimensional case performed by the multigrid design algorithm is represented in Fig. 14.7. It corresponds to a compression arch which holds a three span deck. The first and the third spans are cantilevers which

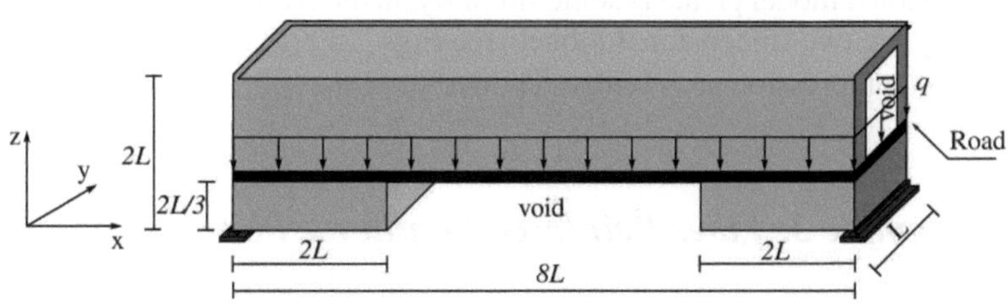

Fig. 14.6 Compression bridge domain

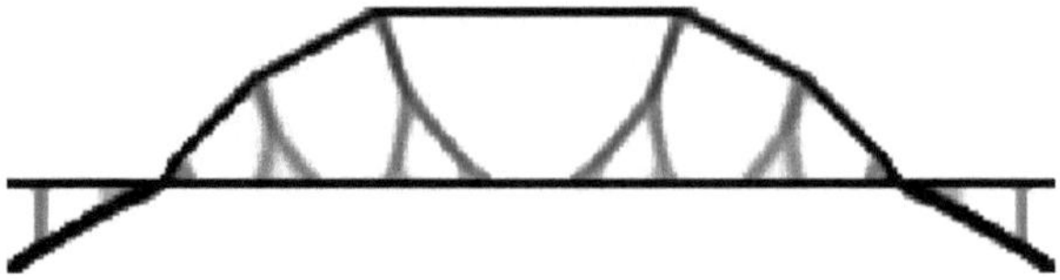

Fig. 14.7 Optimal 2-D topology of compression bridge

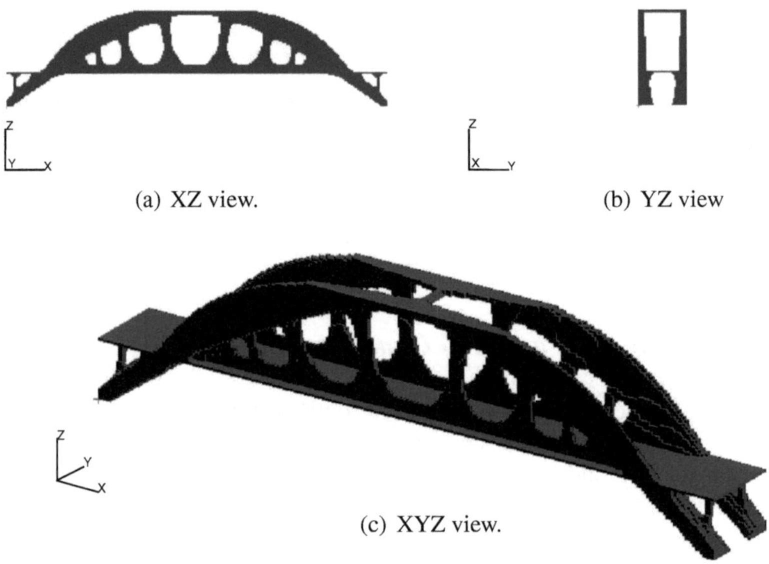

(a) XZ view. (b) YZ view

(c) XYZ view.

Fig. 14.8 Optimal 3-D topology of compression bridge

are supported each by a compression member, whereas the central span is suspended via a series of tension members. Different views for the three-dimensional version of the topology of the bridge are shown in Fig. 14.8. The topology obtained with the three-dimensional model presents some similarly, in the XZ plane, with the topology generated by the two-dimensional model (see Figs. 14.8(a) and 14.7) and with the design of the compression arch bridge reported in [28].

14.6.3 Example 3: Fiber Reinforce Cantilever Plate

To demonstrate the inclusion of the fibre-angle orientation in combined topology optimization environment, the in-plane design of cantilever plates with different material volume fractions is studied. The continuation method [29] is used in this

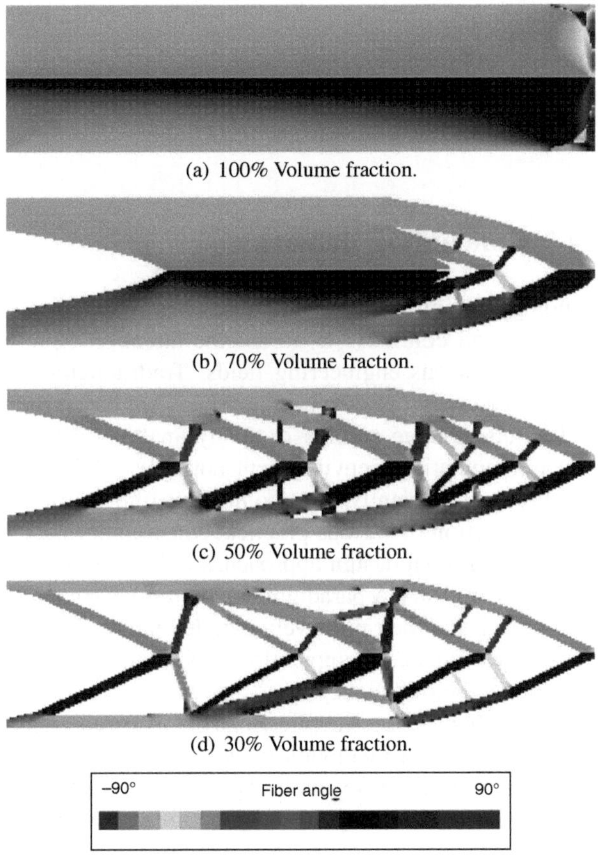

Fig. 14.9 Optimal topology of symmetric cantilever plate (325 × 82 cells) from [14]

study, with the penalization parameter p increasing gradually from 1.0 to 3.0 to avoid local minima. The following material are used:

$$E_1 = 135.2 \text{ GPa}, \; E_2 = 9.241 \text{ GPa},$$
$$G_{12} = 6.276 \text{ GPa}, \; \nu_{12} = 0.318.$$

The symmetric cantilever plate in Fig. 14.4 with an aspect ratio of 4 is modeled with a regular lattice of 325 × 82 cells. The topology optimization problem is solved using HCA scheme. Figures 14.9a through 14.9d show the topology of the optimal designs along with the color-coded fiber orientation angles for different volume fractions (for color version of this figure refer to [14]). These designs, as expected, are quite similar to classical optimal topologies of isotropic material (see example 1).

Corresponding to the designs shown in the figure, normalized compliances with respect to a 0° fiber design are tabulated in Table 14.2. These figures show that with the present choice of density interpolation scheme checkerboards are readily suppressed. Besides, for lower volume fractions, fibers are aligned with thin members similar to Mitchell type of structures.

Table 14.2 Normalized compliance of the symmetric cantilever for different volume fractions

Volume fraction	Normalized compliance
100%	0.74
70%	0.88
50%	1.14
30%	2.22

14.7 Concluding Remarks

Topology optimization of structures has matured enough to be often applied in industry, and continues to attract the attention of researchers and software companies in various engineering fields. Traditionally, most available algorithms for solving topology optimization problems are based on the global solution approach and require a large number of costly analyses. The CA paradigm offers a highly novel computational environment not only solving the topology design optimization problem efficiently, but also in terms of providing a flexible platform for design implementation of various practical constraints easily, which would otherwise render the traditional design approaches computationally infeasible. The main advantages of using the CA paradigm in structural design are the local analysis and design resolutions, and their massively parallel nature. The CA methodology can also take advantage of modern computational tools such as the multigrid acceleration method to improve their efficiency.

In this chapter, some applications of CA paradigm for structural design have been presented. The CA methodology was successfully applied to truss type and continuum structures. Some examples have been treated that illustrate the successes of the CA technique in solving topology optimization problems. Moreover, the multigrid accelerated CA scheme was shown to be an interesting candidate for solving topology optimization for continuum structures in a computationally efficient manner.

References

1. E. Kita, T. Toyoda, Structural design using cellular automata. Struct. Multidiscip. Optim. **19**, 64–73 (2000)
2. Y.M. Xie, G.P. Steven, A simple evolutionary procedure for structural optimization. Comput. Struct. **49**, 885–896 (1993)
3. C. Zhao, G.P. Steven, Y.M. Xie, Effect of initial non-design domain on optimal topologies of structures during natural frequency optimization. Comput. Struct. **62**, 119–131 (1997)
4. Z. Gürdal, B. Tatting, Cellular automata for design of truss structures with linear and non linear response. In *41st AIAA/ASME/ASCE/AHS/ASC Structures, Structural Dynamics and Materials Conference*, Atlanta, GA, 2000
5. R.T. Haftka, Z. Gürdal, *Elements of Structural Optimization*. (Kluwer, Dordrecht, 1993)
6. B. Tatting, Z. Gürdal, Cellular automata for design of two-dimensional continuum structures. In *8th AIAA/USAF/NASA/ISSMO Symposium on Multidisciplinary Analysis and Optimization*, Long Beach, CA, 2003
7. M.M. Abdalla, Z. Gürdal, Structural design using cellular automata for eigenvalue problems. Struct. Multidiscip. Optim. **26**, 200–208 (2004)

8. M.M. Abdalla, Z. Gürdal, Structural design using optimality based cellular automata. In *43th AIAA/ASME/ AHS/ASC Structures, Structural Dynamics and material Conference*, Denver, Co, April 2002
9. M. P. Bendsøe. Optimal shape design as a material distribution problem. Struct. Multidiscip. Optimi. **1**, 193–200 (1989)
10. M. Zhou, G.I.N. Rozvany, The COC algorithm, part II: Topological, geometry and generalized shape optimization. Comput. Meth. Appl. Mechan. Eng. **89**, 197–224 (1991)
11. G.I.N. Rozvany, M. Zhou, T. Birker, Generalized shape optimization without homogenization. Struct. Multidiscip. Optim. **4**, 250–252 (1992)
12. G.I.N. Rozvany, Aims, scope, methods, history and unified terminology of computer-aided topology optimization in structural mechanics. Struct. Multidiscip. Optim. **21**, 90–108 (2001)
13. O. Sigmund, A 99 line topology optimization code written in matlab. Struct. Multidiscip. Optim. **21**, 120–127 (2001)
14. S. Setoodeh, M.M. Abdalla, Z. Gürdal, Combined topology and fiber path design of composite layers using cellular automata. Struct. Multidiscip. Optim. **30**, 413–421 (2005)
15. R. Zakhama, M.M. Abdalla, H. Smaoui, Z. Gürdal, Topology design of geometrically nonlinear 2D elastic continua using CA and an equivalent truss model. In *11th AIAA/ISSMO Symposium on Multidisciplinary Analysis and Optimization*, Portsmouth, VA 2006
16. D.J. Slotta, B. Tatting, L.T. Watson, Z. Gürdal, S. Missoum, Convergence analysis for cellular automata applied to truss design. *Eng. Comput.* **19**, 953–969 (2002)
17. S. Setoodeh, D.B. Adams, Z. Gürdal, L.T. Watson, Pipeline implementation of cellular automata for structural design on message-passing multiprocessors. Math. Comput. Model. **43**, 966–975 (2006)
18. S. Kim, M.M. Abdalla, Z. Gürdal, M. Jones, Multigrid accelerated cellular automata for structural design optimization: A 1-D implementation. In *45th AIAA/ASME/ASCE/AHS/ASC Structures, Structural Dynamics and Materials Conference*, Palm Springs, CA, 2004
19. R. Zakhama, M.M. Abdalla, H. Smaoui, Z. Gürdal, Multigrid implementation of cellular automata for topology optimization of continuum structures. *CMES: Computer Modeling in Engineering and Sciences, to appear*, 2010
20. P. Wesseling, *An introduction to multigrid methods.* (Wiley, Chichester, 1992)
21. A. Tovar, N.M. Patel, G.L. Niebur, M. Sen, J.E. Renaud. Topology optimization using a hybrid cellular automaton method with local control rules. J. Mechan. Des. **128**(6), 1205–1216 (2006)
22. A. Tovar, N.M. Patel, A.K. Kaushik, J.E. Renaud, Optimality conditions of the hybrid cellular automata for structural optimization. AIAA J. **45**(3), 673–683 (2007)
23. S. Missoum, M.M. Abdalla, Z. Gürdal, Nonlinear topology design of trusses using cellular automata. In *44th AIAA/ASME/AHS/ASC Symposium on Structural Dynamics and Material Conference*, Norfolk, VA, 2003
24. S. Missoum, M.M. Abdalla, Z. Gürdal, Nonlinear design of trusses under multiple loads using cellular automata. In *5th World Congress in Structural and Multidisciplinary Optimization*, Lido diJesolo, Italy, 2003
25. S. Setoodeh, Z. Gürdal, L.T. Watson. Design of variable-stiffness composite layers using cellular automata. Comput. Meth. Appl. Mechan. Eng. **195**, 836–851 (2006)
26. N.V. Banichuk, Optimization of anisotropic properties of deformable media in plane problems of elasticity. *Mesh Solids*, **14**, 63–68 (1979)
27. P. Pedersen. Bounds on elastic energy in solids of orthotropic materials. Struct. Optim. **2**, 55–63 (1990)
28. M. Beckers, Topology optimization using a dual method with discrete variables. Struct. Optim. **17**, 14–24 (1999)
29. M.P. Bendsøe, O. Sigmund, *Topology Optimization, Theory, Methods and Applications* (Springer-Verlag, 2003)

Part III
Cellular Automata Software

Chapter 15
Parallel Cellular Programming for Emergent Computation

Domenico Talia and Lev Naumov

15.1 Introduction

In complex systems, global and collective properties cannot be deduced from its simpler components. In fact, global or collective behavior in a complex system emerges from evolution and interaction of many elements. Therefore programming emergent systems needs models, paradigms, and operations that allow for expressing the behavior and interaction of a very large number of single elements.

Because of their inherent parallelism, cellular automata (CA) can be exploited to model large scale emergent systems on parallel computers. In this scenario parallel cellular models and languages provide useful tools for programming emergent computations that model complex phenomena in many application domains from science and engineering to economics and social sciences.

The programming of emergent phenomena and systems based on traditional programming tools and languages is hard and it results in long and complex code. This occurs because these programming approaches are based on the design of a system as a whole. Design and programming do not start from basic elements or system components, but represent a system by modeling its general features. On the contrary, it is better to design emergent and complex systems by means of paradigms that allow for expressing the behavior of the single basic elements and their interactions. The global behavior of these systems then emerges from the evolution and interaction of a massive number of simple elements; hence it does not need to be explicitly coded.

Parallel architectures such as multicore, clusters, and multicomputers are well suited for implementing inherently parallel computing abstract models such as cellular automata, neural networks, and genetic algorithms that represent new mathematical models for describing complex scientific phenomena and systems with emergent properties. All cells of a cellular automaton are updated in parallel. Thus the state of the entire automaton advances in discrete time-steps and the global behavior of

D. Talia (✉)
DEIS, University of Calabria, Rende, Italy
e-mail: talia@deis.unical.it

A.G. Hoekstra et al. (eds.), *Simulating Complex Systems by Cellular Automata*, Understanding Complex Systems, DOI 10.1007/978-3-642-12203-3_15,

the system is determined by the evolution of the states of each cell as a result of multiple local interactions. Cellular automata provide a global framework for the implementation of parallel applications that represent natural solvers of dynamic complex phenomena and systems based on the use of discrete time and discrete space.

CA are intrinsically parallel and they can be efficiently mapped onto parallel machines because the communication flow between processors can be kept low, communication patterns are regular and involve only neighbor cells. Inherent parallelism and restricted communication are two key points for the efficient use of CA for high performance simulation [36].

The cellular automata theory was invented half a century ago. The exact author of this area cannot be named definitely. In 1948 John von Neumann [1] gave a lecture entitled "The General and Logical Theory of Automata", where he presented his ideas of universal and self-reproducing machines. According to his own statement, his work was inspired by Stanislaw Ulam [2]. Konrad Zuse [3] also suggested that the universe could be a cellular automaton. Zuse used this idea for developing computing machines. At the same time, some members of the scientific society regard the paper by Wiener and Rosenblueth [4], or the mathematical work that was done in early 1930s in Russia as the start of the field [5].

However more recently CA emerged as a significant tool for modeling and simulation of complex systems. This occurred thanks to the implementation of cellular automata on high-performance parallel computers. Parallel cellular automata models are successfully used in fluid dynamics, molecular dynamics, biology, genetics, chemistry, road traffic flow, cryptography, image processing, environment modeling, and finance. To explain this approach, we discuss the main features of cellular automata parallel software environments and how those features can support the solution of large-scale problems.

To describe in detail how the marriage of the cellular automata theory with parallel computing is very fruitful, we will discuss some leading examples of high-performance cellular programming tools and environments such as CAMELot and CAME&L. Moreover we will discuss programming of complex systems in CA languages such as CARPET. Those parallel cellular automata environments have been used to solve complex problems in several areas of science, engineering, computer science, and economy. They offer a well structured way to facilitate the development of cellular automata applications, providing transparent parallelism and reducing duplication of effort by implementing a programming environment once and making it available to developers. We discuss the basic principles of parallel CA languages and describe some practical programming examples in different areas designed by means of those parallel CA systems.

15.2 Cellular Automata Systems

In the past decade, several cellular automata environments have been implemented on current desktop computers. For large size two or three dimensional cellular automata the computational load can be enormous. There are two main alternatives that allow to achieve high performance in the implementation of CA. The first one is the design of special hardware devoted to the execution of CA. The second alternative is based on the use of commercially-available parallel computers for developing parallel CA software tools and environments.

CA software and hardware systems belong to the class of problem-solving environments (PSE). The community has formulated the following common recommendations for a general PSE:

1. It should **reduce the difficulty** of the simulation [6].
2. It should **reduce costs and time** of complex solutions development [6].
3. It should allow to perform experiments **reliably** [6].
4. It should have a **long lifetime** without getting obsolete [6].
5. It should **support the plug-and-play paradigm** [6].
6. It should **exploit** the paradigm of the **multilevel abstractions** and complex properties of science [6].
7. User should be able to **use the environment without any specialized knowledge** of the underlying computer hardware or software [7].
8. It should be pointed at the **wide scope** of problems [7].
9. It should be able to coordinate **mighty computational power** to solve a problem [7].
10. It should be **complete**, containing and providing all computational facilities for solving a problem in a target domain [8].
11. **Extensibility** of the environment will provide the ability to enlarge the target problem domain, to enrich the set of supported tools and provided features. This can be achieved with the help of a component-based design. A component approach also complies with the trend that modern distributed problem-solving facilities should be based on web and grid services [6] or Common Object Request Broker Architecture (CORBA) objects [9].

Basing on common considerations, the software or hardware facility, which allows to perform experiments using CA should have the following attributes:

1. It should **hide the complexity** of used computational architecture, operating system or networking mechanism. The language which a researcher should use to control the environment has to be related to basic cellular automata concepts and to the target problem domain.
2. It should allow to **setup and tune** a cellular automaton for the computational experiment. The degree of freedom which is granted to a user here may play the key role.
3. It should give an opportunity to **run and control** a computational experiment. A good solver should use all the benefits provided by the computational architecture and utilize as much parallelizable aspects of the experiment's iteration as it is possible to improve the throughput.
4. It should support **visualization**, because this feature plays one of the key roles in understanding the phenomenon, especially, when modeling spatial-distributed systems.

5. It should provide a set of tools to **analyze** the computational experiment's intrinsic characteristics and their tendencies, current state of the automaton's grid or any other data, which is possible to obtain.
6. It should provide the **reproducibility** and allow to share the description of the way one have done the certain computational experiment. Donald Knuth have declared this feature as a required one for a scientific method [10].

The following list unites two previous ones and consists of concrete required features for a CA-based PSE. At the end of each statement there are references to attributes stated above, given in italics. References to the first list are preceded by "*PSE:*", whereas references to the second one are preceded by "*CA:*". Moreover, the "Workshop on Research Directions in Integrating Numerical Analysis, Symbolic Computing, Computational Geometry, and Artificial Intelligence for Computational Science" [7, 11] have produced "Findings" and "Recommendations" for PSEs. They will not be listed here, but will be referenced in the "*PSE:*" section as "F_n" or "R_n" respectively, where n is the number of distinct finding or recommendation stated in [7].

1. The environment should be as **universal and customizable** as possible. The support of miscellaneous grids, types of neighborhoods and boundary conditions is desirable or even necessary. Environment should allow to choose the type of the automaton to be modeled and the parameters of the experiment from the widest possible spectrum of variants *(PSE: 2, 8, 10; CA: 2)*.
2. **Extensibility** is a contemporary and actively used property of software and hardware systems. The ability to incorporate novel functionalities and algorithms may be one of the most advantageous for the PSE *(PSE: 2, 4, 5, 6, 8, 9, 10, 11,* F_6, R_1*; CA: 1)*.
3. The environment should support **modern parallel or distributed computational technologies**. For software CA system this means that it should involve cluster or Grid computing, Message Passing Interface (MPI), Parallel Virtual Machine (PVM), OpenMP or other technologies. Hence the software has to emulate homogeneous parallel architecture of cellular automaton, but not counterfeit it. Nevertheless without the use of high-performance parallel hardware the CA model would be of no practical use for solving real world problems *(PSE: 1, 2, 3, 9, 10,* F_1, F_2, F_3*; CA: 1, 3)*.
4. The environment should provide a visually attractive, handy and clear **user interface**. It has to preserve the interactivity even when performing long experiments, preserving the reliable control *(PSE: 1, 2, 7,* F_2, F_4, F_7, R_7*; CA: 1, 2, 3)*.
5. The experiment's **description language** has to be close to the language of the target problem domain and as far from the implementation as possible. Description should be independent of the computational architecture, level of resources, and operating system. The ability to involve such dependencies is definitely considered as a powerful option *(PSE: 1, 2, 6, 7, 10,* F_1, F_4, R_7*; CA: 1, 2)*.
6. Grid state **visualization** is a most straightforward way of experiment's representation. Nowadays the scientific visualization seems to be a separate industry [12]. So there is no need for the CA-based PSE to be concurrent with top-level tools in

this area. Nevertheless the environment should contain the basic set of features and preferably be compatible with the specialized visualization software on the level of data-files *(PSE: 2, 10, R_7; CA: 4)*.

7. The environment should support **analysis** functionality to monitor the quality of the experiment, study the progress and the final results for making conclusions and producing new scientific knowledge *(PSE: 2, 6, 10, R_7; CA: 5)*.
8. The environment should allow to **reproduce** experiment made once on the same or another computational system. This will give an opportunity to share the knowledge and the experience between researchers, eliminate ambiguous computations, postpone the simulation, reanalyze, and revisualize or generally reuse the data. This can be achieved by providing the ability to store/restore full computational experiment setup and automaton's grid state to/from a file *(PSE: 1, 2, 4, 10, F_1, F_4, R_6; CA: 6)*.

Such tight connection of properties listed above with general recommendations for the PSEs design and features list for CA modeling facilities allows to conclude that these eight properties are close to be common requirements for cellular automata based modeling environments.

The number of the software created for cellular automata modeling is impressive [13, 14]. The apogee of this boom was at the 1990s. Many projects have been already outdated, but some new successful prototypes appeared.

The comparative survey of the existing software and hardware facilities is summarized in Table 15.1. The first column contains the name of the project with references. The second one presents the target platform and the third – the year of the latest known release or of the last publication devoted to the instrument. Further eight columns contain pluses if project satisfies the requirement with corresponding number (see the previous list) and minuses otherwise.

It is impossible and useless to overview all the existing projects, which were created ever. Those of them which last known version had been released in the

Table 15.1 The comparative survey of existing cellular automata modeling environments. The first raw contains project's name with references, the second is used for the information about the target platform, the third stores the year of current release. The rest raws contain pluses or minuses depending on the conformance to the corresponding requirements (see the previous list)

Name	Platform	Release	1	2	3	4	5	6	7	8
CAGE [15]	Windows	2004	+	−	−	+	+	+	−	+
CAM [16, 17]	iX86 or Sun	1994	+	+	+	+	+	+	−	+
CAMEL [18–20]	UNIX/Linux	1996	+	−	+	+	+	+	−	+
$CAME_{\&}L$ [14, 21, 23]	Windows	2010	+	+	+	+	+	+	+	+
CAMELot [24]	UNIX/Linux	2001	+	−	+	+	+	+	+	+
Cellular [25]	UNIX/Linux, CygWin	1999	+	+	−	+	+	+	−	+
JCASim [26]	Java	2001	+	+	+	+	+	+	+	+
MCell [27]	Windows	2001	+	+	−	+	+	+	−	+
ParCeL-6 [28, 29]	UNIX/Linux, Windows	2004	+	+	+	−	+	−	+	+
SIMP/STEP [30, 31]	UNIX/Linux, Windows	2003	+	+	+	+	+	+	+	+
Trend [32]	UNIX/Linux, Java	2002	+	−	−	+	+	+	−	+

twentieth century were mostly excluded from the study. Only several of them are listed, because of their significant historical value. Also some relatively new projects are deliberately not presented if they are deemed to be unsuitable for the research and scientific modeling.

Projects are listed in the alphabetical order. In the following there are short, one-paragraph reviews, which briefly describe each project in more detail. The name of the project which is subjected to the review is shown with bold when it appears for the first time.

The name **CAGE** stands for "Cellular Automata General Environment". The tool does not support any parallel or distributed computational technology, but this is compensated by the universality. Authors have generalized the notion of the "cellular automaton" and used their vision of it for the software design. The environment supports multilayered grids, rich means for the neighborhoods' formation (including the query-based one) and an ability to use irregular grids. The spectrum of the functionality is extremely rich. Transition rules are to be defined using the C-like language with the help of built-in visual programming means. Written rules are being translated to C++ sources and compiled into the executable code for better computational throughput. Despite of the functional richness, all grid's layers seem to be 2D only.

CAM means "Cellular Automata Machine" and represents a single instruction multiple data (SIMD) hardware implementation of the modeling environment. CAM-6 [16] is a PCI-device which should be plugged into iX86 workstation governed by PC-DOS operating system. It supports 256×256 2D grids with Moore, von Neumann and Margolus neighborhoods. CAM-8 [17] is a device which works in tandem with Sun workstations via CBus and should be controlled by the accompanying STEP software (a predecessor of SIMP/STEP project which is also present in this survey). Eighth version of the machine supports 3D grids. There is an ability to extend the grid by using multiple device specimens. Visualization is performed by the XCAM utility. Transition rules are to be programmed using a dialect of Forth language supplemented with necessary routines. For each transitions function the machine compiles the full lookup table. However, multiple instruction multiple data (MIMD) architectures are more flexible than SIMD machines for implementing CA, as they allow to deal with irregularities on a microscopic level of the algorithm asynchronously and to efficiently simulate also heterogeneous systems. On a higher level of abstraction it is possible to synchronize the parallel components of a MIMD system explicitly as this is the only way to maintain global invariance of CA.

The project's name **CAMEL** stands for "Cellular Automata environMent for systEms modeLing". This software was designed to perform computations on the net of transputers or using MPI. It supports grids of up to three dimensions and complex neighborhoods. The cell's state can be represented by the instance of a data structure composed of basic types. For the CA definition it uses a specialized language CARPET ("CellulAR Programming EnvironmenT") [19] which will be discussed in Sects. 15.3 and 15.4. The program written using this language traditionally consists of the declarative part and the statements. The language is clear and successfully hides implementation issues coming from a parallel computer's architecture complexity, allowing to describe automata and rules in general terms. The additional program IVT (comes from "Interactive Visualization Tool") has been added to CAMEL software to improve the data visualization. In twenty-first century the same group has switched to the development of CAMELot project [24].

By coincidence the name of the project **CAME$_{\&}$L** is very similar to the previous one. Nevertheless in this case it stands for "Cellular Automata Modeling Environment & Library". The ampersand in the abbreviation appeared exactly for it to be distinguishable from the CAMEL. This project will be discussed in more detail in Sect. 15.5, but will be reviewed briefly. The key idea was to create a universal and extensible facility, which supports parallel and distributed computing without any target problem domain specialization. This was achieved by usage of the CA based computational experiment decomposition (see Sect. 15.5.1). The software allows synchronous, asynchronous, probabilistic, inhomogeneous, and any other kind of CA with arbitrary grids, neighborhoods or type of cells' state. Even if the particular functionality or distinct automata type implementation is not included into the standard software package, one may add it and make new solution immediately available for the community. So CAME$_{\&}$L users can be divided into two interconnected and mixed groups: researchers who are just building solutions from the bricks they have, and developers who enhance the set of bricks for themselves and everyone. Ideal situation will be reached when anyone will get the ability to perform arbitrary cellular automata based experiments without the need to create new bricks.

As a descendant of CAMEL, **CAMELot** ("CAMEL Open Technology") also uses CARPET language [19] for the experiment description and MPI for the simulation execution. This project will be discussed in Sect. 15.4 and here it will be overviewed briefly. The software represents the environment for programming and seamlessly parallel execution of cellular automata. It has a graphical user interface for experiment setup, control and visualization. It also includes the customizable tool to produce traces of the simulation in a specified format thus allowing to post-process the output of the experiment by means of the external utilities. Moreover it supports profiling capabilities. The simulator is flexible with regard to cellular space size and dimension (form 1D to 3D), cell's state structure, neighborhood and rules. The program, written using CARPET is translated and compiled into UNIX/Linux executable file. The experiment setup preparation consists of editing of a text file.

Cellular software consists of the programming language (Cellang 2.0), associated compiler (`cellc`), virtual machine for the execution (`pe-scam`) and the viewer (`cellview`). A program written with Cellang 2.0 consists of two parts: the description and the set of statements. The description determines dimensionality of a grid, data-fields, which are contained in each cell, and ranges of acceptable values for each field. There are two possible statements: an assignment and a conditional test. The only possible data type is integer. The viewer is independent of the Cellang 2.0 language and the compiler. The input format for the viewer is identical to the output format of Cellang 2.0 programs. The software supports different grids of arbitrary dimensionality, non-trivial neighborhoods, several kinds of boundary conditions.

JCASim represents a general-purpose system for simulating CA on Java platform. It includes the standalone application and the applet for web presentations.

The cellular automaton can be specified in Java, CDL [33] or using the interactive dialogue. It supports 1D, 2D (square, hexagonal, or triangular) and 3D grids, different neighborhoods, boundary conditions (periodic, reflective or constant), and can display cells using colors, text, or icons. Initially CDL was designed to describe the hardware, which simulates homogeneous structures, but it can also be applied in software as a powerful and expressive tool. JCASim allows any constructions acceptable in CDL. For example, like in CDL, cell's state can be represented with theoretically unlimited amount of integer and floating-point variables. With the package `CAComb` the software allows to simulate CA on several machines in parallel. `CAAnalysis` package incorporates automatic analysis (the mean-fields and similar approximations will be calculated automatically).

MCell or "Mirek's Cellebration" is a very small and simple Windows application which supports 2D grids and no parallel or distributed computing. But despite of this its effort is great, because it can easily show the simplicity, beauty and power of a cellular automata to people who are far from this field of science. This is possible due to successful graphical user interface which is clear for non-specialists and a wide library of examples. Transition rules can be defined using the interface means or by creation of external dynamic-link library.

Project **ParCeL-6** represents the multi-layer cellular computing library for multiprocessor computers, clusters and Grids. The goal of its creation was to decrease the development time for the fine-grained applications. It is implemented in C language and can be linked to C and C++ programs. There are two subversions of the software: ParCeL-6.1 for architectures supporting the memory sharing paradigm and ParCeL-6.2 for architectures supporting the message passing approach. The cluster version of ParCeL-6 was developed in the framework of the Grid-eXplorer project. High level generic and parallel neural model of computations allows smart programming for numerous computing units. ParCeL offers the extended cellular programming model and maps "small" computing units on the "big" processors of parallel machines. When a cell is created, host processor is pointed out and unique registration number is associated with the cell. This number allows to identify it in a cellular network. Finally the cell is created directly on its host processor and executes the computing cycle on it. The software is also able to perform the automatic parallelization of the source code for the multiprocessor machines.

SIMP/STEP is a general-purpose software platform, which includes the language for cellular automata, lattice gases, and a "programmable matter" definitions. It is based on the Python programming language and suites for the wide range of problems. The software consists of two parts: SIMP is the user environment built on STEP, the applications programming interface, which separates conceptual components from implementation details, optimization routines etc. The software supports parallel computing technologies, has visualization and analysis capabilities. SIMP supports 2D rendering, but there are some experimental hooks for the 3D rendering, using VTK [12].

Trend is the 2D cellular automata programming environment with the integrated simulator and the compiler, which produces the virtual machine code for the evaluation module (under UNIX/Linux only) or Java machine. It has several interesting features: the simulation backtracking, conflicts catching, flexible template design and others. The Trend language allows user-defined terms, symmetrically rotatable statements and other constructions specific for the cellular automata programming. The software supports arbitrary neighborhoods within 11×11 region around the center cell. Each cell can be in the state which is coded by unsigned integer variable. The project does not support any parallel computing technologies.

15.3 Parallel CA Languages

For developing cellular automata on parallel computers two main approaches can be used. One is to write programs that encode the CA rules in a general-purpose parallel programming language such as HPF, Erlang, Java, Linda or CILK or still using a high-level sequential language like C++, Fortran or Phyton with one of the low-level toolkits/libraries currently used to implement parallel applications such as MPI, PVM, or OpenMP. This approach does not require a parallel programmer to learn a new language syntax and programming techniques for cellular programming. However, it is not simple to be used by programmers that are not experts in parallel programming and code consists of a large number of instructions even if simple cellular models must be implemented.

The other possibility is to use a high-level language specifically designed for CA, in which it is possible to directly express the features and the rules of CA, and then use a compiler to translate the CA code into a program executable on parallel computers. This second approach has the advantage that it offers a programming paradigm that is very close to the CA abstract model and that the same CA description could possibly also be compiled into different code for various parallel machines. Furthermore, in this approach parallelism is transparent from the user, so programmers can concentrate on the specification of the model without worrying about architecture related issues. In summary, it leads to the writing of software that does express in a natural manner the cellular paradigm, and thus programs are simpler to read, change, and maintain. On the other hand, the regularity of computation and locality of communication allow CA programs to achieve good performance and scalability on parallel architectures.

In recent years, several cellular automata environments have been implemented on current desktop computers as well (see Sect. 15.2). Sequential CA-based systems can be used for educational purposes and very simple simulations, but real world phenomena simulations generally take very long time, or in some cases cannot be executed, on this class of systems because of memory or computing power limits. Therefore, massively parallel computers are the appropriate computing platform for the execution of CA models when real life problems must be solved. In fact, for two and three dimensional cellular automata of large size the computational load can be enormous. Thus, if CA are to be used for investigating large complex phenomena, their implementation on high performance computers composed of several processors is a must.

In particular, general-purpose distributed-memory parallel computers offer a very useful architecture for a scalable CA machine both in terms speed-up, programmability, and portability. These systems are based on a large number of interconnected processing elements (PE) which perform a task in parallel. According to this approach, in the recent years several parallel cellular software environments have been developed.

The main issues that influence the way in which CA languages support the design of applications on high performance architectures are

- The *programming approach*: the unit of programming is the single cell of the automaton.
- The *cellular lattice declaration*: it is based on definition of the lattice dimension and the lattice size.
- The *cell state definition and operations*: cell state is defined as single variable or a record of typed variables; cell state access and update operations are needed.
- The *neighborhood declaration and use*: neighborhood concept is used to define interaction among cells in the lattice.
- The *parallelism exploitation*: the unit of parallelism is the cell and parallelism, like communication, is implicit.
- The *cellular automata mapping*: data partitioning and process-to-processor mapping is implicit at the language level.
- The *output visualization*: automaton global state, as the collection of the cell states, is showed as it evolves.

By addressing these issues we illustrate how this class of languages can be effectively used to implement high-performance applications in science and engineering using the massively parallel cellular approach.

15.3.1 Programming Approach

When a programmer starts to design a parallel cellular program she/he must define the structure of the lattice that represents the abstract model of a computation in terms of cell-to-cell interaction patterns. Then she/he must concentrate on the unit of computation that is a single cell of the automaton. The computation that is to be performed must be specified as the transition function of the cells that compose the lattice. Therefore, differently form other approaches, a user does not specify a global algorithm that contains the program structure in an explicit form.

The global algorithm consists of all the transition functions of all cells that are executed in parallel for a certain number of iterations (steps). It is worth to notice that in some CA languages it is possible to define transition functions that change in time and space to implement inhomogeneous CA computations. Thus, after defining the dimension (e.g., 1D, 2D, 3D) and the size of the CA lattice, she/he needs to specify, by the conventional and the CA statements, the transition function of the CA that will be executed by all the cells. Then the global execution of the cellular program is performed as a massively parallel computation in which implicit communication occurs only among neighbor cells that access each other state.

15.3.2 Cellular Lattice Declaration

As was mentioned above, the lattice declaration defines the lattice dimension and the lattice size. Most languages support two-dimensional rectangular lattices only (e.g., CANL and CDL). However, some of them, such as CARPET and Cellang, allow the definition of 1D, 2D, and 3D lattices. Some languages allow also the explicit definition of boundary conditions such as CANL that allows *adiabatic* boundary conditions where absent neighbor cells are assumed to have the same state as the center cell. Others implement *reflecting* conditions that are based on mirroring the lattice at its borders. Most languages use standard boundary conditions such as *fixed* and *toroidal* conditions.

15.3.3 Cell State

The cell state contains the values of data on which the cellular program works. Thus the global state of an automaton is defined by the collection of the state values of all the cells. While low-level implementations of CA allow to define the cell state as a small number of bits (typically 8 or 16 bits), cellular languages such as CARPET, CANL, DEVS-C++ and CDL allows a user to define cell states as a record of typed variables as follows:

```
cell = (direction :int ;
        mass       : float;
        speed      : float);
```

where three substates are declared for the cell state. According to this approach, the cell state can be composed of a set of sub-states that are of *integer*, *real*, *char* or *boolean* type and in some case (e.g., CARPET) arrays of those basic types can also be used. Together with the constructs for cell state definition, CA languages define statements for state addressing and updating that address the sub-states by using their identifiers, e.g. `cell.speed`.

15.3.4 Neighborhood

An important feature of CA languages that differentiate them from array-based languages and standard data-parallel languages is that they do not use explicit array indexing. Thus, cells are addressed with a name or the name of the cells belonging to the neighborhood. In fact, the neighborhood concept is used in the CA setting to define interaction among cells in the lattice.

In CA languages the neighborhood defines the set of cells whose state can be used in the evolution rules of the central one. For example, if we use a simple neighborhood composed of four cells we can declare it as follows

```
neigh cross = (up, down, left, right);
```

and address the neighbor cell states by the identifiers used in the above declaration (e.g., `down.speed, left.direction`). The neighborhood abstraction is used to define the communication pattern among cells. It means that at each time step, a cell send to and receive from the neighbor cells the state values. In this way implicit communication and synchronization are realized in cellular computing. The neighbor mechanism is a concept similar to the *region* construct that is used in the ZPL language [37] where regions replace explicit array indexing making the programming of vector- or matrix-based computations simpler and more concise. Furthermore, this way of addressing the lattice elements (cells) does not require compile-time sophisticated analysis and complex run-time checks to detect communication patterns among elements.

15.3.5 Parallelism Exploitation

CA languages do not provide statements to express parallelism at the language level. It turns out that a user does not need to specify what portion of code must be executed in parallel. In fact, in parallel CA languages the unit of parallelism is a single cell and parallelism, like communication and synchronization, is implicit. This means that in principle the transaction function of every cell is executed in parallel with the transaction functions of the other cells.

In practice, when coarse grained parallel machines, like clusters or multi-core, are used, the number of cells N is greater than the number of available processors P, so each processor executes a block of N/P cells that can be assigned to it using a domain decomposition approach.

15.3.6 CA Mapping

Like parallelism and communication, also data partitioning and process-to-processor mapping is implicit in CA languages. The mapping of cells (or blocks of them) onto the physical processors that compose a parallel machine is generally done by the run-time system of each particular language and the user usually intervenes in selecting the number of processors or some other simple parameter.

Some systems that run on multicomputers (MIMD machines) use load balancing techniques that assign at run-time the execution of cell transition functions to processors that are unloaded or use greedy mapping techniques that avoid some processor to become unloaded or free during the CA execution for a long period.

15.3.7 Output Visualization and Monitoring

A computational science application is not just an algorithm. Therefore it is not sufficient to have a programming paradigm for implementing a complete application. It is also as much significant to dispose of environments and tools that help a user in all the phases of the application development and execution. Most of the CA languages we are discussing here provide a development environment that allows a user not only to edit and compile the CA programs. They also allow to monitor the program behavior during its execution on a parallel machine, by visualizing the output as composed of the states of all cells. This is done by displaying the numerical values or by associating colors to those values. Examples of these parallel environments are CAMEL for CARPET, PECANS for CANL, and DEVS for DEVS-C++.

Some of these environments provide dynamical visualization of simulations together with monitoring and tuning facilities. Users can interact with the CA environment to change values of cell states, simulation parameters and output visualization features. These facilities are very helpful in the development of complex scientific applications and make possible to use those CA environments as real problem solving environments (PSEs).

Many of these issues are taken into account in parallel CA systems and similar or different solutions are provided by parallel CA languages. In Sect. 15.4 we outline some of the listed issues by discussing the main features of CAMELot, a general-purpose system that can be easily used for programming emergent systems using the CARPET cellular programming language according to a massively parallel paradigm and some related parallel CA environments and/or languages.

15.4 Cellular Automata Based Problem-Solving Environment Case Study: CAMELot and CARPET

CAMELot (CAMEL open technology) is a parallel software system designed to support the parallel execution of cellular algorithms, the visualization of the results, and the monitoring of cellular program execution [38]. CAMELot is an MPI-based portable version of the CAMEL system based on the CARPET language. CARPET offers a high-level cellular paradigm that offers to a user the main CA features to assist her/him in the design of parallel cellular algorithms without apparent parallelism [20].

A CARPET programmer can develop cellular programs describing the actions of many simple active elements (implemented by cells) interacting locally. Then, the CAMELot system executes in parallel cells evolution and allows a user to observe the global complex evolution that arises from all the local interactions. CARPET uses a C-based grammar with additional constructs to describe the rules of the transition function of a single cell. In a CARPET program, a user can define the basic rules of the system to be simulated (by the cell transition function), but she/he does not need to specify details about the parallel execution. The language includes

- a declaration part (cadef) that allows to specify:
- the dimension of the automaton (dimension);
- the radius of the neighborhood (radius);
- the type of the neighborhood (neighbor);
- the state of a cell as a record of substates (state);
- a set of global parameters to describe the global characteristics of the system (parameter).
- a set of constructs for addressing and updating the cell states (e.g., update, GetX, GetY, GetZ).

In a two-dimensional automaton, a very simple neighborhood composed of four cells can be defined as follows:

```
neighbor Stencil[2] ([-1,0]Left, [1,0]Right, [0,1]Up,
                    [0,-1]Down);
```

As mentioned before, the state (`state`) of a cell is defined as a set of typed substates that can be *shorts, integers, floats, char,* and *doubles* or *arrays* of these basic types. In the following example, the state consists of three substates.

```
state(float speedx, speedy, energy);
```

The *mass* substate of the current cell can be referenced by the predefined variable `cell_mass`. The neighbor declaration assigns a name to specified neighboring cells of the current cell and allows such to refer to the value of the substates of these identified cells by their name (e.g., `Left_mass`). Furthermore, the name of a vector that has as dimension the number of elements composing the logic neighborhood it must be associated to neighbor (e.g., `Stencil`). The name of the vector can be used as an alias in referring to the neighbor cell. Through the vector, a substate can be referred as `Stencil[i]_mass`.

To guarantee the semantics of cell updating in cellular automata the value of one substate of a cell can be modified only by the `update` operation, for example

```
update(cell_speedx, 12.9);.
```

After an `update` statement, the value of the substate, in the current iteration, is unchangeable. The new value takes effect at the beginning of the next iteration. Furthermore, a set of global parameters (`parameter`) describes the global characteristics of the system (e.g., the permeability of a soil). CARPET allows to define cells with different transition functions (inhomogeneous CA) by means of the `GetX`, `GetY`, `GetZ` functions that return the value of the coordinate X, Y, and Z of the cell in the automaton. Varying only a coordinate it is possible to associate the same transition function to all cells belonging to a plane in a three dimensional automaton.

The language does not provide statements to configure the automata, to visualize the cell values or to define data channels that can connect the cells according to different topologies. The configuration of a cellular automaton is defined by the graphical user interface (UI) of the CAMELot environment. The UI allows, by menu pops, to define the size of the cellular automata, the number of the processors onto which the automata must be executed, and to choose the colors to be assigned to the cell substates to support the graphical visualization of their values. The exclusion from the language of constructs for configuration and visualization of the data allows executing the same CARPET program with different configurations. Further, it is possible to change from time to time the size of the automaton and/or the number of the nodes onto which the automaton must be executed. Finally, this approach allows selecting the more suitable range of the colors for the visualization of data.

15.4.1 Examples of Cellular Programming

In this section we describe two examples of emergent systems expressed through cellular programming using the CARPET language. The first example is a typical CA application that simulates excitable systems. The second program is the classical Jacobi relaxation that shows how it is possible to use CA languages not only for simulate complex systems and artificial life models, but that they can be used to implement parallel programs in the area of fine grained applications such as finite elements methods, partial differential equations and systolic algorithms that are traditionally developed using array or data-parallel languages.

15.4.1.1 The Greenberg-Hastings Model

A classical model of excitable media was introduced 1978 by Greenberg and Hastings [39]. This model considers a two-dimensional square grid. The cells are in one of a *resting* (0), *refractory* (1), or *excited* (2) state. Neighbors are the eight nearest cells. A cell in the resting state with at least s excited neighbors (in the program we use $s = 1$) becomes excited itself, runs through all excited and resting states and returns finally to the resting state. A resting cell with less than s excited neighbors stays in the resting state.

Excitable media appear in several different situations. One example is nerve or muscle tissue, which can be in a resting state or in an excited state followed by a refractory (or recovering) state. This sequence appears for example in the heart muscle, where a wave of excitation travels through the heart at each heartbeat. Another example is a forest fire or an epidemic model where one looks at the cells as infectious, immune, or susceptible.

Figure 15.1 shows the CARPET program that implements the two-dimensional Greenberg-Hastings model. It appears concise and simple because the programming level is very close to the model specification. If a Fortran+MPI or C+MPI solution is adopted the source code is extremely longer with respect to this one and, although it might be a little more efficient, it is very difficult to program, read and debug.

```
#define resting 0
#define refractory 1
#define excited 2

cadef
{
  dimension 2;
  radius 1;
  state (short value);
  neighbor Moore[8] ([0,-1]North, [1,-1]NorthEast,[1,0]East,
                    [1,1]SouthEast,[0,1]South,[-1,1]SouthWest,
                    [-1,0]West, [-1,-1]NorthWest);
}
 int i, exc_neigh=0;
{
 for (i=0; (i<8) && (exc_neigh==0); i++)
   if (Moore[i]_value == excited) exc_neigh = 1;
 switch (cell_value)
 {
   case excited    : update(cell_value, recovering); break;
   case recovering : update(cell_value, resting); break;
   default         : /* cell is in the resting state */
                     if (exc_neigh == 1)
                       update(cell_value, excited);
 }
}
```

Fig. 15.1 The Greenberg-Hastings model written in CARPET

15.4.1.2 The Jacobi Relaxation

As a second example, we describe the four-point Jacobi relaxation on a $n \times n$ lattice in which the value of each element is to be replaced by the average value of its four neighbor elements. The Jacobi relaxation is an iterative algorithm that is used to solve differential equation systems. It can be used, for example, to compute the heat transfer in a metallic plate on which boundaries there is a given temperature. At each step of the relaxation the heat of each plate point (cell) is updated by computing the average of its four nearest neighbor points. Figure 15.2 shows a CARPET implementation. The initial `if` statement is used to set the initial values of cells that are taken to be 0.0 except for the western edge where boundary values are 1.0.

The Jacobi program, although it is a simple algorithm, is another example of how a CA language can be effectively used to implement scientific programs that are not properly in the original area of cellular automata. This simple case illustrates the high-level features of the CA languages that can be also used for implement applications that are based on the manipulation of arrays such as systolic algorithms and finite elements methods.

For the Jacobi algorithm we present some performance benchmarks that have been obtained by executing the CARPET program using different grid sizes and processor numbers. Table 15.2 shows the execution times for 100 relaxation steps for three different grid sizes (100×200, 200×200 and 200×400) on 1, 2, 4, 8 and

```
cadef
{
 dimension 2;
 radius 1;
 state ( float elem );
 neighbor Neum[4]([0,-1]North,[-1,0]West,[0,1]South,[1,0]East);
}
   int sum;
{
 if (step == 1 )
   if (GetY == 1)
     update (cell_elem, 1.0);
   else
     update (cell_elem, 0.0);
 else
   {
    sum = North_elem+South_elem+East_elem+West_elem;
    update (cell_elem, sum/4);
   }
}
```

Fig. 15.2 The Jacobi iteration program written in CARPET

Table 15.2 Execution time (in) of 100 iterations for the Jacobi algorithm

Grid sizes	1 Proc	2 Procs	4 Procs	8 Procs	10 Procs
100×200	1.21	0.65	0.37		0.25
200×200	3.62	1.25	0.67	0.42	0.37
200×400	8.22	3.65	1.26	0.74	0.62

10 processors of a multicomputer. From the figure we can see that as the number of used processors increases, there is a corresponding decrease of the execution time. This trend is more evident when larger grids are used; while smaller CA do not use efficiently the processors. This means that, because of the algorithm simplicity, when we run an automaton with a small number of cells we do not need to use several processing elements. On the contrary, when the number of cells in the lattice is high, the algorithm benefits from the use of a higher number of computing resources. This can be also deduced from Table 15.3 that shows the relative speed up results for the three different grids. In particular, we can observe that when a 200×400 lattice of cells is used we obtain a superlinear speed up in comparison to the sequential execution mainly because of memory allocation and management problems that occur when all the 80,000 cells are allocated on one single processing element.

Table 15.3 Relative speed up of the Jacobi algorithm

Grid Sizes	1 Proc	2 Procs	4 Procs	8 Procs	10 Procs
100×200	1	1.86	3.27		4.84
200×200	1	2.89	5.40	8.62	9.78
200×400	1	2.25	6.52	11.10	13.25

15.5 Cellular Automata Based Problem-Solving Environment Case Study: CAME$_{\&}$L

The environment CAME$_{\&}$L [14, 21, 23] resulted from a collaboration between the Saint-Petersburg State University of Information Technologies, Mechanics and Optics (Russian Federation) and the Section Computational Science of the University of Amsterdam (The Netherlands).

15.5.1 Cellular Automata Based Computational Experiment Decomposition

The initial idea of "CAME$_{\&}$L" was to distribute the implementation of a computational experiment among the functional parts. Any simulation should be assembled as a set of interacting components of definite types. A researcher will be able to use them in miscellaneous combinations to add arbitrary functionality to the experiment. Components could be taken from the standard set or created by a user to fulfil the target problem requirements.

Consequently the CA based computational experiment decomposition [14, 21] was offered. It was decided to distinguish five types of components. Names of these types are shown with bold in the following list.

- The **grid** implements the visualization of automaton's state and the navigation among cells. It does not actually store cells states. This component's main task should be drawing and interacting with user.
- The **datum** provides cells states storage, exchange and some aspects of the data visualization. Namely it can define
  - the association of cells' states with colors, which will be used for their displaying;
  - the custom single cell drawing routine.
- The **metrics** provides the relationship of neighborhood, coordinates for each cell and distance measurement functions. Implementation of metrics as a separate component instead of entrusting its functions to the grid or the datum allows, for example, to use non-standard coordinate systems, like generalized coordinates [34].
- The **rules** describes computations and controls the iteration. In the introduced ideology terms "rules" and "transition function" are not synonyms. Components of this type define much more: the method of parallelization, methods of computations' optimization (if any are used), many other aspects and the transition function among the rest. This component also should allow
  - to handle experiment's start up (proceed the initialization);
  - to determine and check the criteria of experiment's completion;
  - to handle experiments finish (proceed the finalization);

- to define special tools for checking, changing, pre- and postprocessing;
- to define important experiment's properties for further studying with the help of analyzer components (see below).

- The **analyzer** allows to keep an eye on definite properties of the experiment, draw graphs, create reports, monitor values and all of this kind.

The union of compatible components of first three types totally define a "functionless" cellular automaton. Addition of a component of the fourth type will form a cellular automaton that can perform the computational experiment. Only single instances of the grid, the datum, the metrics and the rules are able to participate in the simulation, but it can involve arbitrary amount of analyzers (even none).

Components are continuously interacting during the whole computational experiment to do the work together. Obviously, each component cannot cooperate with arbitrary another component, but only with one, which is suitable for this. Such compatibility conditions for analyzers are trivially based on the examination of the analyzable parameter's variable data type. For the rest four types there should be a special language of logical expressions to describe their properties and requirements. In this case requirements should represent conditions imposed on properties.

Each component should have specific user interface: the declared set of available parameters, which allow to setup the component for the particular problem and for the accordance to user's needs and preferences.

15.5.2 Software Design

Taking everything, said in Sect. 15.5.1 into account it was decided to implement the software using C++ language. All basic statements, listed above, can be provided with the help of the object-oriented programming paradigm. Windows was chosen as a target operating system. Consequently each component should be represented as a dynamic-link library, developed in the framework of the predefined programming interface. The component's library have to contain the class, which implements the functionality corresponding to one of five types, listed in Sect. 15.5.1.

As a result, CAME$_{\&}$L software consists of three conceptually and functionally interconnected parts:

- **CADLib** or "Cellular Automata Development Library" is the C++ class library, which is designed to present an easy-to-use and rich set of instruments for implementing computational experiments according to given regulations and using definite abstractions. It provides basic classes for all types of components, parameters and for other concepts.
- **Standard components** are most common building blocks of computational experiments, which can be considered as both: ready-made solutions and examples for studying when one is going to create his own component. They also can be reused and extended to fulfil the needs of the researcher.

- The **environment** is the application with rich user interface for simulations and research with the help of cellular automata. It provides the access to tools for the simulation control, studying and analysis, cluster arrangement, workstations management and many other purposes. Important note is that the environment itself contains no computational functionality, but allows to execute components' libraries in the definite software surrounding.

One may say that the ability to use C++ is a too complicated skill to demand it from the researcher. This is true, but at the same time this is totally in the ideology of the extensible environment: the scope of the rules basic class is much wider than just the transition function definition. So one can create a rules component, which represents the parser for the automaton's iteration description from the specific language. This means that one rules component is able to implement not just the single transition function, but the class of such functions. This ideology allows to incorporate arbitrary amount of specific computations description languages into one software and provide specialists from distinct field of the research with the component, which supports necessary abstractions from the given subject field. The code snippets, the rich set examples, scripts and the CADLib itself are provided to make the components creation simpler.

15.5.3 Usage Example. Tumor Growth Modeling

Now $CAME_{\&}L$ is intensively used for the 3D tumor growth modeling. In this section a very schematic example will illustrate the common approach to using this software for a simulation. The example is related to the tumor growth simulation, but is free of plunging into the biological background. Computational Oncology is an active area of research with many promosing results. For instance, Sottoriva et al. [22] report on extensive simulations to reveal Cancer's stem cell driven tumor growth using such models.

To implement the cellular automaton, which will perform modeling, one should select the set of at least four components (grid, datum, metrics and rules), which will arrange the experiment. If particular component is not presented in the set of standard components then it should be created.

Usually, there is no need to create user analyzers, because they are much less problem domain dependant than any others. That is why only grid, datum, metrics and rules components are considered in the list below. The component type's name is shown with bold.

- For performing the computational experiment of 3D tumor growth, the standard **grid** component "Basic 3D Grid" will be suitable. It supports many functions, which are extremely useful for the model of such solid clot: drawing sections, stubs and slices to take a look inside the tumor.
- In the experiment, each automaton's cell is to represent single biological cell, which should be described with distinct user developed data structure. Let's

assume that it is called `BioCell` (there is no need to discuss what it consists of). There is no standard **datum** component implementing 3D storage for cells which contains instances of the `BioCell` structure. This component should be created with the help of CADLib as a descendant of the `CADatum` class. Library makes it extremely easy, providing `CABasicCrts3DDatum` class template, which automatically implements the majority of needed functions. Primitive, but functional class declaration should look like shown of Fig. 15.3. Numbers, given in brackets at the left, are used for further referring to appropriate lines or sections (sets of lines from one number to another) of the code and have no attitude to the source.

On line (1) and following one the parent class template is used with the specific values of parameters: first one is the data type to be stored in each cell, second – the class of the user interface dialog (may be none), used to edit the values of a stored data type, third – the resource identifier of the dialog template. Section, started from line (2) contains constructor and destructor declarations. The main task, which is entrusted to the constructor, is the initialization of component's parameters. Section, started from line (3) presents component's self-introduction functions, treating macrodefinitions, provided by CADLib. These declarations contains (in the same order) components short name, longer description, resource identifier of the corresponding icon, requirements for the properties of other components to be compliant to this one and properties, implemented by this components.

Last two statements worth special discussion. Attributes and requirements specifications are formulated using the trivial language of consequently adjustable properties. The self-characteristics, given on the last line of section (3) should be understood as the declaration of the fact that the component implements the property "Data". Then it is refined: data is "composite". Moreover, composite data is attributed as "biocell". In the same manner requirements represent

```
     class CACrtsCell3DDatum:public
(1)      CABasicCrts3DDatum
         <BioCell, CBioCellDlg, IDD_BIOCELL> {
     public:
(2)      CACrtsCell3DDatum();
         virtual ~CACrtsCell3DDatum();

(3)      COMPONENT_NAME(BioCells for Cartesians 3D)
         COMPONENT_INFO(3D storage for cellular (biological)
           data for cartesian metrics)
         COMPONENT_ICON(IDI_ICON)
         COMPONENT_REQUIRES(Metrics.3D.cartesian.*)
         COMPONENT_REALIZES(Data.composite.biocell)

(4)      virtual inline COLORREF GetCellColor(CACell c); (5)
(5)      virtual inline void SetDefValue(CACell c);
     };
```

Fig. 15.3 The declaration of the datum component for the tumor growth modeling in CAME&L

the conditions over properties, allowing wildcards and logical operations. This component needs to collaborate with another one, which should implement the property "Metrics". The property should be attributed as "3D" and, moreover, "cartesian". The asterisk means that any amount of deeper refining subproperties will fit. There is no strict rule for properties naming. The properties conformance checkup is case-insensitive. The union of components will not form a proper cellular automaton if at least one component has unsatisfied requirements.

There is no need to overload any additional members of the `CADatum` class, because all the required functionality is basically implemented by the `CABasicCrts3DDatum` class template. Nevertheless most likely one will decide to overload two functions, shown on lines (4) and (5).

First of all, note that `CACell` class represents the universal cell identifier, which allows to refer any given cell in the arbitrary metrics. From the technical point of view, it represents the 64 bits integer value. For example, when dealing with standard 2D cartesian metrics the universal cell identifier stores cell's absciss in first 32 bits and the ordinate in rest 32 bits. For standard 3D cartesian metrics the universal cell identifier is divided into three unequal parts: 22 bits for absciss, 21 for ordinate and 21 for applicate. When using generalized coordinates [34] as, for example, Peano-curve-based metrics [35], the universal cell identifier is interpretted as a solid unsigned integer number. So, $\text{CAME}_{\&}\text{L}$ can govern the cellular automaton of up to 2^{64} cells.

The function, overloaded on line (4), is to return the color, which should be used to visualize the value, stored in the cell `c`. The function on line (5) should put the "default value" to the cell `c`. This value will be used for the grid initialization and as the out-of-bounds value for constant boundary conditions.

Finally, the component's library should contain the class declaration, the implementation (in the case of this component, four functions should be implemented: constructor, destructor and two, declared on lines (4) and (5)) and component's library access functions, which can be easily created with following two lines of code:

```
COMPATIBLE_DATUM(1.1)
DATUM_COMPONENT(CACrtsCell3DDatum)
```

First one implements the authentication function for the library, which says that the component was built to be compatible with CADLib version 1.1. Second one adds the creation and the destruction functions for the component, implemented by the `CACrtsCell3DDatum` class.

- It is logical to perform modeling in Cartesian **metrics**, which is implemented by one of the standard components. The name of this component is "Cartesians 3D".
- Each **rules** component should be implemented by the descendant of the `CARules` class. This type of components was designed to allow the full control over the simulation and to support a lot of features. Nevertheless for the plain implementation of algorithm in most cases it's enough to overload its `SubCompute` function only. This function represents the transitions' laws, which are applied to some zone of the grid. The description of the zone is given by the object of

```
     bool CATG3DRules::SubCompute(Zone& z)
     {
(1)      *** Prestep ***

(2)      int i,j,k;
         for(i=(int)z.a1; i<=(int)z.b1; i++) {
(3)          pEnv->SetProgress(((double)i-z.a1)/(z.b1-z.a1+1))
             for(j=(int)z.a2; j<=(int)z.b2; j++)
                 for(k=(int)z.a3; k<=(int)z.b3; k++) {
(4)                  *** Transition for the cell (i;j;k) ***
                     *** Compute the analyzable values ***
                 }
         }

(5)      *** Poststep ***

(6)      *** Compute the analyzable values ***
         *** Assign the values to the analyzable parameters ***

(7)      pEnv->SetProgress(1.0);

(8)      return (*** Criteria of the completion ***); }
```

Fig. 15.4 The schematic representation of the tumor growth modeling algorithm implemented in CAME$_{\&}$L

CADLib's `Zone` class passed as the parameter to the function. If a variable `z` describes the zone, then `z.a1` and `z.b1` are the lower and the higher boundaries of the zone along a first axis, `z.a2` and `z.b2` – along a second one and `z.a3` and `z.b3` – along a third one. All boundaries should be included. Axes are just enumerated, but not named here as the "absciss", the "ordinate" and the "applicate", because zonal mechanism is to be metrics independent and in general situation the meaning of the particular axis is unknown. So, it would be wrong to conclude that the object of the `Zone` class always describes the parallelepiped.
The environment will call the `SubCompute` function with the correct value of the zone description. In most cases, and in the case of tumor growth modeling also, the main structure of the implementation of this function should look like shown in Fig. 15.4 (verbal descriptions of the functionality which replace the code are given between three asterisks, bracketed numbers are used for referencing, as above).

This function will be called once for each timestep. So its beginning (line (1)) is the appropriate place for the prestep routines (initialization of variables, pre-computing values, which will be used later, etc.). On the line (2) three variables, which will run over three Cartesian coordinated are declared. The running is provided by the following loop operators. Inside the loop (line (4) and the next one) the main part of an algorithm should be placed: for each cell its new state should be determined. Moreover, values of the analyzable parameters, which can be influenced by each single cell, should be updated. It is strongly recommended not to reassign the values to such parameters many times, but to deal with the temporary variables until step will be finished. Line (5) is the appropriate place for postprocessing step results. On line (6), after the analyzable values, which are influenced by the simulation step in general (not by any single cell) were calculated, the parameters can get their values for the current iteration (line after (6)).

Member variable `pEnv` allows the rules component to exchange the information with the environment. Its member function `SetProgress` is used to declare which part of the time-consuming process have been accomplished (from 0.0 (nothing have been done) to 1.0 (the process is finished)). The line (3) is the suitable place to report about the progress not excessively often, but adequately. Before finishing the iteration progress should be set to 100% (line (7)).

The value returned by `SubCompute` function (line (8)) plays the role of the computational experiment's completion criterium. Simulation will go on while function returns `true`.

In all the rest a rules component's library should contain the same principal parts as a datum component's library, considered above. Necessary component's library access functions can be also created by two lines of the code.

The situation, which has been considered is quite typical: in the overwhelming majority of cases, excluding purely educational purposes, a user has to create the rules component. In some cases, but not so often she/he has to implement the datum component also. The chance that one will need the non-standard metrics is very low and most likely attitudes to the special metrics-related research. A necessity of new grids or analyzers creation may rise even more rare. Standard analyzers can treat all basic types (boolean, integer and floating-point variables) and standard grids are suitable for 1D, 2D and 3D modeling. Moreover, the visualization can be slightly influenced or customized on the level of datum components (see the description of the `DrawCell`, `GetCellColor`, and `GetPlaceColor` member functions of the `CADatum` class [14, 23]).

From the opposite side, lets sort types of components in the order from the most simple to the most complicated one from the developers point of view. In this case the creation of new datum components will be the simplest. Then rules components follow. It looks not so simple, but in most cases the only thing, which researcher has to do is overloading the `SubCompute` member function. The rest three types are to be created form scratch and number of functions have to be overloaded. The next from the simplicity point of view are analyzer components. Their idea is quite clear and general, it contains less specifics and can be implemented easier than the next type – grid components. Metrics components are the most complex and hard to debug, because they make no visual output, but with the help of CADLib even this can be done without getting stuck.

To run the tumor growth simulation a researcher has to install two created components with the help of the "components manager" built into the environment ("Tools" | "Components Manager..." in the main menu). Then new document should be created ("File" | "New" in the main menu) and four components mentioned above

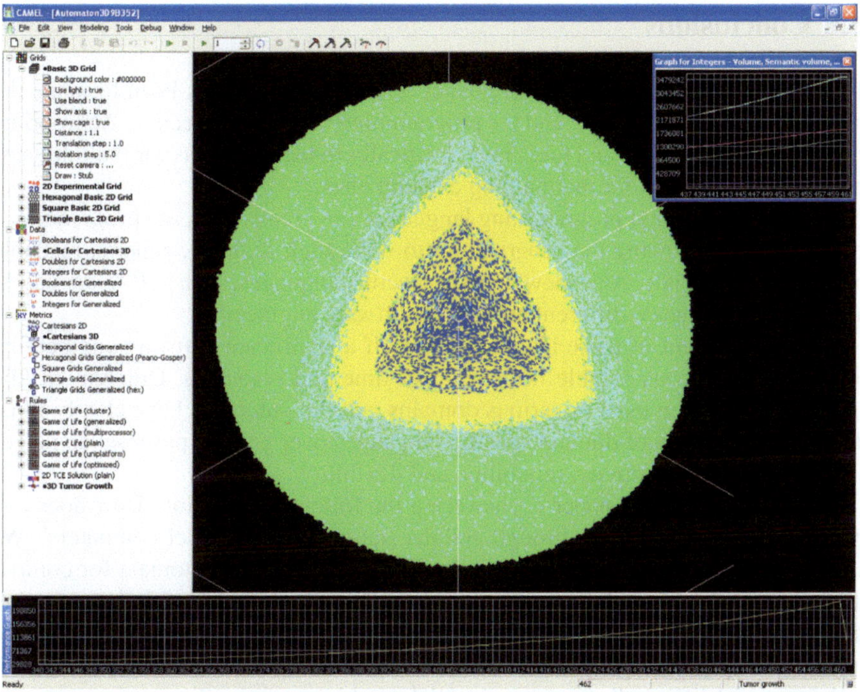

Fig. 15.5 Screenshot of CAME$_{\&}$L, running tumor growth computational experiment, being studied with the help of two analyzers

(two standard and two created) should be chosen. After this the simulation can be executed with the help of "Go" button ("Modeling" | "Go" in the main menu). The screenshot of the environment, running the tumor growth computational experiment is shown on Fig. 15.5.

The experiment's window is divided into two parts. The left one displays the components tree. All components except analyzers are presented there and grouped by types. Tree's leaves of the first level are types' names, on the second level there are the components, and their parameters are on the third one. This tree is handy for fast switching between the components. Those of them, which were selected, are marked out with the small circle in the beginning of the name. Components, which are compatible with the currently chosen instances, are shown with bold font.

In the right part of the experiment's window the grid component is visualizing the simulation. In the shown case the multicellular tumor spheroid is represented as a "stub". This means that cells with positive values of all three coordinates are not drawn, to allow looking inside the formation.

At the bottom and at the upper-right corner there are two analyzer graphs: the performance one and the plot of key tumor growth characteristics (volume, amounts of proliferating, quiescent and dead cells).

15.6 Conclusions

The main goal of programming languages and tools has always been to make the programmer more productive and the programming task more effective. Appropriate programming languages and tools may drastically reduce the costs for building new applications as well as for maintaining existing ones.

It is well known that programming languages can greatly increase programmer's productivity by allowing the programmer to write high-scalable, generic, readable and maintainable code. Also, new domain specific languages, such as CA languages, can be used to enhance different aspects of software engineering.

The development of these languages is itself a significant software engineering task, requiring a considerable investment of time and resources. Domain-specific languages have been used in various domains and the outcomes have clearly illustrated the advantages of domain specific-languages over general purpose languages in areas such as productivity, reliability, and flexibility.

The main goal of the paper is answering the following question: How does one program emergent systems through cellular automata on parallel computers? We think that it is very important for an effective use of cellular automata for computational science on parallel machines to develop and use high-level programming languages and tools that are based on the cellular computation paradigm. These languages may provide a powerful tool for researchers and engineers that need to implement real-life applications on parallel machines using a fine-grain approach. This approach allows designers to concentrate on "how to model a problem" rather than on architectural details as occurs when people use low-level languages that have not been specifically designed to express fine-grained parallel cellular computations.

In a sense, parallel cellular languages provide a *high-level paradigm* for fine-grain computer modeling and simulation. While efforts in sequential computer languages design focused on *how* to express sequential data, objects and operations, here the focus is on finding out *what* parallel cellular objects and operations are the ones we should want to define. Parallel cellular programming emerged as a response to these needs.

References

1. J. von Neumann, *Theory of Self-Reproducing Automata*, ed. by A. W. Burks (University of Illinois Press, Urbana, IL, 1966)
2. S. Ulam, *Random Processes and Transformations / Proceedings of the International Congress of Mathematicians*, vol. 2 (American Mathematical Society Providence, RI 1952).
3. K. Zuse, *Calculating Space*. (Massachusetts Institute of Technology Technical Translation AZT-70-164-GEMIT (Project MAC)). (MIT Cambridge, MA, 1970)
4. N. Wiener, A. Rosenbleuth, The mathematical formulation of the problem of conduction of impulses in a network of connected excitable elements, specifically in cardiac muscle. Archi. Insti. Cardiol. Mex. **16**, 202–265 (1946)
5. P.M.A. Sloot, A.G. Hoekstra, *Modeling Dynamic Systems with Cellular Automata, Chapter 21*. ed. by P.A. Fishwick, Handbook of Dynamic System Modeling. (Chapman & Hall/CRC, London/Boca Raton, FL, 2007)

6. E. Houstis, E. Gallopoulos, J. Bramley, J.R. Rice, Problem-solving environments for computational science. IEEE Comput. Sci. Eng. **4**, 18–21 (1997)
7. E. Gallopoulos, E. Houstis, J.R. Rice, Computer as Thinker/Doer: Problem-solving environments for computational science. IEEE Comput. Sci. Eng. **1**, 11–23 (1994)
8. M. Abrams, D. Allison, D. Kafura, C. Ribbens, M.B. Rosson, C. Shaffer, L. Watson, *PSE Research at Virginia Tech: An Overview. Technical Report: TR-98-21.* (Virginia Polytechnic Institute & State University, Blacksburg, VA, 1998)
9. D.W. Walker, M. Li, O.F. Rana, M.S. Shields, Y. Huang, The software architecture of a distributed problem-solving environment. Concur. Pract. Exp. **12**, 1455–1480 (2000)
10. D.E. Knuth, *Literate programming. Center of the Study of Language and Information.* (Stanford, CA 1992)
11. E. Gallopoulos, E.N. Houstis, J.R. Rice, *Future Research Directions in Problem Solving Environments for Computational Science: Report of a Workshop on Research Directions in Integrating Numerical Analysis, Symbolic Computing, Computational Geometry, and Artificial Intelligence for Computational Science. Technical Report 1259. Center for Supercomputing Research and Development.* (University of Illinois, Urbana-Champaign, IL, 1992)
12. W. Schroeder, K. Martin, B. Lorensen, *Visualization Toolkit: An Object-Oriented Approach to 3D Graphics*, 4th edn. (Kitware, New York, NY, 2006)
13. T. Worsch, Programming Environments for Cellular Automata. *Proceedings of 2nd Conference on CA in Research and Industry (ACRI 96)* (Springer, Heidelberg, 1996)
14. L. Naumov, Generalized coordinates introduction method and a tool for computational experiments software design automation, based on cellular automata. PhD Thesis, SPbSU ITMO, Saint-Petersburg, 2007
15. I. Blecic, A. Cecchini, G. Trunfio, A generalized rapid development environment for cellular automata based simulations. *Cellular Automata: 6th International Conference on Cellular Automata for Research and Industry (ACRI-2004).* (Springer, Heidelberg, 2004) pp. 851–860
16. T. Toffoli, N. Margolus, *Cellular Automata Machines: A New Environment For Modeling.* (MIT Press, Cambridge, MA, 1987)
17. N. Margolus, CAM-8: A Computer Architecture Based on Celluar Automata. *Physics of Computation Seminar* (MIT, Cambridge, MA, 1993)
18. M. Cannataro, S. Di Gregorio, R. Rongo, W. Spataro, G. Spezzano, D. Talia, A parallel cellular automata environment on multicomputers for computational science. Parallel Comput. **21**, 803–823 (1995)
19. G. Spezzano, D. Talia, CARPET: a programming language for parallel cellular processing. *Proceedings 2nd European School on PPE for HPC.* (Alpe d'Huez, France, 1996) pp. 71–74
20. G. Spezzano, D. Talia, A high-level cellular programming model for massively parallel processing. *2nd International Workshop on High-Level Programming Models and Supportive Environments (HIPS97).* IEEE Computer Society Press, LOS Alamitos, CA, pp. 55–63
21. L. Naumov, CAME&L – Cellular Automata Modeling Environment & Library. *Cellular Automata: 6th International Conference on Cellular Automata for Research and Industry (ACRI-2004).* (Springer, Heidelberg, 2004) pp. 735–744
22. A. Sottoriva, J.J.C. Verhoeff, T. Borowski, S.K. McWeeney, P.M.A. Sloot, L. Vermeulen, Modelling cancer stem cell driven tumor growth reveals invasive morphology and increased phenotypical heterogeneity. Cancer Res. **70**, 46–56
23. CAMEL Laboratory – http://camellab.spb.ru/. Accessed date 23 Feb 2005
24. G. Spezzano, D. Talia CAMELot: A parallel cellular environment for modelling complexity. AI*IA Notizie **2**, 9–15 (2001)
25. J.D. Eckart, A cellular automata simulation system: Version 2.0. ACM SIGPLAN Notices **27**(8), 99–106 (1992)
26. U. Freiwald, J.R. Weimar, JCASim a Java system for simulating cellular automata. *Theoretical and Proctical Issues on Cellular Automata (ACRI 2000).* (Springer, Heidelberg, 2001) pp. 47–54
27. Mirek's Cellebration – http://www.mirekw.com/ca/. Accessed date 27 Apr 2010
28. M. Ifrim, Contribution to ParCeL-6 Project: Design of Algorithms Mixing Memory Sharing and Message Passing Paradigms for DSM and Cluster Programming, 2005

29. O. Menard, S. Vialle, H. Frezza-Buet, Making cortically-inspired sensorimotor control realistic for robotics: Design of an extended parallel cellular programming model. *In International Conference on Advances in Intelligent Systems - Theory and Applications*. (IEEE Computer Society, Luxembourg, 2004)
30. T. Bach, T. Toffoli, SIMP, a laboratory for cellular automata and lattice gas experiments. *International Conference on Complex Systems*, (Boston, MA, 2004)
31. T. Toffoli, T. Bach, A common language for "Programmable Matter" (Cellular Automata and All That). Bull. Ital. Assoc. Artif. Intell., **2**, 23–31 (2001)
32. H. Chou, W. Huang, J.A. Reggia, The trend cellular automata programming environment. Simulation **78**(2), 59–75 (2002)
33. C. Hochberger, R. Hoffmann, CDL – a language for cellular processing. *Proceedings of the 2nd International Conference on Massively Parallel Computing Systems*, IEEE, Ischia. pp. 41–46 (1996)
34. L. Naumov, Generalized Coordinates for Cellular Automata Grids. Computational Science – ICCS 2003. Part 2. (Springer, Heidelberg, 2003) pp. 869–878
35. H. Sagan, *Space-Filling Curves*. (Springer, Heidelberg, 1994)
36. D. Talia, Cellular processing tools for high-performance simulation. Computer **33**(9), 44–52 (2000)
37. B.L. Chamberlain, S-E. Choi, S.J. Deitz, L. Snyder, The high-level parallel language ZPL improves productivity and performance. *In: Proceedings of the IEEE International Workshop on Productivity and Performance in High-End Computing* (2004), Madrid
38. G. Spezzano, D. Talia, Programming cellular automata for computational science on parallel computers. Future Gen. Comput. Syst. **16**(2–3), 203–216 (1999)
39. J.M. Greenberg, S.P. Hastings, Spatial patterns for discrete models of diffusion in excitable media. SIAM J. Appl. Math. **34**, 515–523 (1978)

MIX
Papier aus verantwortungsvollen Quellen
Paper from responsible sources
FSC® C105338
FSC
www.fsc.org

If you have any concerns about our products,
you can contact us on
ProductSafety@springernature.com

In case Publisher is established outside the EU,
the EU authorized representative is:
Springer Nature Customer Service Center GmbH
Europaplatz 3, 69115 Heidelberg, Germany

Printed by Libri Plureos GmbH
in Hamburg, Germany